Lecture Notes on Data Engineering and Communications Technologies 181

Series Editor

Fatos Xhafa, *Technical University of Catalonia, Barcelona, Spain*

The aim of the book series is to present cutting edge engineering approaches to data technologies and communications. It will publish latest advances on the engineering task of building and deploying distributed, scalable and reliable data infrastructures and communication systems.

The series will have a prominent applied focus on data technologies and communications with aim to promote the bridging from fundamental research on data science and networking to data engineering and communications that lead to industry products, business knowledge and standardisation.

Indexed by SCOPUS, INSPEC, EI Compendex.

All books published in the series are submitted for consideration in Web of Science.

Zhengbing Hu · Ivan Dychka · Matthew He
Editors

Advances in Computer Science for Engineering and Education VI

Set 2

Editors
Zhengbing Hu
Faculty of Applied Mathematics
National Technical University of Ukraine
"Igor Sikorsky Kiev Polytechnic Institute",
Ukraine
Kyiv, Ukraine

Ivan Dychka
Faculty of Applied Mathematics
National Technical University of Ukraine
"Igor Sikorsky Kiev Polytechnic Institute",
Ukraine
Kyiv, Ukraine

Matthew He
Halmos College of Arts and Sciences
Nova Southeastern University
Fort Lauderdale, FL, USA

ISSN 2367-4512   ISSN 2367-4520 (electronic)
Lecture Notes on Data Engineering and Communications Technologies
ISBN 978-3-031-36117-3   ISBN 978-3-031-36118-0 (eBook)
https://doi.org/10.1007/978-3-031-36118-0

This Springer imprint is published by the registered company Springer Nature Switzerland AG
The registered company address is: Gewerbestrasse 11, 6330 Cham, Switzerland

Preface

Modern engineering and educational technologies have opened up new opportunities in computer science, such as conducting computer experiments, visualizing objects in detail, remote learning, quickly retrieving information from large databases, and developing artificial intelligence systems. Governments and science and technology communities are increasingly recognizing the significance of computer science and its applications in engineering and education. However, preparing the next generation of professionals to properly use and advance computer science and its applications has become a challenging task for higher education institutions due to the rapid pace of technological advancements and the need for interdisciplinary skills. Therefore, higher education institutions need to focus on developing innovative approaches to teaching and providing opportunities for hands-on learning. Additionally, international cooperation is crucial in facilitating and accelerating the development of solutions to this critical subject.

As a result of these factors, the 6th International Conference on Computer Science, Engineering, and Education Applications (ICCSEEA2023) was jointly organized by the National Technical University of Ukraine "Igor Sikorsky Kyiv Polytechnic Institute," National Aviation University, Lviv Polytechnic National University, Polish Operational and Systems Society, Warsaw University of Technology, and the International Research Association of Modern Education and Computer Science on March 17–19, 2023, in Warsaw, Poland. The ICCSEEA2023 brings together leading scholars from all around the world to share their findings and discuss outstanding challenges in computer science, engineering, and education applications.

Out of all the submissions, the best contributions to the conference were selected by the program committee for inclusion in this book.

March 2023

Zhengbing Hu
Ivan Dychka
Matthew He

Organization

General Chairs

Felix Yanovsky   Delft University of Technology, Delft, Netherlands

Ivan Dychka   National Technical University of Ukraine "Igor Sikorsky Kyiv Polytechnic Institute", Ukraine

Online conference Organizing Chairs

Z. B. Hu   National Technical University of Ukraine "Igor Sikorsky Kyiv Polytechnic Institute", Ukraine

Solomiia Fedushko   Lviv Polytechnic National University, Lviv, Ukraine

Oleksandra Yeremenko   Kharkiv National University of Radio Electronics, Kharkiv, Ukraine

Vadym Mukhin   National Technical University of Ukraine "Igor Sikorsky Kyiv Polytechnic Institute", Ukraine

Yuriy Ushenko   Chernivtsi National University, Chernivtsi, Ukraine

Program Chairs

Matthew He   Nova Southeastern University, Florida, USA

Roman Kochan   University of Bielsko-Biała, Bielsko-Biala, Poland

Q. Y. Zhang   Wuhan University of Technology, China

Publication Chairs

Z. B. Hu   National Technical University of Ukraine "Igor Sikorsky Kyiv Polytechnic Institute", Ukraine

Matthew He   Nova Southeastern University, USA

Ivan Dychka   National Technical University of Ukraine "Igor Sikorsky Kyiv Polytechnic Institute", Ukraine

Publicity Chairs

Sergiy Gnatyuk	National Aviation University, Kyiv, Ukraine
Rabah Shboul	Al-al-Bayt University, Jordan
O. K. Tyshchenko	University of Ostrava, Czech Republic

Program Committee Members

Artem Volokyta	National Technical University of Ukraine "Igor Sikorsky Kyiv - Polytechnic Institute", Kyiv, Ukraine
Rabah Shboul	Al-al-Bayt University, Jordan
Krzysztof Kulpa	Warsaw University of Technology, Warsaw, Poland
Bo Wang	Wuhan University, China
Anatoliy Sachenko	Kazimierz Pułaski University of Technology and Humanities in Radom, Poland
Ihor Tereikovskyi	National Technical University of Ukraine "Igor Sikorsky Kyiv Polytechnic Institute", Kyiv, Ukraine
E. Fimmel	Mannheim University of Applied Sciences, Germany
G. Darvas	Institute Symmetron, Hungary
A. U. Igamberdiev	Memorial University of Newfoundland, Canada
Jacek Misiurewicz	Warsaw University of Technology, Warsaw, Poland
X. J. Ma	Huazhong University of Science and Technology, China
S. C. Qu	Central China Normal University, China
Ivan Izonin	Lviv Polytechnic National University, Lviv, Ukraine
Y. Shi	Bloomsburg University of Pennsylvania, USA
Feng Liu	Huazhong Agricultural University, China
Sergey Petoukhov	MERI RAS, Moscow, Russia
Z. W. Ye	Hubei University of Technology, China
Oleksii K. Tyshchenko	University of Ostrava, Czech Republic
C. C. Zhang	Feng Chia University, Taiwan
Andriy Gizun	National Aviation University, Kyiv, Ukraine
Yevhen Yashchyshyn	Warsaw University of Technology, Warsaw, Poland
Vitaly Deibuk	Chernivtsi National University, Chernivtsi, Ukraine

J. Su   Hubei University of Technology, China
G. K. Tolokonnikova   FNAT VIM of RAS, Moscow, Russia

Conference Organizers and Supporters

National Technical University of Ukraine "Igor Sikorsky Kyiv Polytechnic Institute",
Ukraine
National Aviation University, Ukraine
Lviv Polytechnic National University, Ukraine
Polish Operational and Systems Society, Poland
Warsaw University of Technology, Poland
International Research Association of Modern Education and Computer Science, Hong
Kong

Contents

Perfection of Computer Algorithms and Methods

Advances in Technological and Educational Approaches

Perfection of Computer Algorithms and Methods

Improving Piezoceramic Artificial Muscles for Flying Insect-Sized Mini Robots

Improving Piezoceramic Artificial Muscles for Flying Insect-Sized Mini Robots

Sergey Filimonov, Constantine Bazilo$^{(\boxtimes)}$, Olena Filimonova, and Nadiia Filimonova

Cherkasy State Technological University, Shevchenko Blvd, Cherkasy 46018006, Ukraine
{s.filimonov,n.filimonova}@chdtu.edu.ua, b_constantine@ukr.net

Abstract. The relevance of the work is associated with the development of artificial muscles of flying insect-sized mini robots. The development of such robots is due to a significant decrease in the population of insects, namely bees. The purpose of the article is to improve the design of the piezoceramic artificial muscle of flying insect-sized mini robot by determining the rational dimensions of the actuator and developing its design. 3D-dimensional numerical simulation of the process was carried out using the COMSOL Multiphysics software package to study and determine the maximum mechanical vibrations of the piezoactuator. On the basis of the obtained data the graphical dependencies were constructed to determine the rational parameters of the piezoelectric actuator. In addition, a new design of an artificial muscle based on a cruciform piezoelectric actuator has been created and studied. The research results showed an increase in the oscillation amplitude by 7 times compared with the basic design. These results will reduce the power consumption by about 7 times by reducing the amplitude of the control voltage of artificial muscles based on piezoactuators. In addition, this will reduce the weight of the entire structure and make it more autonomous and lighter. The obtained results can be used in the design of piezoelectric actuators for insect-sized mini robots.

Keywords: Air mini robot · Piezoelectric element · Actuator · Artificial muscles · Robobee

1 Introduction

Today, the loss of bees, to which we owe 1/3 of the entire crop, cannot pass without consequences. Having lost the population of bees, humanity will lose not only honey [1]. Since most of the crop depends on bees. Simultaneously with the search for a solution to stop the increase in bee mortality, scientists are looking for a replacement for them. In the modern world, many processes are automated [2, 3], and it is possible to automate the process of pollination of plants as well. One of the analogues to replace the bee is a flying insect-sized mini robot RoboBee.

In addition, robots of this type can be used for numerous applications, from search and rescue operations to environmental monitoring [4] and biological research.

Bee-sized robots (-100 mg) impose greater restrictions on the use of the components. One of the most important components in flying insect-sized mini robots is the element

© The Author(s), under exclusive license to Springer Nature Switzerland AG 2023
Z. Hu et al. (Eds.): ICCSEEA 2023, LNDECT 181, pp. 527–538, 2023.
https://doi.org/10.1007/978-3-031-36118-0_47

that sets the wings in motion. As is known, common drives used in larger robots drive the propellers in most drones. Such electric motor is not advantageous to scale down to an insect in terms of efficiency and power density [5]. This is due to the fact that depending on the surface area, losses, such as Coulomb friction and electrical resistance, take on a greater value, the smaller is scale [6]. Flying insect-sized mini robots mostly use piezoceramic actuators (artificial muscles) to create wing movement. However, the main disadvantage of these artificial muscles is the high control voltage to generate enough wing movement to get airborne. In addition, obtaining a high voltage in autonomous mode leads to a heavier circuitry. All these factors greatly limit the carrying capacity.

To solve the problem, it is necessary to sequentially perform a number of tasks:

- to analyse the design features of modern flying insect-sized mini robots and the main results of research on working processes in them;
- to simulate a piezoelectric actuator (artificial muscle) and determine its rational dimensions;
- to develop a new and improved design of an actuator (artificial muscle) for flying insect-sized mini robot and conduct an experimental study.

Thus, the development of a new type and improvement of known types of piezoceramic actuators is an important task.

The object of research is the processes of bending mechanical oscillations of bimorph piezoelectric element of piezoactuator.

The subject of the study is a piezoelectric actuator based on a bimorph piezoelectric element (artificial muscle).

The aim of the work is to improve the design of the piezoceramic artificial muscle of flying insect-sized mini robot by determining the rational dimensions of the actuator and developing its design.

2 Literature Review

Figure 1 shows the design of one of the analogues of flying mini robots (small-sized flying systems – quadcopters), which uses electromagnetic motors [7–9].

Further reduction of such structures requires a reduction in the dimensions of electromagnetic motors, which in turn leads to a decrease in the efficiency [10], and this is not acceptable. One of the most relevant ways to reduce the size of flying mini robots is the use of piezoelectric actuators [11, 12].

Researchers from the Wyss Institute at Harvard University (USA) are developing an insect-sized mini robot RoboBee (Fig. 2) [13, 14]. First of all, in the framework of this work it is important for us to note the design features of RoboBee piezoelectric actuators which is the main analogue of almost all designs of flying insect-sized mini robots.

The main element for lifting into the air in RoboBee are "unimorph" piezoelectric actuators [15]. The use of actuators of this type allows to reduce the dimensions and weight, and also increases the reliability and durability of the structure [16].

Additional modifications allow some models of RoboBee to switch from swimming underwater to flying, as well as to "hook" on surfaces using static electricity [17, 18]. However, piezoelectric actuators remain the main elements for lifting into the air.

It should be noted that currently RoboBee is autonomous only under certain conditions (bright light with a certain wavelength).

The next example of an insect-sized mini robot is the development of engineers from the University of Washington (USA). They presented their version of the 74-mg miniature flying robot RoboFly (Fig. 3) [19]. As in the previous design, piezoelectric actuators are used for lifting into the air.

Fig. 1. Mini robots based on electric motors "flying monkey"

Fig. 2. Mini robot RoboBee: 1 – piezoelectric actuator, 2 – reducer mechanism, 3 – wings

Another type of mini robot design based on bimorph elements is the Four Wings design (Fig. 4) [20]. It differs from the previous ones in the number and location of piezoelectric actuators.

Having considered the designs of mini robots RoboBee, RoboFly and Four Wings, which are based on a piezoelectric actuator, we can say that their construction is almost the same. Some models differ only in execution (bimorph or monomorph), location and number of piezoelectric actuators in operation. However, options for choosing the geometry, the location of the piezoelectric material on the surface of a metal plate or other geometric structure are not given. Another disadvantage of piezoelectric actuators is the high control voltage, about 200 V. All these shortcomings lead to the impossibility of further improvement of the technical characteristics of piezoelectric actuators, except for adding the number of actuators in one node. However, this leads to an increase in the mass and dimensions of the air mini robot.

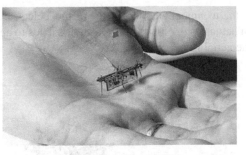

Fig. 3. Mini robot RoboFly

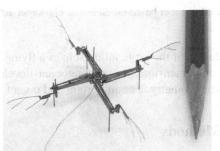

Fig. 4. Mini robot Four Wings

Professor YuFeng Chen from Massachusetts Institute of Technology has developed insect-sized drones with unprecedented agility. Aerial robots are equipped with a new class of soft drives, allowing them to withstand the physical stress of flight (Fig. 5). The actuators of this flying mini robot can make up to 500 strokes per second, making the mini robot resistant to external influences. The robot weighs only 0.6 g, which is approximately the same as the mass of a large bumblebee [21].

Despite all the advantages of this design, it has certain disadvantages. The disadvantages of this mini robot are the high control voltage, which is 2000 V, and the great complexity of manufacturing, because elements of nanotechnology are used in the process of creation. High control voltage leads to certain PCB requirements, increased weight and power consumption.

Thus, having considered the designs of flying insect-sized mini robots, we can draw next conclusions. The main element of the lifting force of the mini robot is the actuator. Actuators are divided into piezoelectric and based on dielectric elastomer materials. The main disadvantages of actuators based on elastomer dielectric materials are mentioned above. Having considered the designs of mini robots Robobee, RoboFly and Four Wings, which are based on a piezoelectric actuator, we can say that their construction is almost the same. Some models differ only in execution (bimorph or unimorph), location and number of piezoelectric actuators in the robot. However, options for choosing the geometry, the location of the piezoelectric material on the surface of a metal plate or other geometric structure are not given. Another disadvantage of piezoelectric actuators is the high control voltage, of about 200 V. All these shortcomings lead to the impossibility of further improving the technical characteristics of piezoelectric actuators, except for adding the number of actuators in one node. In turn, this leads to an increase in the mass and dimensions of the aerial mini robot.

Fig. 5. Mini robot based on dielectric elastomer actuators

So, improving the design of the artificial muscle of a flying insect-sized mini robot by determining the rational dimensions of the actuator and developing its design, as well as creating a new actuator geometry is an important and urgent task.

3   Materials and Methods

Considering the technical features of piezoelectric actuators, which complicate the experimental determination and selection of the correct form of their oscillations, it is optimal to use for this purpose numerous calculation methods implemented by specialized CAD systems [22, 23].

To study the influence of the design parameters of the piezoelectric actuator, numerical simulation of the piezoelectric element operation process was carried out using the COMSOL Multiphysics software package [9], which makes it possible to carry out numerical simulation of three-dimensional models of piezoelectric actuators with the necessary parameters and limiting conditions.

In the process of modelling Lagrange finite elements with elementary basis functions of the second order (Lagrange-Quadratic) are used.

Firstly, we determine the rational location of the piezoelectric element on a metal plate. Secondly, we determine the rational length of the piezoelectric plate, taking into account the first study. Certain results will make it possible to obtain the maximum oscillation amplitude of the bimorph piezoceramic actuator, which in turn can reduce the amplitude of the control voltage.

Initial boundary conditions of bimorph piezoelectric element. The metal plate is a console that is rigidly fixed on one side and the other side is free. The piezoelectric element is symmetrically located relative to the axis of symmetry of the metal plate, and its bottom side is located at a distance of 1 mm from the place where the metal plate is fixed. An input voltage of 100 V is applied to the top and bottom surfaces of the piezoceramic plate. The dimensions of the piezoelectric element are $16 \times 4 \times 0.1$ mm, and the dimensions of the metal plate are $38 \times 6 \times 0.1$ mm. During the experiment, the dimensions of the metal plate remain stable, while the dimensions of the piezoelectric plate change (Fig. 6).

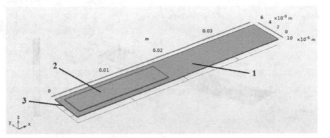

Fig. 6. Basic model of a bimorph piezoelectric actuator: 1 – metal plate; 2 – piezoelectric element; 3 – the place of the cantilever fixing of the metal plate

The principle of operation of a bimorph piezoceramic actuator is as follows. When an alternating electrical voltage is applied, namely, the signal output is connected to the free side of the piezoceramic element 2 and the common output to the metal plate 1, the piezoceramic actuator begins to deform. Since the actuator is fixed on one side, the maximum vibrations will be on the free side. In the general design of a flying mini robot, the free side is connected through a gear mechanism to the wings, which in turn begin to perform movements.

4 Numerical Experiments and Results

Using the functionality of the COMSOL Multiphysics software package, we obtain the amplitude-frequency dependence of bimorph actuator (Fig. 7).

The largest oscillation amplitude of the bimorph piezoelectric actuator corresponds to the frequency of 70 Hz and 280 Hz. A visual display of the oscillation form of a bimorph piezoactuator at these frequencies is shown in Fig. 8.

Figure 8 shows that the most favourable form of oscillations corresponds to a frequency of 70 Hz. It has only the first mode of bending vibrations, but under certain conditions, vibrations at a frequency of 280 Hz can also be used. Thus, our oscillation frequency search range can be narrowed down to 300 Hz.

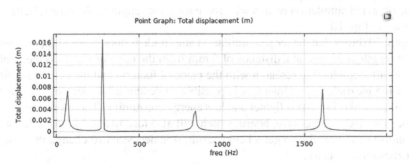

Fig. 7. Amplitude-frequency dependence of bimorph piezoelectric actuator

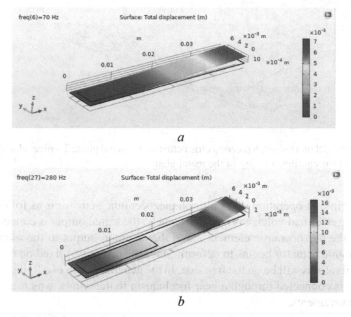

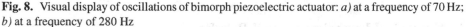

Fig. 8. Visual display of oscillations of bimorph piezoelectric actuator: *a)* at a frequency of 70 Hz; *b)* at a frequency of 280 Hz

To obtain maximum oscillations of the actuator, it is necessary to determine the rational location of the piezoceramic element on the metal plate. To do this, we will change the distance between the edge of the piezoelectric element and the place of fixing with a step of 5 mm. The results of the study (simulation) are shown in Fig. 9.

Figure 9 shows that the largest oscillation amplitude is obtained when the piezoelectric element is located at a distance of 5 mm from the fixed edge of the metal plate.

The next step is to change the length of the piezoelectric plate upward in increments of 3 mm, but taking into account the preliminary results (the initial state of the location from the edge of the fixing is 5 mm).

The obtained simulation results with increasing the length of the piezoelectric plate are shown in Fig. 10.

Figure 10 shows that the largest oscillation amplitude is obtained with a piezoelectric element length of 25 mm at a distance of 5 mm from the edge of the fixed metal plate.

Thus, after conducting research with the place of fixation and the geometric dimensions of the piezoelectric element, it is possible to design a piezoelectric actuator for insect-sized mini robot that is lighter and less energy consuming. Therefore, it is possible to reduce the size of the control board, which will also lead to a reduction in the weight of the entire structure. But to get a better result, it is necessary to change the design of the piezoelectric actuator.

After carefully analyzing the design of piezo actuators used in insect-sized mini robots, the authors proposed an improved design based on a bimorph cruciform actuator, the model of which is shown in Fig. 11.

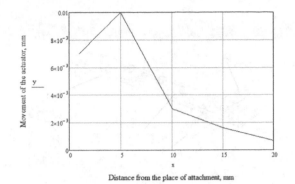

Distance from the place of attachment, mm

Fig. 9. Dependence of the oscillation amplitude on the distance of the piezoelectric element relative to the fixed edge of the metal plate

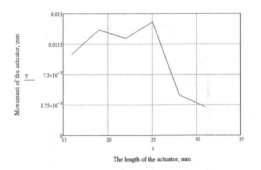

The length of the actuator, mm

Fig. 10. Dependence of the oscillation amplitude on the change in the length of the piezoelectric element relative to the fixed edge of the metal plate

A feature of the proposed design is an increase in the amplitude of oscillations due to the displacement of the fixing point. Thanks to this, we got a kind of lever, in which there are two arms, the ratio of which leads to an increase in the amplitude of oscillations and an increase in mobility.

Initial boundary conditions for a bimorph cruciform piezoelectric actuator. The metal plate is connected by a metal jumper, and is free on the other sides. The metal jumper has a rigid fixing on the edge sides. The piezoelectric element is symmetrically located relative to the axis of symmetry of the metal plate, and its bottom side is located at a distance of 5 mm from the lower edge of the metal plate fixing. An input voltage of 100 V is applied to the top and bottom surfaces of the piezoceramic plate. The dimensions of the piezoelectric element are $25 \times 4 \times 0.1$ mm, and the dimensions of the metal plate are $38 \times 6 \times 0.1$ mm.

Using the COMSOL Multiphysics software package, the amplitude-frequency dependence of this actuator is obtained (Fig. 12).

Figure 12 shows that the largest oscillation amplitude of the bimorph cruciform piezoelectric actuator corresponds to a frequency of 410 Hz, which significantly increases the resistance to external factors, for example, air.

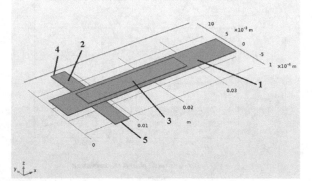

Fig. 11. Bimorph cruciform piezoelectric actuator: 1 – metal plate; 2 – metal jumper; 3 – piezo-electric element; 4 and 5 – places of rigid fixing of the metal jumper

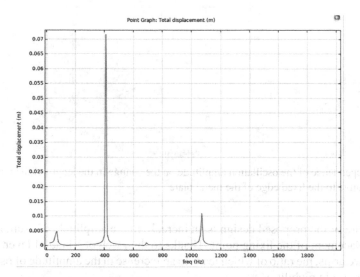

Fig. 12. Amplitude-frequency dependence of a bimorph cruciform piezoelectric actuator

A visual display of the oscillation form of a bimorph cruciform piezo actuator is shown in Fig. 13.

The obtained results are approximately 7 times higher than the results of the basic bimorph piezoelectric actuator design presented above.

To test the proposed design, an experimental sample was developed (Fig. 14).

The test results of the experimental sample confirmed the operability of the developed cruciform design.

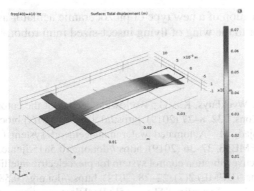

Fig. 13. Visual display of the oscillations of bimorph cruciform piezoelectric actuator at a frequency of 410 Hz

Fig. 14. Experimental sample of a bimorph cruciform piezoelectric actuator

5   Conclusions

The main disadvantages of artificial muscles (piezoelectric actuators), which are the main elements for lifting into the air, have been established.

3D-dimensional modelling was carried out to determine the most rational location of the piezoelectric element on the metal plate, as well as to determine the rational length of the piezoelectric element at its specific location on the metal plate.

A new design of a bimorph cruciform piezoelectric actuator was proposed. The research results showed an increase in the amplitude of vibrations by 7 times compared to the basic design. Thus, the results obtained will reduce energy consumption by about 7 times by reducing the amplitude of the control voltage of artificial muscles based on piezoelectric actuators. In addition, this will reduce the weight of the entire construction and make it more autonomous and lighter. These results accelerate development in this direction.

The obtained results can be used not only in designing flying insect-sized mini robots, but also in designing crawling and floating mini robots. Further research is planned to be directed to the creation of a new type of piezoceramic actuator assembly with a new type of force transfer to the wing of flying insect-sized mini robot.

References

1. Sánchez-Bayo, F., Wyckhuys, K.A.G.: Worldwide decline of the entomofauna: a review of its drivers. Biol. Cons. **232**, 8–27 (2019). https://doi.org/10.1016/j.biocon.2019.01.020
2. Ahemed, R., Amjad, M.: Automated water managemenrt system (WMS). Int. J. Educ. Manage. Eng. (IJEME) **3**, 27–36 (2019). https://doi.org/10.5815/ijeme.2019.03.03
3. Li, J.: Design of active vibration control system for piezoelectric intelligent structures. Int. J. Educ. Manage. Eng. (IJEME) **2**(7), 22–28 (2012). https://doi.org/10.5815/ijeme.2012.07.04
4. Abutu, I.M., Imeh, U.J., Abdoulie, T.M.S., et al.: Real time universal scalable wireless sensor network for environmental monitoring application. Int. J. Comput. Netw. Inf. Secur. (IJCNIS) **6**, 68–75 (2018). https://doi.org/10.5815/ijcnis.2018.06.07
5. Wood, R.J., Finio, B., Karpelson, M., et al.: Progress on 'pico' air vehicles. Int. J. Robot. Res. **31**(11), 1292–1302 (2012). https://doi.org/10.1177/0278364912455073
6. Trimmer, W.S.: Microrobots and micromechanical systems. Sens. Actuat. **19**(3), 267–287 (1989)
7. Mulgaonkar, Y., Araki, B., Koh J., et al.: The flying monkey: a mesoscale robot that can run, fly, and grasp. In: 2016 IEEE International Conference on Robotics and Automation (ICRA) (2016). https://doi.org/10.1109/ICRA.2016.7487667
8. Kushleyev, A., Mellinger, D., Kumar, V.: Towards a swarm of agile micro quadrotors. In: Proceedings of Robotics: Science and Systems Conference (RSS), Sydney, Australia (2012). https://doi.org/10.15607/RSS.2012.VIII.028
9. Bazilo, C., Filimonov, S., Filimonova, N., Bacherikov, D.: Determination of geometric parameters of piezoceramic plates of bimorph screw linear piezo motor for liquid fertilizer dispenser. In: Hu, Z., Petoukhov, S., Yanovsky, F., He, M. (eds.) ISEM 2021. LNNS, vol. 463, pp. 84–94. Springer, Cham (2022). https://doi.org/10.1007/978-3-031-03877-8_8
10. Spanner, K., Vyshnevskyy, O., Wischnewskiy, W.: New linear ultrasonic micro motors for precision mechatronic systems. In: Proceedings of the 10th International Conference on New Actuators, pp. 439–443 (2006)
11. Sharapov, V.: Piezoceramic sensors. Springer (2011).https://doi.org/10.1007/978-3-642-153 11-2
12. Aladwan, I.M., Bazilo, C., Faure, E.: Modelling and development of multisectional disk piezoelectric transducers for critical application systems. Jordan J. Mech. Indust. Eng. **16**(2), 275–282 (2022)
13. Chen, Y., Zhao, H., Mao, J., et al.: Controlled flight of a microrobot powered by soft artificial muscles. Nature **575**, 324–329 (2019). https://doi.org/10.1038/s41586-019-1737-7
14. Mam, K.Y., Chirarattananonm, P., Fullerm, S.B., Woodm, R.J.: Controlled flight of a biologically inspired, insect-scale robot. Science **340**(6132), 603–607 (2013). https://doi.org/10.1126/science.1231806
15. Yang, X., Chen, Y., Chang, L., et al. Bee+: A 95-mg four-winged insect-scale flying robot driven by twinned unimorph actuators. IEEE Robot. Autom. Lett. 4270–4277 (2019). https://doi.org/10.1109/LRA.2019.2931177
16. Jafferis, N.T., Graule, M.A., Wood, R.J.: Non-linear resonance modeling and system design improvements for underactuated flapping-wing vehicles. In: IEEE International Conference on Robotics and Automation (ICRA), pp. 3234–3241. Stockholm, Sweden (2016). https://doi.org/10.1109/ICRA.2016.7487493

17. Graule, M.A., Chirarattananon, P., et al.: Perching and takeoff of a robotic insect on overhangs using switchable electrostatic adhesion. Science **352**(6288), 978–982 (2016). https://doi.org/10.1126/science.aaf1092

18. Chen, Y., Wang, H., Helbling, F., et al.: A biologically inspired, flapping-wing, hybrid aerial-aquatic microrobot. Sci. Robot. **2**(11), eaao5619 (2017). https://doi.org/10.1126/scirobotics.aao5619

19. Chukewad, Y.M., Singh, A.T., James, J.M., Fuller, S.B.: A new robot fly design that is easier to fabricate and capable of flight and ground locomotion. In: IEEE/RSJ International Conference on Intelligent Robots and Systems (IROS), pp. 4875−4882. Madrid, Spain (2018). https://doi.org/10.1109/IROS.2018.8593972

20. Fuller, S.B.: Four wings: an insect-sized aerial robot with steering ability and payload capacity for autonomy. IEEE Robot. Autom. Lett. **4**(2), 570−577 (2019). https://doi.org/10.1109/LRA.2019.2891086

21. Chen, Y., Xu, S., Ren, Z., Chirarattananon, P.: Collision resilient insect-scale soft-actuated aerial robots with high agility. IEEE Trans. Rob. **37**(5), 1752–1764 (2021). https://doi.org/10.1109/TRO.2021.3053647

22. Halchenko, V.Y., Filimonov, S.A., Batrachenko, A.V., Filimonova, N.V.: Increase the efficiency of the linear piezoelectric motor. J. Nano- Electron. Phys. **10**(4), 04025 (5pp) (2018). https://doi.org/10.21272/jnep.10(4).04025

23. Sharapov, V.M., Filimonov, S.A., Sotula, Z.V., Bazilo, K.V., Kunitskaya, L.G., Zaika, V.M.: Improvement of piezoceramic scanners. In: 2013 IEEE XXXIII International Scientific Conference Electronics and Nanotechnology (ELNANO), Kiev, pp. 144–146 (2013). https://doi.org/10.1109/ELNANO.2013.6552063

Platform Construction and Structural Design of Passionfruit Picking and Sorting Robot Under the Guidance of TRIZ

Caigui Huang[1,2]($\boxtimes$)

[1] School of Intelligent Manufacturing, Nanning University, Guangxi 530200, China
hcaigui@163.com
[2] Intelligent Manufacturing and Virtual Simulation Research Center of Nanning University, Nanning University, Guangxi 530200, China

Abstract. The picking and sorting of passionfruit is mainly manual currently, which has problems such as high labor intensity and low efficiency. Aiming at this problem, this paper, guided by TRIZ theory, discusses the principles and methods of solving the invention problem, probes the solution matrix of contradictions and conflicts, analyzes the improvement parameters and deterioration factors, and puts forward the invention principle of solving the design conflict of robot platform. Based on the "material - field" theory of TRIZ, the description model of the passionfruit picking and sorting robot platform is established to identify and resolve the technical conflict, and the passionfruit sorting robot platform is designed, which includes CCD visual recognition system, sorting manipulator body, grab manipulator end actuator, translation module, automatic control system and collection module. According to the principle of the invention, the structure of the robot platform is map out and the end actuator structure is designed to meet the functional requirements of the picking and sorting of passionfruit.

Keywords: Passionfruit · Pick · Sorting · Robot · TRIZ

1 Introduction

Fruit picking and picking is the most common, basic and necessary operation in the harvest season. Using advanced automation technology and equipment can greatly improve the efficiency of fruit harvest [1–3]. The robot designed to complete these operations highlights its unique advantages [4, 5].

Passion fruit originates from the United States and Brazil, has a strong fragrance and high medicinal value, and is rich in vitamins. China's main producing areas are Guangxi, Guangdong, Yunnan and other temperate regions [6–8]. Passion fruit is gradually being used as a fruit crop to get rid of poverty and become rich in southern China, but the current planting, grabbing and sorting are basically realized by manual means. At present, there is little research on the mechanical device of Passion fruit at home and abroad, so the mechanical harvesting and sorting of Passion fruit is particularly important at present.

© The Author(s), under exclusive license to Springer Nature Switzerland AG 2023
Z. Hu et al. (Eds.): ICCSEEA 2023, LNDECT 181, pp. 539–549, 2023.
https://doi.org/10.1007/978-3-031-36118-0_48

The Theory of Inventive Problem Solving, abbreviated as TRIZ, is derived from the invention problem solving theory summarized by Archishuler and his team from the Soviet Union in 1946 in order to solve millions of patents in the invention problem analysis patent database, and has been recognized as a widely popular topic of innovation in the technical field [9, 10]. TRIZ provides a structured problem-solving method, which overcomes unreliable decision-making methods and finds real direct solutions for complex problems. It is a well-structured and innovative problem-solving method in technical and non-technical fields [11, 12]. The combination application analysis of TRIZ and other tools is also common, such as the integration of TRIZ and axiomatic design [13]. TRIZ theory has been widely used in aerospace science and technology, military science and technology and other fields in the former Soviet Union, which has achieved significant development, solved some scientific research problems, and improved the efficiency of solving problems. After more than half a century of development, through continuous research and practice, it is found that TRIZ theory has obvious effect in solving scientific research problems in many countries, such as Europe and the United States, and is known as an internationally recognized innovation method [14, 15]. Due to the universality of TRIZ theory and the objective existence of scientific principles and rules in the process of solving invention problems, in recent years, TRIZ theory has also been constantly tried to apply to other fields of innovation, such as education, management, medicine, agriculture, social relations, etc. to solve related problems [16, 17]. Compared with the traditional innovation method of brainstorming, TRIZ theory is easier to operate, and it is systematic and process-based in solving problems. However, it depends on the personal experience, knowledge, inspiration, etc. of designers, which is suitable for most people to make bold innovation [18].

As an advanced design method, TRIZ aims to define and overcome some key problems that may affect product development through potential innovative solutions [19]. This paper analyzes the problems of low efficiency, low safety, high labor intensity, and the integrity of the fruits in the sorting process of passionfruit, puts forward the conflict resolution matrix of the platform design of passionfruit sorting robot based on TRIZ theory, and establishes the problem solution of structural design through the proposed invention principle.

2   TRIZ Theory Content and Solution to Invention Problems

2.1   TRIZ Theory Content

The main contents of TRIZ theory include 40 invention principles, 39 engineering technical characteristic parameters, and the conflict resolution matrix established by using 39 general parameters and 40 invention principles, as well as the material field analysis theory and the invention problem solving algorithm (ARIZ) [20]. Among them, 40 principles of invention, also known as the conflict resolution principle, were discovered and proposed by Archishuler on the basis of the analysis and research of patents around the world, including 40 principles such as segmentation, refining, local change and asymmetry. In addition, based on this, TRIZ proposed a solution to the separation principle, including the separation of space, time, condition-based, whole and part, and the corresponding relationship with the invention principle is shown in Table 1.

Table 1. Correspondence between separation principle and invention principle

Separation principle	Principle of invention
Spatial separation	1, 2, 3, 4, 7, 13, 17, 24, 26, 30
Time separation	9, 10, 11, 15, 16, 18, 19, 20, 21, 29, 34, 37
Separate parts from whole	12, 28, 31, 32, 35, 36, 38, 39, 40
Conditional separation	1, 7, 25, 27, 5, 22, 23, 33, 6, 8, 14, 25, 35, 13

In TRIZ theory, the conflicting characteristics in different fields are highly summarized and abstracted into 39 technical characteristic parameters, which can uniformly and clearly describe the various conflicts contained in different problems, mainly including 39 engineering technical parameters such as the mass of moving objects, the mass of stationary objects, speed, productivity, etc. In the process of solving system problems, to improve a certain technical feature will often lead to the deterioration of other technical parameters, that is, technical conflict [21]. The conflict resolution matrix establishes the contradiction relationship between the improved technical parameters and the deteriorated technical parameters, and lists the invention principles corresponding to the conflict resolution, as shown in Table 2.

Table 2. Conflict Resolution Matrix (partial)

Principle of invention		Deteriorated technical parameters				
		1	2	3	...	39
Improved technical parameters	1	/	—	15,8,29,34	...	35,3,24,37
	2	—	/	—	...	1,28,15,35
	3	8, 15 29,34	—	/	...	14,4,28,29
	...	....	....	....	...	...
	...	...	...	...	...	...
	39	35,26,24,37	28,27,15,3	18,4,28,38	...	/

2.2 Methods to Solve the Invention Problem

TRIZ theory can solve the problem of invention. Its core is the principle of technological system evolution and the principle of conflict resolution. It also establishes the method of conflict resolution, and uses the method of general solution to solve the technical conflict and then solve the problem of invention. The method flow of TRIZ to solve the invention problem is shown in Fig. 1, which can be carried out in four steps [22]; In the process of using TRIZ theory to solve the invention problem, the invention problem to be solved should be clarified first, and then the problem to be solved by the object to be designed should be converted into a general problem expression by means of matter-field analysis and other methods, that is, the general model of the problem should be established, and then the general solution of the general problem should be found by using the 40 invention principles and tools provided by TRIZ theory. According to the tips of the general problem-solving methods obtained, and referring to the existing design experience and knowledge, the innovative solutions to the specific problems of the structural design are determined.

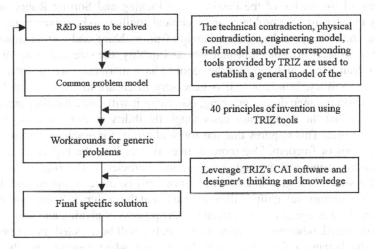

Fig. 1. TRIZ's solution to the invention problem

3   Scheme Design of Robot based on TRIZ

Fruit grabbing and sorting plays an important role in agricultural production, with high labor intensity and seasonality, and high cost of grabbing and sorting. The fruit grabbing and sorting robot has been applied earlier in the United States, Japan, the Netherlands and other countries, and its technology maturity is high. At present, it has realized the grabbing and sorting of grapes, oranges, strawberries, cucumbers, tomatoes and other common fruits and vegetables, with great development potential. There are various robot platforms for grasping and sorting, which are mainly based on machine vision for positioning and recognition, and are used for grasping and sorting by motion control manipulators and end actuators [23, 24]. Based on TRIZ theory, this paper designs the overall structure and system of the thymus fruit grabbing and sorting robot platform to meet the various requirements of grabbing and sorting.

3.1 Analysis of Problems to Be Solved and Establishment of General Problem Model

In order to make the design of the grab and sort robot platform more creative and adaptive, this paper uses TRIZ theory to guide the design, but the key is to do a good job in the pre-and post-processing of specific problem analysis. The process is as follows: first, analyze the specific engineering problems to be solved in R&D and design, establish a general problem model according to TRIZ, analyze the improved and deteriorated engineering technical parameters, determine the corresponding invention principle in the technology conflict matrix, and obtain the general solution, and finally determine the best specific solution by combining engineering knowledge and experience.

In view of the design of the Passion Fruit Picking and Sorting Robot, it is first necessary to determine the problems and technical conflicts to be solved in the process of picking and sorting mechanization. Passion fruit grabbing and sorting robot should first be able to recognize the fruit maturity and quality, the size and color difference of Passion fruit, etc., which requires the robot to have more extensive adaptability and recognition accuracy. In addition, it is necessary to ensure that the manipulator has sufficient working range during grasping, otherwise it will cause missing picking, and it is required that the manipulator has enough flexibility to work so as to be closer to the fruit target. This requires that the robot platform for grabbing and sorting has enough degrees of freedom. The more degrees of freedom, the higher the flexibility of the working range, and the higher the grabbing efficiency. But this means that the structural design of the grabbing robot platform will be more complex, and the more complex the structural reliability will be reduced, and the design difficulty will increase. In addition, the damage of the end effector in the process of picking and sorting should also be considered, otherwise the appearance quality will be affected. In the process of grasping, the damage of fruit skin should be reduced, which requires that the holding force of the end effector of the grasping robot should not be too large in the process of grasping, and the gripped mechanical finger should not be rigid, or the flexible clamping structure should be added, which increases the overall design difficulty and processing manufacturing difficulty, and the cost will also increase.

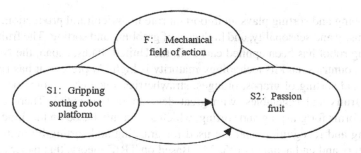

Fig. 2. "Material-field" analysis model

The "material field" analysis of TRIZ theory is used to establish the description model of the passionfruit grabbing and sorting robot platform to determine the technical conflict. As shown in Fig. 2, the material-field analysis model shows that the object of grasping and sorting is Passion Fruit. It is assumed that the grasping and sorting robot platform can achieve efficient and rapid grasping and sorting functions, but it also has harmful effects that make the structure of the grasping robot platform (material) more complex, manufacturing difficulties and high costs, which also leads to technical conflicts. 39 general technical parameters are used to analyze and describe the above technical conflicts. The parameters for improvement and deterioration are shown in Table 3.

Table 3. Parameter analysis of improvement and deterioration

Content	Improved technical parameters	Deteriorated technical parameters
1. The structure of the robot platform is simple and stable	Structural stability (13): the integrity of the system and the relationship between its components	Productivity (39): Useful value created per unit time
2 The grabbing robot can accurately identify and avoid obstacles	Reliability (27): the ability of the system to complete the required functions in the expected manner and state	Productivity (39): Useful value created per unit time
3. The grasping process does not damage the fruit and is flexible	Harmful factors generated by objects (31): Harmful factors will reduce the efficiency of objects or systems or the quality of completing functions	Device complexity (36): number and diversity of components in the system
4. It has a wide working range and good adaptability	Applicability and versatility (35): the ability of an object or system to respond to external changes or to apply under different conditions	Manufacturability (32): The ability to process, manufacture and assemble easily

3.2 Ways Reaching Solutions

According to the above technical conflicts, the technical parameters that need to be improved and the deterioration parameters that will occur in the design of the thymus fruit grabbing and sorting robot platform are analyzed, and all the corresponding TRIZ invention principles are listed in the technical conflict matrix in Table 2. Since the method to solve the problem is not unique, the invention principle obtained by referring to the technical conflict matrix table in Table 2 is not single, and the TRIZ invention principle obtained represents the possible direction to solve the problem, not the specific solution, so not all invention principles can solve the problem, which requires us to make specific analysis according to the actual situation, Select the most suitable scheme to apply to the design of the grab and sort robot platform structure. According to Table 2, the serial numbers of technical parameters that need to be improved are 13, 27, 31 and 35, and the serial numbers of technical parameters that cause deterioration are 32, 33, 36 and 39. Combined with Table 3, the corresponding invention principle serial numbers in TRIZ's 40 invention principles can be obtained, and the invention principle serial numbers to solve the problem are shown in Table 4.

Table 4. Invention principles for solving problems

Improve technical parameters	Deteriorating technical parameters	Solved invention principle	Screening appropriate invention principles
13	39	23, 35, 40, 3	1 (segmentation); 13 (reverse); 17 (dimension change), 23 (feedback), 35 (parameter change), 40 (composite material)
27	33	27, 17, 40	
31	36	19, 1, 31	
35	32	1, 13, 31	

4 Solution for the Design of Thymus Fruit Grabbing and Sorting Robot

For the design of the thymus fruit grabbing and sorting robot, this paper mainly describes the design of the overall structure of the grabbing and sorting robot platform, the design of the terminal actuator structure responsible for completing the grabbing action, and the machine vision grabbing and grading control scheme responsible for feedback and adjustment.

4.1 Design of Grab and Sort Robot Platform Structure

The principle of "segmentation" is adopted to increase the independence of the object and facilitate disassembly and assembly. Specifically, the robot platform for passionfruit grabbing can be modular designed, and the robot platform structure is simple and stable.

The "reverse" principle is adopted to make the moving part of the object stationary and the stationary part moving. Drive and control the robot platform base, manipulator, and end effector separately. If you want to grab and sort, you need to keep one or two of the modules stationary, and then the rest of the modules are controlled by motion to grab and sort.

The principle of "dimension change" is adopted to convert one-dimensional space motion into two-dimensional or three-dimensional space motion. The robot platform is designed as a multi-degree of freedom operation platform. In this paper, the robot operation platform is set as a four-degree of freedom to make the robot operation more flexible. The grasping robot can accurately identify and avoid obstacles, and the multi-degree of freedom can improve the working range of the manipulator.

According to the guidance of the above invention principles, the overall design of the thymus fruit grabbing and sorting robot platform can obtain the overall structure shown in Fig. 3, including the CCD visual recognition system, the main body of the sorting manipulator, the end effector of the grabbing manipulator, the translation module, the automatic control system and the collection module. The main component of CCD visual recognition system is CCD industrial camera, the main body of sorting manipulator includes adjusting arm, motor, gear and bracket, etc., and the translation module includes base, pulley, synchronous belt, motor, connecting frame, guide mechanism, etc. The end grabbing and sorting actuator is installed at the front end of the telescopic arm. At the same time, an industrial camera is installed at this position for fixation. The main body of the sorting manipulator is installed on the mobile platform module.

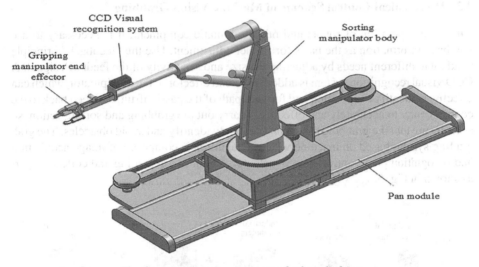

Fig. 3. Overall design structure of robot platform

4.2   Structure Design of End Actuator

Use the principle of "parameter change" to change the flexibility of the object. The joint and contact plate of the gripper finger of the end actuator are made of soft rubber, and the spring is set as a buffer to increase flexibility and solve the problem of fruit damage during grasping. In addition, based on the principle of "composite materials" and different functional requirements, materials with single material are replaced with composite materials. Here, the key stressed finger joints are made of lightweight high-strength aviation materials or aluminum alloy materials. The structure of the end actuator is shown in Fig. 4, including joint, contact plate, transmission mechanism and support part.

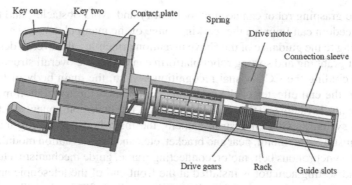

Fig. 4. Structure diagram of end actuator

4.3 Hierarchical Control Scheme of Machine Vision Grabbing

During the operation of robots and other automatic equipment, it is necessary to use feedback information as the basis for further adjustment. Use the "feedback" principle to adapt to different needs by adjusting the size and sensitivity of the feedback signal. A CCD visual recognition system is added to the end effector of the manipulator, which can effectively identify the position and forward path of the passionfruit, and feed back to the control center to accurately and effectively carry out the grabbing and sorting action, so as to ensure that the grabbing robot can accurately identify and avoid obstacles. The grab grading system based on machine vision recognition is composed of image acquisition and recognition part, controller, grab actuator, transmission grading and collection part as shown in Fig. 5.

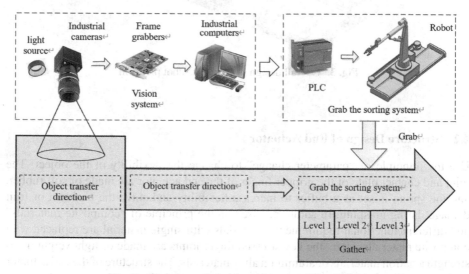

Fig. 5. Workflow of visual grabbing grading scheme

First, turn on the light source to adjust the contrast to assist the camera's recognition, and process the acquired image for contrast recognition, and then control the operation of grabbing and grading.

5 Conclusion

In view of the vacancy of thymus fruit grabbing and sorting mechanization, this paper designs the overall structure of thymus fruit grabbing and sorting robot platform based on TRIZ theory. The main work of this paper is as follows:

1) Analyze the specific engineering problems to be solved in the development and design of the thymus fruit grabbing and sorting robot platform, and establish a general problem model according to TRIZ.
2) Analyze 4 improved engineering technical parameters and 4 deteriorated engineering technical parameters in the design process.
3) According to the conflict matrix, the invention principle suitable for the design is obtained. Finally, the invention principle of "segmentation, reverse, dimension change, feedback, parameter change, composite material" is selected to design the overall structure of the passionfruit grabbing and sorting robot.
4) According to the design results, the structure of the passionfruit grabbing and sorting robot platform is complete, which can meet the set functional requirements and meet the design requirements.

Acknowledgment. This thesis is supported by (1) the project of the Yongning District Scientific Research and Technological Development Plan of Nanning City "Research on Passion Fruit Harvesting Device Based on Computer Vision Recognition Technology" (20180206A) and (2) Nanning University "Machinery Design and Manufacturing and Automation Professional Certification Construction" (ZYRZ03).

References

1. Wang, Z., Xun, Y., Wang, Y., Yang, Q.: Review of smart robots for fruit and vegetable picking in agriculture. Int. J. Agric. Biol. Eng. **15**(1), 33–54 (2022)
2. Bogue, R.: Fruit picking robots: has their time come?. Ind. Robot. **47**(2), 141–145 (2020)
3. Dewi, T., Mulya, Z., Risma, P., Oktarina, Y.: BLOB analysis of an automatic vision guided system for a fruit picking and placing robot. Int. J. Comput. Vis. Robot. **11**(3), P315–P327 (2021)
4. Polishchuk, M., Opashnianskyi, M., Suyazov, N.: Walking mobile robot of arbitrary orientation. Int. J. Eng. Manufac. **8**(3), 1–11 (2018)
5. Marie, M.J., Mahdi, S.S., Tarkan, E.Y.: Intelligent control for a swarm of two wheel mobile robot with presence of external disturbance. Int. J. Mod. Educ. Comput. Sci. **11**(11), 7–12 (2019)
6. Fang, Y.: Key points of high-yield cultivation technology of passionfruit. Agric. Develop. Equip. **09**, 162–163 (2021). (in Chinese)

7. Liang, Q., Li, Y., Long, M., et al.: Research progress in chemical constituents and pharmacological activities of passionfruit. Food Indust Sci. Technol. **39**(20), 343–347 (2018). (in Chinese)
8. Yao, L., Chen, Y., Zhao, Y., et al.: Cultivation technology of mountain passionfruit hedge. Mod. Agric. Sci. Technol.(18), 86–87+90 (2021). (in Chinese)
9. Mann, D.: An introduction to TRIZ: the theory of inventive problem solving. Creat. Innov. Manage. **10**(2), 123–125 (2001)
10. Chaoton, S., Lin, C., Chiang, T.: Systematic improvement in service quality through TRIZ methodology: an exploratory study. Total Qual. Manag. Bus. Excell. **19**(3), 223–243 (2008)
11. Sauli, S.A., Ishak, M.R., Mustapha, F., et al.: Hybridization of TRIZ and CAD-analysis at the conceptual design stage. Int. J. Comput. Integrat. Manufac. **32**(9), 890–899 (2019)
12. Lin, C., Chaoton, S.: An innovative way to create new services: applying the triz methodology. J. Chin. Inst. Indust. Eng. **24**(2), 142–152 (2007)
13. Duflou, J.R., Dewulf, W.: On the complementarity of TRIZ and axiomatic design: from decoupling objective to contradiction identification. Procedia Eng. **9**, 633–639 (2011)
14. Ye, C., Wang, M., Zhu, Y.: Innovative design of feeding equipment based on TRIZ theory. Manufac. Technol. Mach. Tool **11**, 47–52 (2018). (in Chinese)
15. Akay, D., Demıray, A., Kurt, M.: Collaborative tool for solving human factors problems in the manufacturing environment: the Theory of Inventive Problem Solving Technique (TRIZ) method. Int. J. Prod. Res. **46**(11), 2913–2925 (2008)
16. Lau, D.K.: The role of TRIZ as an inventive tool in technology development and integration in China. In: International Conference on the Business of Electronic Product Reliability and Liability, pp. 157–161 (2004)
17. Kim, T.-Y., Kim, J.-H., Park, Y.-T.: Improving the inventive thinking tools using core inventive principles of TRIZ. J. Korean Soc. Qual. Manage. **46**(2), 259–268 (2018)
18. Zhang, C., Zhao, Z.: Mechanical Innovation Design (Fourth Edition). Machinery Industry Press, Beijing (2021). (in Chinese)
19. Frizziero, L., Francia, D., Donnici, G., et al.: Sustainable design of open molds with QFD and TRIZ combination. J. Ind. Prod. Eng. **35**(1), 21–31 (2018)
20. Wang, L., Chen, G., Zhao, H.: Blueberry grabbing robot design based on TRIZ theory. J. Northeast Forest. Univ. **37**(06), 45–47 (2009). (in Chinese)
21. Wang, J., Ge, Z., Zhu, H.: Research on fruit grabbing robot based on TRIZ theory. Agric. Mechan. Res. (06), 34–36+59 (2007). (in Chinese)
22. Donnici, G., Frizziero, L., Francia, D., et al.: TRIZ method for innovation applied to an hoverboard. Cogent Eng. **5**(1), 1–24 (2018)
23. Mahfouz, A.A., Aly, A.A., Salem, F.A.: Mechatronics design of a mobile robot system. Int. J. Intell. Syst. Appl. **5**(3), 23–36 (2013)
24. Masrour, T., Rhazzaf, M.: A new approach for dynamic parametrization of ant system algorithms. Int. J. Intell. Syst. Appl. **10**(6), 1–12 (2018)

Development and Application of Ship Lock System Simulation Modules Library

Yuhong Liu[✉] and Qiang Zhou

College of Transportation and Logistics Engineering, Wuhan University of Technology,
Wuhan 430063, China
333240@whut.edu.cn

Abstract. System simulation is a crucial technology for complex engineering system, while current simulation systems lack of library for ship lock systems. In this paper, a new library of simulation modules for ship lock systems is proposed. Firstly, based on the object-oriented modeling technology and system simulation method, the composition, association relationship between the subsystems and system configuration of the ship lock logistics system are analyzed, and then the object model and dynamic model of the lock system is established. After that, the universality of the development of ship lock system simulation module library is expounded, and the ship lock system simulation module library is constructed based on the object model and dynamic model. Meanwhile, the structure and usage instructions of the module library are introduced in detail. Finally, the simulation model of Three Gorges Lock is quickly built by using the ship lock system simulation module library and the simulation test is carried out to verifies the engineering application value of the module library.

Keywords: Ship lock system · Simulation module library · Quickly built

1 Introduction

System simulation technology is one of the most effective tool to analyze complex engineering systems, and attracting much attention of many researchers in recent decade [1, 2]. Many works have used system simulation technology to solve the real problems of discrete event dynamic systems in transportation, mathematics, engineering application [3–5]. Some researchers focus on taking advantage of system simulation on ship lock. As a typical discrete event dynamic system, the lock system has complex characteristics such as nonlinearity and uncertainty, so many scholars emphasis on analyzing the passing ability and reliability of the lock system through simulation, so as to conduct more in-depth research on the ship lock system [6–10]. For instance, Li simulated the hydraulic system of the lock opening and closing machine in Weicun Hub based on AMESim, and puts forward ideas for the detailed design of the subsequent hydraulic system [11]. Chen et al. used Simulink to build a simulation model and found that the bilateral herringbone gate follow-up control system can greatly improve the rapidity of system response without affecting the stability and accuracy of the system [12]. Liu Ying

Z. Hu et al. (Eds.): ICCSEEA 2023, LNDECT 181, pp. 550–559, 2023.
https://doi.org/10.1007/978-3-031-36118-0_49

et al. established a simulation model of the Three Gorges Lock based on the simulation software SIVAK, analyzed the waiting time of the ship, and studied the lock scheduling strategy [13].

However, when building the simulation model of the lock system, the system boundary, system construction, and logical flow of the lock should be analyzed first, and then the model is built, which requires not only familiarity with the logistic activities and operational operations of the lock, but also proficiency in using different simulation software. Therefore, the modeling period is often long. In view of this, this paper deeply analyzes the complex relationship of the lock system, expounds the development ideas of the system simulation module library with wide applicability, and develops a set of object-oriented lock system simulation module library, which users can use to quickly build the lock model through simple steps such as "dragging in the module "→" establishing logical connection "→" entering basic data" on the simulation platform, and in-depth study of several problems in the lock system.

2 Module Library Development Ideas

The object-oriented modeling method is an effective method to study complex problems in complex systems by constructing software systems, in which the object-oriented object modeling technique is a programming-independent graphical representation method. The development of the lock system simulation module library is carried out.

2.1 Object Model of the Ship Lock System

The object model of the ship lock system is mainly based on the object modeling. It represents the static structure of the system by describing the objects in the system, the relationships between objects, attributes and describing the attributes and operations of each object class [14]. The main objects of the extracted lock system are: anchorage, dispatch center, approach channel, waiting area, lock door, lock chamber, and filling and emptying system. The specific descriptions are as follows.

1) Anchorage: for ships to wait at anchor, with two states of idle and occupied.
2) Scheduling Center: Scheduling the lock chamber according to the scheduling plan, generating the lock chamber scheduling chart and issuing scheduling instructions to the vessels at berth.
3) Approach channel: Has dimensions and can be occupied or vacant.
4) Waiting area: When the water level adjustment inside the gate is not completed, vessels can wait in the waiting area.
5) Lock door: Open or close.
6) Lock chamber: With dimensions for ships to berth waiting for water level adjustment, vacant or occupied.
7) Filling and emptying system: Regulate the water level elevation inside the gate by watering and sluicing water.

The object model of the ship lock system is shown in Fig. 1.

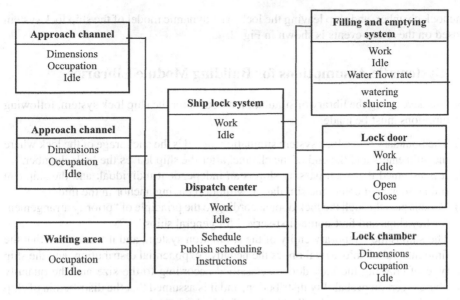

Fig. 1. Object model of the ship lock system

2.2 Dynamic Model of the Lock System

As a typical complex discrete-event dynamic system, the dynamic model of the ship lock system is mainly used to describe the instantaneous and behavioral control characteristics of the system. The state diagram is used as a descriptive tool to establish the dynamic model of the ship lock system, which represents the state of the object and the change of state between objects (event operation). The main events in the lock system are: ships declaration, ships entering to anchorage, dispatch center generating the lock chamber scheduling chart, ship entering to approach channel, filling and emptying system regulating the water level, lock door opening, lock door closing, ships entering to

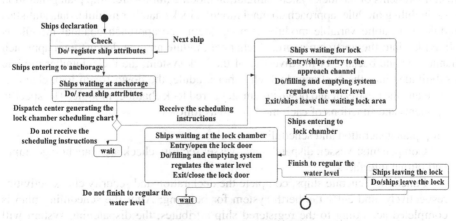

Fig. 2. Dynamic model of the ship lock system

the lock chamber, and ship leaving the lock. The dynamic model of the ship lock system based on the above events is shown in Fig. 2.

3 Systematic Assumptions for Building Module Libraries

Before developing the library of simulation modules for the ship lock system, following assumptions must be made:

1) The boundary of the lock system simulation model is the anchorage of the lock where the ship enters and the end of the channel after the ship leaves the lock chamber.
2) It is assumed that each passing ship is an independent individual, and the ship can run as soon as it enters the simulation without breaking anchor in the middle.
3) Vessels pass through the lock door according to the principle of "priority arrangement for key ships and first-come-first-served for general ships".
4) The ship is the temporary entity of the simulation system, and it is assumed that the time interval of ship arrival obeys the negative exponential distribution, and the ship type of crossing the lock door is classified according to the size and the quantity obeys a certain probability distribution, and it is assumed that the distribution of ship type is consistent with the distribution of ship deadweight tonnage.
5) When the ship is moored in the lock chamber, there are safety distances between the ships in the longitudinal and transverse directions, which are reflected in the mooring length and berthing width of the lock chamber, that is, the two are calculated by subtracting the safety distance from the original size of the lock chamber.

4 Systematic Construction of the Module Library and Description of the Module Structure

4.1 Systematic Construction of Module Libraries

The object model and dynamic model based on the object-oriented modeling method are the important basis for the division of the lock system simulation module library, and the main systems of the lock system simulation module library are: ship plan generation and scheduling module, approach channel module, lock chamber module, data statistics module and public variable module. Among them, the approach channel module is subdivided into the upstream approach channel module and the downstream approach channel module because of the diversity of the lock system; the lock chamber module is subdivided into the upstream lock chamber module, the downstream lock chamber module and the upstream and downstream staggered lock chamber module. The specific components and functions of each module are as follows.

1) Ship plan generation and scheduling module.

Components: Vessel, dispatch plan, vessel security check area, anchorage, input and output interfaces.

Function: Generate ships, complete the declaration and security check activities respectively, and enter the berth system for berthing; After the scheduling plan is completed according to the registered ship attributes, the dispatching system will issue a lock door crossing command to the ship to realize the scheduling function.

2) Approach channel module.

Components: approach channel, waiting area, input and output interfaces.

Function: The navigation route of the ship from the anchorage to the lock door is specified, and the ship can queue up at the waiting area near the lock door to prepare for entering the lock door.

3) Lock chamber module.

Components: locks, chambers, filling and emptying systems, input and output interfaces.

Function: The water level inside the lock chamber is adjusted by the filling and emptying system, so that the water level in the two adjacent levels of the lock chamber is equal and the ship completes its passage. At the same time, the ship can moor and wait inside the lock chamber during the water level adjustment.

4) Data statistics module.

Components: histograms, pie charts, variables and other tools used to count each indicator.

Function: In the simulation process, the data such as the passing capacity of the lock system, ship waiting time, ship crossing time, and the occupancy rate of the lock chamber area are counted, and quantitative data are obtained to analyze the problems studied and find ways to solve them.

5) Public variables module.

Components: public variables and attribute variables for each module entity element.

Function: As an auxiliary module, it mainly consists of common variables and attribute variables of each module.

4.2 Module Structure and Instructions for Use

The main purpose of establishing a library of ship lock system simulation modules is to enable the rapid construction of a ship lock model and to obtain the required statistical parameters for in-depth study of the system. The main structural components of each module in the lock system module library are: module input interface, module logic body and module output interface.

Among them, the module input interface is represented in the system module library as the input interface sub-module of the module, and the main component element is the variable, which can be used to receive the basic parameters input by external users, and the user only needs to import the data parameters to the input interface as required after completing the rapid construction of the model with the module library.

The module output interface is represented in the system module library as the output interface sub-module of the module. On the one hand, it is used to output statistical parameters to the data statistics module for statistics; on the other hand, it is connected with the module input interface, and the output ship is received by the module input interface and then input to the logical body part of the module for the transfer of entity flow and information flow between models.

5 Application Validation of the Module Library

In the following, the Three Gorges double-line continuous V-stage locks are used as the application object of the simulation module library, and this module library is used to build the simulation model of Three Gorges locks on the WITENSS simulation platform to verify the usability and convenience of this module library.

5.1 Module Selection and Model Building

The Three Gorges locks are double-line continuous five-stage locks, and the basic components of each line of locks are: anchorage, approach channel, lock chamber, lock door, and filling and emptying system. The layout of the Three Gorges Locks system model is built based on the Fig. 3 [15].

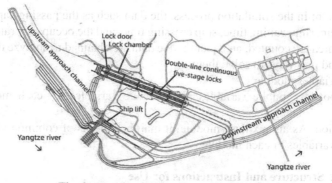

Fig. 3. Layout of Three Gorges lock system

5.1.1 Module Selection

Each line of locks consists of anchorage, approach channel and 5 levels of lock chambers. The selected basic modules and the required quantities are shown in Table 1.

5.1.2 Model Building

On WITNESS simulation platform, drag in modules in the order of public variable module, data statistics module, ship plan generation and scheduling module, upstream approach channel module, upstream first stage lock chamber module, upstream intermediate stage lock chamber module, upstream last stage lock chamber module, then establish the logical relationship between modules, import data according to the data parameters of Three Gorges ship lock system, and complete the simulation model of north line ship lock. The process of building the south line locks is similar. In addition to the infrastructure data of the Three Gorges Locks, the ship crossing scheduling follows the principle of "priority arrangement for VIP ships and first-come-first-served for general ships", and the ship type distribution is set according to the "Standard Ship Type

Table 1. Number of base modules

Module Name	Quantity
Ship plan generation and scheduling module	2
Upstream approach channel module	1
Upstream first stage lock chamber module	1
Upstream intermediate stage lock chamber module	3
Upstream last stage lock chamber module	1
Downstream approach channel module	1
Downstream first stage lock chamber module	1
Downstream Intermediate stage lock chamber Module	3
Downstream last stage lock chamber module	1
Data Statistics Module	1
Public Variables Module	1

Main Scale Series for Inland River Transportation Ships - Yangtze River System, and the interval of the scheduling plan is set to 85 min according to the scheduling plan issued by the Yangtze River Three Gorges Navigation Administration [16]. The screenshot of the established Three Gorges Lock simulation model in WITNESS is shown in Fig. 4.

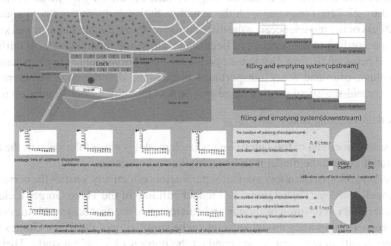

Fig. 4. Simulation model of Three Gorges lock system

5.2   Model Simulation Verification

According to the navigation data statistics released by the Yangtze River Three Gorges Navigation Administration, taking October 2022 as an example, excluding 3.12 h of suspension of the Three Gorges South Line due to high winds, 4.2 h of suspension of the Three Gorges North Line locks due to high winds, and 5 h of suspension of maintenance, the design model simulation time is 739 h, 44340 min. From the following statistical parameters to carry out the validation of the model built with the module library.

Table 2. Simulation data statistics of Three Gorges Locks

Statistical parameters	Model Simulation Data	Realistic statistics
Number of ships(ships)	3631	3606
Passing capacity (million tons)	1669	1360
Lock door opening times (times)	938	936

From Table 2, we can analyze.

1) Number of ships

In terms of the number of ships, the difference between the model simulation data and the real statistics is 0.7%, which indicates that the model simulation is very similar to the actual operation.

2) Passing capacity

It can be seen from the table that the cargo volume obtained by the model simulation increased by 22.72% compared with the actual statistics, but this does not explain the incorrectness of the model, because the distribution of ship types set in the model simulation is set according to the standard of large-scale ships, and the load factor is uniformly distributed between (0.7, 1). But in the operation of the Three Gorges Lock, because the freight volume of inland water transportation is unbalanced or there are still more small ships, the actual statistical cargo volume will be less than the model simulation data.

3) Lock door opening times

The simulation data of the Three Gorges Lock model is almost the same as the actual reality statistics, and the model is in line with the real situation.

The comparison of the above statistical parameters effectively verifies the correctness of the model; moreover, in the process of model building of the Three Gorges locks with the module library, it is time-consuming and efficient, and the user only needs to complete a few fixed steps to quickly model and realize the simulation of the Three Gorges locks, obtain the statistical parameters, and conduct the bottleneck problem study. Therefore, it can be determined that the library of simulation modules for ship lock system based on object-oriented development is not only usable, but also convenient for users to build the required models quickly.

6 Conclusion

The lock system simulation module library based on object-oriented development can quickly build lock models, reduce the construction time of models required for research, and more systematically and intuitively explore the logistics technology problems in the planning ship lock systems or existing ship lock systems, reveal the mechanism of various complex factors on the passing ability and reliability of lock systems. The module library also has excellent scalability and portability, which can facilitate users to modify the parameters and module structure according to special needs. The simulation application case of Three Gorges Lock shows that this module library has important practical value and will greatly improve the research efficiency of locks.

The development of this lock system simulation module library is based on the principle of "priority for key ships and first-come-first-served for general ships" for ship crossing, and in the subsequent research process, different principles of ship crossing can be set to give priority to the utilization rate of the lock chamber area, and the scheduling plan can be made optional, which will make the simulation more complete.

References

1. Ma, G., Xinmei A., He, L.E.I.: The numerical manifold method: a review. Int. J. Comput. Meth. **7**(01), 1–32 (2010)
2. Machaček, J., Wichtmann, T., Zachert, H., et al.: Long-term settlements of a ship lock: measurements vs. FE-prediction using a high cycle accumulation model. Comput. Geotechn. **97**, 222–232 (2018)
3. Numerical study on hydrodynamic interaction between a berthed ship and a ship passing through a lock. Ocean Eng. **88**, 409–425 (2014). (in Chinese)
4. Smith, L.D., Sweeney, D.C.: Campbell. Simulation of alternative approaches to relieving congestion at locks in a river transportion system. J. Oper. Res. Soc. (2009)
5. Richter, J., Verwilligen, J., Dilip Reddy, P., et al.: Analysis of full ship types in high-blockage lock configurations. Masin **2012**, 1–9 (2012)
6. Shang, J., Liu, C., Tang, Y., Guo, Z.: Review of lock passing capacity. Water Transp. Eng. **07**, 103–108 (2018). (in Chinese)
7. Hu, X.: Study on the passing capacity of cruise ship locks under the Minjiang River. Southwest Jiaotong University, Chengdu (2011). (in Chinese)
8. Huang, H., Zhang, W., Li, X.: Research on the passing capacity of Beijing-Hangzhou Canal locks based on queuing theory. J. Wuhan Univ. Technol. (Transp. Sci. Eng.) **33**(03), 604–607 (2009). (in Chinese)
9. Zhang, W., Gu, D., Wang, Q.: Navigation analysis and simulation study of Yangzhou section of Yanshao Line. Water Transp. Eng. (05), 122–127 (2015). (in Chinese)
10. Liu, Z.: Reliability analysis of herringbone gate lock hydraulic system of Subei Canal. China Water Transp. (Second Half Month) **18**(11), 85–86+89 (2018). (in Chinese)
11. Li, J.: Design and simulation of hydraulic system of lock opening and closing machine based on AMESim. China Water Transp. **08**, 72–74 (2022). (in Chinese)
12. Chen, K., Gao, S., Chen, X.: Simulation study on follow-up control of double-sided herringbone gates of Three Gorges Locks. Yangtze River **52**(08), 235–238 (2021). (in Chinese)
13. Liu, Y., Mou, J.: Simulation of passage capability of Three Gorges Lock based on SIVAK. J. Dalian Maritime Univ. **41**(04), 37–41 (2015). (in Chinese)

14. Zhang, H., Liu, X.: Research on object-oriented object modeling technology and its application. Software **03**, 66–68 (2011). (in Chinese)
15. Cheng, M.: Three gorges ship lock capacity evaluation and impact factor research. Chongqing Jiaotong university, Chongqing (2019). (in Chinese)
16. JTS 196-6-2012. Technical regulations for navigation scheduling of Three Gorges Locks. Industry standard – transportation, Wuhan (2012)
17. Yin, Y.: Study on main scale series of standard ship types of typical inland waterway transportation vessels. China Maritime **05**, 33–35 (2021). (in Chinese)

Hydrodynamics of Inhomogeneous Jet-Pulsating Fluidization

Bogdan Korniyenko[(✉)], Yaroslav Kornienko, Serhii Haidai, and Andrii Liubeka

National Technical University of Ukraine «Igor Sikorsky Kyiv Polytechnic Institute», Kyiv 03056, Ukraine
bogdanko@gmx.net

Abstract. The method of creating inhomogeneous (non-uniform) fluidization in the self-oscillating (auto-oscillating) mode by using a special gas distribution device (GDD) with a slit-type construction and a granulator chamber is presented. The impact of the design parameters of the gas distribution device on the quality of the hydrodynamics of the jet-pulsating mode of fluidization was determined, due to the reduction of the risk of formation of stagnant zones on the horizontal working surface of the gas distributing device for the case when the height of the vertical gas jet is three or more times less than the initial height of bed of solid particles (granules) in the granulator chamber. This condition follows from the determination of the heat and mass exchange surface during the granulation process of heterogeneous liquid systems in a fluidized bed.

Keywords: Granulation · Fluidization · Inhomogeneous Fluidization · Jet-Pulsating Mode

1 Introduction

To obtain granulated organic-mineral fertilizers of a new generation that contain nutrients of mineral and organic origin, deoxidizing and stimulating impurities it is most appropriate to use fluidization technique, which provides obtaining a granulated product with a uniform distribution of components throughout the volume of granule with a heat utilization coefficient of more than 50% [1–4].

For further intensification of transfer processes, it is proposed by the authors [5–7] to carry out the processes of dehydration and granulation with inhomogeneous fluidization.

In the works [8–12] was carried out a study of the non-uniform (inhomogeneous) jet-pulsating mode of fluidization at the ratio of the gas jet breakdown height z_f to the height of the initial stationary bed of solid granulated material H_0 in $z_f/H_0 = 0.33$.

The use of such a fluidization mode during the granulation of a liquid heterogeneous systems that based on an ammonium sulfate solution with sunflower ash impurities made it possible to obtain granulated organic-mineral fertilizers of the composition [Humates]:[K]:[S]:[N]:[Ca]:[Mg]:[P] = 1.5:21.5:13.8:9.1:4.6:3.2:1.8 with a layered structure [13–15]. At the same time, the granulation coefficient was $\psi \geq 90\%$ and the intensity of moisture removal from a unit surface of the bed of solids was 1.5 times greater than in the bubbling mode [16].

An increase in the productivity of the device based on evaporated moisture while preserving the layer-by-layer granulation mechanism is associated with an increase in the total surface of the granular material, which is accompanied by an increase in the total height of the bed of solids.

The purpose of the article is to determine the influence of the design parameters of the gas distribution device (GDD) on the intensity of movement of solid granular material along the working surface of the gas distribution device without the formation of stagnant zones.

2 Analysis of Scientific Papers

In the works [17–19] it is proposed to use a pulsating supplying of the heat carrier when processing heat-resistant materials. In particular, the authors [20] carried out the pulsating mode of supplying the heat carrier to the fluidized bed is using a mechanical pulsator, that is, the reliability of the device depends on the functioning of the mechanical pulsator operating in the environment with the high-temperature heat carrier. In this case, there is a time interval in the working cycle when the supplying of the liquefying agent to the device is completely stopped. This leads to the formation of stagnant (low-moving) zones on the working surface of gas distribution device (GDD). This increases the risk of melting of the material during the granulation of heat-labile substances when the heat carrier is supplied, the temperature of which exceeds the melting point temperature of the components of solid particles.

To eliminate this disadvantage by the authors [21] is proposed a method of inhomo-geneous (non-uniform) jet-pulsating fluidization by using of original construction of the gas distribution device (GDD).

An innovative method of interaction between the gaseous heat carrier and the solid granular material is proposed, in which the self-oscillating jet-pulsating mode of fluidiza-tion is realized without the formation of stagnant zones on the working surfaces of the gas distribution device, which contributes to the intensification of diffusion-controlled processes. For this, two jets in orthogonal planes are injected into the granulator chamber through the slit type gas distribution device. The basic jet is injected horizontally along the curved surface of the gas distributing device (GDD) and vertical jet (fountaining jet) is injected at a certain distance. The speed of the horizontal jet is selected in such a way that its merging with the vertical jet can be achieved.

During the dehydration and granulation processes, the total surface of the bed of solids is determined from the conditions of mass transfer, therefore the height of the initial bed of solid granular material H_0 is three times greater than the height of the breakdown height of the vertical jet z_f. Therefore, a gas bubble intensively forms on the upper part of the jet, which, after reaching a critical size, begins to rapidly move vertically upwards.

The potential energy accumulated by the bubble causes the emission of solid granular material into the space above bed of solids, which, thanks to the guiding insert, is intensively moved to the left part of the chamber of the apparatus in the area affected by the base slit. This leads to an instant increase in the hydraulic resistance of the gas ejecting from the horizontal slit, which is accompanied by a decrease in the speed of the dispersed system.

Such design of gas distributing device (GDD) in combination with structural changes in the granulator chamber ensures the realization of inhomogeneous (non-uniform) jet-pulsating fluidization, which under certain conditions goes into the self-oscillating mode of fluidization [14–20].

In works [9–13] it is proposed to evaluate the quality of inhomogeneous jet-pulsating fluidization in the absence of stagnant zones on working surfaces of gas distributing device (GDD) with continuous supplying of the fluidizing agent. The physical model of hydrodynamics in the zone of gas distributing device (GDD) is shown in Fig. 1 [7–9].

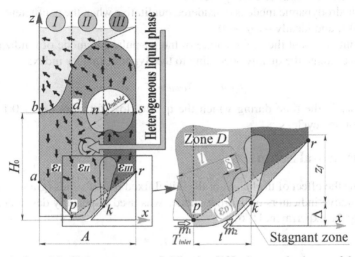

Fig. 1. Physical model of inhomogeneous fluidization [19], a) general scheme of the granulator chamber b) features of hydrodynamics in zone D, 1 – granulator chamber; 2 – slit-type GDD; 3 – guiding insert; 4 – mechanical disperser; 5 – slow moving granules

The peculiarity of the hydrodynamic mode of fluidization is that the liquefying agent is injected into the chamber of the granulator with gas distributing device (GDD) of slit type 2 through two slits, at points p and k, Fig. 1, in the horizontal (m_1) and vertical (m_2) directions.

The distance between the slits t is determined by the horizontal range of the gas jet h_{hor}, and the shape of the working surface of the gas distributing device (GDD) plate repeats the shape of the gas jet, which determines the need for the location of the second slit at a height Δ relative to the first slit. At the point k, the two jets merge, which leads to the formation of a combined jet with the breakdown height of z_f. Conventional planes drawn through points p and k divide the apparatus chamber into three zones of equal width $A/3$.

It was established that the height of the initial fixed bed of solids $H_0 = 0.32$ m is three times greater than the height of the breakdown of the gas jet $z_f/H_0 \le 0.33$. Therefore, a gas bubble begins to form cyclically at the top of the jet, which, upon exiting the bed of solids, causes the inertial removal of solid granular material into the space above bed of solids of zones *II* and *III*. After contact with the guiding insert 3, solid particles move to zone *I*, after which they quickly return to the initial volume of the bed of solids.

In the case when the energy of the horizontal gas jet coming out of the slit in (point *p*) and moving horizontally over the working surface of the gas distributing device (GDD) is insufficient, stagnation of granular material is formed in zone *D*.

$$L_D = K_1\big([\varepsilon_D] - \varepsilon_{D(i)}\big)^2 + K_2\left(\frac{[\delta] - \delta_{(i)}}{l}\right)^2, \tag{1}$$

where K_1 and K_2 – are the coefficients of proportionality ($K_1 = 0.3$ and $K_2 = 0.7$);

$\varepsilon_{D(i)}$ – experimentally determined the current value of porosity in zone *D*; $[\delta] = 0.01l$, m; l – chord length of GDD plate, m.

Such a hydrodynamic mode is considered qualitative when the coefficient of quality loss $L_D \leq 0.1$, and ideally $– L_D \to 0$.

Taking into account the cyclic nature of the jet-pulsating mode of fluidization, it is proposed to evaluate the quality according to the dynamic quality index:

$$i_{quality} = \tau_{quality}/\tau_{cycle} \tag{2}$$

where $\tau_{quality}$ is the time during which the quality loss function $L_D \leq 0.1$, s; τ_{cycle} – duration of one cycle, s.

3 Experimental Setup

To determine the effect of the height of the initial fixed bed of solids (granular material) on the qualitative indicators of hydrodynamics was used a specially developed method of providing the experiment [1–6].

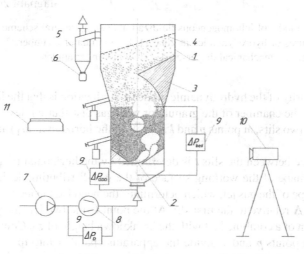

Fig. 2. Schematic representation of the experimental setup, 1 – fluidized bed granulator chamber; 2 – slit type gas distribution device (GDD); 3 – guiding insert; 4 – elastic mesh-bumper; 5 – cyclone; 6 – container for collecting dust; 7 – gas blower; 8 – chamber diaphragm; 9 – differential manometers with pressure drop sensors; 10 – video camera; 11 – weights.

Studies of the hydrodynamic mode of fluidization were carried out on a pilot plant with the dimensions of the granulator chamber $A \times B \times H = 0.3 \times 0.11 \times 1.5$ m, Fig. 2.

The slit-type gas distribution device is located in the lower part of the granulator chamber [9], and the upper one has a guiding insert and an elastic mesh-bumper. The cross-section coefficient of the gas distributing device (GDD) φ varies from 3.2% to 4%. For video and photo analysis, the front wall of the granulator chamber is transparent. The pressure drop in the bed of solids was measured using pressure drop sensors with an accuracy of ± 0.1 Pa.

4 Materials

To determine the coefficients of hydraulic resistance, was used a two-slit gas distributing device (GDD) with two values of the cross-section coefficient of GDD $\varphi = 4\%$ and $\varphi = 3.2\%$.

As granular material was used granulated ammonium sulfate with humate impurities with equivalent diameter $d_e = 2.07$ mm, density $\rho_{solids} = 1450$ kg/m^3. The mass of the bed of solid granules loaded into the apparatus chamber varied from 11.32 to 17.46 kg, which determined the height of the bed of solids H_0 (0.42, 0.52 and 0.6 m) and the nominal hydrostatic pressure ΔP_{nom}. With porosity $\varepsilon_0 = 0.4$, the height of the injection of the first jet is shifted by $\Delta = 40$ mm.

5 Results and Discussion

On Fig. 3 is given the comparison of hydraulic resistance coefficients of gas distributing device (GDD) when the cross-section coefficient of GDD φ changes from 3.2 to 4.9%, where $A = w_{slits}\, \rho_{solids}/2$.

The obtained results indicate that the reducing of the ratio of the cross-section coefficient φ by 34.6% leads to an adequate increase in hydraulic resistance value of gas distributing device (GDD) by 32.4%, Fig. 3.

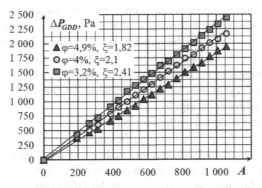

Fig. 3. Determination of hydraulic resistance coefficient of gas distributing device (GDD)

The fluidization curve was taken for the initial height of the bed of solids $H_{0(1)} = 0.42$ m at the GDD cross-section coefficient $\varphi = 4\%$, Fig. 4, where $w_{average}$ is the average velocity of fluidizing agent in granulator chamber, m/s, w_{slits} is velocity of gas flow in the slits of GDD, m/s, K_w is the fluidization number ($K_w = w_{average}/w_{fluidization}$), $w_{fluidization}$ is the average velocity of fluidizing agent at which fluidization begins, m/s, $\Delta P_{nominal}$ is the nominal hydraulic resistance of fluidized bed equal to the hydrostatic pressure of the fixed bed of solids, Pa. Jet breakdown height ratio is $z_f /H_0 = 0.1/0.42 = 0.24$.

The obtained fluidization curve can be conditionally divided into four zones. In zone 1, when the fluidization number is $0 \le K_w < 0.54$, gas movement in the apparatus occurs in filtration mode.

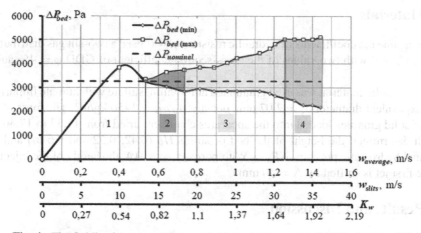

Fig. 4. The fluidization curve ($H_{0(1)} = 0.42$ m; $w_{fluidization} = 0.735$ m/s; $\varphi = 4\%$)

In zone 2 at values of the fluidization number $0.54 \le K_w < 1.0$ occurs the breakdown of gas jets that come out of two slits and merge into one with an upward motion, as a result of which a pulsation mode is realized in the apparatus.

In zone 3 at values of the fluidization number is within $1.0 \le K_w < 1.71$ the system operates in the bubbling mode of fluidization, which is characterized by an insufficient volumetric mixing index and the formation of stagnant zones on the GDD surface.

In zone 4 where $K_w \ge 1.71$ and the speed in the GDD slits is 31.4 m/s the device implements a high-quality self-oscillating jet-pulsating mode of fluidization, which ensures the absence of stagnant zones on the working surface of GDD.

Photo fixation of cycle of the self-oscillating jet-pulsating fluidization in the granulator chamber is given on Fig. 5.

After analyzing the obtained cyclogram was determined the dynamics of changes in the porosity of the bed of solids in the zone D, Fig. 6a, and was established that during the entire pulsation cycle $\varepsilon_D > 0.85$, and the quality loss function $L_D < 0.1$, Fig. 6b.

That is, with the value of $K_w = 1.71$, the index of the dynamic quality of hydrodynamics is $i_{quality} \rightarrow 1$ while the gas velocity in the GDD slits is $w_{slits} = 31.4$ m/s and the ratio of the gas jet breakdown height to the initial height of the granular material bed $z_f/H_0 = 0.24$.

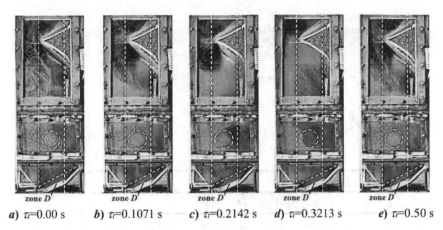

zone D   zone D   zone D   zone D   zone D

a) τ_i=0.00 s   *b)* τ_i=0.1071 s   *c)* τ_i=0.2142 s   *d)* τ_i=0.3213 s   *e)* τ_i=0.50 s

Fig. 5. Photo fixation of the state of the bed of solid granular material in the granulator chamber ($H_{0(1)} = 0.42$ m; $K_w = 1.71$; $w_{slits} = 31.4$ m/s; $f = 2$ Hz; $\tau_{cycle} = 1/f = 0.5$ s)

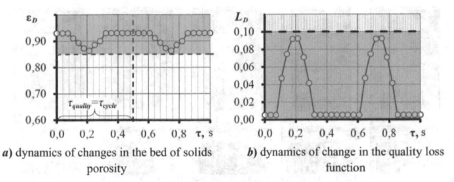

a) dynamics of changes in the bed of solids porosity

b) dynamics of change in the quality loss function

Fig. 6. Evaluation of the hydrodynamics quality in zone D ($H_{0(1)} = 0.42$ m; $\varphi = 4.0\%$; $f = 2$ Hz; $\tau_{cycle} = 1/f = 0.5$ s)

Increasing the initial height of the bed of solids by 23% to $H_{0(2)} = 0.52$ mm led to a decrease in the index of dynamic quality of hydrodynamics since $z_f/H_0 = 0.1/0.52 = 0.19$.

In order to ensure $i_{quality} \rightarrow 1$ when increasing the initial height of the bed of solids there was installed a gas distributing device (GDD) with a cross-section coefficient $\varphi = 3.2\%$ in the apparatus, which ensured an increase in the breakdown height of the gas jet z_f to $z_f = 0.12$ m, then $z_f/H_0 = 0.12/0.52 = 0.23$.

For this case, the fluidization curve was taken for the initial height of the bed of solids $H_{0(2)} = 0.52$ m, Fig. 7.

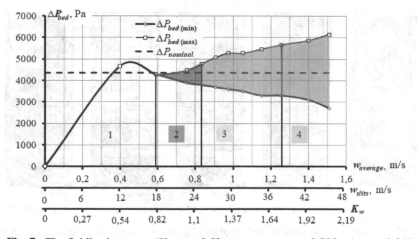

Fig. 7. The fluidization curve ($H_{0(2)} = 0.52$ m; $w_{fluidization} = 0.735$ m/s; $\varphi = 3.2\%$)

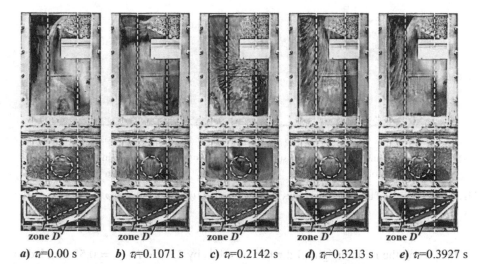

a) τ_i=0.00 s   *b)* τ_i=0.1071 s   *c)* τ_i=0.2142 s   *d)* τ_i=0.3213 s   *e)* τ_i=0.3927 s

Fig. 8. Photo fixation of the state of the bed of solid granular material in the granulator chamber ($H_{0(2)} = 0.52$ m; $K_w = 1.71$; $w_{average} = 1.26$ m/s; $w_{slits} = 38.7$ m/s; $f = 2.5$ Hz; $\tau_{cycle} = 1/f = 0.4$ s)

Photo fixations of the state of the bed of solids for the self-oscillating jet-pulsating mode of fluidization for a bed of solids with $H_{0(2)} = 0.52$ m, that was achieved at $K_w = 1.71$ and $w_{slits} = 38.73$ m/s are given on Fig. 8.

The dynamics of changes in the porosity of the bed of solids in zone D and the the quality loss function of hydrodynamics are shown in Fig. 9.

From the graphs it follows that at the values of $K_w = 1.71$ and of gas velocity in the GDD slits $w_{slits} = 38.73$ m/s a high-quality self-oscillating jet-pulsating mode of fluidization is ensured during the entire cycle time $\tau_{quality} = \tau_{cycle}$. The porosity in the

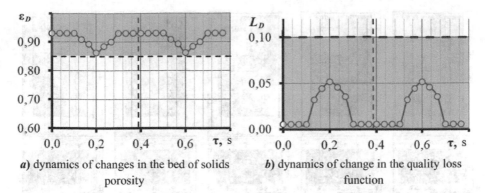

a) dynamics of changes in the bed of solids porosity

b) dynamics of change in the quality loss function

Fig. 9. Evaluation of the hydrodynamics quality in zone D ($H_{0(2)} = 0.52$ m; $\varphi = 3.2\%$; $f = 2.5$ Hz; $\tau_{cycle} = 1/f = 0.4$ s)

zone D is $\varepsilon_D > 0.85$ and the quality loss function $L_D < 0.1$, $i_{quality} \to 1$, which eliminates the risk of formation of stagnant zones on the GDD surface.

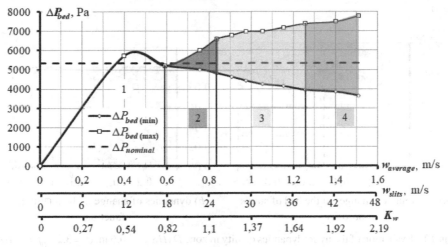

Fig. 10. The fluidization curve ($H_{0(3)} = 0.6$ m; $w_{fluidization} = 0.735$ m/s; $\varphi = 3.2\%$)

Similarly, the fluidization curve was determined with a further increase in the height of the fixed bed of solids by 15% to $H_{0(3)} = 0.6$ m. While maintaining the cross-section coefficient of GDD $\varphi = 3.2\%$ was taken the fluidization curve, Fig. 10, with photo fixations of the state of the bed of solid granular material in the high-quality self-oscillating mode of fluidization, Fig. 11.

The dynamics of changes in the porosity of the bed of solid granular material in the zone D corresponds to the conditions of the qualitative hydrodynamic mode $\varepsilon_D > 0.85$ and the quality loss function $L_D < 0.1$, Fig. 12, and so $i_{quality} \to 1$.

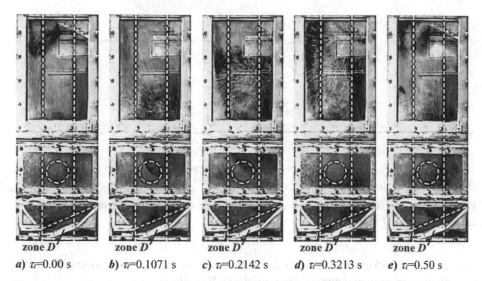

zone D zone D zone D zone D zone D

a) $\tau=0.00$ s   **b)** $\tau=0.1071$ s   **c)** $\tau=0.2142$ s   **d)** $\tau=0.3213$ s   **e)** $\tau=0.50$ s

Fig. 11. Photo fixation of the state of the bed of solid granular material in the granulator chamber ($H_{0(3)} = 0.6$ m; $K_w = 1.71$; $w_{average} = 1.26$ m/s; $w_{slits} = 38.73$ m/s; $f = 2$ Hz; $\tau_{cycle} = 1/f = 0.5$ s)

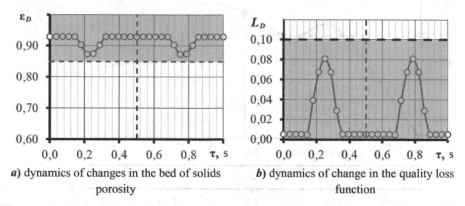

a) dynamics of changes in the bed of solids porosity

b) dynamics of change in the quality loss function

Fig. 12. Evaluation of the hydrodynamics quality in zone D ($H_{0(3)} = 0.6$ m; $\varphi = 3.2\%$; $f = 2$ Hz; $\tau_{cycle} = 1/f = 0.5$ s)

Based on the results of the researches, it was found that reducing the live cross-section coefficient of GDD allows increasing of the gas velocity in the GDD slits and increasing the kinetic energy of the gas jet, which ensures a high-quality hydrodynamic mode in the zone D ($i_{quality} = 1.0$) at the ratio $z_f/H_0 = 0.12/0.6 = 0.2$.

A decrease in the cross-section coefficient of GDD contributed to an increase in the height of the breakdown of the gas jet, Fig. 13a. At the same time, the diameter of the gas bubble D_{bubble} increases, Fig. 13b, as a result of which the inertial removal of a bed of solid granular material increases, Fig. 14, as a result of which stable kinetics of the granulation process is not ensured.

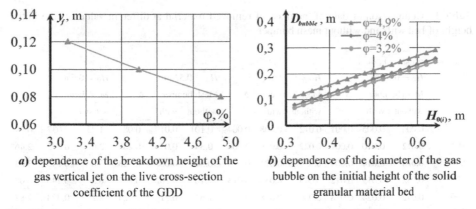

a) dependence of the breakdown height of the gas vertical jet on the live cross-section coefficient of the GDD

b) dependence of the diameter of the gas bubble on the initial height of the solid granular material bed

Fig. 13. Determination of the influence of the cross-section coefficient of GDD on the geometric dimensions of the vertical gas jet and the diameter of gas bubble

Impulse removal of granular material into the space above bed of solids causes inertial removal of granular material from the apparatus. To eliminate this disadvantage, it was proposed to install an elastic bumper in the separation zone of the granulator. The results of the experiments are shown on Fig. 14 and a comparison of this data is given Table 1.

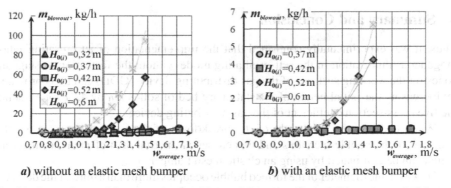

a) without an elastic mesh bumper

b) with an elastic mesh bumper

Fig. 14. Dependence of the mass of removed material $m_{blowout}$ from the average speed of the gas flow in the apparatus $w_{average}$

Thus, from the Table 1 it follows that the expediency of using an elastic mesh bumper was confirmed, as it made it possible to eliminate inertial removal of solids at the initial height of the bed of granular material of 0.37 m, and to reduce it from 94.28 kg/h to 6.24 kg/h i.e. by 15.1 times for the maximum investigated bed of solids height.

Table 1. Comparison of data of the mass of removed material at different values of the initial height of bed with and without mesh bumper

$w_{average}$, m/s	$m_{blowout}$, kg/h											
	$H_0 = 0.37$ m			$H_0 = 0.42$ m			$H_0 = 0.52$ m			$H_0 = 0.6$ m		
	Mesh bumper		Δ	Mesh bumper		Δ	Mesh bumper		Δ	Mesh bumper		Δ
	without	with		without	with		without	with		without	with	
0.77	0.006	0.005	0.001	0.012	0.008	0.004	0.101	0.019	0.081	1.03	0.037	0.99
0.91	0.012	0.010	0.001	0.022	0.016	0.006	0.34	0.056	0.285	2.6	0.095	2.46
0.97	0.018	0.014	0.004	0.024	0.021	0.003	0.58	0.088	0.488	3.8	0.143	3.63
1.05	0.032	0.019	0.012	0.045	0.032	0.013	1.16	0.162	0.997	6.3	0.246	6.08
1.09	0.042	0.024	0.018	0.060	0.038	0.022	1.64	0.219	1.424	8.2	0.323	7.87
1.15	0.064	0.032	0.032	0.095	0.051	0.044	2.77	0.344	2.429	12.1	0.485	11.59
1.20	0.090	0.040	0.050	0.139	0.065	0.074	4.29	0.501	3.787	16.7	0.681	16.00
1.26	0.137	0.054	0.084	0.219	0.087	0.132	7.24	0.790	6.449	24.6	1.023	23.58
1.30	0.182	0.065	0.117	0.297	0.106	0.191	10.26	1.070	9.193	31.9	1.341	30.52
1.40	0.366	0.104	0.263	0.635	0.172	0.463	24.54	2.260	22.278	60.8	2.642	58.20
1.49	0.688	0.158	0.530	1.258	0.267	0.991	53.81	4.460	49.347	94.28	6.242	88.04
1.59	1.387	0.253	1.134	2.688	0.434	2.255						
1.71	3.217	0.445	2.772	6.685	0.776	5.908						

6 Summary and Conclusion

Thus, it was experimentally confirmed that the implementation of jet-pulsating inhomogeneous fluidization in the self-oscillating mode without the formation of stagnant zones on the working surfaces of the gas distributing device (GDD) can be is provided by increasing the initial height of the stationary bed of solids H_0 by 42% by reducing the live cross-section coefficient of the gas distributing device.

When φ decreases, the height of the breakdown of the gas jet increases but the diameter of the bubble increases and this causes an inertial drift of granular material, which can be eliminated by using an elastic mesh bumper.

However, the diameter of the formed bubble occupies more than a third of the device's width, $D_{bubble}/A > 1/3$, which makes it impossible to ensure the necessary volumetric mixing of the bed of solids in the middle of the apparatus. To eliminate this disadvantage, it is necessary to provide a corresponding increase in the length of the device.

References

1. Kornienko, Y.M., Haidai, S.S., Sachok, R.V., Liubeka, A.M., Korniyenko, B.Y.: Increasing of the heat and mass transfer processes efficiency with the application of non-uniform fluidization. ARPN J. Eng. Appl. Sci. **15**(7), 890–900 (2020)
2. Korniyenko, B., Kornienko, Y., Haidai, S., Liubeka, A., Huliienko, S.: Conditions of non-uniform fluidization in an auto-oscillating mode. In: Hu, Z., Petoukhov, S., Yanovsky, F., He, M. (eds.) ISEM 2021. LNNS, vol. 463, pp. 14–27. Springer, Cham (2022). https://doi.org/10.1007/978-3-031-03877-8_2

3. Korniyenko, B., Kornienko, Y., Haidai, S., Liubeka, A.: The heat exchange in the process of granulation with non-uniform fluidization. In: Hu, Z., Petoukhov, S., Yanovsky, F., He, M. (eds.) ISEM 2021. LNNS, vol. 463, pp. 28–37. Springer, Cham (2022). https://doi.org/10.1007/978-3-031-03877-8_3

4. Tuponogov, V., Rizhkov, A., Baskakov, A., Obozhin, O.: Relaxation auto-oscillations in a fluidized bed. Thermophys. Aeromech. 15(4), 603–616 (2008)

5. Shevchenko, Y.M., Kornienko, Y.M., Haidai, S.S., Denisenko, V.R.: Gas distributing device of the apparatus of fluidized bed. Patent UA 136196 U Ukraine, IPC B01J 8/44, 18 February 2019, Bulletin № 15 (Ukr.)

6. Kornienko, Ya.N., Podmogilnyi, N.V., Silvestrov, A.N., Khotyachuk, R.F.: Current control of product granulometric composition in apparatus with fluidized layer. J. Autom. Inform. Sci. 31(12), 97–106 (1999)

7. Korniyenko, B., Ladieva, L.: Mathematical modeling dynamics of the process dehydration and granulation in the fluidized bed. In: Hu, Z., Petoukhov, S., Dychka, I., He, M. (eds.) ICCSEEA 2020. AISC, vol. 1247, pp. 18–30. Springer, Cham (2021). https://doi.org/10.1007/978-3-030-55506-1_2

8. Korniyenko, B., Ladieva, L., Galata, L. Control system for the production of mineral fertilizers in a granulator with a fluidized bed. In: ATIT 2020 - Proceedings: 2020 2nd IEEE International Conference on Advanced Trends in Information Theory, № 9349344, pp. 307–310 (2020). https://doi.org/10.1109/ATIT50783.2020.9349344

9. Korniyenko, B., Galata, L., Ladieva, L.: Research of Information Protection System of Corporate Network Based on GNS3. In: 2019 IEEE International Conference on Advanced Trends in Information Theory, ATIT 2019 - Proceedings, № 9030472, pp. 244–248 (2019). https://doi.org/10.1109/ATIT49449.2019.9030472

10. Korniyenko, B., Galata, L., Ladieva, L.: Mathematical model of threats resistance in the critical information resources protection system. CEUR Workshop Proc. 2577, 281–291 (2019)

11. Kornienko, Y.M., Liubeka, A.M., Sachok, R.V., Korniyenko, B.Y.: Modeling of heat exchangement in fluidized bed with mechanical liquid distribution. ARPN J. Eng. Appl. Sci. 14(12), 2203–2210 (2019)

12. Korniyenko, B.Y., Ladieva, L.R., Galata, L.P.: Mathematical model of heat transfer process of production of granulated fertilizers in fluidized bed. ARPN J. Eng. Appl. Sci. 16(20), 2126–2131 (2021)

13. Korniyenko, B., Ladieva, L., Galata, L., Nesteruk, A., Matviichuk-Yudina, O.: Information control system for the production of mineral fertilizers in the granulator with a fluidized bed. In: 2021 IEEE 3rd International Conference on Advanced Trends in Information Theory, ATIT 2021 - Proceedings, pp. 250–255 (2021)

14. Korniyenko, B., Zabolotnyi, V., Galata, L.: The optimization of the critical resource protection system of a mineral fertilizers manufacturing facility. In: Proceedings of the 11th IEEE International Conference on Intelligent Data Acquisition and Advanced Computing Systems: Technology and Applications, IDAACS 2021, vol. 1, pp. 172–178 (2021)

15. Kravets, P., Shymkovych, V.: Hardware Implementation Neural Network Controller on FPGA for Stability Ball on the Platform 2nd International Conference on Computer Science, Engineering and Education Applications, ICCSEEA 2019, Kiev, Ukraine, 26 January 2019 - 27 January 2019 (Conference Paper), vol. 938, pp. 247–256 (2019)

16. Malekzadeh, M., Khosravi, A., Noei, A.R., Ghaderi, R.: Application of adaptive neural network observer in chaotic systems. Int. J. Intell. Syst. Appl. 6(2), 37–43 (2014). https://doi.org/10.5815/ijisa.2014.02.05

17. Bhagawati, K., Bhagawati, R., Jini, D.: Intelligence and its application in agriculture: techniques to deal with variations and uncertainties. Int. J. Intell. Syst. Appl. 8(9), 56–61 (2016). https://doi.org/10.5815/ijisa.2016.09.07

18. Wang, W., Cui, L., Li, Z.: Theoretical design and computational fluid dynamic analysis of projectile intake. Int. J. Intell. Syst. Appl. **3**(5), 56–63 (2011). https://doi.org/10.5815/ijisa. 2011.05.08

19. Patnaik, P., Das, D.P., Mishra, S.K.: Adaptive inverse model of nonlinear systems. Int. J. Intell. Syst. Appl. **7**(5), 40–47 (2015). https://doi.org/10.5815/ijisa.2015.05.06

20. Babak, V.P., Babak, S.V., Myslovych, M.V., Zaporozhets, A.O., Zvaritch, V.M.: Methods and models for information data analysis. In: Diagnostic Systems For Energy Equipments. SSDC, vol. 281, pp. 23–70. Springer, Cham (2020). https://doi.org/10.1007/978-3-030-44443-3_2

21. Polishchuk, M., Tkach, M., Parkhomey, I., Boiko, J., Eromenko, O.: Experimental studies on the reactive thrust of the mobile robot of arbitrary orientation. Indonesian J. Electric. Eng. Inform. **8**(2), 340–352 (2020)

Approaches to Improving the Accuracy of Machine Learning Models in Requirements Elicitation Techniques Selection

Denys Gobov[1][✉] and Olga Solovei[2]

[1] National Technical University of Ukraine "Igor Sikorsky Kyiv Polytechnic Institute", Kyiv, Ukraine
d.gobov@kpi.ua

[2] Kyiv National University of Construction and Architecture, Kyiv, Ukraine
solovey.ol@knuba.edu.ua

Abstract. Selecting techniques is a crucial element of the business analysis approach planning in IT projects. Particular attention is paid to the choice of techniques for requirements elicitation. One of the promising methods for selecting techniques is using machine learning algorithms trained on the practitioners' experience considering different projects' contexts. The effectiveness of ML models is significantly affected by the balance of the training dataset, which is violated in the case of popular techniques. The paper aims to analyze the efficiency of the Synthetic Minority Over-sampling Technique usage in Machine Learning models for elicitation technique selection in case of the imbalanced training dataset and possible ways for positive feature importance selection. The computational experiment results confirmed the effectiveness of using the proposed approaches to improve the accuracy of machine learning models for selecting requirements elicitation techniques. Proposed approaches can be used to build Machine Learning models for business analysis activities planning in IT projects.

Keywords: Requirement elicitation technique · machine learning · decision tree · over-sampling technique · binary classification problem

1 Introduction

The choice of techniques for effectively identifying requirements in developing IT solutions is essential in planning business analysis work. A thorough understanding of the variety of techniques available, their advantages and disadvantages assists the business analyst in adapting to a particular project context [1]. The business analyst must create a combination of techniques to guarantee the success of software requirements identification activities, as it is impossible to fulfill all project stakeholders' needs using just one technique [2]. One approach to solving this problem is to use machine learning models, where training samples are formed based on practitioners' experience or recommendations. For example, in studies [3, 4], a machine learning model was built that recommends the usage of elicitation techniques depending on the combinations of

© The Author(s), under exclusive license to Springer Nature Switzerland AG 2023
Z. Hu et al. (Eds.): ICCSEEA 2023, LNDECT 181, pp. 574–584, 2023.
https://doi.org/10.1007/978-3-031-36118-0_51

factors. However, the use of machine learning models for commonly used techniques and the Accuracy of their work is associated with several difficulties. If we select the most frequently used techniques as a target class, the gathered dataset got imbalanced, i.e., most observations belong to a target class with a positive value equal to "1," which indicates that the technique was used in the project.

The mentioned model from [3] was constructed with Decision Jungle Tree (DJT) algorithm that was empirically selected as the most efficient. DJT learner, like a binary decision tree learner, creates a tree that is biased to the majority class when a dataset is imbalanced [5]. The reason causes are the following: while recursively partitioning the dataset so that the observations with similar target values are grouped together, it qualifies the candidate split of the node m using the parameters that minimize the impurity.

$$Q = argmin_t G(Q_m, t) \tag{1}$$

where

$$G(Q_m, t) = \frac{r_m^{left}}{r_m} H^{left}\left(Q_m^{left}(t)\right) + \frac{r_m^{left}}{r_m} H^{right}\left(Q_m^{right}(t)\right) \tag{2}$$

$$H^{left}\left(Q_m^{left}(t)\right) \tag{3}$$

$$H^{right}\left(Q_m^{right}(t)\right) \tag{4}$$

are gini or entropy measures of split's impurity for the classification task. The lower the value of

$$H^{left}\left(Q_m^{left}(t)\right) \tag{5}$$

$$H^{right}\left(Q_m^{right}(t)\right) \tag{6}$$

the better the split. When the dataset is imbalanced, there is a significant probability that the majority of the class samples are included in the same nodes, which makes entropy close to zero. As a result, the created tree is biased toward the majority class [6, 7].

The mentioned problem negatively impacts the model's performance which is visible in Table "Accuracy metrics' values" in the study [3]: the high value of Accuracy (>0.9) for techniques: "Interview", "Document Analysis" and relatively low value (~0.7) of the area under the ROC curve (AUC). According to studies [8, 9], AUC is an accurate indicator of the model's prediction ability when a dataset is imbalanced because it shows the ability of the classifier to rank the positive instances relative to the negative ones. The ROC curve graph is based upon true positive rate (tpr) and false positive rate (ftp), so it does not depend on the class's distributions' changes. In contrast, Accuracy measures the minimization of the overall error to which the minority class contributes very little [10].

To tackle the problem and improve the performance, in the current work, we will propose to enhance the built ML model by adding a data preprocessing step, which is to balance the dataset before training a machine learner algorithm. The prediction's quality will be measured based on the same metrics used in work [3] to obtain the predictions' comparable results.

Another point is a "Positive feature importance" algorithm used in [3] – according to which features' importance is identified based on the build-in capability of the decision tree learner when the feature used in the node's split most often receives the higher score. Such an approach depends on the learner's prediction Accuracy and can result in a wrong feature's score when a created decision tree is biased toward the majority class. In the current work, we propose to use machine learner-independent methods to identify the feature's importance score. The paper is structured as follows. Section 2 reviews the related works on approaches for handling imbalanced datasets in ML models. Section 3 contains the description of the input data and the experiment's scheme: dataset characteristics, data pre-processing procedure, train and test procedures. Section 4 is devoted to the experiments' results of using proposed improvements for the ML model. Section 5 concludes the paper with the main study findings and future work.

2   Related Work

In the study [11], to improve classification models' results for imbalance datasets are considered the following data preprocessing strategies: 1) random resampling: 1.1) resampling the small class at random until it contains the number of samples that is equal to the majority class; 1.2) focused resampling the small class only with data occurring close to the boundaries with a factor equal to 0.25 between the concept and its negation. 2) downsizing: 2.1) random down-sizing – eliminating at random the elements of the over-sized class until it matches the size of the majority class; 2.2) focused down-sizing - eliminating only elements further away from the boundaries; 3) learning by recognition - to use unsupervised machine learning algorithms and ignore target class in dataset. The study concluded that resampling and down-sizing methods are very effective compared to the recognition-based approach. However, it was left without a recommendation, which is a preferable method – random or focused resampling/down-sizing. Furthermore, it was specified that by random down-sizing of the majority class, there is a chance to exclude meaningful information; therefore, the preferences to be given to resampling or focused down-sizing.

In the study [12] it was proposed an over-sampling approach – Synthetic Minority Over-sampling Technique (SMOTE) in which the minority class is over-sampled by creating "synthetic" examples that are identified as k minority class nearest neighbors. If the amount of over-sampling needed is 100%, then the five nearest neighbors are chosen. The authors specified that the SMOTE approach could improve the accuracy of classifiers because, with an over-sampled dataset, the classifier builds larger decision regions. After all, the created tree isn't biased. Except for the Accuracy, oversampling algorithm SMOTE can improve the CV score, F1 Score, and recall of classifiers [13]. Further enhancements to SMOTE were proposed in the study [14] – the new method was called SMOTEBoost – according to which SMOTE is applied on each boosting round. With SMOTEBoost, the higher prediction Accuracy of the classifier was achieved.

The study [15] proposed to combine the sampling technique and ensemble idea, i.e., to make the dataset balanced by down-sizing or over-sampling and after to grow each tree. The Random Forest algorithm was used in experiments with methods: SMOTE, SMOTEBoost (called Balanced Random Forest), and the results are compared with the Weighted Random Forest algorithm. The conclusions were that both Weighted Random Forest and Balanced Random Forest outperformed other learners, and there was no winner between those two learners. The comparison was made by applying the ROC convex hull method [16]. The better the "more northwest" ROC convex hull because it corresponds to the classifier with a lower expected cost.

Several filter methods are proposed in studies [17] to have the most predictive feature selected independently from machine learning algorithms. The appropriate filter's method is selected depending on the parameters: a type of machine learning task; in case of classification – several unique values in target class (binary or multi-class); type of predictive features – continuous/discrete or categorical; type of data in the dataset – flat feature; structure feature; linked data; multi-source; multi-view; streaming feature; streaming data. However, the selection of the best-fit method for the concrete dataset was not formalized; therefore, an empirical test is required.

In the current work, based on the results of studies [11–16], to balance a dataset, we will apply the over-sampling method SMOTE. We also use a learning algorithm called Random Forest Classifier (RFC) to build a model. The mentioned enhanced methods: SMOTEBoost and Weighted Random Forest, won't be used to avoid over complication of our model. We will use the ROC convex hull method to evaluate the built model's ability to recommend an elicitation technique. However, classification task metrics like precision and recall will be calculated to compare the effect of the proposed enhancements with the results achieved in the work [3]. Because the study [17] shows no recommendations on how systematically identify the filter technique which fits best the concrete dataset, in the current study, we will empirically find an optimal for our dataset a filter method.

3 Methodology

To apply and evaluate the effects of the proposed enhancements, we will include additional blocks in the traditional supervised machine learning work cycle ("blue" rectangles in Fig. 1). A new block, "Data Balancing," will generate synthetic samples that will diminish the class-imbalance problem with method SMOTE. The applied technique will get a new sample

$$S = x + k \cdot \left(x^R - x \right) \tag{7}$$

as linear combinations of two samples from the minority class (x^R and x) with $0 \le k \le 1$; x^R is randomly chosen among the 5-minority class nearest neighbors of x.

In the added new "Feature scoring" block, we use the filter methods, independent from the learning algorithm, to estimate the score of each feature. Different strategies can be used to get feature scores without knowing which filter method fits the dataset the best: 1) to average the feature's scores received by applying different filters; 2) to choose the feature's highest score from the scores received by applying different filters; 3) to apply learning algorithm with feature selected by different methods and select the best

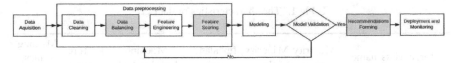

Fig. 1. Supervised machine learning work cycle with added blocks.

filter method to get features' scores based on the analysis of ROC convex hull. The last strategy will be used in the current work. In a new "Recommendations Forming" block with the model's forecasts and feature scores as input information, recommendations regarding recommended techniques will be formed.

4 Experiment

We execute experiments to assess the effectiveness of the proposed enhancements. The effectiveness of the built machine learning model was measured via Accuracy, AUC, precision, and recall metrics. The received metrics' values are compared with the results achieved in work [3], and validations of the proposals are based on the comparison results.

Our databases' characteristics and imbalanced ratios calculated as majority-to-minority samples are specified in Table 1. The features that are included in the dataset are two types: 1). Features to describe the project's context; 2) features to list all elicitation techniques used in the project [18]. The following features belong to the first type: Country, ProjectSize, Industrial Sector, Company Type, Company Size, System/Service Class, Team Distribution, Experience, Way of Work, Project Category, BA Activities (BA Only Role), and Certified. The following features belong to the second type: Benchmarking and Market Analysis, Brainstorming, Business rules analysis, Collaborative games, Data mining, Design Thinking, Interface analysis, Interviews, Observations, Process analysis, Prototyping, Reuse database and guidelines, Stakeholders list, map or Personas, Survey or Questionnaire, Document Analysis, Workshops and focus groups, Mind Mapping.

The features' information that is included in the dataset with target class names: "Elicitation", "Document Analysis", "Interface Analysis", and "Process Analysis" is the same with the following exception when the feature's name is equal to the target class, then this feature isn't included in features' list.

The data pre-processing procedure has included: 1) removing features whose values are not unique; 2) applying SMOTE to over-sample a minority class to match 100% of the majority class.

Train and test procedures included: 1) Random Forest tree classifier is trained on the sub-dataset, which is received as a result of a random split with the proportion 80/20 for train and test correspondingly. 2) Trained learner is tested with the test subset.

Identification of feature's importance score: based on the dataset's parameters from Table 1 to identify the feature's score can be used methods: 1) Chi-squared stats - measures dependency between non-negative feature and class, so that irrelevant for classification feature is set a low score [19]; 2) Analysis of variance (ANOVA f-test) - measures the ratio of explained variance to unexplained variance, the higher values mean

Table 1. The characteristics of datasets

Target class name	Majority class	Minority class	Imbalance ratio	Machine learning task	Feature type	Missing values, Y/N?
Interviews	282	41	6.9			
Document Analysis	276	47	5.9	Binary classification	Discrete	N
Interface Analysis	232	91	2.5			
Process Analysis	213	110	1.9			

more feature's importance to a target class [20]; 3) mutual information (MI)– measures the amount of shared information between independent feature and target class. It can detect non-linear dependencies, which is its main strength, and it can be applied when the dataset includes continuous, discrete, or both types of data [21, 22].

Because the types of the results of the mentioned filters are different, to understand which filter best fits datasets from Table 1, we will do the following: 1) create ROC curves calculated by Random Forest with balance data and feature selected by methods: chi-squared stats, mutual information, Anova f-value; 2) select the best-fit method based on ROC convex hull graphs analysis.

5   Results and Discussions

Figure 2 illustrates how the split's impurity measured by entropy increased after the dataset was balanced by applying SMOTE, which indicates that more samples from different target classes are included in the node; hence both classes are presented before the decision in favor of the majority, or minority class is made, and the created tree isn't biased. The gap between mean entropies is $\approx$ 10% for datasets with imbalance ratios of more than 5 (Fig. 2- pictures (a) and (b)), and it is 5% and 4% for datasets with imbalance ratios of 2.5 and 1.9, correspondingly. Therefore, the graphs in Fig. 2 display the direct dependency between the decision tree's bias and the dataset's imbalance ratio.

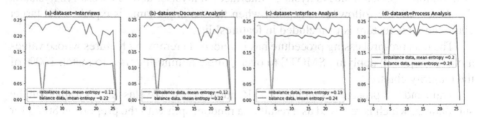

Fig. 2. Split's impurity of imbalance and balance datasets calculated by entropy

The values of the performance metrics of Random Forest learners with a balanced test subset are recorded in Table 2. The area under ROC curves for the balanced dataset increased significantly compared to the imbalanced dataset, and accuracy increased as well.

Similar improvements are recorded in Fig. 3, where SMOTE-Random Forest computed ROC convex hull colored by orange, and Random Forest computed ROC convex hull shown by blue color with the imbalanced dataset. The SMOTE-Random Forest ROC convex hull dominates the ROC convex hull without SMOTE, making SMOTE a more optimal classifier.

Table 2. Accuracy and AUC for balanced by SMOTE and original dataset

Techniques	Accuracy			AUC		
	Imbalanced	Balanced	Improving	Imbalanced	Balanced	Improving
Interviews	0,938	0.94	0%	0,720	0.97	35%
Document Analysis	0,901	0.91	1%	0,728	0.97	33%
Interface Analysis	0,790	0.84	6%	0,676	0.9	33%
Process Analysis	0,778	0.82	5%	0,757	0.88	16%

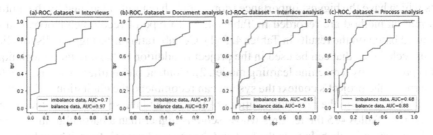

Fig. 3. ROC convex hull of imbalance and balance datasets calculated by RFC

The p-value of paired t-test for precision and recall received for the balanced dataset and imbalanced dataset from [3] are recorded in Table 3. p-value = 0.018 indicates a statistically significant difference in the precision of the Random Forest classifier with a balanced dataset compared to imbalance which means that the ability of the classifier not to label as positive a sample that is negative is affected when the negative's class distribution's changes because of over-sampling. The p-value = 0.087 indicates the statistically insignificant difference in the recall, i.e., the classifier's ability to correctly find all positive samples when the positive class is a majority class has not been affected by applying the oversampling of the minority class.

Table 3. p-value of paired t-test of precision and recall

Dataset	Interviews	Document Analysis	Interface Analysis	Process Analysis	p-value
Precision, balanced data	0.89	0.88	0.81	0.78	0.018
Precision, imbalanced data	0.936	0.899	0.831	0.818	
Recall, balanced data	1	0.96	0.9	0.88	0.087
Recall, imbalanced data	1	1	0.922	0.9	

To select the best feature selection method, we used ROC convex hull graphs presented in Fig. 4. For each dataset from Table 1, graphs show the "more northwest" ROC curve when features are selected by the mutual information method. The ten features which have the biggest importance score for each dataset calculated by the mutual information method are recorded in Tables 4, 5. To type of recommendations can be produced based on the results in Tables 4, 5: 1). Collaborating filtering - the list of elicitations techniques that can be used in the project in addition to the elicitation technique that is predicted by a machine learning model. 2). Content-based filtering – based on the similarity of the project's context the system can recommend the elicitation technique to be used. Due to form the recommendations of the first and the second types, the system selects from Tables 4, 5 the features whose score is higher than a particular number and which correspondingly belong to the first and the second feature's type. The value of the score, which the system uses to select the feature, is proposed to be a user's choice.

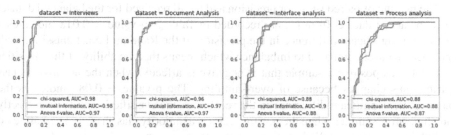

Fig. 4. ROC convex hull calculated by Random Forest with balance data and feature selected by methods: chi-squared, mutual information, Anova f-value

Table 4. Features' importance score. Interviews and Document analysis.

Interviews		Document analysis	
Feature	Score	Feature	Score
Project Size	0.3	Experience	0.32
Experience	0.28	Project Size	0.27
WoW	0.27	WoW	0.25
Prototyping	0.25	Industrial Sector	0.22
Project Category	0.23	Company Size	0.2
Company Type	0.21	Project Category	0.2
Process analysis	0.19	Process analysis	0.19
Industrial Sector	0.18	Company Type	0.19
Company Size	0.18	Interface analysis	0.18
Interface analysis	0.17	Observations	0.17

Table 5. Features' importance score. Interface analysis and Process Analysis

Interface analysis		Process Analysis	
Feature	Score	Feature	Score
Experience	0.22	Business rules analysis	0.15
Project Size	0.15	Project Category	0.15
WoW	0.15	Company Type	0.14
Company Type	0.15	Industrial Sector	0.13
Project Category	0.14	WoW	0.12
Team Distribution	0.14	Experience	0.11
Observations	0.13	Workshops and focus groups	0.11
Prototyping	0.12	Stakeholders list, map or Personas	0.1
Document analysis	0.12	Benchmarking and Market Analysis	0.1
Company Size	0.12	Team Distribution	0.09

6 Conclusions

The current study proposes several enhancements to the machine learning model for getting recommendations on elicitation techniques usage based on the combinations of factors. The first proposal is to balance the dataset by applying SMOTE methods before the training and testing classifier. As a result, improvement in the "Accuracy" indicator was obtained in the range from 0% to 6%, and the "AUC" indicator from 16% to 35%. The dependency between the value of the imbalance ratio and created tree biased to the majority class was visual. The results can be used in future works to formalize whether applying SMOTE to balance a dataset is necessary, as an increased dataset will demand more operation time for machine learning. The second proposal is to calculate the features' importance score independently from the machine learner classifier. A mutual information method was identified as the best fit for the considered datasets. The produced list of the most predictable features is proposed to use in recommended systems of both types: collaborating filtering and content-based filtering. The results obtained allow the construction of more precise models for the recommendation of requirements elicitation techniques contingent upon the project context. Further research can be conducted to explore additional BA tasks to determine correlations and make suggestions for selecting techniques for requirement specification and modeling, validation, and verification based on balanced datasets.

References

1. Rehman, T., Khan, M., Riaz, N.: Analysis of requirement engineering processes, tools/techniques and methodologies. Int. J. Inform. Technol. Comput. Sci. **5**(3), 40–48 (2013). https://doi.org/10.5815/ijitcs.2013.03.05
2. Manzoor, M., et al.: Requirement elicitation methods for cloud providers in IT industry. Int. J. Mod. Educ. Comput. Sci. **10**, 40–47 (2018). https://doi.org/10.5815/ijmecs.2018.10.05
3. Gobov, D., Huchenko, I.: Influence of the software development project context on the requirements elicitation techniques selection. Lect. Notes Data Eng. Commun. Technol. **83**, 208–218 (2021)
4. Gobov, D., Huchenko, I.: Software requirements elicitation techniques selection method for the project scope management. CEUR Workshop Proc. **2851**, 1–10 (2021)
5. Derya, B.: Data Mining in Banking Sector Using Weighted Decision Jungle Method. Data Mining-Methods, Applications and Systems. IntechOpen (2020)
6. Visa, S., Ralescu, A.: Issues in mining imbalanced data sets-a review paper. In: Proceedings of the Sixteen Midwest Artificial Intelligence and Cognitive Science Conference, vol. 2005, pp. 67–73 (2005)
7. Krawczyk, B.: Learning from imbalanced data: open challenges and future directions. Progress Artific. Intell. **5**(4), 221–232 (2016). https://doi.org/10.1007/s13748-016-0094-0
8. Fawcett, T.: An introduction to ROC analysis. Pattern Recogn. Lett. **27**(8), 861–874 (2006)
9. Flach, P.A.: ROC analysis. In: Sammut, C., Webb, G.I. (eds.) Encyclopedia of machine learning and data mining, pp. 1–8. Springer US, Boston, MA (2016). https://doi.org/10.1007/978-1-4899-7502-7_739-1
10. Orallo, H., et al.: The 1st workshop on ROC analysis in artificial intelligence (ROCAI-2004). ACM SIGKDD Explor. Newsl **6**(2), 159–161 (2004)
11. Japkowicz, N.: Learning from imbalanced data sets: a comparison of various strategies. AAAI workshop on learning from imbalanced data sets, vol. 68, pp. 10–15 (2000)
12. Chawla, N., et al.: SMOTE: synthetic minority over-sampling technique. J. Artific. Intell. Res. **16**, 321–357 (2002)
13. Chakraborty, V., Sundaram, M.: An efficient smote-based model for dyslexia prediction. an efficient smote-based model for dyslexia prediction. Int. J. Inform. Eng. Electron. Bus. **13**(6), 13–21. (2021). https://doi.org/10.5815/ijieeb.2021.06.02
14. Chawla, N., et al.: SMOTEBoost: improving prediction of the minority class in boosting. Lect. Notes Comput. Sci. **2838**, 107–119 (2003)
15. Díez-Pastor, J., et al.: Random balance: ensembles of variable priors classifiers for imbalanced data. Knowl. Based Syst. **85**, 96–111 (2015)

16. Bettinger, R.: Cost-sensitive classifier selection using the ROC convex hull method. SAS Institute, pp. 1–12 (2003)
17. Li, J., et al.: Feature selection: a data perspective. ACM Comput. Surv. **50**(6), 1–45 (2017)
18. Gobov, D., Huchenko, I.: Requirement elicitation techniques for software projects in Ukrainian IT: an exploratory study. In: Proceedings of the Federated Conference on Computer Science and Information Systems, pp. 673–681 (2020). https://doi.org/10.15439/2020f16
19. Su, C.T., Hsu, J.H.: An extended chi2 algorithm for discretization of real value attributes. IEEE Trans. Knowl. Data Eng. **17**(3), 437–441 (2005)
20. Lindman, H.R.: Analysis of Variance in Experimental Design. Springer Science & Business Media (2012)
21. Gao, W., Kannan, S., Oh, S., Viswanath, P.: Estimating mutual information for discrete-continuous mixtures. Adv. Neural Inform. Process. Syst. **30** (2017)
22. Ross, B.C.: Mutual information between discrete and continuous data sets. PLoS ONE **9**(2), e87357 (2014)

Machine Learning Based Function for 5G Working with CPU Threads

Maksim Iavich[✉]

Caucasus University, 1 Paata Saakadze St., 0102 Tbilisi, Georgia
miavich@cu.edu.ge

Abstract. A huge amount of information is transmitted over wireless networks. Moreover, the volume of transmitted information is constantly increasing, one of the many factors in this is that new mobile devices are continuously communicate with each other through the network, and the number of multimedia applications, streaming, video conferencing, social networks and other things are growing behind them. To meet current and expected needs, Transition to 5G networks is taking place at great pace. As of 2021 and 2022, approximately one million minutes of video per second was transmitted over the Internet worldwide, and as data grows rapidly, the 4G system needs to be replaced with a more powerful 5G system with increased bandwidth and improved quality of service to ensure a secure and stable connection. 5G have the security problems. During our research, we were able to identify various 5G security vulnerabilities and then study them in detail. On the basis of which we have implemented a new machine learning based cybersecurity model with its Firewall and IDS/IPS, which we describe in this article. We have analyzed the use of different machine learning algorithms for our task. Finally, decision tree algorithm was chosen. We have implemented this algorithm using CPU threads to increase the efficiency. The main significance of the offered security function is that it is very efficient and works very quickly; therefore, it can be considered for real time usage. The offered security function is trained with attack patterns created in a simulation lab. We have tested this mechanism in the simulated 5G network. The experiment was carried out in our laboratory, where we used 10 Raspberry Pi modems to simulate attacks on the server. A similar approach will be useful for future versions of 5G.

Keywords: 5G · machine learning · networks · CPU threads

1 Introduction

A large amount of data is transmitted over wireless networks. Moreover, the size of transmitted data is increasing. New mobile devices are continuously added to the network, and the multimedia applications, streaming, video conferencing, social networks and other things are growing behind them. To meet ever-increasing demand, the telecommunications industry is rapidly deploying 5G and beyond [1–4]. At 2021–2022 more than one million minutes of video per second was transmitted over the internet, the size of data grows rapidly. Therefore, the 4G system needs to be replaced with a more powerful

Z. Hu et al. (Eds.): ICCSEEA 2023, LNDECT 181, pp. 585–594, 2023.
https://doi.org/10.1007/978-3-031-36118-0_52

5G system with increased bandwidth and improved quality. This will ensure a secure and stable connection and communication. The introduction of 5G requires the introduction of new data storage and processing technologies, which will certainly set new tasks for us in terms of the functionality and security of 5G systems. The world's leading scientists are actively researching the problems related to the security of 5G systems. Research shows that 5G security is not yet at the desired level. This is confirmed by a number of successful attacks on these networks, such as vulnerabilities of MNmap, MiTM, and a battery drain attack that can lead to malicious code infiltrating the system. So, it is needed to look for new ways to secure 5 and future 6G networks by creating new architectures and new algorithms based on AI/ML [5–7].

5G networks must be able to transmit very large amounts of information with very low latency. All this can be achieved by deploying high-frequency towers at short distances. In addition, there will be a need for a new network architecture, new data storage technologies with great capabilities, the introduction of which will inevitably lead to new cybersecurity problems. Infrastructure will soon be fully migrated to 5G system, so to ensure security, strong security systems must be established, Which requires comparing the long-established 4G systems with 5G and studying the differences between them. Weaknesses in 5G systems also need to be identified and ways to address them need to be found. It must be proposed the efficient security function to protect 5G and beyond networks. The major research objectives is to offer the security function, which will be very efficient, using machine learning approach.

As the machine learning approach we use decision tree algorithm. There are the related works, which use decision tree algorithms, but they don't describe the parallel approaches using CPU threads. Our approach is also unique, because it uses the manual approach, which increases the detection precision. The main limitation of the existing approaches is that they cannot be used in the real time. Our main goal is to increase the efficiency in order to make it usable in the real time.

2   About 5G Security Features

The LTE systems providing communication and security of 4g and 5g systems are similar. Therefore, we can divide the 5g security features into two categories, and we can take care of the device and the radio node to ensure safety. The network access security devices, which ensure security of connection between radio node and device, and also devices ensuring security of access various services for users are included in first category (Fig. 1).

WAMS data and SCADA data can be stable between two short periods. Before determining the correlation coefficient, the upper and lower bounds of the correlation coefficient are required. The Pearson correlation coefficient proposed in this paper is widely used to measure sequence correlation.

In a 5G system, access authentication is at the forefront. The operation is completed at the starting stage of identification, if authentication completes successfully the session keys are determined, which are used for network – device communication.

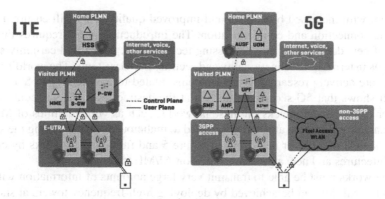

Fig. 1. LTE/4G and 5G security architecture

This the operation is executed in compliance with 3GPP 5G security regulations to protect EAP – Extended Authorization Protocol, whose purpose is to use data, certificates, public key passwords, and usernames – credentials stored on SIM cards. Identification and key agreement mechanisms are used to pass authorization procedures of device and network. As a result of this process, The KSEAF keys generated by this process are sent to the secure channel – SEAF service network. Thus, KSEAF is used in situations where re-authentication is not required. In 3GPP network authentication, keys are generated using 3GP and UE functions. 5G uses SUPI, which assigns a globally unique, permanent subscriber ID to each user. Permanent caller identifier supports IMSI-international mobile subscriber identifier and NAI – network access identifier. It is important that SUPI remains secret when connected to a specific mobile device, unlike 3 and 4G systems. in which the IMSI identification of a specific item on the network is exposed. Please note that the SUCI-secret signature identifier should be used until verification the authenticity of the network and device completes. After completing identification, the home network SUPI can be opened to service network to ensure that subscriber's identity is not stolen. To avoid this a device is connected to RBS while trying to sniff unencrypted traffic. To conceal SUPI out of SUCI the SIDF – Subscription ID disclosure feature should be used. Which one uses privacy key from a privacy-related home network public/private key pair that is securely stored on the home network of operator. Access parameters in SIDF should allow only the home network to request SIDF. Disclosure must go through so called UDM – the UniFi Dream Machine. Port settings should be set so that only the home network should be able to request the port [8–10].

To ensure the security of messages sent through the N32 interface around the perimeter of the public land mobile network (PLMN), the 5G architecture includes a proxy security boundary protection proxy. All network function (NF) service layers drop messages to SEPP, which are being kept till they are sent through the N32 interface. On the other way each message is sent to SEPP by N32 interface, at his turn SEPP checks data and transmits back it to the suitable network functional level. SEPP protects all information from application layer that moves among two NFs in two various PLMNs. In order to achieve the highest best security level, 5G uses a network slicing method. The network segment is responsible for an corresponding quality of service. Thus, we can divide the whole network into slices so that they do not overlap. This mechanism allows you to segment the network and securely manage the corresponding services. Thus, the system is easy to manage and moreover, it is more secure.

3   5G Security Issues

Our research gives us reason to say that 5G security is still not perfect and faces challenges. The presented article analyzes the problems related to 5G security:

1. Various types of attacks are constantly being carried out on the 5G network, it has much more entry points for attackers, because the 5G network architecture allows them to exploit numerous weak points, flaws and bugs.
2. With the introduction of 5G networks we need to introduce new functionality, and with new functionality comes new vulnerabilities. Attackers can attack base points and control functions.
3. Since mobile operators are dependent on providers may cause the development of new attack routes and may greatly increases attack damage.
4. After the introduction of 5G systems, almost all vital IT applications will use it, so we may encounter security issues related to the integrity and availability.
5. There will be a lot of devices on the 5G network, which can provoke DoS and DDoS attacks.
6. Network segmentation also may cause some security clauses, as attackers can attempt to connect a certain device using a desired network segment.

It should be noted that an attacker can use the flaws found in 5G with the help of which attacker can perform illegal actions. We can group these attacks as follows:

- MNmap

To conduct an experiment, our researchers sent a message to the network, the purpose of this action was to reveal fake network. So they fabricated base station and were able to get all the information about the devices that were connected to the networks. Finally managed to determine device type, model, manufacturer, operating system, and version.

- MiTM

The design of 5g systems also allows for Man-in-the-Middle (MiTM) attacks. By means of which it Bidding and battery drain attacks may be performed possible. As it's possible to slow down 5G speeds by disabling Multiple Input–Output (MIMO). Thus, the speed may fall to the of 2G/3G/4G networks speed.

- Battery drain attack

During this attack, small information packets are sent to the victim device, which drains the battery. The attacker adapts PSM, while the victim device is connecting network the hacker offers it a desired network, thereby gaining control of the device.

It is worth noting that for the new design of 5G and future 6G networks, it is necessary to develop new AI/ML-based algorithms that provide a appropriate level of cybersecurity and appropriate defense of mobile subscribers, industry and government.

4 Advantages of Our Approach

Our proposal is to create a security module where IDS/IPS and firewall are integrated on the same server and used on each base station (Fig. 2).

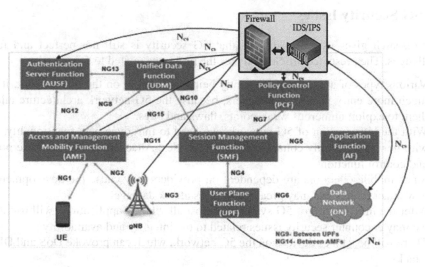

Fig. 2. Cybersecurity module

In the article, we have presented materials proving that we can still hack 5G security with Probe and Dos or software attacks. Our IDS initial version is based on M/L. Effective protection against these attacks can be performed by using different datasets in the training process. Most often, NSL-KDD kits are used for such studies. Which is widely used for creating IDS prototypes and conducting various M/L based experiments. It should be noted that some of the data collected in this set is outdated and no longer corresponds to today's realities. We propose to build our own template and teach IDS on these templates. The laboratory consists of a server that includes a network protocol analyzer – TSark and 20 pieces of a RASPBERRY PI device. We use a network analyzer to create four datasets, each consisting of different data patterns ranging in size from 1 to 2 gigabytes. We performed brute force and DoS/DDOS attacks (ICMP, slow Loris, UDP flood, SYN flood, volume-based and Ping of Death attacks). The first set of data includes: UDP and SYN flood attack patterns. Second: ICMP and Slow Loris attacks. Volume Based and Ping of Death attacks the third one and Brute Force attack is the fourth set of data.

5 Machine Learning Algorithms

We have checked our system using different machine learning algorithms in order to choose the most suitable one. We have tried decision tree, Deep Learning and random forest classifier algorithms [11–15].

For every experiment, we split our data into training and test datasets. The experiment results for decision tree logic are illustrated in the Table 1. And the experiment results for Deep Learning and random forest classifier algorithms are illustrated on Tables 3 and 4 respectively. The best results were achieved when using Decision tree algorithm. The

Table 1. Decision tree without threads

Training data size	Test data size	Accuracy	Time spent for predicting and calculating	Time spent for learning data	Full session time
0.9	0.1	0.9847	0.71(s)	35(s)	404.73(s)
0.8	0.2	0.9821	1.39(s)	34.57(s)	
0.7	0.3	0.978	2.10(s)	38.59(s)	
0.6	0.4	0.975	2.81(s)	34.19(s)	
0.5	0.5	0.9715	3.56(s)	36.54(s)	
0.4	0.6	0.9689	4.22(s)	32.47(s)	
0.3	0.7	0.9664	4.96(s)	31.67(s)	
0.2	0.8	0.9621	5.62(s)	30.26(s)	
0.1	0.9	0.9561	6.27(s)	28.34(s)	
0.05	0.95	0.9326	6.62(s)	27.44(s)	
0.01	0.99	0.8337	6.89(s)	28.15(s)	

Table 2. Decision tree with threads

Training data size	Test data size	Accuracy	Time spent for predicting and calculating	Time spent for learning data	Full session time
0.9	0.1	0.9847	8.61(s)	269.6(s)	350.55(s)
0.8	0.2	0.9821	9.22(s)	270.03(s)	
0.7	0.3	0.978	5.24(s)	272.42(s)	
0.6	0.4	0.975	5.11(s)	271.32(s)	
0.5	0.5	0.9715	15.61(s)	265.17(s)	
0.4	0.6	0.9689	13.41(s)	266.84(s)	
0.3	0.7	0.9664	11.01(s)	250.17(s)	
0.2	0.8	0.9621	20.46(s)	258.02(s)	
0.1	0.9	0.9561	6.77(s)	62.6(s)	
0.05	0.95	0.9326	9.83(s)	56(s)	
0.01	0.99	0.8337	41.18(s)	206.56(s)	

best split of dataset was 90% for training and 10% for testing. The received accuracy was 0.9847. We have implemented the same algorithm using the threads, the results are illustrated in Table 2.

Table 3. Deep learning

Training data size	Test data size	Accuracy	Time spent for predicting and calculating	Time spent for learning data	Full session time
0.9	0.1	0.9847	0.37(s)	108.52(s)	1065.39(s)
0.8	0.2	0.9821	0.70(s)	103.03(s)	
0.7	0.3	0.978	11.69(s)	131.42(s)	
0.6	0.4	0.975	1.39(s)	125.50(s)	
0.5	0.5	0.9715	14.22(s)	69.32(s)	
0.4	0.6	0.9689	5.99(s)	79(s)	
0.3	0.7	0.9664	1.66(s)	61.96(s)	
0.2	0.8	0.9621	1.94(s)	63.04(s)	
0.1	0.9	0.9561	2.79(s)	65.76(s)	
0.05	0.95	0.9326	42.68(s)	42.26(s)	
0.01	0.99	0.8337	17.72(s)	67.27(s)	

Table 4. Random forest classifier

Training data size	Test data size	Accuracy	Time spent for predicting and calculating	Time spent for learning data	Full session time
0.9	0.1	0.987640	3.25(s)	434.49	1870.54 (s)
0.8	0.2	0.986984	6.27(s)	216.34(s)	
0.7	0.3	0.983707	9.59(s)	193.83(s)	
0.6	0.4	0.982442	11.52(s)	168.52(s)	
0.5	0.5	0.980148	14.34(s)	144.25(s)	
0.4	0.6	0.978150	16.49(s)	121.83(s)	
0.3	0.7	0.976001	20.23(s)	140.72(s)	
0.2	0.8	0.971416	20.61(s)	96.22(s)	
0.1	0.9	0.963210	23.63(s)	67.71(s)	
0.05	0.95	0.947818	24.04(s)	60.52(s)	
0.01	0.99	0.882751	22.25(s)	53.69(s)	

6   The Offered Function

Our proposal is the following IDS model. The TShark network protocol analyzer and our new smart IDS are installed on the main server. IDS is trained using decision tree algorithm. A decision tree algorithm is a supervised non-parametric learning algorithm, it is used in both regression and classification tasks [16–18]. The approach has a hierarchical, tree structure, and it has the root node, the branches, leaf nodes and the internal nodes. In order to choose the best parameter for our decisional tree, we choose the attributes with the minimal entropy values. We calculate the entropy using the following formula:

$$E(D) = - \sum_{cl \in CL} p(C) \log_2 pr(C) \qquad (1)$$

D – is the data set.

C – are the classes in the set, D

pr(c) – is the proportion of the data points of cl according to the number of total points of data in D.

After the IDS learning process is completed, the system waits for data input. To store all traffic patterns, we have to add new server, log server. Each successful attack is analyzed manually and supplemented to our database. Our IDS system will be permanently trained using the updated datasets as described and will be pushed to main server. We offer to use decision tree algorithm for the training purposes. To increase its efficiency, we use CPU treads. Thus, we can break down our IDS model into the following stages:

1) Data training on the main server
2) Traffic analysis on the main server
3) Pushing data to the log server
4) Data training on the log server
5) The trained model updating on the main server
6) In case an attack is detected, information is immediately sent to IPS
7) In the event attack succeeds, the data is analyzed manually and supplemented to the log server. The process is repeated endlessly.

7   Tests

We have generated the attacks vectors in the simulation lab. Using these attack patterns we have trained IDS and checked it in Google Colab. We have got the following results (Table 5):

Table 5. Results of the experiment

Attack name	Performed attacks	Identified attacks
UDP flood	20000	19979
SYN flood	20000	19896
ICMP, Ping of Death	20000	19870
Slowloris	20000	18980
Brute force attacks	20000	19960
Volume Based Attacks	20000	18982

Based on the analysis of the obtained results, we can claim that the obtained IDS is quite effective and can be proposed as a real IDS prototype.

It must be also mentioned that the whole work of the system was rather effective and the whole session time was 350.55(s), that was much quicker than in the case of other machine learning algorithms. The parallel programming using threads also made the speed up of the algorithm. The system was checked in Google Colab. The results collide with the expectations of the experiment.

8 Conclusion and Plans for the Future

The model presented in this article covers most of the registered attacks that threaten 5G systems. The approach we propose is in principle different from the strategies developed before, where IDS training is mainly carried out only on the KDD99 dataset.

Presented approaches are purely academic and cannot be applied in practice. Our approach is rather efficient and is uses CPU threats. Conducted experiments show that proposed mechanism is able to reveal majority of the listed attacks and gradually will become more precise. Accordingly, IDS detection results will become subtle. Besides, new attack vectors are emerging. The work advances the field from the present state of knowledge, because the proposed IDS is constantly updated and is in permanent learning regime. Respectively, proposed IDS model will be applied in 5G systems. In the future, – it is interesting to work on automating the analysis process of attacks detected by our IDS system.

Acknowledgment. This work was supported by Shota Rustaveli National Science Foundation of Georgia (SRNSF) [STEM-22-1076].

References

1. Bingham, J.A.C.: Multicarrier modulation for data transmission: an idea whose time has come. IEEE Commun. Mag. **28**(5), 5–14 (1990). https://doi.org/10.1109/35.54342
2. Weinstein, S., Ebert, P.: Data transmission by frequency-division multiplexing using the discrete Fourier transform. IEEE Trans. Commun. Technol. **19**(5), 628–634 (1971). https://doi.org/10.1109/TCOM.1971.1090705

3. Doelz, M.L., Heald, E.T., Martin, D.L.: Binary data transmission techniques for linear systems. Proc. IRE **45**(5), 656–661 (1957). https://doi.org/10.1109/JRPROC.1957.278415
4. Chang, R., Gibby, R.: A theoretical study of performance of an orthogonal multiplexing data transmission scheme. IEEE Trans. Commun. Technol. **16**(4), 529–540 (1968). https://doi.org/10.1109/TCOM.1968.1089889
5. Iavich, M., Gnatyuk, S., Odarchenko, R., Bocu, R., Simonov, S.: The novel system of attacks detection in 5G. In: Barolli, L., Woungang, I., Enokido, T. (eds.) AINA 2021. LNNS, vol. 226, pp. 580–591. Springer, Cham (2021). https://doi.org/10.1007/978-3-030-75075-6_47
6. Pan, F., Wen, H., Song, H.H., Jie, T., Wang, L.Y.: 5G security architecture and light weight security authentication. In: 2015 IEEE/CIC International Conference on Communications in China – Workshops (CIC/ICCC), pp. 94–98 (2015). https://doi.org/10.1109/ICCChinaW.2015.7961587
7. Park, S., Kim, D., Park, Y., Cho, H., Kim, D., Kwon, S.: 5G Security threat assessment in real networks. Sensors **21**, 5524 (2021). https://doi.org/10.3390/s21165524
8. Iwamura, M.: NGMN view on 5G architecture. In: 2015 IEEE 81st Vehicular Technology Conference (VTC Spring), pp. 1–5 (2015). https://doi.org/10.1109/VTCSpring.2015.7145953
9. Gupta, A., Jha, R.K.: A survey of 5G network: architecture and emerging technologies. IEEE Access **3**, 1206–1232 (2015). https://doi.org/10.1109/ACCESS.2015.2461602
10. Agyapong, P.K., Iwamura, M., Staehle, D., Kiess, W., Benjebbour, A.: Design considerations for a 5G network architecture. IEEE Commun. Mag. **52**(11), 65–75 (2014). https://doi.org/10.1109/MCOM.2014.6957145
11. Safavian, S.R., Landgrebe, D.: A survey of decision tree classifier methodology. IEEE Trans. Syst. Man Cybern. **21**(3), 660–674 (1991). https://doi.org/10.1109/21.97458
12. Shrestha, A., Mahmood, A.: Review of deep learning algorithms and architectures. IEEE Access **7**, 53040–53065 (2019). https://doi.org/10.1109/ACCESS.2019.2912200
13. Wang, H., Chen, S., Xu, F., Jin, Y.-Q.: Application of deep-learning algorithms to MSTAR data. In: 2015 IEEE International Geoscience and Remote Sensing Symposium (IGARSS), pp. 3743–37452015). https://doi.org/10.1109/IGARSS.2015.7326637
14. Devetyarov, D., Nouretdinov, I.: Prediction with confidence based on a random forest classifier. In: Papadopoulos, H., Andreou, A.S., Bramer, M. (eds.) AIAI 2010. IAICT, vol. 339, pp. 37–44. Springer, Heidelberg (2010). https://doi.org/10.1007/978-3-642-16239-8_8
15. Kulkarni, V.Y., Sinha, P.K.: Pruning of random forest classifiers: a survey and future directions. In: 2012 International Conference on Data Science and Engineering (ICDSE), pp. 64–68 (2012). https://doi.org/10.1109/ICDSE.2012.6282329
16. Rifat-Ibn-Alam, M.: A comparative analysis among online and on-campus students using decision tree. Int. J. Math. Sci. Comput. **8**(2), 11–27 (2022). https://doi.org/10.5815/ijmsc.2022.02.02
17. Maharjan, M.: Comparative analysis of data mining methods to analyze personal loans using decision tree and naïve Bayes classifier. Int. J. Educ. Manage. Eng. **12**(4), 33–42 (2022). https://doi.org/10.5815/ijeme.2022.04.04
18. Nutheti, P.S.D., Hasyagar, N., Shettar, R., Guggari, S., Umadevi, V.: Ferrer diagram based partitioning technique to decision tree using genetic algorithm. Int. J. Math. Sci. Comput. **6**(1), 25–32 (2020). https://doi.org/10.5815/ijmsc.2020.01.03

Analysis of Threat Models for Unmanned Aerial Vehicles from Different Spheres of Life

Hanna Martyniuk[1]([✉]), Bagdat Yagaliyeva[2], Berik Akhmetov[2], Kayirbek Makulov[2], and Bakhytzhan Akhmetov[3]

[1] Mariupol State University, Kyiv 03037, Ukraine
ganna.martyniuk@gmail.com
[2] Yessenov University, Aktau 130000, Kazakhstan
[3] Abay University, Almaty 050010, Kazakhstan

Abstract. Unmanned aerial vehicles are currently used in various spheres of life: they are used for the agricultural industry, for photo and video filming, for rescue operations, in military operations, etc. Depending on the areas of application of UAVs, they can be used both for attacks on various objects and for obtaining some information from objects. The purpose of this work is to classify UAVs and consider the cybersecurity problems that arise when using them. Based on the above classification, a threat model for UAVs in various fields of activity is considered and analyzed. There is a large number of works related to the creation of a threat model. However, such works do not consider the scope of drones. It should be noted that the threat model for military UAVs should be different from the threat model for UAVs in the agricultural sector. In this regard, the authors set themselves the goal of constructing and analyzing a UAV threat model based on their classification depending on the scope of application.

Keywords: Cybersecurity · UAV · Threat model

1 Instruction

Unmanned aerial vehicles (UAVs) are becoming more and more popular in the world. The scope of their application is expanding: from military measures to the operations of rescuers, doctors, firefighters, agriculture and simply the supply of small-sized cargo. UAVs are capable of conducting aerial reconnaissance and surveillance, transmitting photo and video information in real time, being carriers and targets, operating in extreme conditions, in particular in areas subjected to radiation, chemical or biological contamination, in areas of disasters and intense fire countermeasures.

In modern conditions, the use of UAVs and robotics has become the usual norm. With their help, the tasks of ensuring military security, as well as issues in research, security and other areas are solved [1–8]. In the fight against terrorism, drones and other types of robotics are becoming an increasingly effective and regularly used tool [9–11].

However, it should be noted that today there are problems with the protection of such UAVs from unauthorized hacking and use. Thus, a number of publications [12–17]

© The Author(s), under exclusive license to Springer Nature Switzerland AG 2023
Z. Hu et al. (Eds.): ICCSEA 2023, LNDECT 181, pp. 595–604, 2023.
https://doi.org/10.1007/978-3-031-36118-0_53

describe the main problems of UAV cybersecurity. Based on this analysis, the authors decided to compile a threats model, depending on the possible methods of attacking aircraft.

At the same time, terrorist organizations are trying to keep up with progress and are actively using unmanned vehicles in their destructive activities [19–22]. Therefore, consideration of issues on the use of UAVs and countering them must be carried out in parallel.

Only the general application of UAVs is considered in the literature for compiling a threat model. At the same time, it is impossible to equally assess the threats and risks that affect the confidentiality of these drones in the civil and military spheres. So, for example, the loss of data confidentiality during aerial reconnaissance of the state of crops will not cause such damage as the loss of data confidentiality during aerial reconnaissance of enemy sites in war. In this regard, the work will consider a model of UAV security threats depending on their areas of application.

2   Classification of Unmanned Aerial Vehicles by Features

An unmanned aerial vehicle is a device that is controlled remotely without the direct participation of the pilot. The UAV consists of the following components: an air platform with a special landing system, a power plant, a power supply system for all components, and avionics equipment. The general classification of UAVs is given in Table 1.

This classification is not complete, but it is quite enough to determine the cybersecurity problems that exist today with UAVs. For creating and analyzing threat models authors decided to use only usage method like classification feature,

3   Methodology of Building a Threats Model for a UAV

As already mentioned, many authors have paid attention to UAV safety issues. Summing up, we can say that attackers can attack software, a communication network, or directly hardware. An illustration of such threats is shown in Fig. 1.

All such attacks are used to violate the confidentiality or integrity of data, or in the worst case, take control of the UAV in their hands. In order to assess the damage that can be caused by UAVs in various areas of human activity, the authors proposed to use threat model. In general, a threat model is a formalized description of methods and means of implementing threats to information. According to [23], there are four main criteria for assessing the security of an information and telecommunication system (ITS):

- confidentiality criterion - unauthorized access to information;
- availability criterion - violation of the possibility of using the UAV or the information being processed;
- integrity criterion - unauthorized modification of information transmitted by the UAV (forgery, distortion of information);
- observation criterion - refusal to identify, authenticate and log actions.

The following threats to UAVs have been identified in [19]: sniffing, spoofing, DoS-attack. Using [16], threats can be classified like active interfering, backdoor malware, collisions, de-authentication etc.

Table 1. Classification of UAVs by characteristic features

Classification feature	UAVs
Usage method	military
	civil
	antiterrorist
Way of solving the problem	tactical (flight range up to 80 km)
	operational-tactical (flight range up to 300 km)
	operational-strategical (flight range up to 700 km)
Weight	small-sized (up to 200 kg)
	medium-sized (up to 2000 kg)
	oversized (up to 5000 kg)
	heavy (over 5000 kg)
Flight duration	short duration (up to 6 h)
	medium duration (up to 12 h)
	large duration (over 12 h)
Classification feature	**UAVs**
Maximum altitude	low-altitude (up to 1 km)
	medium-altitude (up to 4 km)
	high-altitude (up to 12 km)
	stratospheric (over 12 km)

The authors also analyzed and divided the main threats in [12, 16, 17, 19] depending on the criteria listed above (Table 2).

For a better understanding of the types of threats, their description is given below.

Eavesdropping – attackers listen to unencrypted messages over a communication channel.

Sniffing – intercepting data that is delivered within the observed network in the form of packets.

Malware Infection – most UAV controllers are based on the use of mobile phones, laptops or wireless systems. Their use makes it possible to illegally install malicious software on UAV controllers, which leads to violations of the ability to use aircraft.

Manipulation – allows you to change the specified route of the UAV, which can lead to the loss of the drone or the cargo it is carrying.

Jamming – the introduction of interference signals, which disrupts the connection of the UAV with the controller and distorts the transmitted information.

Spoofing – by intercepting an encrypted message, the attacker subsequently disguises himself as the legitimate sender of messages.

De-authentication attack – this type of threat prevents the user from gaining access to their UAV.

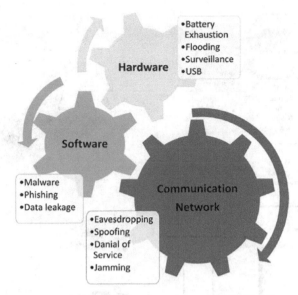

Fig. 1. Classification of attacks used on UAVs [1]

Table 2. Classification of UAV threats depending on the criteria of the threats model

Criterion name	Type of threats
confidentiality criterion	Eavesdropping
	Sniffing
availability criterion	Malware Infection
	Manipulation
integrity criterion	Jamming
	Spoofing
observation criterion	De-authentication attack

3.1 Threat Model for Civilian UAVs

To date, the scope of UAVs for civilian purposes is very extensive [1–3, 8] For example, amateur drone technology is being used to acquire high-resolution imagery data of remote areas, islands, mountain peaks, and coastlines. UAVs have made performance mapping, construction monitoring and site inspection efficient, simple and fast. UAVs can be used in precision agriculture to collect data from ground sensors (water quality, soil properties, moisture, etc.), pesticide spraying, disease detection, irrigation planning, weed detection, and crop monitoring and management. The traffic monitoring system is an area where the integration of UAVs has also generated great interest. Some possibilities of using UAVs for civilian purposes are shown in Fig. 2.

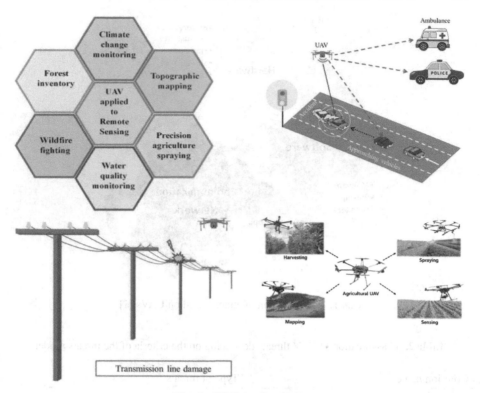

Fig. 2. UAV Applications

Taking into account the areas of application of UAVs for civilian purposes, the authors proposed an assessment of the risks and damages from attacks on UAVs (Table 3). As estimates, it is proposed to use such a scale:

1. low damage/risk;
2. medium damage/risk;
3. high damage/risk.

The authors believe that threats such as "Manipulation" bring the greatest harm to civilian UAVs. This is due to the fact that with this type of threat, it is possible to lose valuable cargo or disrupt the growth of plants in agriculture, etc. The least harm will be caused by threats related to the loss of confidentiality, since in most cases the information transmitted by UAV data can be obtained by any other legal means.

3.2 Threat Model for Military UAVs

Unmanned aerial vehicles play an important role in military surveillance missions. Several countries have added UAVs to their defense strategic plans [1, 9, 10, 18, 22]. Countries use these flying robotic vehicles for enemy detection, anti-poaching, border control, and maritime surveillance of critical shipping lanes. Inexpensive, reliable and versatile UAVs are now making a significant contribution to aerial surveillance, monitoring and

Table 3. Calculation of risks and damage for civilian UAVs

Type of threats	Risk assessment	Damage assessment	Overall Threat
Eavesdropping	1	1	2
Sniffing	1	1	2
Malware Infection	2	2	4
Manipulation	3	3	6
Jamming	2	3	5
Spoofing	2	3	5
De-authentication attack	1	3	4

inspection of any specific area to prevent any illegal activity. For example, surveillance of any threat can be detected by drones and they can be used to track any movement in any restricted area. The UAV can provide these services by receiving automatic alerts with minimal manual effort.

In addition, using the example of Ukraine, in the conditions of a full-scale war in Ukraine, the issue of conducting aerial reconnaissance by means of unmanned aircraft is an important and relevant task [9, 22]. The collection and transmission of operational information requires the solution of a number of tasks, among which a separate place is occupied by the processing of data from the payload cameras directly on board the UAV, the preparation of such data for transmission and direct transmission to a ground work-station. Accordingly, the transmitted information must be protected from unauthorized access, depending on the degree of access restriction.

Taking into account the areas of application of UAVs for military purposes, the authors proposed an assessment of the risks and damages from attacks on UAVs (Table 4).

Table 4. Calculation of risks and damage for military UAVs

Type of threats	Risk assessment	Damage assessment	Overall Threat
Eavesdropping	3	3	6
Sniffing	3	3	6
Malware Infection	1	2	3
Manipulation	1	2	3
Jamming	2	3	5
Spoofing	3	3	6
De-authentication attack	1	1	2

When conducting military operations, including aerial reconnaissance, the loss of confidential data poses the greatest threat, since not only the military, but also the civilian population can suffer from this. In this regard, the authors decided that the highest threat to military UAVs is the methods of "Eavesdropping", "Sniffing" and "Spoofing".

3.3   Threat Model for UAVs Used as a Means of Destruction of Targets

In this section, attention will be paid to UAVs that can directly harm civilian or military targets. These aircraft include [19] reconnaissance UAVs, which are used to collect information about a certain telecommunications network with a subsequent attack on this network. This also includes UAVs for electronic warfare. This type of UAV is capable of affecting the performance of the telecommunications network, introducing distortions and creating false base and subscriber stations. It is also necessary to note strike UAVs (Fig. 3), which are capable of causing material damage and incapacitating the entire object of information activity.

Taking into account the considered UAVs, the authors proposed an assessment of the risks and damages from attacks on UAVs (Table 5).

Table 5. Calculation of risks and damage for UAVs as means of destruction targets

Type of threats	Risk assessment	Damage assessment	Overall Threat
Eavesdropping	3	1	4
Sniffing	3	1	4
Malware Infection	1	1	2
Manipulation	1	1	2
Jamming	3	3	3
Spoofing	3	3	6
De-authentication attack	1	3	4

For such UAVs, the loss of confidentiality will not lead to a violation of the integrity or functional features of information activity objects. The accessibility criterion for such aircraft is also not in the first place. At the same time, breaching the integrity of information has the highest priority for attackers, as it will compromise the integrity of the information and result in the most damage. In this regard, the authors decided that for UAVs that are used as a means of destruction targets, the greatest total threat is posed by attacks such as "Jamming" and "Spoofing".

Fig. 3. Types of strike UAVs [17]

4 Summary and Conclusion

Cybersecurity issues and methods of attacks on UAVs have received much attention in the literature. There are a number of publications devoted to threats models. However, the threats models are of a general nature and do not depend on the scope of the UAV in practice. In this paper, the authors carried out work to distinguish between attacks and the damage that these attacks can cause, depending on the scope of the UAV. The authors proposed to consider the threats model for aircraft used for civil and military purposes. In addition, a threats model for UAVs is given, which are used to cause damage at the objects of information activity. These types of attacks on UAVs pose the greatest threat:

1) "Manipulation" threats do the most harm to civilian UAVs;
2) the highest threat to military UAVs are the methods of "Eavesdropping", "Sniffing" and "Spoofing";
3) the greatest total threat is posed by attacks such as "Jamming" and "Spoofing" when using UAVs as a means of destroying objects.

The authors believe that such detailing of threat models, depending on the areas of application of UAVs, will make it possible to build the best protection for them.

References

1. Mohsan, S.A.H., Khan, M.A., Noor, F., Ullah, I., Alsharif, M.H.: Towards the Unmanned Aerial Vehicles (UAVs): A Comprehensive Review. Drones **6**, 147 (2022). https://doi.org/10.3390/drones6060147
2. Bauk, S., Kapidani, N., Sousa, L., Lukšić, Ž., Spuža, A.: Advantages and disadvantages of some unmanned aerial vehicles deployed in maritime surveillance. The 8th Maritime Conference - MT2020 At: Barcelona, pp. 1–12 (2020)
3. Chan, K., Nirmal, U., Cheaw, W.: Progress on drone technology and their applications: a comprehensive review. AIP Conf. Proc. **2030**, 020308 (2018). https://doi.org/10.1063/1.5066949

4. Hryb, O.G., Karpaliuk, I.T., Shvets, S.V., Zakharenko, N.S.: Increasing the reliability of the power supply system through unmanned aerial vehicles. Zapysky VNTU **1**, 1–6 (2020). (in Ukrainian)
5. Macrina, G., Pugliese, L.D.P., Guerriero, F., Laporte, G.: Drone-aided routing: a literature review. Transp. Res. Part C Emerg. Technol. **120**, 102762 (2020)
6. Khan, M.A., Alvi, B.A., Safi, A., Khan, I.U.: Drones for good in smart cities: a review. In: Proceedings of the International Conference on Electrical, Electronics, Computers, Communication, Mechanical and Computing (EECCMC). Tamil Nadu, India (28–29 January 2018)
7. Alshbatat, A.I.N.: Fire extinguishing system for high-rise buildings and rugged mountainous terrains utilizing quadrotor unmanned aerial vehicle. Int. J. Ima. Graph. Sig. Proce. (IJIGSP) **10**(1), 23–29 (2018). https://doi.org/10.5815/ijigsp.2018.01.03
8. Ramesh, K.N., Chandrika, N., Omkar, S.N., Meenavathi, M.B., Rekha, V.: Detection of rows in agricultural crop images acquired by remote sensing from a UAV. Int. J. Ima. Graph. Sig. Proce. (IJIGSP) **8**(11), 25–31 (2016). https://doi.org/10.5815/ijigsp.2016.11.04
9. Prystavka, P., Sorokopud, V., Chyrkov, A., Kovtun, V.: Automated complex for aerial reconnaissance tasks in modern armed conflicts. CEUR Workshop, vol. 2588, 1–10
10. Unmanned Aerial System DEVIRO "Leleka-100", http://deviro.ua/leleka-100, last accessed 16 October 2019. (in Ukrainian)
11. Chyrkov, A., Prystavka, P.: Suspicious Object Search in Airborne Camera Video Stream. In: Hu, Z., Petoukhov, S., Dychka, I., He, M. (eds.) ICCSEEA 2018. AISC, vol. 754, pp. 340–348. Springer, Cham (2019). https://doi.org/10.1007/978-3-319-91008-6_34
12. SavasIlgi, G., Ever, Y.K.: Chapter Eleven - Critical analysis of security and privacy challenges for the Internet of drones: a survey Drones in Smart-Cities. Security and Performance 207–214 (2020)
13. Bakkiam, D.D., Fadi, A.-T.: Chapter One - Aerial and underwater drone communication: potentials and vulnerabilities. Drones in Smart-Cities. Security and Performance 1–26 (2020)
14. Ly, B., Ly, R.: Cybersecurity in unmanned aerial vehicles (UAVs). J. Cyb. Secu. Technol. **5**(2), 120–137 (2021). https://doi.org/10.1080/23742917.2020.1846307
15. Gudla, C., Rana, S., Sung, A.H.: Defense techniques against cyber attacks on unmanned aerial vehicles. Int'l Conf. Embedded Systems, Cyber-physical Systems, & Applications, ESCS'18, pp. 110–116
16. Tsao, K.-Y., Girdler, T., Vassilakis, V.G.: A survey of cyber security threats and solutions for UAV communications and flying ad-hoc networks. Ad Hoc Networks, 1–39 (2022)
17. Siddiqi, M.A., Iwendi, C., Jaroslava, K., Anumbe, N.: Analysis on security-related concerns of unmanned aerial vehicle: attacks, limitations, and recommendations. MBE **19**(3), 2641–2670 (2021)
18. Hartmann, K., Giles, K.: UAV Exploitation: A New Domain for Cyber Power. In: Pissanidis, N., Rõigas, H., Veenendaal, M. (eds.) 2016 8th International Conference on Cyber Conflict Cyber Power, pp. 205–241. 2016 © NATO CCD COE Publications, Tallinn
19. Voitenko, S., Druzhynin, V., Martyniuk, H., Meleshko, T.: Unmanned aerial vehicles as a source of information security threats of wireless network. International Journal of Computing **21**(3), 377–382 (2022)
20. The Drone Cyberattack That Breached a Corporate Network. Accessed December 2022. https://blogs.blackberry.com/en/2022/10/the-drone-cyberattack-that-breached-a-corporate-network
21. Yeo, M.: Electronic warfare in the land domain - The threats continue to increase. Asia-Pacific Defence Reporter **45**(7), 20–22 (2019)

22. Kjellèn, J.: Drone-based jamming (REB-N). Russian Electronic Warfare. The role of Electronic Warfare in the Russian Armed, Swedish Defence Research Agency, p. 105

23. RD TPI 2.5-004-99. Criteria for assessing the security of information in computer systems from unauthorized access (in Ukrainian)

Identification of Soil Water Potential Sensors Readings in an Irrigation Control System Using Internet-of-Things (IoT): Automatic Tensiometer and Watermark 200SS

Volodymyr Kovalchuk[1]([✉]), Oleksandr Voitovich[1], Pavlo Kovalchuk[1], and Olena Demchuk[2]

[1] Institute of Water Problems and Land Reclamation NAAS, Kyiv 03022, Ukraine
volokovalchuk@gmail.com
[2] National University of Water and Environmental Engineering, Rivne 33028, Ukraine
o.s.demchuk@nuwm.edu.ua

Abstract. The analysis of world practice showed that various authors identified the relationship between soil water potential, Watermark 200SS gypsum block resistance and soil temperature $P = P(R, T)$ for light soils. However, as our research shows, such dependencies cannot be applied to other types of soils. The method of identifying the parameters of the presented nonlinear model is also not given in the works of other authors. In the development of the Internet of Things (IoT), a method for identifying the dependences of soil water potential on the resistance of the Watermark 200SS gypsum block and soil temperature has been developed. In a laboratory experiment, the soil water potential (using a tensiometer) and the resistance value of a Watermark 200SS gypsum block are measured in parallel at certain temperatures using monoliths of an undisturbed structure of heavy loamy chernozem soil. Non-linear dependence is reduced to a linear relationship. The parameters of the linear model, which are used to identify the nonlinear model, are found by the method of least squares. Based on the non-linear dependence, the values of soil water potential are calculated based on Watermark 200SS indicators and soil temperature. Based on the soil water potential, watering timings are determined in irrigation control systems using the Internet of Things sensor system.

Keywords: Sensor calibration · irrigation control · parameter identification · soil water potential · least squares method

1 Introduction

1.1 Formulation of the Problem

In the context of climate change, the rational use of irrigation water is the basis for the sustainable development of agriculture. It is impossible without the use of the Information System for Operational Irrigation Control, which is based on field networks of various

Z. Hu et al. (Eds.): ICCSEEA 2023, LNDECT 181, pp. 605–614, 2023.
https://doi.org/10.1007/978-3-031-36118-0_54

sensors, that is, on the use of the Internet of Things (IoT) [1–7]. Many researchers use Watermark 200SS sensors to control irrigation in their field moisture sensor networks. Irrometer Inc. Developed and patented a whole series of such sensors [8].

However, the given formula for the relationship between soil water potential and Watermark 200SS resistance and soil temperature $P = P(R,T)$ is recommended for soils of light mechanical composition. The problem is to develop a method for conducting experimental studies and identifying dependencies based on experimental data for an arbitrary type of soil.

1.2 Literature Review

Modern systems for measuring soil moisture content use the physical properties of the soil, such as volumetric moisture and soil water potential. The fundamental mathematical model of the relationship between the volumetric soil moisture and the soil water potential $\Theta = f(P)$ was proposed by Van Genuchten, M.T. [9]. Functional features of the relationships model, depending on the soil mechanical composition, are calculated by the Rosetta program [10]. In this way, the measured soil moisture or soil water potential can easily be converted from to each other and used in the irrigation control system to determine watering.

In a previous publication [5] we recommended to use an automatic tensiometer and Watermark 200SS for building field networks of sensors that measure soil water potential in irrigation control systems.

Our recommendations are supported by a significant body of literature, where Watermark 200SS sensors [11–15] or automatic tensiometers [1, 14, 16–18] are also used for these purposes.

The Watermark 200SS measures moisture indirectly, namely by measuring the variable resistance of a wet gypsum block [8], as water exchange between the soil and the sensor changes the electrical resistance between the electrodes in the sensor. Sensor readings are identified in units of soil water potential, and electrical resistance depends on soil moisture and temperature [11]. The authors [11, 12] identified the Watermark 200SS readings for soils of light mechanical composition based on research. Studies [19] revealed a significantly lower correlation coefficient for soils of heavy mechanical composition when obtaining the dependence between the soil moisture and soil water potential. The work [14] indicates an overestimated value of the dryness of the Watermark 200SS readings for heavy soils compared to other measurement methods. Therefore, it is necessary to calibrate such sensors of soil water potential for soils of heavy mechanical composition for their use in irrigation control systems.

It is known that a mechanical [20] or modern automatic tensiometer [16–18, 21] directly measures the vacuum pressure, which is equal in modulus to the soil water potential. The tensiometer consists of a sealed system: a sensitive ceramic porous sensor element buried in the soil, a tube that goes to the surface and a hydraulically connected mechanical indicator (vacuum meter) or an automatic sensor for vacuum measuring. This equipment (tensiometer) is also often used in irrigation control systems [1, 5, 16–18] to monitor soil water potential and make decisions about watering.

The accuracy and reliability of the tensiometer in determining the soil water potential is high [1, 14, 15, 20, 21] and largely depends on the accuracy of vacuum pressure measuring instruments. To a lesser extent, the accuracy depends on the quality of the ceramic probe immersed in the soil layer [20].

Since the measurement data using this sensor does not require any additional interpretation, it is used to identify other sensors readings with indirect measurement methods [1, 12, 14, 15], such as the Watermark 200SS [12, 14, 15].

1.3   The Purpose of the Study

The purpose of this study is to develop a method for identifying the relationship between soil water potential and Watermark 200SS readings and soil temperature values based on data from laboratory studies of soil samples for various soils types. The identified dependencies are used to calculate the soil water potential and assign watering in irrigation control systems using the IoT sensor system.

2   Research Methods

The following there is the identification method based on experimental studies and mathematical modeling of the relationship between soil water potential and electrical resistance measured by Watermark 200SS at different temperatures.

2.1   Experimental Equipment for Data Obtaining

Experimental equipment (Fig. 1), which is able to maintain the soil temperature in the soil monoliths of an undisturbed structure within the specified limits using the controller (5) Arduino Nano V3.0 AVR ATmega328P with a relay that controls the aquarium heater (4) was designed and manufactured. The transfer of commands by the user to the server and from the server to the relay through the Arduino controller is carried out remotely using the GPRS module. The water temperature and the temperature inside the soil monoliths are measured by digital sensors (3) Dallas DS18B20, located inside a stainless flask using thermally conductive paste.

The data of vacuum pressure and electrical resistance of the Watermark are received and collected using the Arduino Nano V3.0 AVR ATmega328P controller (6) (Fig. 1) and transmitted to the server using Wi-Fi.

Watermark 200SS sensors (2) were calibrated using automatic tensiometers (Fig. 1), which were developed at the Institute of Water Problems and Land Reclamation (IWPLR) of the National Academy of Agrarian Sciences of Ukraine (UA utility model patent No. 132271 [21]) based on a BMP280 digital vacuum gauge (1). And although the BMP280 has a significant drawback, it is afraid of water getting on the sensitive element; its advantage is high accuracy in measuring vacuum pressure.

It can be argued that both sensors measure the water potential of soil moisture: the automatic tensiometer directly measures the vacuum pressure equal to the modulus of the soil water potential, and the Watermark 200SS is calibrated in potential units.

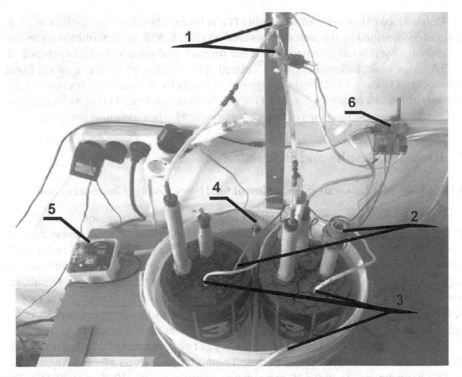

Fig. 1. Experimental equipment for measurement of readings of soil water potential sensors and Watermark 200SS (the designations are given in the text of the publication)

2.2 Model Identification Method

At present, the relationship between the parameters P — P(R,T) of Watermark 200SS sensors is determined by the formula [11]:

$$P = \frac{-a - bR}{1 - cR - dT}.$$ (1)

where P is the soil water potential (module of vacuum pressure), kPa; R is the resistance, kΩ; T is the temperature, °C; a, b, c, d are desired coefficients. This formula was proposed by American scientists [11], verified [12] for soils of light mechanical composition. They obtained the following calibration coefficients: $a = 3.213$; $b = 4.093$; $c = 0.009733$; $d = 0.01205$. However, for soils of heavy mechanical composition, the identification of dependence (1) is required on the basis of experimental studies and mathematical modeling.

To identify the coefficients of non-linear Eq. (1), we rewrite it in the form:

$$S(1 - cR - dT) = -a - bR,$$ (2)

or.

$$S = -a - bR + cSR + dST.$$ (3)

Dependence (3), in contrast to equality (1), is linear. Therefore, the coefficients a, b, c, d can be identified by the least squares method [22]. S, R, T are determined according to the data of the laboratory experiment. Then for the initial value S, the input values $S*R$, $S*T$, R of linear dependence (3) are calculated. The coefficients a, b, c, d of the linear dependence (3), calculated from the experimental data by the least squares method, are substituted into the nonlinear formula (1). We obtain dependence (1) for a given type of soil, which is used to bind the readings of both sensors when assigning watering.

3   Research Results and Discussion

3.1   Results of a Laboratory Experiment for Heavy Loamy Chernozem Soil

Monoliths of the undisturbed structure of heavy loamy chernozem soil were selected at the experimental production site of the State Enterprise "Experimental farming Andriyivske" of IWPLR in Odessa region from the arable (10–25 cm) and underarable (30–45 cm) horizons. In the process of laboratory research, there was a problem with the cracking of the monolith with the soil of the underarable horizon and, as a result, the constant loss of vacuum by the soil medium and the tensiometer. The physical properties of both soils are similar. Therefore, despite some loss of information content of the experiment, the results are given according to the studies of one monolith.

The monoliths were tested in the laboratory at two temperature thresholds of about 20.6 °C and about 24.65 °C inside each soil monolith (value was averaged over Dallas DS18B20 readings [23], ± 0.5 °C temperature accuracy from −10 °C to + 85 °C). The monoliths were saturated close to full capacity, and then dried naturally. The vacuum pressure with a tensiometer (based on the BMP280 sensor element [24], pressure resolution is 0.16 Pa), the gypsum block resistance in the Watermark 200SS sensor and the temperature of the water around the soil monoliths were automatically recorded.

A comparison of the dynamics of vacuum pressure according to the readings of the automatic tensiometer and the calculated values of the readings from Watermark according to formula (1) with the coefficients [11] for light soils at different values of the gypsum blocks electrical resistance of the sensor was carried out (Fig. 2). Section 2 has a temperature threshold of about 20.6 °C, and Sect. 3 has a temperature threshold of about 24.65 °C.

Visual analysis of dependencies in Fig. 2 shows the systematic deviations between the readings of the automatic tensiometer (vacuum in kPa) and the calculated vacuum values (soil water potential) according to the electrical resistance R data from Watermark on all four sections of the curves. The relative error in Sects. 2 and 3 of the systematic deviations to calibration was calculated using the formula:

$$M[\Delta X]_{relative} = \frac{1}{n} \sum_{i=1}^{n} \frac{|X_{act}^i - X_{calc}^i|}{|X_{act}^i|} * 100\%, \tag{4}$$

where X_{act}^i are the experimental values of the potential (vacuum pressure); X_{calc}^i are calculated values of potential (vacuum pressure) according to formula (1) with coefficients [11]. The error between the X_{act} values for deep drying of the soil monolith and the X_{calc}^i

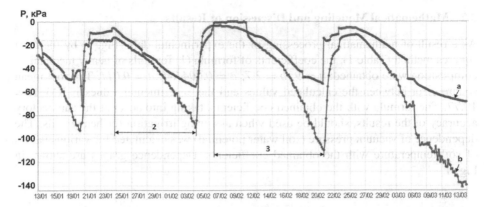

Fig. 2. Comparison of vacuum pressure dynamics based on (a) automatic tensiometer readings and (b) calculated values based on electrical resistance data from Watermark according to formula (1)

values is about 80%. This confirms the need to identification of the sensor Watermark readings according to the experimental data for soils of heavy mechanical composition.

According to the experiment plan, after the initial unstable period and soil monolith maximum saturation reaching from 01/24/2022 15:37, its natural drying began at an average temperature of 20.6 °C, which was detected by the sensors (Fig. 2, curve's Sect. 2). The maximum vacuum value was reached on 02/04/2022 12:06. Another drying cycle (Sect. 3) after reaching the maximum saturation of the monolith with water lasted from 02/07/2022 09:19 to 02/21/2022 07:31. The sensors recorded the average temperature for this cycle of 24.65 °C. Based on the obtained data, two curves of the dependence R(P) were plotted at a certain constant value of the soil temperature (Fig. 3).

As can be seen from Fig. 3, in the area close to the full moisture capacity (P = 0...8–10 kPa.), there is an uncertainty in the electrical resistance of Watermark 200SS, equal to a certain constant value.

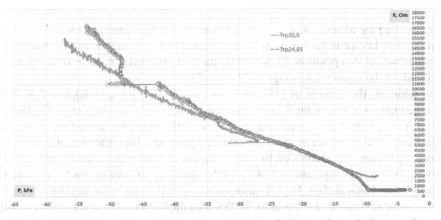

Fig. 3. Graphs of experimental dependences R(P) for further mathematical processing

3.2 Mathematical Modeling and Discussion of Results

As a result of mathematical processing of the experimental data (Fig. 3) by the least squares method (Table 1), the coefficients of formula (1) for soils of heavy mechanical composition were obtained: $a = 9$; $b = 2.7$; $c = 0.004484$; $d = 0.011$. The root mean square error between the calculated values and the experimental values is RMSE = 10.77. The formula with the obtained coefficients, taking into account the satisfactory accuracy of the results, should be used when assigning irrigation for heavy soils. The dependence of vacuum pressure (soil water potential) was calculated for various values of soil temperature with the obtained coefficients. Dependence graphs are shown in Fig. 4.

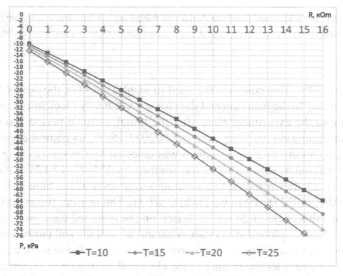

Fig. 4. Graph of dependence between the Watermark 200SS electrical resistance, temperature and vacuum pressure (soil water potential)

Regarding the identification results of Watermark 200SS readings by other authors, the most well-known is the verification of formula (1) by the authors of [12] for soils of light mechanical composition. In the study, the authors used a number of temperature thresholds in the temperature range of 21–30 °C.

In addition, the authors of [13] attempted to describe the dependence P(R,T) by several equations, dividing the R curve into several sections and comparing the obtained data at T = 24 °C with the calculated values of the potential according to the Van Genuchten-Mualem model by the USDA Rosetta program [10] without using the control measuring instrument. Studies on the identification of Watermark 200SS readings [15] were carried out at a temperature of 24 °C for the sand silty soil, which can be attributed to the middle mechanical class. In [25] calibration was carried out in the field by comparing the measured resistance of Watermark 200SS with the known electrical resistance of the resistors.

Table 1. An example of applying the least squares method to calculate the coefficients of the model (3)

N	S	R	T	Ai1 (S*R)	Ai2 (S*T)	Ai3 (-R)	Bi (S)	Bi'	(Bi-Bi')2
1	−1	0	20,6	0	−20,6	0	−1	−0,87357	0,015985
2	−10	2	20,6	−20	−206	−2	−10	−9	1
3	−10	1,8	24,65	−18	−246,5	−1,8	−10	−10,691	0,477505
4	−20	4,6	20,6	−92	−412	−4,6	−20	−18,1129	3,561268
5	−20	4,2	24,65	−84	−493	−4,2	−20	−21,492	2,226012
6	−30	7	20,6	−210	−618	−7	−30	−27,2343	7,648925
7	−30	6,5	24,65	−195	−739,5	−6,5	−30	−32,3133	5,35136
8	−40	6	20,6	−240	−824	−6	−40	−35,867	17,08146
9	−40	9,5	24,65	−380	−986	−9,5	−40	−43,276	10,73236
10	−45	3,9	20,6	−175,5	−927	−3,9	−45	−39,9256	25,74981
11	−45	10,9	24,65	−490,5	−1109,25	−10,9	−45	−48,758	14,12287
12	−50	11,8	20,6	−590	−1030	−11,8	−50	−45,5823	19,5157
13	−50	10,6	24,65	−530	−1232,5	−10,6	−50	−53,976	15,8083
14	−52	12,3	20,6	−639,6	−1071,2	−12,3	−52	−47,4281	20,90221
15	−52	11,1	24,65	−577,2	−1281,8	−11,1	−52	−56,1635	17,33447

Therefore, we think that the use of an experimental equipment for several temperature thresholds with the ability to constantly maintain it, in comparison with the calibration examples described above, makes it possible to obtain experimental data at various temperature thresholds (similar to [12]), to highlight in detail the effect of temperature on the Watermark 200SS resistance and on soil water potential for any soil type. In particular, this is done for soils of heavy mechanical composition. Results of Watermark 200SS sensor calibration can be performed both for a single field soil and other soil types. They are extrapolated for other temperature values and for any soils of heavy mechanical composition. This makes it possible to effectively use the Watermark 200SS in irrigation management systems for watering assigning.

4   Summary and Conclusion

As the analysis showed, in irrigation control systems many farmers use Watermark 200SS sensors for field networks for measuring soil moisture by indirect methods. However, in order to watering assigning through measurements by these sensors, it is necessary to identify their readings of the soil water potential. There is such dependence only for light mechanical composition soils.

Therefore, in our study, the problem of developing a method for finding a functional dependence between the soil water potential, the Watermark 200SS gypsum block resistance and the soil temperature P(R,T) was formulated and solved. The method makes it possible to obtain experimental data on the relationship between electrical resistance and soil water potential for an arbitrary soil sample on experimental equipment. On the basis of experimental data, a method of identifying the parameters of linear dependence is presented, which ensures the identification of a nonlinear model of the relationship P = P(R,T).

The experimental approach and the identification of non-linear models provided a constructive implementation of the method for identifying the non-linear relationship between water potential, resistance and soil temperature P = P(R,T) for soils of heavy mechanical composition. In the future, it is possible to implement the method for creating a knowledge base of model's coefficients in automated irrigation control systems during watering.

References

1. Maddah, M., Olfati, J.A., Maddah, M.: Perfect irrigation scheduling system based on soil electrical resistivity. Int. J. Veg. Sci. **20**(3), 235–239 (2014). https://doi.org/10.1080/19315260.2013.798755
2. Lozoya, C., Mendoza, C., Aguilar, A., Román, A., Castelló, R.: Sensor-based model driven control strategy for precision irrigation. Journal of Sensors (2016). https://doi.org/10.1155/2016/9784071
3. Payero, J.O., Mirzakhani-Nafchi, A., Khalilian, A., Qiao, X., Davis, R.: Development of a Low- Cost Internet-of-Things (IoT) System for Monitoring Soil Water Potential Using Watermark 200SS Sensors. Advances in Internet of Things **7**, 71–86 (2017). https://doi.org/10.4236/ait.2017.73005
4. Okine, A., Appiah, M., Ahmad, I., Asante-Badu, B., Uzoejinwa, B.: Design of a green automated wireless system for optimal irrigation. Int. J. Comput. Netw. Inf. Secur. **12**(3), 22–32 (2020). https://doi.org/10.5815/ijcnis.2020.03.03
5. Kovalchuk, V., Voitovich, O., Demchuk, D., Demchuk, O.: Development of Low-Cost Internet-of-Things (IoT) Networks for Field Air and Soil Monitoring Within the Irrigation Control System. In: Hu, Z., Petoukhov, S., Dychka, I., He, M. (eds.) ICCSEEA 2020. AISC, vol. 1247, pp. 86–96. Springer, Cham (2021). https://doi.org/10.1007/978-3-030-55506-1_8
6. Akwu, S., Bature, U.I., Jahun, K.I., Baba, M.A., Nasir, A.Y.: Automatic plant irrigation control system using Arduino and GSM module. Int. J. Eng. Manuf. (IJEM) **10**(3), 12–26 (2020). https://doi.org/10.5815/ijem.2020.03.02
7. Anusha, K., Mahadevaswamy, U.B.: Automatic IoT based plant monitoring and watering system using Raspberry Pi. Int. J. Eng. Manuf. (IJEM) **8**(6), 55–67 (2018). https://doi.org/10.5815/ijem.2018.06.05
8. Irrometer Inc. WATERMARK Soil Moisture Sensor. https://www.irrometer.com/200ss.html
9. Van Genuchten, M.T.: A closed-form equation for predicting the hydraulic conductivity of unsaturated soils. Soil Sci. Soc. Am. J. **44**(5), 892–898 (1980)
10. Rosetta Version 1.0 (Free downloaded program). U.S.Salinity Laboratory ARSUSDA. www.ussl.ars.usda.gov. Retrieved from: http://www.ussl.ars.usda.gov
11. Shock, C.C., Barnum, J.M., Seddigh, M.: Calibration of Watermark Soil Moisture Sensors for Irrigation Management. Malheur Experiment Station, Oregon State University (1998). Retrieved from https://www.researchgate.net/profile/Clinton-Shock/2
12. Chard, J.: Watermark soil moisture sensors: characteristics and operating instructions. Utah State University (2002). https://www.researchgate.net/publication/237805713
13. Radman, V., Radonjić, M.: Arduino-based system for soil moisture measurement. Proc. 22nd Conference on Information Technologies IT. vol. 17 (2017)

14. Kumar, J., Patel, N., Rajput, T.B.S., Kumari, A., Rajput, J.: Performance evaluation and calibration of soil moisture sensors for scheduling of irrigation in brinjal crop (Solanum melongena L. var. Pusa Shyamla). Journal of Soil and Water Conservation **19**(2), 182–191 (2020). https://doi.org/10.5958/2455-7145.2020.00025.9
15. Vettorello, D.L., Marinho, F.A.: Evaluation of time response of GMS for soil suction measurement. In: MATEC Web of Conferences, vol. 337, p. 01014. EDP Sciences (2021). https://doi.org/10.1051/matecconf/202133701014
16. Thalheimer, M.: A low-cost electronic tensiometer system for continuous monitoring of soil water potential. J. Agri. Eng. **44**(3), XLIV: e16. (2013). https://doi.org/10.4081/jae.2013.e16
17. Matus, S.K.: Information and measurement system of data collection and control of soil moisture reserves. Bulletin of the National University of Water and Environmental Engineering. Technical Sciences (2), 198–208 (2014). (in Ukrainian)
18. Pereira, R.M., Sandri, D., Rios, G.F.A., Sousa, D.A.: Automation of irrigation by electronic tensiometry based on the arduino hardware platform. Revista Ambiente & Água 15 (2020). https://doi.org/10.4136/ambi-agua.2567
19. Jabro, J., Evans, R., Kim, Y.: Estimating in situ soil-water retention and field water capacity in two contrasting soil texture. Irrig. Sci. **27**, 223–229 (2009). https://doi.org/10.1007/s00271-008-0137-9
20. Romashchenko, M.I., Koriunenko, V.M.: Recommendations for operational control of the crops' irrigation regime using the tensiometric method. K.: DIA Ltd. (2012). (in Ukrainian)
21. Kovalchuk, V.P., Voitovich, O.P., Demchuk, D.O.: Ukrainian Patent for Utility Model UA132271 (25 February 2019). https://sis.ukrpatent.org/en/search/detail/1223767/
22. Lawson, Ch., Henson, R.: Numerical solution of problems by the method of least squares. — M.: Nauka (1986). (in Russian)
23. Maxim Integrated Products, Inc. DS18B20 Programmable Resolution 1-Wire Digital Thermometer: Data sheet. Retrieved from: https://www.analog.com/media/en/technical-documentation/data-sheets/DS18B20.pdf
24. Adafruit. Digital Pressure Sensor BMP280: Data sheet. Retrieved from: https://cdn-shop.adafruit.com/datasheets/BST-BMP280-DS001-11.pdf
25. Fisher, D.K.: Automated collection of soil-moisture data with a low-cost microcontroller circuit. Appl. Eng. Agric. **23**(4), 493–500 (2007)

Machine Learning Algorithms Comparison for Software Testing Errors Classification Automation

Liubov Oleshchenko[✉]

National Technical University of Ukraine "Igor Sikorsky Kyiv Polytechnic Institute",
Kyiv 03056, Ukraine
oleshchenkoliubov@gmail.com

Abstract. Automatic classification of software errors is an important tool for ensuring the quality of software being tested. By automating this process, testers can save time and effort, improve accuracy, gain a better understanding of errors, and improve communication between stakeholders. Software error clustering is an important concept in software testing that involves identifying and analyzing patterns in software errors or defects. The goal of error clustering is to understand the relationships between defects and identify the root causes of software errors, so that they can be prevented or corrected. The article provides a comprehensive survey of various techniques for software error clustering, including clustering algorithms. The author note that many of these techniques require significant expertise in software engineering and data analysis, and that there is a need for more user-friendly tools for software error clustering. This article presents an empirical research of software defect clustering, in which the author analyzes a large dataset of defects from a software development projects to identify several patterns in the defects, including clusters of related defects and common root causes.

The proposed software method uses stack traces to cluster data about software testing errors. The method uses a kNN algorithm to analyze test results, allowing the user to assign the text of the software test result to specified categories. The kNN algorithm has a high accuracy rate of 0.98, which is better than other clustering methods like Support Vector and Naive Bayes. The proposed method has several advantages, including a single repository for test results, automatic analysis of software testing results, and the ability to create custom error types and subtypes for error clustering. The developed new software method allows multiple launches to be combined. If there are a large number of test suites, they are divided into smaller groups as they cannot be included in a single run. The test suites can then be combined into a single run to present the data on dashboards and generate reports. The method is integrated into software using a Docker container, which reduces the time and human resources required for error analysis. Overall, the method provides real-time monitoring of project status for managers and customers.

Keywords: Machine learning algorithms · software defect · software errors classification · clustering · automation

© The Author(s), under exclusive license to Springer Nature Switzerland AG 2023
Z. Hu et al. (Eds.): ICCSEEA 2023, LNDECT 181, pp. 615–625, 2023.
https://doi.org/10.1007/978-3-031-36118-0_55

1 Introduction

Automatic classification of software errors is an important aspect of software testing, as it helps ensure that all errors are identified and categorized correctly. Software error clustering refers to the phenomenon where software defects are not randomly distributed across software components, but rather tend to occur in clusters within specific modules or parts of the system. This can have important implications for software testing and quality assurance, as it suggests that testing efforts should be focused on these high-risk areas of the system.

Consider the several key benefits to automating this process.

1. Improved testing efficiency: automating the classification of software errors can save time and effort that would otherwise be spent manually identifying and categorizing errors. This can help speed up the testing process and make it more efficient.
2. Enhanced accuracy: automated error classification can help reduce human errors that may occur during the manual classification of software errors. This can help ensure that all errors are correctly identified and categorized, which is important for ensuring the overall quality of the software.
3. Better understanding of errors: by automatically categorizing errors, testers can gain a better understanding of the types of errors that are occurring and the frequency with which they occur. This can help them identify patterns and trends that may be useful in improving the software and preventing similar errors in the future.
4. Improved communication: automated error classification can help improve communication between testers, developers, and other stakeholders. By categorizing errors in a standardized way, everyone involved in the software development process can more easily understand the issues that need to be addressed.

There are several machine learning algorithms that can be used for software error classification during software testing. K-means is a popular clustering algorithm that can be used to group software errors based on their similarity. The algorithm partitions the errors into k clusters, where k is a user-defined parameter. Each error is assigned to the nearest cluster based on its features or characteristics. Hierarchical clustering is another popular algorithm that can be used to group errors based on their similarity. The algorithm builds a tree-like structure of clusters, where the leaf nodes represent individual errors and the internal nodes represent groups of errors that share some common characteristics. Density-based clustering algorithms such as DBSCAN (Density-Based Spatial Clustering of Applications with Noise) can be used to identify clusters of errors that are dense in a feature space. These algorithms can be particularly useful when the errors are distributed in a non-uniform manner. Fuzzy clustering algorithms such as Fuzzy C-means can be used to assign errors to multiple clusters with varying degrees of membership. This can be useful when errors have characteristics that overlap with multiple clusters. It's important to note that the choice of clustering algorithm depends on the specific characteristics of the errors being classified, as well as the goals of the software testing process.

1.1 Analysis of Recent Research and Related Work

The study [1] provides a systematic review of software error clustering techniques, focusing on recent research from the past decade. The authors identify several key challenges in software error clustering, including the lack of standardized data formats and the difficulty of interpreting clustering results.

The authors also highlight several promising approaches for software error clustering, including machine learning techniques and graph-based methods. They suggest that these approaches can provide more accurate and comprehensive analysis of software errors, and can be applied to a wide range of software development contexts.

Software error clustering is a phenomenon in software testing where a particular software defect or bug is found to occur in multiple instances. In other words, certain errors tend to cluster together, indicating a deeper underlying issue in the software code.

The article [2] presents an empirical study of error clustering in commercial software applications. The study found that a small percentage of software errors were responsible for a large percentage of the total defects in the software. The authors recommend using error clustering to identify high-priority areas for software testing and debugging.

The article [3] presents an analysis of software error clustering in open-source software projects. The study found that errors tended to cluster in certain areas of the software code, such as specific modules or functions. The authors recommend using clustering analysis to identify areas of the software that are particularly prone to errors.

The article [4] presents a study of error clustering in large-scale software systems. The study found that errors tended to cluster in certain parts of the code, such as frequently executed functions or those with complex logic. The authors recommend using clustering analysis to prioritize testing and debugging efforts in large-scale software systems. By analyzing the distribution of errors in software code, software developers and testers can gain insight into the underlying causes of defects and improve the quality of their software.

The authors of article [5] conducted a systematic review of research studies on software error clustering, analyzing the methods used to identify and quantify clustering, the types of software defects that tend to cluster, and the factors that contribute to clustering. The article highlights several key findings from the research studies reviewed. For example, the studies generally found that a small number of modules or components tend to contain the majority of software defects. The studies also found that different types of software defects tend to cluster in different ways; for example, logic errors may cluster differently than memory-related errors. The authors also discuss some of the challenges and opportunities associated with studying and addressing software error clustering. For example, they note that identifying and quantifying clustering can be difficult, particularly in large and complex software systems. They also note that addressing clustering may require a shift in focus from detecting and fixing individual defects to identifying and addressing underlying structural or design issues.

The article [6] also reviews research studies on software error clustering, with a focus on the methods used to identify and measure clustering, as well as the techniques used to address clustering in software testing. The authors highlight several key trends and challenges in the research on software error clustering. For example, they note that different techniques for identifying and measuring clustering have been proposed, but it can be difficult to determine which technique is most appropriate for a particular software system. They also note that addressing clustering may require a combination of testing techniques, such as static analysis, dynamic analysis, and mutation testing.

Overall, the last articles [7–19] highlight the importance of understanding and addressing software error clustering in software testing. By identifying high-risk areas of the software system, testing efforts can be focused more effectively, leading to better software quality and more efficient use of testing resources. These articles provide valuable insights into the challenges and opportunities of software error clustering in software testing. They highlight the importance of using multiple techniques for software error clustering, and suggest that advances in machine learning and data analysis can help improve the accuracy and efficiency of software testing and quality assurance.

1.2 Research Data

To solve the problem of identifying and fixing software errors in a timely manner the author proposes an automated solution that can achieve high accuracy at all stages of software development using the microservice architecture to compare errors found during software testing.

The dataset of 100,000 stack traces from software companies in Ukraine was use and was develop the software method for classifying testing errors using Java programming language and MongoDB as a database. To analyze the test results, author used HTML, CSS, JS technologies, Angular framework, Node. Js and compared Naive Bayes, k-Nearest Neighbors, Support Vector Machines machine learning methods. The research aims to develop an automated software error analysis process that can reduce the time it takes to identify and fix errors during software development.

1.3 Objective

The object of the research in this article is creating the software method to automatically classify software testing errors more accurately and quickly using machine learning algorithms. The goal is to improve the efficiency and accuracy of identifying errors in software testing.

2 Research Method

2.1 Clustering Algorithms Comparison for Software Errors Classification

Naive Bayes (NB) is a probabilistic algorithm that can be used for software error classification during software testing. It is a simple and efficient algorithm that is often used for text classification tasks, but it can also be applied to software error classification. NB algorithm works by calculating the probability of each error belonging to a particular class (i.e., error category). It does this by using Bayes' theorem, which calculates the conditional probability of a class given the error features, assuming that the features are conditionally independent. This assumption of independence is called "naive" and is often violated in real-world applications, but the algorithm still tends to perform well in practice. To apply the NB algorithm to software error classification, we need to first define the error features (i.e., input variables) that we will use to train the model. These features could include the error message, stack trace, error type, error severity, and other relevant information. Once the features have been defined, we need to train the Naive Bayes model on a set of labeled error data. During the training phase, the algorithm learns the conditional probability of each feature given each error class. To classify a new error, the NB algorithm calculates the conditional probability of the error belonging to each class, using the learned probabilities from the training data. The class with the highest probability is then assigned to the error.

Overall, NB algorithm can be a useful approach for software error classification during software testing, especially when the number of error classes is relatively small, and the features are well-defined and easily measurable. It is important to note that the performance of the algorithm depends on the quality and relevance of the error features, as well as the quality and size of the training data.

Support Vector Machines (SVM) is a popular machine learning algorithm that can be used for software error classification during software testing. SVM is a supervised learning algorithm that is often used for classification tasks in which the input data is linearly separable. To apply SVM to software error classification, we first need to define the error features and create a labeled dataset of error instances. The features could include information such as error message, stack trace, error type, and severity. Next, we train an SVM model on the labeled dataset of error instances. During the training process, the SVM algorithm uses the error features to create a hyperplane that separates the errors into different classes. The goal of the algorithm is to find the hyperplane that maximizes the margin (i.e., the distance between the hyperplane and the closest error instances of each class). Once the SVM model has been trained, it can be used to classify new error instances. The SVM algorithm maps the error features to a point in a high-dimensional space, and then uses the hyperplane to classify the error into one of the predefined classes. One advantage of SVM algorithm is that it can handle non-linearly separable data by using kernel functions to transform the input features into a higher-dimensional space where the data becomes separable. Additionally, SVM is relatively insensitive to the presence of irrelevant features, which can be useful when working with error data that contains noise or unimportant variables.

Overall, SVM algorithm can be a useful approach for software error classification during software testing, especially when the error data is well-defined and the classes are linearly or non-linearly separable. However, it is important to carefully select and engineer the features used for classification, as well as tune the hyperparameters of the model to achieve the best performance.

The k-Nearest Neighbors (kNN) algorithm can be used for software error classification during software testing. The kNN algorithm is a non-parametric and lazy learning algorithm that can be used for both classification and regression tasks. To apply kNN to software error classification, we first need to define the error features and create a labeled dataset of error instances. The features could include information such as error message, stack trace, error type, and severity. Next, we need to choose a value for k, which represents the number of nearest neighbors to consider when classifying new error instances. Then, when a new error instance is presented, the kNN algorithm searches for the *k* nearest neighbors in the labeled dataset based on the similarity of their error features to those of the new instance. Once the k nearest neighbors have been identified, the kNN algorithm uses the class labels of those neighbors to assign a class label to the new error instance. This is done by majority vote, where the class with the most instances among the k nearest neighbors is assigned to the new instance. One advantage of the kNN algorithm is that it is simple and easy to implement. It also doesn't make assumptions about the underlying distribution of the data, making it applicable to a wide range of problems. However, the performance of the kNN algorithm can be sensitive to the choice of *k* and the distance metric used to calculate the similarity between error instances.

Overall, the kNN algorithm can be a useful approach for software error classification during software testing, when the classes are well-defined. However, it is important to carefully select and engineer the features used for classification, as well as choose an appropriate value for k and distance metric.

2.2 Proposed Software Error Classification Method

The new proposed method involves forming a stack trace after running tests to identify errors in software. The stack trace contains messages about software errors, and describes the reasons for failed tests and shows where the errors are located in the software files at different levels of the libraries. The proposed method repeats this process for each test, removing repeated text and irrelevant information like dates, and analyzes the unique information specific to each test. This allows for the classification of software errors using clustering algorithms. (Fig. 1).

```
java lang assertionError  invalid upc service navigation
link redirection  expected  true        found  false
org testng assert fail assert java
org testng assert failnotequals assert java
org testng assert asserttrue assert java
my project tests checkLinksareclickable maintest java
```

● Build #1   ● Build #2   ● Build #3   ● Build #4   ● Build #5

Fig. 1. Code text for clustering

After gathering a significant amount of redundant text, the text is analyzed to collect information on the frequency of repeated words for each error category. This information is shown in Fig. 2. As more tests are performed and more text is collected, the accuracy of the kNN algorithm improves.

After obtaining the metrics on the frequency of repeated words for each error category, the proposed method uses two metrics called Term Frequency (TF) and Inverse Document Frequency (IDF). These metrics help to determine the significance of words in documents and assign them to the appropriate class of software errors.

In the dataset that was analyzed, the errors causing failed tests were initially categorized into three groups: Product Bugs, System Issues, and Automation Issues. The

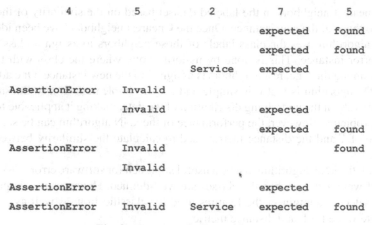

	4	5	2	7	5
				expected	found
				expected	found
		Service	expected	
AssertionError	Invalid			
	Invalid		expected	
			expected	found
AssertionError	Invalid			found
	Invalid			
AssertionError			expected	
AssertionError	Invalid	Service	expected	found

Fig. 2. Frequency indicators example

text from the files was then represented as a mathematical vector in a multi-dimensional space.

2.3 The Software Architecture of the Proposed Method

The software system for error classification was developed using a microservices approach. The client part of the software consists of Logger, Agent, and Client services. HTTP clients are used to send HTTP requests through the API.

The proposed software system for organizing tests is structured in four levels: Launch, Test Suite, Test Case, and Test Step. The Launch level contains sets of tests that were performed during a launch. The Test Suite level groups test cases that are related to one testing module, functionality, priority, or type of testing. Each Test Suite includes multiple Test Cases and is typically performed as a whole in the testing process. The Test Case level describes the algorithm for testing a program to determine the occurrence of a particular situation and source data. The Test Step level describes the steps required to reproduce a bug, with the aim of keeping these steps as short as possible. This level also allows users to identify at which step the error occurred and view the launch history. The result of the test case execution is displayed, including all its steps and software errors that occurred during the launch. Users can switch between different runs and view the errors that occurred in the past. Attachments can also be added to each step, such as images of code fragments, which can be viewed by clicking on the corresponding attachment in the developed software system.

The developed software method allows multiple launches to be combined. If there are a large number of test suites, they are divided into smaller groups as they cannot be included in a single run. The test suites can then be combined into a single run to present the data on dashboards and generate reports. The software implements both linear and deep merging (as shown in Fig. 3). If the user selects the "Linear Merge" option, a new launch is created that includes elements from the merged launches, while the item levels remain the same as in the original launch.

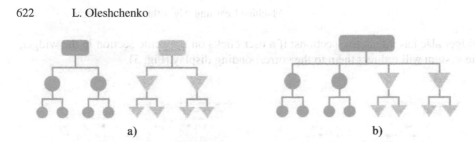

a)   b)

Fig. 3. Test launches visual scheme: a) before merging, b) after linear merger

This passage describes a feature of a software system that displays the result of a test case execution, including any errors that occurred during the launch. The system also saves the last 10 test runs, allowing users to switch between them and see previous errors. Additionally, there is a feature to attach images of code fragments to the corresponding step, which can be opened in a pop-up window. The system also has a Stack trace tab that shows the entire stack trace used in automatic analysis. The root cause of the test failure can be written in the first lines of the log lines, and the analyzer can be configured to consider only the necessary lines.

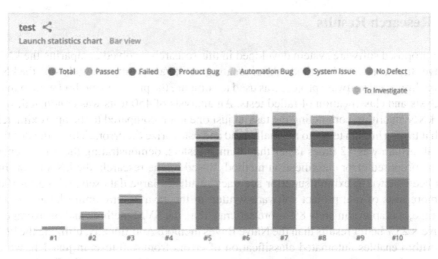

Fig. 4. Launch statistics chart in startup mode of proposed software method

The software system that has been suggested enables users to design and build their own widgets. These widgets have specialized graphical tools that are intended to make it simple to view and evaluate the outcomes of the test automation process. (Fig. 4).

The widget presents an overview of the statistics for every test run, including the total number of tests, as well as the number of tests that passed, failed, or were associated with various types of errors. These types of errors include software product bugs, system issues, automation bugs, defects that require investigation, and those that are not considered defects. Each type of error is displayed as a percentage of the total. The widget also has interactive sections; if a user clicks on a specific section in the widget, the system will redirect them to the corresponding display (Fig. 5).

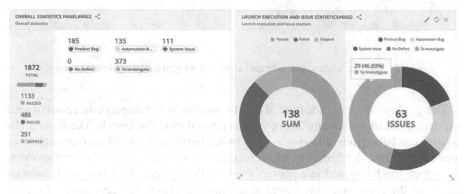

Fig. 5. Statistics panel of proposed software method

3   Research Results

The proposed software system developed in this research involved comparing the kNN, Naive Bayes, and SVM methods. After creating a software testing method using the kNN algorithm, a real software project was used to compare the proposed method with manual analysis and classification of failed tests. An analysis of 450 tests was conducted, with the kNN algorithm completing the task in just one hour, compared to the approximately 12 h it takes a human tester to manually read and categorize the errors. This indicates that the algorithm was 12 times faster than human testers, demonstrating the effectiveness of the proposed error classification method. Based on the research, the kNN algorithm has been shown to exhibit superior accuracy results in large data sets, with 500–1000 or more tests of real project software written in the Java programming language. On average, it is approximately 8% more accurate than the SVM method, and on average, it delivers 25% better results than the Naive Bayes method. Additionally, utilizing the kNN algorithm enables automated classification of errors from 450 tests in just 1 h, which is approximately 12 times faster than a human tester could do manually. Furthermore, the accuracy of clustering with 1000 failed tests using the kNN algorithm was found to be 0.98 (Table 1), which is the best result among the clustering methods considered in this study (kNN, Support Vector, and Naive Bayes). The kNN algorithm provides flexibility to choose the distance when building the kNN model. The Naive Bayesian classifier has low computational costs for training and classification. However, the SVM algorithm requires a lot of memory and has long training times for large datasets, making it computationally complex.

Table 1. Clustering algorithms accuracy comparison in proposed software method

Machine learning algorithm	The failed tests number					
	50	100	200	250	500	1000
kNN	0,73	0,75	0,83	0,85	0,91	**0,98**
SVM	0,76	0,79	0,86	0,86	0,86	**0,92**
NB	**0,83**	0,81	0,77	0,77	0,76	0,75

4   Conclusions

Automatic classification of software errors is an important tool for ensuring the quality of software being tested. By automating this process, testers can save time and effort, improve accuracy, gain a better understanding of errors, and improve communication between stakeholders.

The proposed new software method for classifying software testing errors has a number of competitive advantages, including a central location for test results, integration with other systems, and the ability to analyze test results automatically. It also allows for the creation of custom error categories and subcategories to better organize test results, maintains a history of test launches, and is compatible across multiple platforms. Users can view self-test performance in real time, filter results as needed, and access automation statistics in the form of tables and graphs.

However, there are several areas that require further research and analysis. The study did not consider data-driven features that could enhance the accuracy and dependability of automated systems. Additionally, the evaluation was limited to only three methods, and a broader range of programs should be examined before drawing conclusions. Multiple programs that operate on different operating systems often interact, which means that information from multiple sources may be necessary to identify the root cause of errors. Relatedly, hardware failures can cause errors that are difficult to differentiate from software errors, and this is an area that requires more analysis.

References

1. Kaur, H., Malhotra, R., Singh, H.: Software error clustering: a systematic review. IEEE Access **8**, 98244–98261 (2020). https://doi.org/10.1109/ACCESS.2020.2996594
2. Ramalingam, S., Arumugam, S., Kannan, S.: An empirical study on error clustering in commercial software applications. Int. J. Comp. Sci. Netw. Secu. **14**(7), 27–38 (2014)
3. Aggarwal, A., Williams, L., Nagappan, N.: An analysis of software error clustering in open-source software. Empir. Softw. Eng. **23**(1), 582–625 (2018). https://doi.org/10.1007/s10664-017-9526-y
4. Xu, P., Fu, J., Su, Z.: Identifying error clustering in large-scale software systems: a study of 23 million lines of code. IEEE Trans. Software Eng. **42**(4), 349–367 (2016). https://doi.org/10.1109/TSE.2015.2482201
5. Jalali, R., Wohlin, C.: A systematic review of software error clustering: trends, challenges, and opportunities. J. Syst. Softw. **85**(6), 1213–1227 (2012). https://doi.org/10.1016/j.jss.2012.01.060

6. Sedighi, S., Rezaei, M.: Software error clustering: A systematic literature review. J. Syst. Softw. **165**, 110591 (2020). https://doi.org/10.1016/j.jss.2020.110591
7. Truong, A., Hellström, D.: Clustering and Classification of Test Failures Using Machine Learning. Master's thesis work. Department of Computer Science Faculty of Engineering LTH (2018)
8. Dang, Y., Wu, R., Zhang, H., Zhang, D., Nobel, P.: ReBucket: a method for clustering duplicate crash reports based on call stack similarity. In: 2012 34th International Conference on Software Engineering (ICSE), pp. 1084–1093. IEEE. Conference on Software Engineering (ICSE). IEEE (2012, June)
9. Moran, K., Linares-Vásquez, M., Bernal-Cárdenas, C., Vendome, C., Poshyvanyk, D.: Automatically discovering, reporting and reproducing android application crashes. In: 2016 IEEE international conference on software testing, verification and validation, pp. 33–44. IEEE (2016)
10. Abbineni, J., Thalluri, O.: Software defect detection using machine learning techniques. In: 2nd International Conference on Trends in Electronics and Informatics, IEEE. pp. 471–475 (2018)
11. Buchgeher, B.R.R., Klammer, G., Pfeiffer, M., Salomon, C., Thaller, H., Linsbauer, L.: Improving defect localization by classifying the affected asset using machine learning. In: International Conference on Software Quality, pp. 125–148 (2019)
12. Durelli, V.H.S., et al.: Machine Learning Applied to Software Testing: A Systematic Mapping Study. IEEE Transactions on Reliability, 1–24 (2019)
13. Karim, S., Leslie, H., Spits, H., Abdurachman, E., Soewito, B.: Software metrics for fault prediction using machine learning approaches. In: IEEE International Conference on Cybernetics and Computational Intelligence, pp. 19–23 (2017)
14. Alsmadi, I., Alda, S.: Test cases reduction and selection optimization in testing web services. I.J. Information Engineering and Electronic Business **5**, 1–8 (2012)
15. Nasar, M.d., Johri, P., Chanda, U.: Software testing resource allocation and release time problem: a review. I.J. Modern Education and Computer Science **2**, 48–55 (2014)
16. Mohapatra, S.K., Prasad, S.: Finding Representative Test Case for Test Case Reduction in Regression Testing. Int. J. Intell. Sys. Appli. (IJISA) **7**(11), 60–65 (2015)
17. Ansari, G.A.: Detection of infeasible paths in software testing using UML application to gold vending machine.Int. J. Edu. Manage. Eng. (IJEME) **7**(4), 21–28 (2017)
18. Mahmood, H., Sirshar, M.: A case study of web based application by analyzing performance of a testing tool. Int. J. Edu. Manage. Eng. (IJEME) **7**(4), 51–58 (2017)
19. Nawaz, Z.: Proposal of Enhanced FDD Process Model. Int. J. Edu. Manage. Eng. (IJEME) **11**(4), 43–50 (2021)

Machine Learning for Unmanned Aerial Vehicle Routing on Rough Terrain

Ievgen Sidenko[1](✉), Artem Trukhov[1], Galyna Kondratenko[1], Yuriy Zhukov[2], and Yuriy Kondratenko[1,3]

[1] Petro Mohyla Black Sea National University, 68th Desantnykiv Street, 10, Mykolaiv 54000, Ukraine
ievgen.sidenko@chmnu.edu.ua
[2] C-Job Nikolayev, Artyleriyska Street, 17/6, Mykolaiv 54006, Ukraine
[3] Institute of Artificial Intelligence Problems, Mala Zhytomyrs'ka Street, 11/5, Kyiv 01001, Ukraine

Abstract. The paper considers the main methods of machine learning for unmanned aerial vehicle (drone) routing, simulates an environment for testing the flight of a drone, as well as a model with a neural network for the unmanned routing of a drone on rough terrain. The potential use of unmanned aerial vehicles is limited because today the control of drone flight is carried out in a semi-automatic mode on the operator's commands, or in remote mode using a control panel. Such a system is unstable to the human factor because it depends entirely on the operator. The relevance of the work is to use machine learning methods for drone routing, which will provide stable control of the unmanned aerial vehicle to perform a specific task. As a result of the work, a neural network architecture was developed, which was successfully implemented in a test model for routing an unmanned aerial vehicle on rough terrain. The test results showed that the unmanned aerial vehicle successfully avoids obstacles in the new environment.

Keywords: Transport routing problem · Unmanned aerial vehicle · Machine learning · Reinforced learning · Artificial intelligence

1 Introduction

Recently, the development of mobile robots has gained rapid momentum, especially the use of unmanned aerial vehicles in various fields of human activity [1–3].

Among the main reasons and advantages of using unmanned aerial vehicles for the following purposes: high efficiency; no threat to the life and health of staff; economic efficiency due to the relative cheapness of the drone; visual monitoring in real time with high quality images; quick search with target identification [4].

Machine learning, including neural networks, helps unmanned aerial vehicles to move more accurately and positionally on rough terrain, detect and classify objects with a camera and bypass them, adapt to changes in weather conditions based on training models and traffic scenarios, and deliver medicines, essential goods in locations with increased risk to human life, and more [2].

Z. Hu et al. (Eds.): ICCSEEA 2023, LNDECT 181, pp. 626–635, 2023.
https://doi.org/10.1007/978-3-031-36118-0_56

The main problem in this task is the choice of the effective use of machine learning for unmanned aerial vehicles routing in rough terrain. In addition, the problem also lies in the need to create a training environment for training and testing the operation of the UAV, and then modeling obstacle avoidance situations in conditions close to reality. The existing works did not provide a comprehensive solution to the problems of UAV routing on rough terrain. In this work, the authors try to solve these problems using reinforcement learning and game engine Unity3D.

The main purpose of this paper is to describe the process of machine learning and developing system to ensure efficient routing of unmanned aerial vehicles on rough terrain, which will allow more accurate movement of a certain route and increase flight stability.

2 Related Works and Problem Statement

Most of the works are related to object detection and recognition using UAVs [5–7] In the proposed approach [5], firstly, the training for machine learning on the objects is carried out using convolutional neural networks, which is one of the deep learning algorithms. By choosing the Faster-RCNN and YoloV4 architectures of the deep learning method, it is aimed to compare the achievements of the accuracy in the training process.

In this study [7], the methods of deep learning based detection and recognition of threats, evaluated in terms of military and defense industry, using Raspberry Pi platform by UAVs are presented. In the proposed approach, firstly, the training for machine learning on the objects is carried out using convolutional neural networks, which is one of the deep learning algorithms.

The flight controller, based on the performance of sensors and software protection algorithms, can independently change the flight parameters of the drone, without resorting to the help of the operator [6]. The design of the device determines its possible use. Currently, there are many areas where unmanned aerial vehicles have proven themselves well: geodesy and cartography; agriculture in terms of field control and cultivation; farming; normal photo and video shooting; transportation of medicines and essential goods [7]. The huge potential of drones in transporting medications or vaccines helps in situations where the time factor is important. A number of large trade and postal companies have seriously considered investing in the development of drones engaged in the delivery of parcels, mail, medicines, food [1, 3].

Machine learning allows computers to draw conclusions based on data without following strict rules. In other words, the machine can find patterns in complex multiparameter problems (which the human brain cannot solve) to find more accurate answers. The result is a correct forecast [8, 9]. Depending on whether there is a teacher, training is divided into teacher training (controlled training), lack of teacher training (uncontrolled training) and reinforcement training [10].

In this study [11], authors applied reinforcement learning based on the proximal policy optimization algorithm to perform motion planning for UAV in an open space with static obstacles. The test of the trained model shows that the UAV reaches the goal with an 81% goal rate using the simple reward function suggested in this work [11].

The success of deep learning directly depends on the power of technology. At the time of the emergence of neural networks, the power of computers was low, which is why the networks themselves were quite weak. That is why at that time it was impossible to create a large number of layers of neural networks, namely the number of layers depends on the capabilities of the network. But with the advent of GPU and TPU, everything has changed. Modern Deep Learning is able to cope with large network sizes. In addition, for deep learning use special frameworks: Keras, Detectron, TensorFlow, PyTorch, CNTK and others [12–14].

All of the above publications are directly related to the objectives of the study in this work, while all of them lack a comprehensive solution to this problem in conditions close to reality.

3 Methods of Machine Learning for Unmanned Aerial Vehicle Routing

Convolutional neural network (CNN) is one of the most influential innovations in the field of computer vision [11, 15].

Use reinforcement training where the machine is facing a task - there are many possible options for performing the task correctly in the external environment, such as computer games, trade operations and unmanned vehicles [16]. Reinforcement learning (RL) is almost the same as learning with a teacher, but the role of a "teacher" is a real or virtual environment [10, 11, 16–18].

SARSA (state - action - reward - state - action) is a special algorithm of machine learning with reinforcement. It was suggested by Rammeri and Niranjan in a technical note entitled "Modified Connectionist Q-Learning" (MCQ L) [19]. The name of the algorithm reflects the fact that the main function for maximizing the value of Q depends on the current state of the agent S, the action that the agent chooses A, the reward R that the agent receives for choosing an action, the state S2 to which the agent goes after this action, and, finally, the next action A2, which the agent chooses in his new state.

$$Q(s_t, a_t) \leftarrow Q(s_t, a_t) + \alpha \left[r_t + \gamma Q(s_{t+1}, a_{t+1}) - Q(s_t, a_t) \right] \tag{1}$$

The SARSA agent interacts with the environment and updates policies based on actions taken. The value of Q for a certain action and state is updated by the alpha score. The value of Q represents the possible reward obtained in the next time step for performing action A in state S, plus the future reward obtained in the next observation of the action of the state [20].

In comparison, Q-learning updates the estimate of the optimal state-action function based on the maximum reward for available actions, while SARSA learns the value of Q associated with adopting a policy that it itself follows [21–24].

The speed of learning (α) determines the extent to which new information overlaps the old. A coefficient equal to 0 will force the agent to not change anything, and a coefficient equal to 1 will force the agent to consider only the most recent information. The discount factor (γ) determines the importance of future rewards. A coefficient of 0 forces the agent to consider only current rewards, while a coefficient of 1 forces it to seek long-term high rewards.

Q-learning is a learning algorithm used in the field of reinforcement learning. It does not require an environment, and it solves problems with stochastic transitions and rewards without requiring adaptation [21, 22]. Q-learning uses an action-value function for strategy

$$Q(s, a) = E[R|s^t = s, a^t = a],\tag{2}$$

where s is a state, a is an action.

Q-learning in its simplest form stores data in tables. This approach fails when the number of states/actions increases, as the probability that an agent will visit a certain state and perform a certain action becomes increasingly small [24].

Genetic algorithm (GA) is a search algorithm and heuristic method that simulates the process of natural selection, using methods such as mutation and crossover, to create new genotypes in the hope of finding good solutions to this problem [25–28]. Genetic algorithms are mainly used to find solutions in very large, complex search spaces [29].

Different approaches to vehicle route planning using fuzzy logic and fuzzy sets, IoT technologies, neural networks, genetic, evolutionary and heuristic algorithms are also given in [30–35].

Due to the limitations of genetic algorithms and other machine learning methods, it was decided to use reinforcement learning algorithms, because they give better results and allow influencing the training of agents when solving the UAV routing problem. Also, reinforcement learning algorithms have a wide range of settings, which allows controlling the behavior of the agent and determining the correct learning goals, for example, neglecting the short-term reward for the long-term one, which is important for solving this problem. In addition, models trained by reinforcement learning algorithms are more versatile and adapt well in a dynamic environment.

4 Practical Implementation of the Drone Routing

The main libraries for creating artificial intelligence models supported by the Unity game engine are Acord.Net, CNTK, Tensorflow and unity ml-agents. Accord.NET Framework is a library for scientific computing in.NET [36–38]. The source code of the project is available under the terms of the Gnu Lesser Public License version 2.1.

The project was originally designed to expand the capabilities of the AForge.NET Framework, but since then it has included AForge.NET. The new versions combine both frameworks called Accord.NET [38, 39].

Microsoft Cognitive Toolkit [40–46] is a standardized toolkit for designing and developing neural networks of various types, uses artificial intelligence to work with large amounts of data through deep learning, uses internal memory to process sequences of arbitrary length.

The system is implemented using a game engine Unity3D and a plug-in unity-ml-agents. The main components of the plugin are the brain, the agent and the academy [47–50]. The academy is responsible for the learning process and the learning environment. An agent is actually an object that uses the brain to learn and coordinate its actions. It has observational vectors and available actions, and accumulates a reward. The brain is responsible for the decision made by the agent. This final element is being trained [51].

To solve the problem of drone routing, a training environment was created that will force the aircraft to perform movements on all axes. It is shown in Fig. 1.

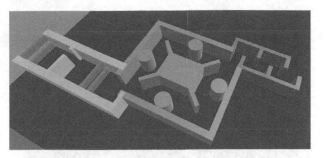

Fig. 1. The training environment

An important task was to force the agent not only to move sideways and forward or backward, but also to rotate the body when necessary, for which the first segment of the environment is responsible (Fig. 2a). Instead, to create the effect of rising and falling, a section with special vertical obstacles was created (Fig. 2b).

a)   b)

Fig. 2. Parts for testing: (a) drone hull turns, (b) drone lifting and lowering

With the help of vectors, the agent will observe the environment, a total of 17 vectors were used: 8 diagonals at an angle of 20 and 45 degrees, forward at an angle of 45, 65, 90, 110 and 155, up, down, left and right. The agent is shown in Fig. 3.

The development of the system includes the creation of a neural network architecture that can be used to control an unmanned aerial vehicle in the environment. Develop an environment for learning and testing the neural network, which includes: creating an agent, determining the desired behavior for him; creating an environment, placing sensors in it, which are used to orient the agent in the environment and to train the model; development of a system of physical interaction of the agent with the environment.

The advantages of the developed system are that the user can start the learning process again with the parameters defined by him, teach the model, if necessary, and transfer the model to another environment, provided new sensors are used to orient the agent in the environment. The created model is saved in ONNX format for use on a real drone. In addition, the created system is well optimized and designed to run on any platform.

The proposed neural network structure marks 19 neurons in the input layer, 128 neurons in 2 fully connected layers and 4 neurons in the output layer. To speed up

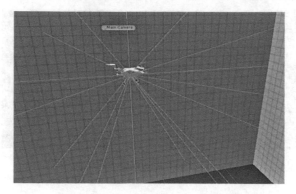

Fig. 3. Agent with observable vectors

learning and improve its quality, four agents have been created who use the same brain to make decisions, but two of them are moving in the opposite direction.

The input are 17 observable vectors, end point position and drone position, the output are discrete numbers, one for each axis of motion.

Since the playing time is relative, it was decided to increase it by 50 times, to speed up the process of training agents. The training process lasted about 4 hours, with 1,500,000 decisions made and the average reward of agents increased from 9.55 to 2441.

System testing begins with planning the strategy by which this process will take place. During testing, the correctness of the drone operation on a certain segment must be checked. Testing should include checking the movement of the drone on all axes and the correctness of their use depending on the situation. It is also necessary to create and test the developed model in another environment where there was no training and where the agent will be for the first time. One of the most important steps is to test behavior in a new environment. In addition to checking the correct behavior of the agent is also tested his ability to achieve a given goal. To move to a more realistic environment, a prototype city with a central park in the middle was created. In order to specify the end point, an empty object was created and located at the end of the route, thus transmitting the exact coordinates of the destination. The next step is to set the drone to the starting position and start the routing with the Play button, the agent starts moving towards the end point and wraps around the obstacles (Fig. 4).

In the test environment, the agent successfully reached the end point without encountering any of the obstacles. During the testing, the agent's ability to navigate in space, perform all basic movements along 4 axes, reach the end point and bypass obstacles in different environments was tested.

It should be noted that the model is designed for specific physical conditions, drone models and observation vectors. Therefore, such a system can be used as initial data to complete the model, but it can not be used for any conditions. Use specific conditions or adjust certain parameters must be done in an artificial environment, then refine the model and transfer to a real device that contains analogues of all devices that provide observation vectors.

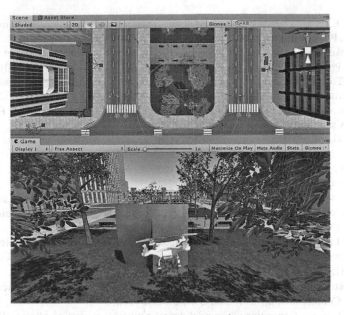

Fig. 4. Flight agent in a test environment

5 Conclusions

During the work, a drone routing system was designed using machine learning methods, which provides high efficiency and stable flight in rough terrain.

The result is a comprehensive system that allows you to perform neural network training with custom parameters for your own purposes or use an already trained and developed model of artificial intelligence, which is provided by default with the drone model.

A neural network architecture with reinforcement learning algorithms was developed, which was successfully implemented in a test model for routing an unmanned aerial vehicle on rough terrain. The test results showed that the unmanned aerial vehicle successfully avoids obstacles in the new environment.

References

1. Sinulingga, L.S.A., Ramdani, F., Saputra, M.C.: Spatial multi-criteria evaluation to determine safety area flying drone. Proc. of the ISyG Int'l Symp. on Geoinformatics, pp. 55–61. Malang, Indonesia (2017)
2. Kharchenko, V., Matiychyk, D., Babenko, A.: Mathematical model of unmanned aerial vehicle control in manual or semiautomatic modes. Proc. of the 4th Int'l Conf. APUAVD, pp. 37–40. Kiev, Ukraine (2017)
3. Zhdanov, S., et al.: The Mathematical Model for Research of the UAV Longitudinal Moving. I. J. Computer Network and Information Security **5**, 29–39 (2021)
4. Alshbatat, A.: Fire extinguishing system for high-rise buildings and rugged mountainous terrains utilizing quadrotor unmanned aerial vehicle. I.J. Image, Graphics and Signal Processing **1**, 23–29 (2018)

5. Bayhan, E., et al.: Deep learning based object detection and recognition of unmanned aerial vehicles. Proc. of the 3rd Int'l Conf. HORA, pp. 1–5. Ankara, Turkey (2021)
6. Timchenko, V.L., Lebedev, D.O.: Control systems with suboptimal models for stabilization of UAV. Proc. of the 5th Int'l Conf. APUAVD, pp. 117–122. Kiev, Ukraine (2019)
7. Ozkan, Z., et al.: Object detection and recognition of unmanned aerial vehicles using raspberry Pi platform. Proc. of the Int'l Conf. ISMSIT, pp. 467–472. Ankara, Turkey (2021)
8. Tania, M., et al.: Image recognition using machine learning with the aid of MLR. I.J. Image, Graphics and Signal Processing 6, 12–22 (2021)
9. Meena, G., et al.: Traffic prediction for intelligent transportation system using machine learning. Proc. of the 3rd Int'l Conf. ICETCE, pp. 145–148. Jaipur, India (2020)
10. Yasir, M., et al.: Machine learning based analysis of cellular spectrum. I.J. Wireless and Microwave Technologies 2, 24–31 (2021)
11. Kim, S., et al.: Motion planning by reinforcement learning for an unmanned aerial vehicle in virtual open space with static obstacles. Proc. of the Int'l Conf. Control, Automation and Systems (ICCAS), pp. 784–787. Busan, Korea (2020)
12. Wu, W., Liao, M.: Reinforcement fuzzy tree: a method extracting rules from reinforcement learning models. Proc. of the 18th Int'l Conf. on Computer and Information Science (ICIS), pp. 47–51. Beijing, China (2019)
13. Zinchenko, V., et al.: Computer vision in control and optimization of road traffic. Proc. of the 3rd Int'l Conf. DSMP, pp. 249–254. Lviv, Ukraine (2020)
14. Striuk, O., et al.: Generative adversarial neural network for creating photorealistic images. Proc. of the 2nd Int'l Conf. ATIT, pp. 368–371. Kyiv, Ukraine (2020)
15. Leizerovych, R., et al.: IoT-complex for monitoring and analysis of motor highway condition using artificial neural networks. Proc. of the 11th Int'l Conf. DESSERT, pp. 207–212. Kyiv, Ukraine (2020)
16. Chornovol, O., et al.: Intelligent Forecasting System for NPP's Energy Production. Proc. of the 3rd Int'l Conf. DSMP, pp. 102–107. Lviv, Ukraine (2020)
17. Bidyuk, P., Gozhvi, A., Kalinina, I.: Modeling military conflicts using bayesian networks. Proc. of the 1st Int'l Conf. SAIC, pp. 1–6. Kyiv, UKraine (2018)
18. Tiwari, A.K., Nadimpalli, S.V.: Augmented random search for quadcopter control: an alternative to reinforcement learning. I.J. Information Technology and Computer Science 11, 24–33 (2019)
19. Tathe, P.K., Sharma, M.: Dynamic actor-critic: reinforcement learning based radio resource scheduling for LTE-Advanced. Proc. of the 4th Int'l Conf. ICCUBEA, pp. 1–4. Pune, India (2018)
20. Kiran, B.R., et al.: Deep reinforcement learning for autonomous driving: a survey. IEEE Trans. Intell. Transp. Syst. 23(6), 4909–4926 (2022)
21. Pandey, D., Pandey, P.: Approximate Q-Learning: An Introduction. Proc. of the 2nd Int'l Conf. on Machine Learning and Computing, pp. 317–320. Bangalore, India (2010)
22. Sun, C.: Fundamental Q-learning Algorithm in Finding Optimal Policy. Proc. of the Int'l Conf. ICSGEA. Changsha, pp. 243–246 (2017)
23. Chen, Y.-H., Wang, H., Chiu, C.-S.: Navigation Application of Q-learning Neural Network*. Proc. of the Int'l Conf. ICSSE, pp. 1–4. Kagawa, Japan (2020)
24. Strauss, C., Sahin, F.: Autonomous navigation based on a Q-learning algorithm for a robot in a real environment. Proc. of the Int'l Conf. on System of Systems Engineering, pp. 1–5. Monterey, CA, USA (2008)
25. Ziwu, R., Ye, S.: A hybrid optimized algorithm based on simplex method and genetic algorithm. Proc. of the 6th World Cong. on Intelligent Control and Automation, pp. 3547–3551. Dalian, China (2006)

26. Guo, P., Wang, X., Han, Y.: The enhanced genetic algorithms for the optimization design. Proc. of the 3rd Int'l Conf. on Biomedical Engineering and Informatics, pp. 2990–2994. Yantai, China (2010)
27. Kondratenko, Y., et al.: Fuzzy and evolutionary algorithms for transport logistics under uncertainty. Proc. of the INFUS Int'l Conf. on Intelligent and Fuzzy Techniques: Smart and Innovative Solutions. Advances in Intelligent Systems and Computing, 1197, pp. 1456–1463. Springer, Cham (2021)
28. Linkens, D.A., Okola Nyongesa, H.: A distributed genetic algorithm for multivariable fuzzy control. Proc. of the Coll. on Genetic Algorithms for Control Systems Engineering, pp. 9/1–9/3. London, UK (1993)
29. Jiang, J., Butler, D.: A genetic algorithm design for vector quantization. Proc. of the 1st Int'l Conf. on Genetic Algorithms in Engineering Systems: Innovations and Applications, pp. 331–336. Sheffield, UK (1995)
30. Kondratenko, G.V., Kondratenko, Y.P., Romanov, D.O.: Fuzzy models for capacitive vehicle routing problem in uncertainty. Proc. of the 17th Int'l DAAAM Symp. on Intelligent Manufacturing and Automation: Focus on Mechatronics & Robotics, pp. 205–206. Vienna, Austria (2006)
31. Werners, B., Kondratenko, Y.: Alternative fuzzy approaches for efficiently solving the capacitated vehicle routing problem in conditions of uncertain demands. In: Berger-Vachon, C., et al. (eds.) Complex Systems: Solutions and Challenges in Economics, Management and Engineering. Studies in Systems, Decision and Control, 125, pp. 521–543. Springer, Cham (2018)
32. Wang, X.: Vehicle Routing. Operational Transportation Planning of Modern Freight Forwarding Companies, pp. 7–23. Produktion und Logistik (2015)
33. Escobar, J.W.: Heuristic algorithms for the capacitated location-routing problem and the multi-depot vehicle routing problem. 4OR-Q Oper. Res. 12, 99–100 (2014)
34. Pereira, F.B., Tavares, J.: Bio-inspired Algorithms for the Vehicle Routing Problem[M]. Springer, Berlin, Heidelberg (2009)
35. Skakodub, O., Kozlov, O., Kondratenko, Y.: Optimization of Linguistic Terms' Shapes and Parameters: Fuzzy Control System of a Quadrotor Drone. Proc. of the 11th Int'l Conf. on Intelligent Data Acquisition and Advanced Computing Systems: Technology and Applications, pp. 566–571. Cracow, Poland (2021)
36. Levchenko, B., Chukhray, A., Chumachenko, D.: Development of Game Modules with Support for Synchronous Multiplayer Based on Unreal Engine 4 Using Artificial Intelligence Approach. Proc. of the Int'l Conf. on Integrated Computer Technologies in Mechanical Engineering, pp. 503–513. Springer, Cham (2020)
37. Koonce, B.: Convolutional Neural Networks with Swift for Tensorflow [M]. Apress, Berkeley, CA (2020)
38. Firdaus, M., Pramono, A., Fibri, M.R.: Optimizing Performance of Real Time Multiplayer 3D Games on Smartphone Using Unsupervised Learning (K-Means). Proc. of the 4th Int'l Conf. ICIC, pp. 1–8. Semarang, Indonesia (2019)
39. Castaño, A.P.: Support Vector Machines. Chapter in Practical Artificial Intelligence, pp. 315–365. Berkeley, CA: Apress (2018)
40. Etaati, L.: Deep Learning Tools with Cognitive Toolkit (CNTK). Chapter in Machine Learning with Microsoft Technologies, pp. 287–302. Berkeley, CA: Apress (2019)
41. Seide, F.: Keynote: The computer science behind the Microsoft Cognitive Toolkit: An open source large-scale deep learning toolkit for Windows and Linux. Proc. of the IEEE/ACM Int'l Symp. CGO. Austin, TX, USA (2017)
42. Banerjee, D.S., Hamidouche, K., Panda, D.K.: Re-Designing CNTK Deep Learning Framework on Modern GPU Enabled Clusters. Proc. of the Int'l Conf. CloudCom. Luxembourg, pp. 144–151 (2016)

43. Sugiyarto, A.W., Abadi, A.M.: Prediction of Indonesian Palm Oil Production Using Long Short-Term Memory Recurrent Neural Network (LSTM-RNN). Proc. of the 1st Int'l Conf. AiDAS, pp. 53–57. Ipoh, Malaysia (2019)

44. Rahhal, J.S., Abualnadi, D.: IOT Based Predictive Maintenance Using LSTM RNN Estimator. Proc. of the Int'l Conf. ICECCE, pp. 1–5. Istanbul, Turkey (2020)

45. Sidenko, I., et al.: Peculiarities of human machine interaction for synthesis of the intelligent dialogue chatbot. Proc. of the Int'l Conf. IDAACS. Metz, France (2019)

46. Siriak, R., Skarga-Bandurova, I., Boltov, Y.: Deep convolutional network with long short-term memory layers for dynamic gesture recognition. Proc. of the Int'l Conf. IDAACS, pp. 158–162. Metz, France (2019)

47. Kuntsevich, V.M., et al. (eds.): Control systems: theory and applications. Book Series in Automation, Control and Robotics. Gistrup, Delft: River Publishers (2018)

48. Opanasenko, V., Kryvyi, S.: Synthesis of multilevel structures with multiple outputs. Proc. of the 10th Int'l Conf. UkrPROG. pp. 32–37 (2016)

49. Kondratenko, Y.P., Simon, D.: Structural and parametric optimization of fuzzy control and decision making systems. In: Zadeh, L., Yager, R., Shahbazova, S., Reformat, M., Kreinovich, V. (eds.) Recent Developments and the New Direction in Soft-Computing Foundations and Applications. Studies in Fuzziness and Soft Computing, 361, pp. 273–289. Springer, Cham (2018)

50. Gerasin, O., et al.: Remote IoT-based Control System of the Mobile Caterpillar Robot. Proc. of the 16th Int'l Conf. ICTERI, vol. I, pp. 129–136. Kharkiv, Ukraine (2020)

51. Youssef, A.E., et al.: Building your kingdom imitation learning for a custom gameplay using unity ML-agents. Proc. of the 10th Int'l Conf. IEMCON, pp. 0509–0514. Vancouver, BC, Canada (2019)

Improvement of the Dispenser of the Liquid Fertilizer Dosing Control System for Root Feeding of Crops

Sergey Filimonov, Dmytro Bacherikov, Constantine Bazilo$^{(\boxtimes)}$, and Nadiia Filimonova

Cherkasy State Technological University, Shevchenko blvd., 460, Cherkasy 18006, Ukraine
{s.filimonov,n.filimonova}@chdtu.edu.ua, b_constantine@ukr.net

Abstract. Agriculture is rapidly using the latest technologies that are applied to manage and optimize agricultural production. Fertilizers are an essential element of agriculture, as their proper use increases yields. The pouring of liquid fertilizers is carried out using special systems. The main element of the pouring system are dispensers of various types. The satisfactory vegetation of the plant will depend on the accuracy of pouring liquid fertilizers, which will affect the future harvest. The main features of modern models of dispensers for pouring liquid fertilizers are determined. Their advantages and disadvantages are revealed. The new design of the dispenser with a screw piezoceramic motor is developed. New original solution for automatic control of the pouring norm is proposed. A stand for research and determination of the parameters of dispensers is developed. The norms of pouring out the minute flow rate for the developed design of the dispenser are experimentally tested and the hydraulic characteristics are obtained. The research results can be used in the design of dispensers for pouring liquid fertilizers.

Keywords: agriculture · dispenser · dosing orifice plate · piezoelectric motor · bimorph piezoelectric element

1 Introduction

Today, modern agricultural production and the agro-industrial complex as a whole is one of the most important sectors of the economy of each country. Advanced implementation of innovative technologies in the modern agricultural sector is not only autopilots, monitoring systems, but also navigation systems that increase the productivity of technological operations. Innovative technologies in the modern agricultural sector also include both modern approaches to tillage technology and equipment for the efficient use of fertilizers [1, 2].

The most common fertilizers are granular ones. Traditionally fertilizers are applied in the form of granules and, firstly, must be dissolved in water. After the transition from one state to another one they become available to plants. All these processes take a certain time, which will negatively affect the development and condition of the plant, and is not very effective.

Z. Hu et al. (Eds.): ICCSEEA 2023, LNDECT 181, pp. 636–644, 2023.
https://doi.org/10.1007/978-3-031-36118-0_57

A more effective fertilizers are liquid ones, namely fertilizers that are on the market in the form of anhydrous ammonia (which can rather be attributed to gaseous form), ammonia water, complex fertilizers (CF) and UAM (urea-ammonia mixture) [3–5]. The liquid form operates at the start of growth, so the plant receives nutrition earlier. Liquid fertilizers have a prolonged action that provides plant nutrition throughout the growing season. Also, the direction of precision farming is becoming relevant, which is currently technically advanced in systems of pouring and accurate dosing. The most famous manufacturers of special equipment and machinery for the application of liquid fertilizers are the following brands: Precision Planting, Raven, Hexagon, JohnDeere, Amazone, Jacto, Fast, Fliegl, Vredo, Hi-Spec, Boguslav. Almost all of the above manufacturers use in their dosing systems the principle of pouring liquids, which depends on the pressure control in the system, this causes certain problems in dosing accuracy. The deviation of the actual application rate from the specified one is 3%. The development and implementation of priority areas of precision farming will contribute to the development of agricultural production and the agro-industrial complex as a whole.

2 Formal Problem Statement

At the moment, a large number of farmers from different countries are upgrading existing models of tillage and sowing complexes and purchasing new units with additional equipment for applying liquid fertilizers. The equipment provides for the installation on the frame of the unit or on the wheels of a special container for the operating solution. At the same time, hoses are connected to each share or coulter, which supply fertilizer using pumps. Thus, the application of liquid fertilizers occurs together with the sowing of seeds, or with another technological operation. These pouring systems allow to accurately dose and comply with the rate set by the operator. This increases the economic effect and the yield in the fields [6–9].

For accurate dosing of liquid fertilizers, the agricultural machinery manufacturers offer the use of special dispensers that are installed on fertilizer dosing units. Also, manufacturers of agricultural machinery offer different types of control. Figure 1 shows the classification of the main methods to control the liquid pouring system.

The disadvantages of these control systems, widely used in applicators, trailers, mounted and self-propelled sprayers, are the control of electric motors and solenoid valves are energy consuming due to significant currents for self-holding, as well as high pressure in the system and the complexity of manufacturing, a significant step of regulation and the complexity of controlling these types of systems [10, 11].

A more promising direction in the field of modern agricultural production and the agro-industrial complex as a whole is the creation and use of motors, the principle of operation of which is based on the reverse piezoelectric effect. The reverse piezoelectric effect is the appearance of mechanical deformations under the influence of an electric field.

Due to the absence of radiating magnetic fields and to resistance to their effects, wide ranges of rotational speeds and torques on the shaft (0.1… 1.0 Nm) the design has a number of technical advantages over electromagnetic motors. The piezoceramic motor has a high positioning accuracy [12, 13] of the order of 0.5 μm, and a large torque on the

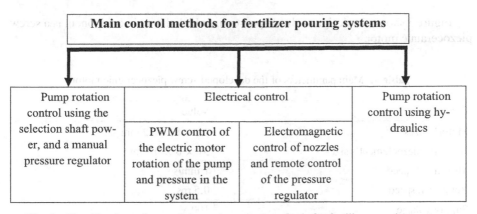

Main control methods for fertilizer pouring systems		
Pump rotation control using the selection shaft power, and a manual pressure regulator	Electrical control		Pump rotation control using hydraulics
	PWM control of the electric motor rotation of the pump and pressure in the system	Electromagnetic control of nozzles and remote control of the pressure regulator	

Fig. 1. Classification scheme of the main control methods for fertilizer pouring systems

shaft of 10 N. Such motors can be used in all industries, and especially in modern agricultural machinery and the agro-industrial complex [14, 15]. The great advantage of these motors is their unpretentiousness in operating and maintenance conditions [16–19]. One of the significant advantages of the piezoelectric motor is the start-stop characteristics, which are characterized by the ability of the motor to instantly start and stop, as well as hold a given position, without applying additional energy or effort.

Thus, the actual task is the development of liquid fertilizer dispensers based on piezoelectric motors.

3   Materials and Methods

The work is based on the new method of dosage control proposed by the authors, which consists not in maintaining a certain pressure in the system, but in changing the area of the orifice, which will allow you to accurately change the flow rate. To implement this method, the design of the developed dispenser uses piezoelectric motor [20, 21] based on bimorph piezoelectric elements [22], which is shown in Fig. 2.

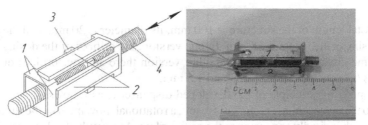

Fig. 2. Design of a piezoelectric motor using bimorph piezoelectric elements: 1 – brass plates; 2 – piezoelectric elements; 3 – four-sided metal nut, 4 – running shaft

Table 1 shows the main parameters of the developed screw piezoceramic motor.

Figure 3 shows the design of the dispenser, in which the main component is a screw piezoceramic motor.

Table 1. Main parameters of the developed screw piezoceramic motor

Parameter	Value
Motor's sizes	36 × 12 × 12 mm
Bimorph piezoelement's sizes	31 × 6 × 0.4 mm
Movement speed	1 mm/s
Rotational speed	0.5 rps
Supply voltage	100 V
Resonant frequency	7000–7300 Hz
Piezoceramic type	PZT-5H

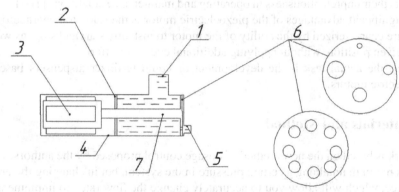

Fig. 3. The design of the dispenser with a screw piezoceramic motor: 1 – liquid supply; 2 – stuffing box; 3 – developed screw piezoceramic motor, 4 – dispenser housing, 5 – liquid outlet, 6 – replaceable dosing orifice plate, 7 – shaft

The total length of the structure is 100 mm, its diameter is 20 mm, and the diameter of the dosing orifice plate is 18 mm. In one version, the orifices of the dosing plate are of the same diameter 0.5 mm, in the other version there are four orifices of different diameters 0.5 mm, 0.75 mm, 1 mm and 1.5 mm.

The operating principle of the developed dispenser is as follows. When voltage is applied to the piezoceramic motor, as a result, a rotational movement of the running shaft occurs, which is rigidly connected to the orifice plate. As a result, the dosing orifice plate rotates and sets its orifice against the liquid outlet hole. In this way, the pouring out of the fertilizer can be controlled.

The developed stand for the study and determination of the parameters of the dispensers is shown in Fig. 4. This stand consists of a tank 3 with a capacity of 50 l, pump

1, filters 2, rubber sleeves 11, pressure regulator 4, manometer 5, rods 8 made of poly-
mer with a diameter of 32 mm, liquid dispenser 6, dispenser holder 7, contain-er for
collecting liquid 9, precision scales 10.

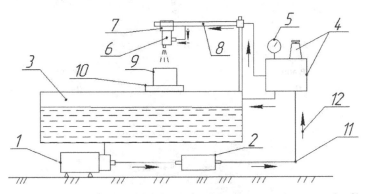

Fig. 4. Functional scheme of the stand for the study of the dispenser: 1 – pump; 2 – filter; 3 – tank
with liquid; 4 – manual pressure regulator; 5 – manometer; 6 – dispenser; 7 – dispenser holder;
8 – rod; 9 – container for collecting liquid; 10 – scales; 11 – sleeve; 12 – fluid movement

The operation of the stand is as follows: pump 1 sucks the solution from tank 3,
and supplies it to pressure regulator 4, which regulates the pressure in rod 8, the liquid
passes through filters 2. Pressure control is carried out thanks to the manometer 5. The
liquid passes through the rod to the liquid dispenser 6 and is distributed into the tank 9.
The distributed liquid in container 9 is weighed on precision scales, for a more accurate
result during the experiments and after weighing it is poured into tank 3. Stand operates
cyclically.

4   Experiments and Results

With the help of the developed stand, experimental studies of the dispenser were carried
out, where the minute flow rate of the liquid was determined. For this, the following
limiting conditions were adopted. The flow is measured in liters per 1 min for each
orifice of dosing plate, the pressure in the tests was (0.25 – 4 atm.), with an error of less
than 1%. The duration of the measurement, determined by a stopwatch with an error of
less than 1 s, must be greater than or equal to 60 s. For a more accurate determination
of the amount of liquid, the tank with liquid was weighed on scales with an error of 1 g.
Figure 5 shows the hydraulic characteristics for dosing plates with different orifices.

Figure 5 clearly shows that if you rotate the shaft with dosing orifice plate in the
developed dispenser, you can set the required pouring rate.

From the obtained graphs, a function that describes the dependence of the liquid
pouring rate at various values of pressure P and diameter d of the orifice of the dosing
plate of the dispenser was constructed.

$$V = a - be^{-cP^f} \tag{1}$$

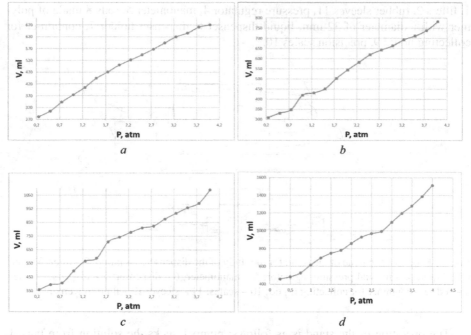

Fig. 5. Hydraulic characteristics of dosing plates with different orifice diameter:a) d = 0.5 mm; b) d = 0.75 mm; c) d = 1.0 mm; d) d = 1.5 mm

where V is the volume of the pouring liquid, P is the pressure in the system, a, b, c, f are the coefficients given in Table 2.

Table 2. The coefficients of the dependence (1)

The coefficients for dispenser 0.5 mm	The coefficients for dispenser 0.75 mm	The coefficients for dispenser 1.0 mm	The coefficients for dispenser 1.5 mm
a: 896.6892	a: 951.10074	a: 1628.8107	a: 5845.618
b: 646.69384	b: 661.13363	b: 1336.7101	b: 5391.4818
c: 0.21584292	c: 0.1761869	c: 0.17188192	c: 0.026710483
f: 1.1506088	f: 1.433036	f: 1.1332526	f: 1.4750516

This dispenser is versatile and contains a dosing plate with various orifices, which allows you to quickly change the pouring rate and allows you to use it with liquids with different liquid fertilizer densities. Also, a big plus of this dispenser is that it has practically no clogging of orifices.

The next experiment was to compare two dosing orifice plate, one of which was in operation and the other was not in operation. Figure 6 shows the hydraulic characteristics of the dispensers, which was in operation (about 2000 motohours), and which was not in operation.

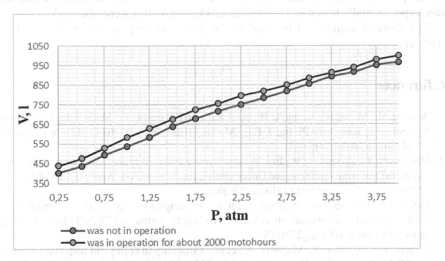

Fig. 6. Hydraulic characteristics of dosing plates with an orifice diameter d = 0.5 mm

Figure 6 shows that the daily flow through the dispenser that was in operation at a pressure of 2 atm is about 751 l, and the daily flow through the dispenser was not in operation is about 710 l. The difference between the values of the dispensers is 41 L. Each dispenser that has been in operation for about 2000 h and is installed on a row has significant deviations from the true value. This means that if it is used later in technological operations, it will lead to non-compliance with the specified pouring rates, and excessive consumption of liquids, which in turn increases the consumption of the tank mixture. If such a dispenser is installed on a liquid fertilizer system that have 24 rows, then the total daily excessive consumption will be 984 L. Thus, from the experiments presented above, it can be seen that the use of a dispenser with one orifice (for one pouring rate) has negative consequences, which con-sist in over-consumption of liquid fertilizers or non-compliance with the specified pouring rate. Therefore, the use of the developed dispenser with the ability to regulate different pouring rates will reduce these negative factors.

In the future, field studies can be carried out, as well as the creation of an equivalent circuit using the method of electromechanical analogies to match with the control circuit.

5 Conclusions

The design features of modern models of dispensers for pouring liquid fertilizers were analyzed. Their advantages and disadvantages were revealed.

A new design of the dispenser with a screw piezoceramic motor has been pro-posed and developed. The design proposed a new original solution for automatic regulation of the pouring norm.

The stand for research and determination of the parameters of dispensers was developed.

The pouring norms of the minute flow rate for the developed design of the dispenser were experimentally tested and the hydraulic characteristics were obtained.

The research results can be used in the design of dispensers for pouring liquid fertilizers.

References

1. Anusha, K., Mahadevaswamy, U.B.: Automatic IoT Based Plant Monitoring and Watering System using Raspberry Pi. Int. J. Eng. Manuf. (IJEM) **8**(6), 55–67 (2018). https://doi.org/10.5815/ijem.2018.06.05
2. Bhagawati, K., Bhagawati, R., Jini, D.: Intelligence and its Application in Agriculture: Techniques to Deal with Variations and Uncertainties. Int. J. Intell. Sys. Appli. (IJISA) **8**(9), 56–61 (2016). https://doi.org/10.5815/ijisa.2016.09.07
3. Parent, S.É., Dossou-Yovo, W., Ziadi, N., et al.: Corn response to banded phosphorus fertilizers with or without manure application in Eastern Canada. Agron. J. **112**(3), 2176–2187 (2020). https://doi.org/10.1002/agj2.20115
4. Kablan, L.A., Chabot, V., Mailloux, A., et al.: Variability in corn yield response to nitrogen fertilizer in Eastern Canada. Agron. J. **109**(5), 2231–2242 (2017). https://doi.org/10.2134/agronj2016.09.0511
5. Scharf, P.C., Wiebold, W.J., Lory, J.A.: Corn yield response to nitrogen fertilizer timing and deficiency level. Agron. J. **94**(3), 435–441 (2002). https://doi.org/10.2134/agronj2002.4350
6. Bermudez, M., Mallarino, A.P.: Yield and early growth responses to starter fertilizer in no-till corn assessed with precision agriculture technologies. Agron. J. **94**(5), 1024–1033 (2002). https://doi.org/10.2134/agronj2002.1024
7. Galpottage Dona, W.H., Schoenau, J.J., King, T.: Effect of starter fertilizer in seed-row on emergence, biomass and nutrient uptake by six pulse crops grown under controlled environment conditions. J. Plant Nutr. **43**(6), 879–895 (2020). https://doi.org/10.1080/01904167.2020.1711945
8. Mallarino, A.P., Bergmann, N., Kaiser, D.E.: Corn responses to in-furrow phosphorus and potassium starter fertilizer applications. Agron. J. **103**(3), 685–694 (2011). https://doi.org/10.2134/agronj2010.0377
9. Kaiser, D.E.M., Mallarino, A.P.M., Bermudez, M.: Corn grain yield, early growth, and early nutrient uptake as affected by broadcast and in-furrow starter fertilization. Agronomy. Journal **97**(2), 620–626 (2005). https://doi.org/10.2134/agronj2005.0620
10. Taghizadeh, M., Ghaffari, A., Najafi, F.: Modeling and identification of a solenoid valve for PWM control applications. Comptes Rendus Mec. **337**(3), 131–140 (2009). https://doi.org/10.1016/j.crme.2009.03.009
11. Xu, X., Han, X., Liu, Y., et al.: Modeling and dynamic analysis on the direct operating solenoid valve for improving the performance of the shifting control system. Appl. Sci. **7**(12), 1266 (2017). https://doi.org/10.3390/app7121266
12. Sharapov, V.M., et al.: Improvement of piezoceramic scanners. In: IEEE XXXIII International Scientific Conference Electronics and Nanotechnology (ELNANO). Kiev, vol. 2013, pp. 144–146 (2013). https://doi.org/10.1109/ELNANO.2013.6552063
13. Bobtsov, A.A., Bojkov, V.I., S.V. Bystrov, V.V., et al.: Actuating devices and systems for micromovements. SPb, ITMO University (2017)

14. Song, S., Shao, S., Xu, M., et al.: Piezoelectric inchworm rotary actuator with high driving torque and self-locking ability. Sens. Actuators, A **282**, 174–182 (2018). https://doi.org/10.1016/j.sna.2018.08.048
15. Kiziroglou, M.E., Temelkuran, B., Yeatman, E.M., Yang, G.Z.: Micro motion amplification – a review. IEEE Access **8**, 64037–64055 (2020). https://doi.org/10.1109/ACCESS.2020.2984606
16. Spanner, K., Koc, B.: Piezoelectric Motors, an Overview. Actuators **5**(1), 6 (2016). https://doi.org/10.3390/act5010006
17. Hunstig, M.: Piezoelectric inertia motors – a critical review of history, concepts, design, applications, and perspectives. Actuators **6**(1), 7 (2017). https://doi.org/10.3390/act6010007
18. Brahim, M., Bernard, Y., Bahri, I.: Modelling, design, and real time implementation of robust H-infinity position control of piezoelectric actuator drive. Int. J. Mechatronics and Automation **6**(4), 151–159 (2018). https://doi.org/10.1504/IJMA.2018.095516
19. Spanner, K., Koc, B.: Piezoelectric Motor Using In-Plane Orthogonal Resonance Modes of an Octagonal Plate Actuators **7**(1), 2 (2018). https://doi.org/10.3390/act7010002
20. Bazilo, C., Filimonov, S., Filimonova, N., Bacherikov, D.: Determination of Geometric Parameters of Piezoceramic Plates of Bimorph Screw Linear Piezo Motor for Liquid Fertilizer Dispenser. In: Hu, Z., Petoukhov, S., Yanovsky, F., He, M. (eds.) ISEM 2021. LNNS, vol. 463, pp. 84–94. Springer, Cham (2022). https://doi.org/10.1007/978-3-031-03877-8_8
21. Li, J.: Design of active vibration control system for piezoelectric intelligent structures. Int. J. Edu. Manage. Eng. (IJEME) **2**(7), 22–28 (2012). https://doi.org/10.5815/ijeme.2012.07.04
22. Bazilo, C.: Modelling of Bimorph Piezoelectric Elements for Biomedical Devices. In: Hu, Z., Petoukhov, S., He, M. (eds.) AIMEE 2019. AISC, vol. 1126, pp. 151–160. Springer, Cham (2020). https://doi.org/10.1007/978-3-030-39162-1_14

A System of Stress Determination Based on Biomedical Indicators

Lesia Hentosh[1]([✉]), Vitalii Savchyn[1], and Oleksandr Kravchenko[2]

[1] Rtificial Intelligence Department, Lviv Polytechnic National University, Lviv 79013, Ukraine
lesia.i.mochurad@lpnu.ua
[2] National Technical University Kharkiv Polytechnic Institute, Kharkiv 61002, Ukraine

Abstract. This paper investigated methods of determining stress in humans and developed of smart service system in medicine to automate this process. This paper evaluates existing research in this area. We conducted a study of the method of determining stress levels based on biomedical indicators. Also, we have developed a system that is relevant for use in many different areas. For example, such a system is convenient to use in office firms to prevent overexertion of workers, to prevent emergencies in jobs with a high level of human impact, where human life is endangered, and also for daily use in health care. The smart service system works with input data based on heart rate variability indices. Neural network training has been launched in 100 epochs. In each epoch, the results of accuracy and loss were recorded. Also, for better reliability of the results, we recorded the data obtained not only from training data but also from variation data. As a result, the classification problem is solved. We get 99% accuracy.

Keywords: Stress level · Psychophysical state · Computer technologies · Machine- learning method · Accuracy

1 Introduction

After studying the phenomenon of stress and its symptoms, it is advisable to find out what are the ways to determine stress. One of the most common methods, which was used before and is still actively used today is to conduct a survey. Based on the answers to the questions posed in the questionnaire, conclude the emotional and stressful state of the person. Another accurate but more complex method than the previous one is the detection of stress using various sensors in the laboratory. For example, electromyography, electrocardiography, etc. are performed, which examine the work of the human musculoskeletal and cardiac systems.

So making decisions under uncertainty and lending strategies are also associated with the occurrence of stress.

Another particularly interesting way to diagnose stress is to diagnose it with a smart-watch and fitness bracelets. Almost all of these devices have built-in heart rate sensors, but not all have a function for detecting stress and a function for determining HRV (heart rate variability). Usually, only more expensive or newer versions of such devices have this functionality, and for many other users the feature is not available.

Also, often use the method of diagnosing stress using Schulte tables [1]. But psychologists do not recommend using Schulte tables in the electronic version to determine stress. This is because their idea is to use peripheral vision, and the computer mouse cursor will distract from the center square of the table. Therefore, researchers recommend using only paper Schulte tables. Devices with a touch screen do not have a mouse cursor, but will distract the finger that clicks on the screen.

The latest in the list and the most innovative way to diagnose stress are intelligent systems for recognizing signs of stress using human biometrics [2]. Typically, such systems work using machine learning algorithms [3–6]. The input data for such a system are various biometric indicators, but more popular and accurate is heart rate variability. To classify a person's stress level, the dynamics of his pulse can be analyzed using, for example, the Random Forest algorithm, Logistic Regression, or other classification methods.

Smartwatches and fitness bracelets also use similar smart systems, but there these systems are directly integrated into the device. Such a system can be designed and used separately from smart devices, but you will have to enter the biomedical data yourself, as they will not be obtained automatically by the device.

The purpose of this study is to develop a method for diagnosing human stress using computer technology and machine learning. To achieve this aim, we need to perform the following tasks:

1) to study the phenomenon of stress and all available methods of diagnosing it in humans;
2) develop a method for diagnosing stress;
3) to develop a smart service system for the diagnosis of stress in humans based on biomedical indicators.

The object of research: stress and psychophysical condition of a person.

The subject of research: smart service system for digital determination and prediction of stress.

Therefore, the main contribution of this chapter can be summarized as follows:

1) it is designed a universal method for diagnosing human stress using machine learning methods, which is based on the use of computational intelligence tools;
2) algorithmic implementations of the designed universal method with the use of both: machine learning algorithm and artificial neural networks are presented; procedures for their training and application within the designed method are described;
3) it is designed a smart service system in medicine; an experimental investigation of the efficiency of work of proposed algorithm; it gets accuracy – 99%.

2 Related Works

Interest in this topic is growing every year, because the timely diagnosis of stress and taking precautions can help people avoid nervous disorders, overwork or negative health consequences. Recently, stress detection systems have begun to integrate into smartwatches and fitness bracelets, and new scientific articles and research in this area have been published. This indicates that at the moment the research topic is still relevant.

The article [7] describes the development of a method for automatically determining stress on two popular scales of depression: the scale of stress, anxiety, and depression – DASS-21 and the scale of psychological stress Kessler – K-10. Six regression models were used for this: normal regression, least squares, generalized linear models, beta-binomial regression, fractional logistic regression model, MM-estimation and censored smallest absolute deviation. This article is useful and interesting because it describes popular stress assessment scales and machine learning algorithms that have been adapted so that they can measure stress on these scales.

The article [8] describes the construction of an information system not only to determine the level of stress, but also to track cognitive impairment. Another feature of this work is that studies have been conducted for infirm elderly people with mild cognitive impairment. To do this, it analyses the collected data from physiological sensors, which for some time were worn by the elderly. The paper also describes a stress detection system based on several machine learning algorithms, and tested their effectiveness on a real data set. It has been found that stress detection algorithms usually provide only the identification of a stressful/non-stressful event. Researchers suggest constantly monitoring physiological parameters in real-time to assess stress levels.

In a study [9], the authors conducted a study of stress in the military or people who participated in hostilities. The research was conducted in Ukraine and the Ukrainian military took part in the survey and testing. Classification and clustering of data were performed using machine learning algorithms. The k-means algorithm was used to identify the number of clusters, the optimal number of clusters was found using the elbow algorithm, and the Random Forest algorithm was used to test the model and its subsequent use.

The article [10] also investigates methods for determining the level of human stress using modern technologies. Researchers say the study is relevant because high levels of stress are a major factor that has a negative impact on heart health, and there is a relationship between stress levels and heart rate. This paper presents a technique where the heart rate of patients is recorded by IoT devices. Using an analytical model, the proposed system shows better accuracy in shorter processing times compared to other approaches to machine learning, and thus it is a cost-effective solution in the IoT system than medical research.

The study [11] investigated the risk of high blood pressure in humans through machine learning. The article also describes the methods of processing human biomedical data using machine learning. To predict the risk of high blood pressure, the algorithm uses additional data such as age, anger, the general level of anxiety or stress in humans, and indicators of obesity and cholesterol in humans. The Random forest algorithm showed a prediction accuracy of 87.5%.

Ways to use classification methods to identify key predictors of the patient, which are considered the most important in the classification of the worst cognitive abilities, which is an early risk factor for dementia, were studied in [12]. During the study, three classifiers were used: decision trees, Bayesian algorithm and random forests.

The impact of the global pandemic COVID-19 on the emotional state and level of stress in humans was analyzed in [13]. A random forest algorithm was used to detect psychological stress during a pandemic. The study found that the prevalence of clinical levels of anxiety, depression and post-traumatic stress was higher during a pandemic.

The article [14] investigates the potential of machine learning in predicting health status after the body has transmitted certain types of infections or diseases. A matrix of percentage similarity between helper genes and Escherichia coli genomes was created, which was later used as input for machine learning algorithms. The paper also compared the efficiency of such algorithms as random forest, vector support machine algorithm (radial and linear core) and gradient gain. This article does not describe how to determine stress, but it describes the application of machine learning algorithms for human biomedical parameters.

Research [15] investigates a method called ensemble classification, which is used to increase the accuracy of weak algorithms by combining several classifiers. Experiments with this tool were performed using a data set on heart disease. A comparative analytical approach was performed to determine how ensemble techniques can be used to improve prediction accuracy in heart disease. The main focus of this paper is not only on improving the accuracy of weak classification algorithms, but also on the implementation of an algorithm with a medical data set to show its usefulness for predicting the disease at an early stage. This article also does not describe the definition or prediction of stress, but here is a useful study of ways to improve classification algorithms based on biomedical data.

An indicator that quantifies the stress experienced by a cyclist on certain sections of the road has been studied in the article [16]. The study proposes a classification that consists of two parts: clustering and interpretation. At the heart of this methodology is a classifier combined with a predictive model to make the approach scalable to massive datasets.

A hybrid approach to stress clustering at the personal level, using basic stress self-reports to increase the success of human-independent models without requiring significant amounts of personal data is described in [17]. Machine learning algorithms have been used to solve this problem. The study is developing a method to quickly detect persistent daily stress to prevent serious health consequences.

The article [18] describes a comparative study of classifiers of tumors and heart disease. A variety of medical methods based on deep learning and machine learning have been developed to treat patients, as cancer is a very serious problem in this era. The researchers compared machine learning algorithms such as logistic regression, K-NN and decision trees. Researchers argue that stress in patients can also cause heart disease. The paper does not describe the topic of stress or ways to determine it, but a comparison of machine learning algorithms, which are trained on biomedical indicators.

In a study [19], the authors developed a mathematical model for the problem of classification and analysis of the psychophysical state of man using a modification of the tapping test as a data source. To conduct the experiment, an application was developed for a smartphone that is a generalization of the classic tapping test. The developed tool collects much more data related to motility. This allowed us to collect a sample, to analyze the structure of each individual experiment, and the sample as a whole. The data were investigated for anomalies and a method of reducing their number was developed on the basis of methods of descriptive statistics and methods of optimization of hyperparameters. The machine learning model was used to predict some features of the corresponding data set.

As a result, the above-mentioned literature sources were considered and analyzed. Some articles have examined the level of stress for a certain category of people or under certain circumstances. For example, the articles analyzed situations for the elderly or the military or in a pandemic. In contrast to the existing approaches, this paper proposes a system that will determine the level of stress for anyone, by entering certain biometric indicators. The proposed smart service system will not require special devices for their measurement. Diagnosis will be performed with HRV or heart rate data obtained in any convenient way. The obtained results have a special practical value for diagnosing stress in students during training or for office workers.

3 Materials and Methods

The approach used in this work focuses on the parallel encryption of the blocks into which the file is divided. This is possible because the encryption process for each block is independent of the others. For example, suppose you need to encrypt a file of 1048576 bytes. This file will be divided into 16-byte blocks. Thus, the AES algorithm will be applied to 65536 blocks. Here you can use parallelization to distribute the work between the available threads depending on the technology used.

Neural networks were chosen to implement the stress diagnosis system. The main components of an artificial neural network are nodes, which in a certain hierarchy transmit to each other information, during the passage of each such node partially changes and the last layer of the neural network is already processed source data. To solve the problem, you need to classify stress into three categories: no stress, time pressure and interruption. To do this, we will use a serial-type neural network that will have an input layer that will input data, one hidden layer, and the last output layer, which will already return the data divided into three categories. So we need to develop a Sequential model with three layers, choose the activation function for these layers, and configure the model training.

In the developed model, on the first two layers, we used the ReLU function, and on the last layer, we used the softmax function.

The ReLU function is determined by the formula:

$$f(x) = \max(0, x),$$

where x – is the input value of the neuron.

This function leaves the input unchanged if it is greater than zero and converts all negative argument values to zero. This feature is popular because it is simple and fast to calculate. There are also variations and improvements to this feature in cases where zeroing negative elements are not suitable for some models. An example of such a function is the Leaky ReLU function, which is calculated by the formula:

$$f(x) = \max(0.1x, x).$$

This type of function is used in the work on the input and hidden layers of the model, because it perfectly copes with the task and is quickly calculated, so it optimizes the neural network. But the third layer cannot be used because the effect of themultilayer will be lost. Therefore, for the last layer, we used Softmax [20]:

$$f(x) = \frac{e^{x_j}}{\sum_i^K \sum_i e^{x_i}},$$

where x_j – the value of the input data, K – the number of classes in the classifier, $j = 1, \ldots, K$.

The advantage of this activation function is that it can already handle several classes. At the output, these classes take the form from 0 to 1.

If we know the number of layers and the number of neurons on each layer, then we can calculate the number of weights that will be counted by the designed neural network. We chose to have three layers. On the first layer – 34 neurons, because it is the input layer, and the studied dataset has 34 input parameters. The last layer is the output layer, so its number of nodes is also known, there will be 3, because the system will classify three states of people. And only the hidden layer remains available for experiments with the number of nodes. For example, we first set the number of nodes to 30. In this case, the number of weights can be calculated as follows:

34 x 30 = 1020 – the number of weights between the first and second layer;
30 x 3 = 90 – the number of weights between the second and third layers;
1020 + 90 = 1110 – the total number of weights.

Also, a necessary step in the design of a neural network is the choice of metrics. Metrics are a thing that will help determine how accurately the network has been trained. Metrics are of the following types: accuracy, recall, precision, and F-measurement. After analyzing the advantages and disadvantages of each of the metrics, for the implementation of the neural network, we chose the accuracy metric:

$$accuracy = \frac{TP + TN}{TP + TN + FP + FN},$$

where TP – True Positive, TN – True Negative, FP – False Positive, FN – False Negative.

Another important part of the design of neural networks is the choice of the function by which the losses will be calculated. We chose the categorical cross-entropy loss function:

$$-\frac{1}{N} \sum_{i=1}^{N} \sum_{c=1}^{C} 1_{y_i \in C_c} \log p_{model}[y_i \in C_c],$$

where N – is the number of observations (records in the dataset), C – is the number of categories of the classifier (classes).

In the above formula, the entry $1_{y_i \in C_c}$ indicates the $i-$ th observation belonging to class C among all N observations. Part of the formula $p_{model}[y_i \in C_c]$ is the probability of the model predicting that the $i-$ th observation belongs to class C.

The Keras API was used to develop the neural network. This high-level API is part of TensorFlow. TensorFlow has developed this library to simplify and speed up the process of building neural networks. Thanks to Keras, we can design a neural network using intuitive abstraction. Keras has most of TensorFlow's capabilities but speeds up the network development and design process.

4 Results

We took the SWELL dataset [21] as input for training the neural network. This dataset contains 369289 records for network training and 41033 records for testing. So in total the data set has 410322 records with data. The data set contains 34 biometric indicators that will be included in the network input. These biometric indicators describe the work of the human heart, namely the indices of the internal heart rate. All data were collected from 25 office workers, who were temporarily monitored and their heart rate variability and other indicators were monitored using special sensors. During the study, these workers were given a variety of tasks and periodically created stressful situations to monitor the body's response to such stressful stimuli. The condition of workers was divided into three states: no stress, time pressure and interruption. These three states are the source data for the network. The first condition is without stress. In this case, employees were given tasks and allowed to work on them for as long as they needed. In the dataset, it is called "no stress". The second type of condition of employees is already stressful for them and was caused by the limited time to complete the task. In the dataset, it is called "time pressure". In the third case, employees were distracted from the task by sudden e-mails. Sent eight letters with different content, with one that is relevant to the task, and one that was not related to the task. In a dataset, this type of stress is called "interruption". During the development of the neural network, we experimented with and trained a neural network with a different number of nodes on the hidden layer. We also recorded the accuracy and number of losses in each of the epochs.

We launched neural network training in 100 epochs. In each epoch, the results of accuracy and loss were recorded. Also, for better reliability of the results, we recorded the data obtained not only from training data, but also from variation data.

Table 1 shows the values of losses and accuracy of the neural network as a result of 100 epochs. For convenience and compactness, the results in the table are given in increments of 10 epochs. The first column in the table indicates the number of epochs for which the results are recorded. The second and third columns indicate the loss and accuracy when using training data. The fourth and fifth columns also indicate loss and accuracy, but for validation data. We used validation data for better independence and reliability of results.

Analysing Table 1, we can observe how the accuracy of model training improved and losses decreased as the number of training epochs increased.

Figure 1 shows a graph of the model learning speed. This is a graph of the dependence of losses on past epochs. Analyzing this graph, we can see that until about the 20th century, the speed of learning was very high, because the values of losses were rapidly declining. This indicates that the model was not yet trained and was just beginning to train. With almost every era, the speed of network training has decreased, which is a sign of a good training process. If at some point in time, the loss function began to increase sharply, it would mean the effect of retraining and training the model at this stage would have to stop.

Analysing Fig. 1 in more detail, we can see that almost along its entire length there are small fluctuations in the values of the function, similar to noise. This is because validation data is used to plot the graph. Due to the validation data in the figure, many

Table 1. The results of neural network training

№ epoch	Training data		Validation data	
	Loss	Accuracy	Loss	Accuracy
1	0.8105	0.6360	0.7059	0.6760
10	0.2549	0.9067	0.2431	0.9117
20	0.1302	0.9565	0.1302	0.9555
30	0.0804	0.9734	0.0739	0.9757
40	0.0534	0.9835	0.0514	0.9843
50	0.0388	0.9885	0.0371	0.9897
60	0.0301	0.9911	0.0261	0.9932
70	0.0243	0.9931	0.0241	0.9933
80	0.0204	0.9941	0.0228	0.9934
90	0.0177	0.9951	0.0189	0.9939
100	0.0156	0.9955	0.0143	0.9964

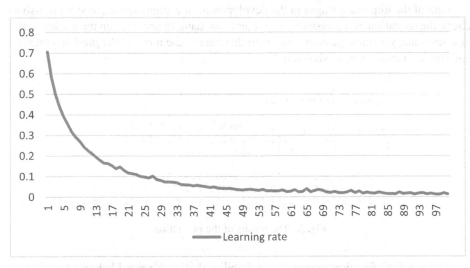

Fig. 1. Changing the learning speed of the model depending on the epoch

of these small local minima and maxima arise. And if you look at the whole chart, it is generally declining. This means that the effect of retraining does not occur.

Very similar to the graph of learning speed, but inverted, is the graph of the accuracy of the model on the number of epochs passed (see Fig. 2).

In contrast to the graph of learning speed, the graph shown in Fig. 2 increases almost throughout.

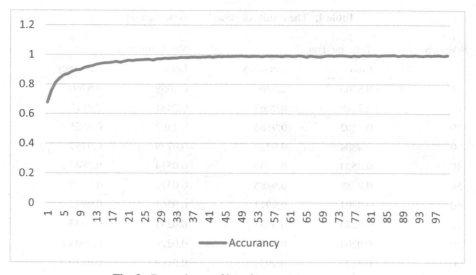

Fig. 2. Dependence of learning accuracy on epochs

One of the important stages in the development of a smart service system is also to check the operation of the system on tests and new data. In order to run the model on the new test data, you must pass an array with this data to the input of the *predict()* method and run it on execution, as shown in Fig. 3.

```
1 model.predict(testSamples)

array([[2.0427158e-13, 1.0000000e+00, 1.8924628e-21],
       [3.8080251e-11, 5.4651661e-08, 1.0000000e+00],
       [8.2618625e-12, 1.0000000e+00, 1.2235136e-08],
       ...,
       [3.9729514e-22, 1.0000000e+00, 3.6257938e-10],
       [5.7275296e-09, 1.0000000e+00, 2.2881335e-14],
       [1.0891031e-04, 1.0470001e-05, 9.9988055e-01]], dtype=float32)
```

Fig. 3. The results of the prediction

The *predict()* function returns the probability that each record belongs to different classes. There are three classification classes, so the function returns three columns with probabilities. Figure 3 shows that for each record the probability of one of the classes is one, or a number close to one, and for other classes, the probability is a number close to zero. To verify the reliability of the results, look at the original values of these classes (Fig. 4).

Figure 3 and Fig. 4 show that in all cases the prediction coincided with the original data, so it can be argued that the system works correctly with an accuracy of more than 99%.

It should be noted that the first column in the output in Fig. 3 and Fig. 4 indicates the likelihood of stress due to employee distraction (interruption). The second column

`1 testLabels`

```
array([[0., 1., 0.],
       [0., 0., 1.],
       [0., 1., 0.],
       ...,
       [0., 1., 0.],
       [0., 1., 0.],
       [0., 0., 1.]])
```

Fig. 4. Original values

indicates a state without stress (no stress). The third column indicates the probability of stress due to pressure in time constraints (time pressure). We have presented the data in Table 2 for the convenience of their analysis.

Table 2. Probabilities of emotional states

Interruption	No stress	Time pressure
2.0427158e-13	1.0000000e + 00	1.8924628e-21
3.8080251e-11	5.4651661e-08	1.0000000e + 00
8.2618625e-12	1.0000000e + 00	1.2235136e-08

It is convenient to analyze the results in the form of a table because we can see the probabilities of all three states at once. If the user is superfluous or not interested in this information, we can immediately get the name of one state, the probability of which is the highest. To do this, we need to use the *predict_classes()* function instead of the *predict()* function.

5   Summary and Conclusion

In this research, we have developed a smart service system to identify signs of human stress based on its biomedical indicators. To do this, we first analyzed the various literature sources in which the study similar problems. We proposed a method for determining stress based on the neural network. We chose the neural network because neural networks have many parameters for tuning and optimization and allow new experiments. As a result of the analysis and research of the subject area, we chose a model and developed an algorithm based on which we implemented a system for determining the signs of stress in humans. The system works with input data based on HRV indices. At the input, it accepts 34 parameters based on which the prediction takes place. The source data for the network are from three different states. The first condition is not stress. The second condition is stress due to time constraints. The third stage is also stress due to the distraction of the person from the task on which he focused. Surveys for recruitment have been conducted among office workers, and a network trained in such a dataset will also be better able to detect the stress of office work. There is a good reason to monitor stress in employees because early detection of stress helps to take precautions, maintain a high level of productivity, and prevent damage to health. The advantage of this system is that it is not necessary to wear special sensors to use human HRV to use it, as such sensors already have some smartwatches. Our proposed method for diagnosing stress showed an accuracy of 99%. The developed smart service system is tested by test data and validation data.

Further research will be carried out in two main areas for making decisions under uncertainty [22] and for risk analysis in lending [23] to identify signs of human stress based on her biomedical indicators.

1) the use of CUDA technology [24];
2) the use of OpenMP technology [25].

References

1. Maksimenko, V.A., et al.: Human personality reflects spatio-temporal and time-frequency EEG structure. PLoS ONE **13**(9), 1–13 (2018)
2. Mochurad, L., Yatskiv, M.: Simulation of a Human Operator's Response to Stressors under Production Conditions. Proceedings of the 3rd International Conference on Informatics & Data-Driven Medicine, November 19–21, pp. 156–169. Växjö, Sweden (2020)
3. Deo, R.C.: Machine Learning in Medicine. Circul. Bas. Sci. Clini. **132**, 1920–1930 (2015)
4. Gao, J., Yang, Y., Lin, P., Park, D.S.: Computer vision in healthcare applications. J. Healthc. Eng. **2018**, 1–5 (2018)
5. Izonin, I., Trostianchyn, A., Duriagina, Z., Tkachenko, R., Tepla, T., Lotoshynska, N.: The combined use of the wiener polynomial and SVM for material classification task in medical implants production. Int. J. Intelli. Sys. Appli. (IJISA) **10**(9), 40–47 (2018)
6. Santra, A., Dutta, A.: A comprehensive review of machine learning techniques for predicting the outbreak of Covid-19 cases. Int. J. Intell. Sys. Appli. (IJISA) **14**(3), 40–53 (2022)
7. Gamst-Klaussen, T., Lamu, A.N., Chen, G., Olsen, J.A.: Assessment of outcome measures for cost-utility analysis in depression: mapping depression scales onto the EQ-5D-5L. BJPsych Open **4**(4), 160–166 (2018)
8. Delmastro, F., Martino, F.D., Dolciotti, C.: Cognitive training and stress detection in MCI frail older people through wearable sensors and machine learning. IEEE Access **8**, 65573–65590 (2020)
9. Pavlova, I., Zikrach, D., Mosler, D., Ortenburger, D., Gora, T., Wasik, J.: Determinants of anxiety levels among young males in a threat of experiencing military conflict-applying a machine-learning algorithm in a psychosociological study. PLoS ONE **15**(10), 1–14 (2020)
10. Akanksha, E.: Framework for propagating stress control message using heartbeat based IoT remote monitoring analytics. Int. J. Elec. Comp. Eng. **10**(5), 4615–4622 (2020)
11. Nimmala, S., Ramadevi, Y., Sahith, R., Cheruku, R.: High blood pressure prediction based on AAA++ using machine-learning algorithms. Cogent Engineering. **5**(1), 1–12 (2018)
12. Rankin, D., et al.: Identifying key predictors of cognitive dysfunction in older people using supervised machine learning techniques: Observational study. JMIR Med. Inform. **8**(9), 1–23 (2020)

13. Prout, T.A., et al.: Identifying predictors of psychological distress during COVID-19: a machine learning approach. Front. Psychol. **11**, 1–14 (2020)
14. Njage, P.M.K., Leekitcharoenphon, P., Hald, T.: Improving hazard characterization in microbial risk assessment using next generation sequencing data and machine learning: predicting clinical outcomes in shigatoxigenic escherichia coli. Int J Food Microbiol. **292**, 72–82 (2019)
15. Latha, C.B.C., Jeeva, S.C.: Improving the accuracy of prediction of heart disease risk based on ensemble classification techniques. Info. Medi. Unloc. **16**, 1–9 (2019)
16. Huertas, J.A., et al.: Level of traffic stress-based classification: a clustering approach for Bogotá, Colombia. Transp. Res. Part D: Transp. Environ. **85**, 1–17 (2020)
17. Can, Y.S., Chalabianloo, N., Ekiz, D., Fernandez-Alvarez, J., Riva, G., Ersoy, C.: Personal stress-level clustering and decision-level smoothing to enhance the performance of ambulatory stress detection with smartwatches. IEEE Access. **8**, 38146–38163 (2020)
18. Khan, T.A., Kadir, K.A., Nasim, S., Alam, M., Shahid, Z., Mazliham, M.S.: Proficiency assessment of machine learning classifiers: an implementation for the prognosis of breast tumor and heart disease classification. Int. J. Adv. Comput. Sci. Appl. **11**(11), 560–569 (2020)
19. Mochurad, L., Ya, H.: Modeling of psychomotor reactions of a person based on modification of the tapping test. Int. J. Comp. **20**(2), 190–200 (2021)
20. Du, K.-L., Swamy, M.N.S.: Neural Networks and Statistical Learning, pp. 1–24. Springer-Verlag, London (2014)
21. Dataset SWELL [Electronic resource] // Access mode: https://www.kaggle.com/qiriro/swell-heart-rate-variability-hrv
22. Nehrey, M., Hnot, T.: Data science tools application for business processes modelling in aviation. In: Cases on Modern Computer Systems in Aviation. IGI Global, pp. 176–190 (2019). https://www.igi-global.com/gateway/chapter/222188
23. Kaminskyi, A., Nehrey, M.: Information technology model for customer relationship management of nonbank lenders: coupling profitability and risk. In: 11th International Conference on Advanced Computer Information Technologies (ACIT), pp. 234–237 (2021)
24. Kalaiselvi, T., Sriramakrishnan, P., Somasundaram, K.: Performance of medical image processing algorithms implemented in CUDA running on GPU based machine. Int. J. Intell. Sys. Appli. (IJISA). **10**(1), 58–68 (2018). Jan
25. Umbarkar, A.J., Rothe, N.M., Sathe, A.S.: OpenMP Teaching-learning based optimization algorithm over multi-core system. Int. J. Intelli. Sys. Appli. (IJISA) **7**(7), 57–65 (2015)

Sub-Gigahertz Wireless Sensor Network for Smart Clothes Monitoring

Andrii Moshenskyi[1], Dmitriy Novak[2], and Liubov Oleshchenko[3]([✉])

[1] National University of Food Technologies, Kyiv 01601, Ukraine
[2] Kyiv National University of Technologies and Design, Kyiv 01011, Ukraine
[3] National Technical University of Ukraine "Igor Sikorsky Kyiv Polytechnic Institute",
Kyiv 03056, Ukraine
oleshchenkoliubov@gmail.com

Abstract. Wireless sensor system is based on a set of sensor modules and custom devices that are installed on clothing to achieve the goal of collecting micro-meteorological data for interlayer spaces research. Last time clothes testing is usage short distance data transfer systems or usage of cellular networks for the coverage extension. Sub-gigahertz systems integration into existed data networks in the research was proposed. Module consists of an Atmel microcontroller, a battery power supply, a battery management module, radio module and a set of temperature and relative humidity sensors. These modules are integrated into the clothing in interlayer spaces. The microcontroller firmware is written in C programming language. The software client side for displaying data collection and statistics processing is written in the Python programming language with the use of Numpy and Matplotlib libraries for the serial port. It was established that for the first subject an increase of the average temperature and relative humidity is due to physical activity and type of uniform. The second subject results were more stable than for the first one. Parameters of changes in temperature and air humidity can be used to predict comfortable conditions and optimize clothing in relation to climatic conditions and the load on the human body. Proposed adopted licensing and data modes make possible to build networks with coverages, which can concurrent with national cell networks but with much lower data rates. Usage ISM on the sub-gigahertz bands data links produce data rates less than 10 Kb/s but the radius increase more then on hundreds of percent (>200%). Amateur radio and the same speed limits produce extension of coverage radius in 10 and more times versus ISM 433. Testing of proposed system is perform in thousands of data transfer sessions. Coverage simulation was very close to the real data transfer area. Proposed network can be recommend for the similar researches of smart systems of local data collecting.

Keywords: Sub-gigahertz · wireless sensor network · software · microcontroller · sensor · radio module · temperature · humidity

Z. Hu et al. (Eds.): ICCSEEA 2023, LNDECT 181, pp. 657–669, 2023.
https://doi.org/10.1007/978-3-031-36118-0_59

1 Introduction

Recently, there has been intensive work on the development of so-called smart textiles. Most textile parameters depend on the fact that it is produced, namely on fibers and filler. In order to achieve the above optimization, there are actual tasks related to the collection of micro-meteorological parameters, namely the temperature and humidity of the air in the interlayer spaces between the human body and clothing, between layers of clothing and between clothing and external spaces.

The integration of radio electronics and textiles enables the development of new materials for industry. Smart textiles are dividing into several subgroups. The first group includes the so-called passive clothing, which allows analyzing the environment or the actions of the user with the help of sensors.

Active smart textiles belonging to the second group can react to external conditions. The third group includes the so-called very smart textiles, they can not only react and analyze the conditions and adapt to them. Sometimes a fourth group is distinguished, the so-called intelligent textiles, which react not only to external conditions and adapt, but also allow be programmed.

The integration of conductive and non-conductive materials from which clothes are made allows to build, as it were, integrated microcircuits, but not according to integral, but according to the so-called hybrid technology. In the form of clothing, integrate some types of sensors, some various devices in the form of electronic circuits that make up elements of construction and design.

The main problem of smart clothes testing is usage short distance data transfer systems or usage of cellular networks for the coverage extension. Sub-gigahertz systems integration into existed data networks was not consider. Sub-gigahertz technology operates in a frequency band less than 1 GHz and can be used for wireless applications requiring low power consumption and long-range transmission, which reduces the locating intermediate base stations cost.

1.1 Analysis of Recent Research

Smart clothes researchers have developed materials that integrate various sensors to study such physiological parameters as temperature, humidity, heart rate, concentration of gases, namely carbon and carbon dioxide. Leading manufacturers use the monitoring of biological parameters of the human body and international gas spaces, namely such manufacturers as Adidas and Numetrex [1, 2] smart clothes in the form are used to detect so-called dangerous situations for the human body and even the environment [3, 4].

In certain works, the temperature limit of 37.80 °C means that monitoring is carried out not every 10 s, but every second. The upper limit of 65.6 degrees Celsius, which leads to the activation of an alarm, looks rather strange. Usually, for a biological object, a temperature exceeding 43 degrees Celsius is dangerous, and a temperature exceeding 53 degrees Celsius leads to painful symptoms. Work [5] provides a set of methods and materials that allows developing clothing intended for industrial purposes and sports purposes. In papers [6–8], detailed monitoring of the inter-layer parameters of clothing, namely temperature and air humidity, is carried out.

In work [9] emergency data network with unstandardized usages of digital radio modules was described, which allows using human based ghost analytics to show the coverage and levels. Ad-Hoc networks [10] can be use in the similar research, but this architecture is too difficult for our subject area, peer-to-peer will be enough.

Materials usage [11, 12] and interlayer microclimatic features were describe, but nothing about sensors that can be uses in the interlayer areas of clothes. WSN security and key generation [13–15] for not secured network for clothes studying can be use in future research, but not on this step. Attack detection and models actuality [16] is urgent task, but real coverages of WSN not described. Adaptive antenna systems for radar and telecommunications applications [17] not adopted for chosen ISM bands.

In review of 15 papers [18] the reconfigurable antennas for body centric communication were focused on wide used band 2.4 GHz, but nothing about long-range sub-gigahertz systems. Robots really can help in lifesaving tasks, but radio control data network is described poor in work [19].

Cloud of Things is very close to IoT Cloud, especially for research of smart things and clothes [20] and unfortunately, that TEQIP-III project decryption is too slow for our task. Sleep-mode energy management [21] in cellular networks management technic cannot be use in peer-to-peer network. In [22] compared measured and calculated signal losses on the low distances, but again only for popular 2.4GHz band only.

In the analyzed works was not described a sub-gigahertz wireless sensor network and a computer system for processing statistical data collected from this network.

1.2   Objective

The objective of the research in this article is sub-gigahertz wireless sensor network development that allows increase coverage area for the manufacture of military troop shirt with the zonal arrangement of temperature and relative humidity sensors with the possibility of their real-time remote monitoring and cloud storing of obtained data.

2   Materials and Research Method

The proposed experimental setup consists of a power supply lithium-ion batteries module with a small capacity of approximately 500 milliamps per hour, microcontroller of the Atmel family of 8 -bit architecture, which can be from mega 88 to 328, temperature and humidity sensor ds18b20 for temperature measuring in hard-to-reach places and sensors temperature humidity, produced by Amsong company am2301 2302. Radio modules produced by Silabs company, namely si4432 or si4463, are used for data transmission over a wireless channel, which allow working in the unlicensed ISM band of 433 MHz with the use of a small power of approximately 10 mW. On the server side, there is a radio module also from the Silabs family, which is connected to a computer via UART (Universal Asynchronous Receiver Transmitter). Further data processing is done by reading the stream from the UART and processing the data using the built-in libraries of the Python language.

Figure 1 shows the internal structure and construction of the sensors, with 1-wire bus in sensor Dallas for use in hard-to-reach areas of clothing.

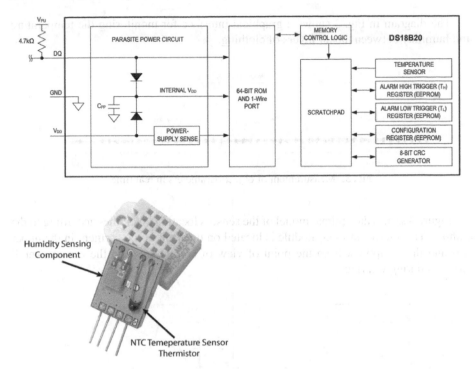

Fig. 1. Temperature and relative humidity sensors that were used in the research

Figure 2 shows the structural diagram of the module with the microcontroller in the center and shows which buses and ports the corresponding sensors and peripheral devices are connected to. It shows that the analog- to- digital converter of the controller is used to monitor the battery status.

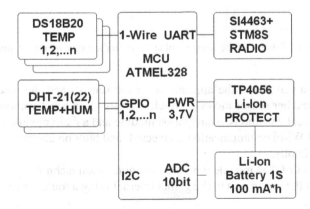

Fig. 2. Scheme of the mobile module

The diagram in Fig. 3 shows a graphical interface for monitoring the temperature and humidity between the air layers of clothing.

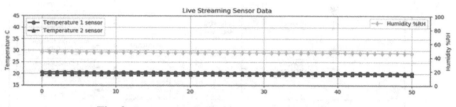

Fig. 3. Measurement of object parameters in real time

Figure 4 shows the optimal model of the sensors location on clothes according to the authors. The microcontroller module is located on the back of the garment in the upper part and this is optimal from the point of view of ergonomics and the use of overall modules during wearing.

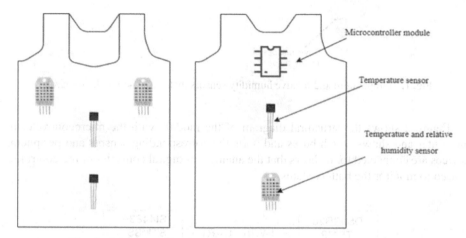

Fig. 4. Scheme of the sensor's placement in the experimental sample

In Fig. 5, part a) shows the appearance of the sensors and the receiving device, part b) shows another sensor module, which consists of an accelerometer, sensors for microclimatic local conditions, a navigation module and a radio module for GSM communication and Wi-Fi communication connected, and built on another microcontroller of the esp8266 family.

Using Wi-Fi or GSM module wireless communication technology, it is possible to directly connect the sensor unit to the global Internet using a router or cellular station as shown in Fig. 6.

The research of the authors is related to the development of smart clothing for military purposes, therefore the range of wireless and Wi-Fi technology was insufficient, and the coverage of cellular networks in some areas, namely military training grounds, was also

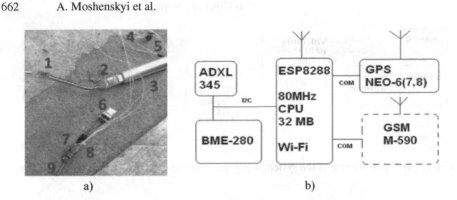

a)   b)

Fig. 5. a) Modifications of sensors for the remote monitoring system of changing in the internal microclimate in the air gap between cloth layers: 1 – base module – control station 1SI4463 based radio module, 2 – USB convertor ch340, 3 – powerbank 1 X 18650 USB port of PC mobile module, sensors on the clothes, 4– AM2302 temperature and humidity sensor, wire connected, 5 –DS18b20 thermometer, 6 – 100mah LiIon battery, 3 h of supply without sleep mode and with cyclic transmissions on 100 mW, 7 – ADXL345 3 axis digital accelerometer, 8 – Atmega328 MCU for sensor request, data processing, 9 – SI4463; b) GSM module diagram

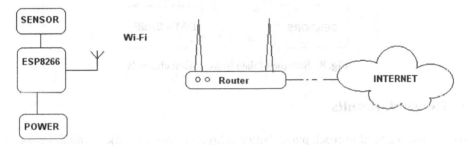

Fig. 6. The block diagram of the moving objects survey using a WiFi network

insufficient, therefore the technology of data transmission over a radio channel on the VHF and UHF was used, which are shown in more detail in the structural diagram of Fig. 7. The left part of Fig. 7 shows a sensor module consisting of a microcontroller, a set of sensors, a power supply module and a radio module with an integrated portable antenna. In the right part of Fig. 7, a radio module connected to a serial interface of a computer connected to the Internet is shown, and an example of using the APRS (Automatic Packet Reporting System) network is given.

Figure 8 shows the structural diagram of the data transmission channel. Data from the sensors are processed by microcontrollers and sent to the transmitter radio module of the portable device. Through the air, data is transmitted to the receiving radio module, processed by a computer and sent to a data storage in the Internet or a local server.

The computer transfers data to the cloud database where it is stored for usage.

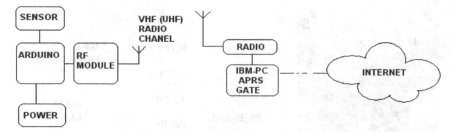

Fig. 7. Structural scheme of the system using the VHF radio network (range up to 100 km)

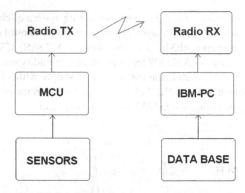

Fig. 8. Scheme of data transmission channels

3 Research Results

For monitoring hard-to-reach places between layers of clothing, the ds18b20 sensor is recommended due to the use of a convenient one-wire bus and the ability to work in 12-bit mode, which allows assess temperature trends with a fairly high resolution. The unique 64 bit address of each sensor cannot impose a limit on the number of sensors on one item of clothing within reason. From the Fig. 9 we could see that average temperature 1 slightly increases from 27.4 to 27.9 °C and relative humidity 1 increases from 44.8 to 46.5% when subject do physical activities (walking for 10 min) after some rest average temperature decreases to 24.1 °C and relative humidity decreases to 37.5%, after another portion of physical activities (running for 15 min) average temperature increases to 26.2 °C and relative humidity to 43%.

Without physical activity temperature in inter-layer clothing space decreases up to 23.1 °C and relative humidity almost does not changes it slightly decreases to 40.8 and increases to 43%. The average temperature 2 almost does not changes and account for about ~ 26 °C and relative humidity decreases from 43.7 to 29.8% for 2 h on the street in the winter with temperature ~ - 10 °C. The research subjects were in different types of troop uniform.

Temperature and humidity sensors manufactured by the Amsong company can be removed from the outer case and deeply integrated into thin interlayer spaces. In this research we obtained the following indicators: operating range of humidity is 0–100%

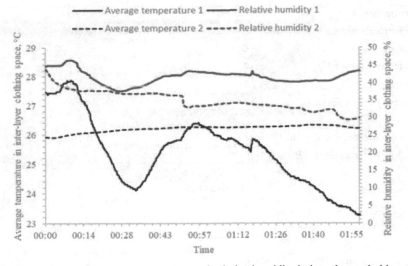

Fig. 9. Dependence of average temperature and relative humidity in inter-layer clothing space

RH; operating range of temperature -40 ~ 80 deg. Celsius; accuracy of related humidity + -2% RH in this range; accuracy of temperature + -0.5 °C in this range. Sensor initialization and library usage is on the Wiring language for replication simplicity.

Figure 10 shows radio coverage parameters modeling of the research.

Figure 11 shows calculation of the radio link losses in the free space. On UHF band we can use Friis equation for free space losses plus 20 dB for compensation of the "green" and surface losses in the first Frenel zone on distances for mile less. Figure 11 shows the calculation of attenuation in the radio channel described by the ideal radio communication formula, the Friis formula. In the band of 430 MHz, at the permissible power of the radio module of 10 mW and the maximum sensitivity of the radio-receiving device, respectively, for a data transfer rate of 9600 bits per second, a communication distance of more than 1 km can be easily achieve.

The Fig. 11 shows the calculation of the path losses and influence the choice of data modem modes. Radio coverage area for the specified frequency, power and the antenna location height of 2 m above the ground in Ukraine, Cherkasy region in the northern part in Fig. 12. To legalize the conduct of experiments, the call sign of the author ut5uuv [9] was used and the power was increased to 100 mW for the permissible limit of 5000 mW of the CEPT (European Conference of Postal and Telecommunication administration) limits.

Testing of proposed system is perform in thousands of data transfer sessions. Coverage simulation was very close to the real data transfer area. Proposed network can be recommend for the similar researches of smart systems of local data collecting. Experiment result comparison shows the great increasing of data transfer coverage areas (Table 1).

Proposed adopted licensing and data modes make possible to build networks with coverages, which can competitive with national cell networks but with much lower data rates. As shown, usage ISM on the sub-gigahertz bands data links produce data rates

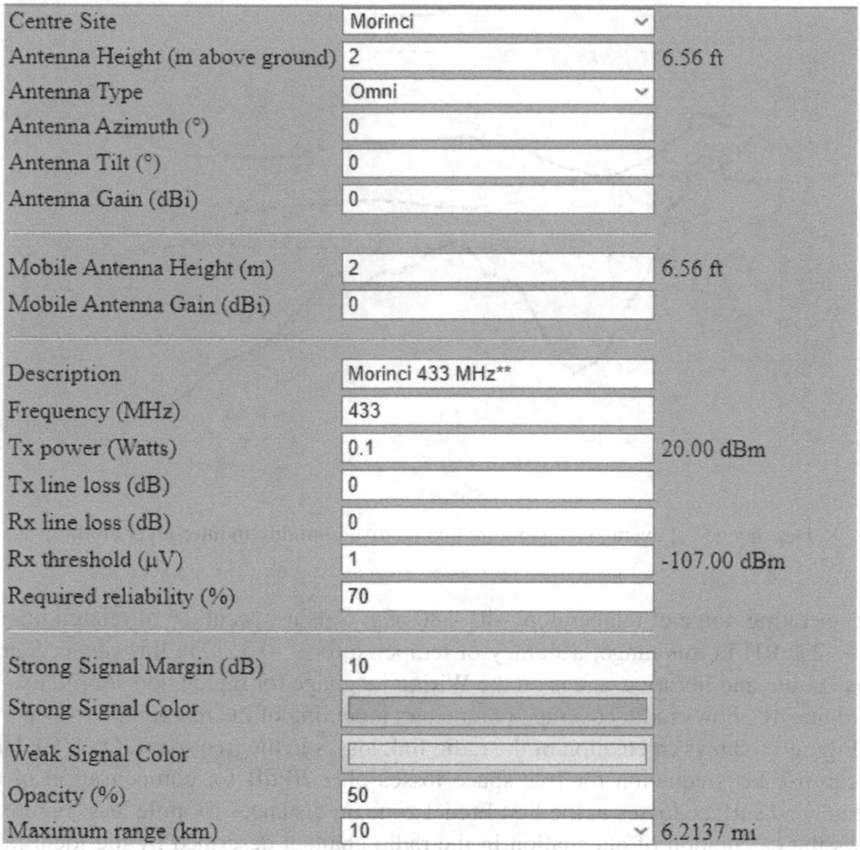

Centre Site	Morinci ⌄	
Antenna Height (m above ground)	2	6.56 ft
Antenna Type	Omni ⌄	
Antenna Azimuth (°)	0	
Antenna Tilt (°)	0	
Antenna Gain (dBi)	0	
Mobile Antenna Height (m)	2	6.56 ft
Mobile Antenna Gain (dBi)	0	
Description	Morinci 433 MHz**	
Frequency (MHz)	433	
Tx power (Watts)	0.1	20.00 dBm
Tx line loss (dB)	0	
Rx line loss (dB)	0	
Rx threshold (μV)	1	-107.00 dBm
Required reliability (%)	70	
Strong Signal Margin (dB)	10	
Strong Signal Color		
Weak Signal Color		
Opacity (%)	50	
Maximum range (km)	10 ⌄	6.2137 mi

Fig. 10. Radio coverage parameters modeling

less than 10 Kb/s but the radius increase more then on hundreds of percent (>200%). Amateur radio and the same speed limits produce extension of coverage radius in 10 and more times versus ISM 433.

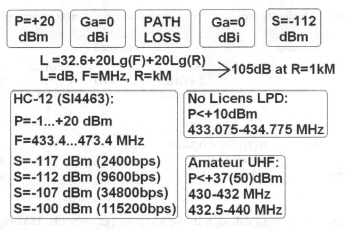

Fig. 11. Link path calculation and speed comparison in proposed system.

Fig. 12. Radio link path loss and wireless sensor network coverage

Superposition of proposed usage of public APRS service and Amateur CEPT license adaptation for low speed data links extends the range over the areas with allocated gates of proposed services to hundreds of kilometers. Digipiter (Digital repeater) on the International Space Station and another LEO Satellites also can be used for proposed system extension.

For increase of the link budget, author uses the LoRa technology on the local and LEO satellites [23]. Some problems has a CEPT license now, so transmission can be

Table 1. Radius of average coverage comparison for different technologies

Research technologies	Standard			Proposed		
	Wi-Fi	Cellular	ISM 2.4 GHz baseband	ISM 433 MHz	Amateur CEPT 433 MHz (5W)	Amateur CEPT 433 MHz (5W) + APRS
Radius of average coverage [km]	0.1	35(71 extended range)	0.3–0.5	>1	>>10	>>100 across the gates

done not under the amateur call sign in the Ukraine, but only like ISM object. Local emission class tables not changed yet.

4 Conclusions

Wireless sensor network developed and tested by the authors, was used to test military clothing in Ukraine. This made it possible to monitor interlayer spaces of clothes in real time in an area that did not have cellular communication and the ability to connect to local Wi-Fi networks.

The accuracy of temperature sensors of 12 bits made it possible to measure the temperature with a resolution of 0.0625 degrees Celsius, which allows determine clearly the tendency to increase or decrease in temperature. At the same time, a real accuracy of plus or minus 0.5 degrees Celsius was ensure in the excessive temperature range from minus 10 to + 85 degrees Celsius.

New combination of proposed equipment and usage of sub-gigahertz band and ISM licensed network extended the radius of WSN to more than 1 km. Usage the same band in combination with amateur radio license and output power 5 watts less increase the coverage to limits of line-of-slight.

First time data transfer protocols and emission classes was adopted to ISM and amateur radio service (CEPT) limits. Integration of the above wireless sensor network into APRS networks made it possible to conveniently use publicly available services and extend the coverage by public receiver gates and tens kilometers radius across them. The new technology of textile materials modification for smart clothes with the use of sensors and microcontroller equipment were proposed.

New dependencies of temperature and relative humidity in inter-layer clothing space from the time were determined. It was establish that for the first subject an increase of the average temperature and relative humidity is due to physical activity and type of troop uniform. The second subject results were more stable than for the first one. Proposed WSN have chances for the real life cycle due the ISM limitations.

Future work will be integration of the LoRa networks for the subject areas and similar research. Chirp modulation and frequency spreading increase the link budget, but the local license limitation for amateur radio did not allow to use wide band modulations, so ISM power limitation create problems for the authors research.

References

1. Langereis, G.R., Bouwstra, S., Chen, W.: Sensors, Actuators and Computing Architecture Systems for Smart Textiles. In: Chapman, R. (ed.) Smart Textiles for Protection, vol. 1, pp. 190–213. Woodhead Publishing; Cambridge, UK (2012)
2. Custodio, V., Herrera, F.J., López, G., Moreno, J.I.: A review on architectures and communications technologies for wearable health-monitoring systems. Sensors. **12**, 13907–13946 (2012)
3. Giddens, H., Paul, D.L., Hilton, G.S., McGeehan, J.P.: Influence of body proximity on the efficiency of a wearable textile patch antenna. Proceeding of the 6th European Conference Antennas & Propagation (EuCAP); Prague, Czech. pp. 1353–1357 (26–30 March 2012)
4. Viry, L., et al.: Flexible three-axial force sensor for soft and highly sensitive artificial touch. Adv. Mater. 2659–2664 (2014)
5. Zhang, L., Wang, Z., Psychoudakis, D., Volakis, J.L.: Flexible Textile Antennas for Body-Worn Communication. Proceedings of IEEE International Workshop on Antenna Technology, pp. 205–208. Tucson, ZA, USA (5–7 March 2012)
6. Yan, S., Soh, P.J.: Wearable dual-band composite right/left-handed waveguide textile antenna for WLAN applications. Electron. Lett. **50**, 424–426 (2014)
7. Salvado, R., Loss, C., Gonçalves, R., Pinho, P.: Textile materials for the design of wearable antennas: A survey. Sensors. **12**, 15841–15857 (2012)
8. Dursun, M., Bulgun, E., Şenol, Y., Akkan, T.: A Smart Jacket Design for Firefighters. Tekstil ve Mühendis **26**(113), 63–70 (2019)
9. Moshenskyi, A.: Private rescue echo beacon with FSK radiomodule. Scientific journal "Science-Based Technologies" **4**(48), 478–483 (2020)
10. Oleshchenko, L., Movchan, K.: AODV Protocol Optimization Software Method of Ad Hoc Network Routing. Lecture Notes on Data Engineering and Communications Technologies **83**, 219–231 (2021)
11. Zhdanova, O., Bereznenko, S., Bereznenko, N., Novak, D.: Textile materials manufacturing features with the use of antimicrobial additives. Fibres and Textiles **24**(4), 41–46 (2017)
12. Kurhanskyi, A., Bereznenko, S., Novak, D., et al.: Effects of multilayer clothing system on temperature and relative humidity of inter-layer air gap conditions in sentry cold weather clothing ensemble. Fibres and Textiles **25**(3), 43–50 (2018)
13. Salehi, M., Karimian, J.: A Trust-based security approach in hierarchical wireless sensor networks. Int. J. Wirel. Microw. Technol. (IJWMT) **7**(6), 58–67 (2017)
14. Kaur, J., Randhawa, S., Jain, S.: A novel energy efficient cluster head selection method for wireless sensor networks. Int. J. Wirel. Microw. Technol. (IJWMT) **8**(2), 37–51 (2018)
15. Pandey, A.K., Gupta, N.: An Energy Efficient Clustering-based Load Adaptive MAC (CLA-MAC) Protocol for Wireless Sensor Networks in IoT. Int. J. Wirel. Microw. Technol. (IJWMT) **9**(5), 38–55 (2019)
16. Nangue, A.: Elie Fute Tagne, Emmanuel Tonye, Robust and Accurate Trust Establishment Scheme for Wireless Sensor Network. Int. J. Comp. Netw. Info. Secu. (IJCNIS) **12**(6), 14–29 (2020)
17. Shcherbyna, O., Zaliskyi, M., Kozhokhina, O., Yanovsky, F.: Prospect for using low-element adaptive antenna systems for radio monitoring stations. Int. J. Comp. Netw. Info. Secu. (IJCNIS) **13**(5), 1–17 (2021)
18. Munawar, H.S.: An Overview of Reconfigurable Antennas for Wireless Body Area Networks and Possible Future Prospects. Int. J. Wirel. Microw. Technol. (IJWMT) **10**(2), 1-8 (2020)
19. Punith Kumar, M.B., Sumanth, S., Savadatti, M.A.: Internet Rescue Robots for Disaster Management. Int. J. Wirel. Microw. Technol. (IJWMT) **11**(2), 13–23 (2021)

20. Bashir, A., Sholla, S.: Resource efficient security mechanism for cloud of things. Int. J. Wirel. Microw. Technol. (IJWMT) **11**(4), 41–45 (2021)
21. Agubor, C.K., Akande, A.O., Opara, C.R.: On-off Switching and Sleep-mode Energy Management Techniques in 5G Mobile Wireless Communications – A Review. Int. J. Wirel. Microw. Technol. (IJWMT) **12**(6), 40–47 (2022). https://doi.org/10.5815/ijwmt.2022.06.05
22. Ng, S.T., Lee, Y.S., Radhakrishna, K.S., Krishnan, T.M.: Study of 2457 MHz WIFI Network Signal Strength at Indoor and Outdoor Enviroment. Int. J. Wirel. Microw. Technol. (IJWMT) **11**(6), 1–9 (2021). https://doi.org/10.5815/ijwmt.2021.06.01
23. UT5UUV Console. https://tinygs.com/station/UT5UUV@320362520

Paradoxical Properties Research of the Pursuit Curve in the Intercepting a Fugitive Problem

Viktor Legeza and Liubov Oleshchenko$^{(\boxtimes)}$

Igor Sikorsky Kyiv Polytechnic Institute, Kyiv 03056, Ukraine
oleshchenkoliubov@gmail.com

Abstract. The actual problem of building a mathematical model of the process of "flat" pursuit with the selection of the optimal angle of inclination of the line of escape in order to avoid capture by the fugitive on the longest segment along the given direction has been solved.

The proposed new formulation of the problem makes it possible to determine new properties of the pursuit curve and characteristics obtain from it, which affect the fugitive's ability to avoid capture at the longest horizontal distance. A quantitative and qualitative analysis of the influence of the characteristics of the pursuit curve on the maximum movement of the fugitive before his capture was carried out. The best case shows the minimum extremum in result comparison, it can show the optimal path.

The obtained optimized model makes it possible to choose the angle of inclination of the line of escape, which ensures exactly such an optimal movement of the escapee. It was established first that for a successful escape, the strategy of choosing an angle of inclination of the escape line equal to zero (that is, escape along the shortest segment connecting two parallel lines   the vertical axis of the ordinate and the "life" line) is not optimal.

First time the dependence of the change in the distance (on the horizontal axis) to the point of capture of the fugitive, depending on the magnitude of the angular coefficient of his movement along an inclined line at a fixed angular coefficient, was established. It is shown that this dependence has a well-defined local maximum, which indicates the presence of a certain non-zero angle of inclination of the fugitive's movement line, which provides him with the opportunity to reach the maximum movement towards the lifeline on the border between countries.

New paradoxical phenomenon was founded: for a successful escape, the strategy of choosing the angle of inclination of a straight line equal to zero (that is, moving along the shortest segment connecting two parallel straight lines) is not correct.The results of the numerical experiment give reasons to recommend the proposed model for practical use. The future development of this research can be use in military task in the Southeast of Ukraine in torpedoes usage.

Keywords: Pursuit curve · local maximum · angle coefficient · selection strategy · slope angle · escape line · "life" line

Z. Hu et al. (Eds.): ICCSEEA 2023, LNDECT 181, pp. 670–681, 2023.
https://doi.org/10.1007/978-3-031-36118-0_60

1 Introduction

The pursuit curve is the line along which the pursuer moves while pursuing the fugitive. The problem of persecution probably dates back to the time of Leonardo da Vinci. He was the first to investigate this problem when the fugitive moved in a horizontal straight line. The general case was studied by the French scientist Pierre Bouguer in 1732. The task was to find the curve of pursuit of a merchant ship by a pirate ship. At the same time, it was assumed that the speeds of the two vessels are always in the same ratio.

The use of differential equations for the pursuit-evasion problem has recently become very relevant for the development of computer games.

Multi variables vas reduced to only two variables in research [1] for the tasks of graphic in pursuit-evasion game. In research [2] authors described pursuit and evasion strategy and used only the deep neural network instruments, but nothing in determinate math.

Works [3–5] describe the same game tasks by the deep reinforcement learning that can make instruments for the decision-making technology without proposed optimal routes for the moved object.

In works [6, 7], the well-known problem of "chasing four mice" in various settings was considered. Suppose four mice are located in each of the four corners of a square table, and each mouse runs to the one to the right of it. We need to find parametric curves that describe the trajectory of each mouse. The solutions of this problem will be spiral trajectories that coincide in the center of the table.

In research [8], deterministic continuous pursuit is considered, in which n ants chase each other in a circle and have predetermined variable speeds. Two discrete analogues are considered, in which a cricket or a frog is cyclically pursued with a constant and equal speed. The possible evolution of this motion as time approaches infinity is explored: collisions, limit points, equilibrium states, and periodic motion.

The paper [9] presents a simple mathematical model of the local interaction of a colony of ants or other natural or artificial creatures with a great "sense of global geometry" to find a direct path from the anthill to food. This task was also considered within the framework of the general task of chasing n ants.

The paper [10] investigates the movement of an arbitrary set of points (or beetles) on a plane chasing each other in cyclic pursuit. It is shown that for regular centrally symmetric configurations, analytical solutions are easily obtained by switching to the corresponding rotating frame of reference. Several cases of asymmetric configurations are discussed. For the case where all beetles have the same speed, a theorem is proved that whenever premature (i.e., non-reciprocal) capture occurs, the collision must be head-on. This theorem is then applied to the case of three- and four-beetle configurations to show that these systems collapse up to a point, i.e., the capture is mutual. Some aspects of these results are generalized to the case of systems with n beetles.

The work [11] considers a well-known problem that continues to be analyzed today. Three (or more) beetles, which are initially located at the vertices of a regular polygon, begin to cyclically chase each other, moving together at the same speed.

It is usually necessary to know the distance traveled by each beetle before mutual capture. The case of four beetles is the simplest. Due to symmetry, the four beetles always remain on the four vertices of the square. Since the fleeing beetle Z_2 always moves at right angles to the pursuing beetle Z_1, the capture speed of these two beetles depends only on the speed of the beetle Z_1. The case for n bugs is only slightly more complicated. This problem is also symmetric, since n bugs are always located at n vertices of a regular n-gon.

The book [12] gives systematic methods of winning in differential pursuit and evasion games. The scope and application of game procedures have been studied. Numerous useful examples illustrate basic and advanced concepts, including capture, strategy, and algebraic theory. Also includes consideration of both linear and non-linear games. Chapters in the book include stroboscopic and isochronous target acquisition, algebraic theory, and selection strategies.

The article [13] considers the pursuit problem with simple movement for the case when the maximum speeds of the players are the same, and the fleeing person moves along a strictly convex smooth n-dimensional surface. It is proven that the end of pursuit is possible from any starting position. It is established that evasion is possible from some initial positions if the hypersurface contains a two-dimensional flat part.

The book [14] is a well-known textbook on differential games. It treats differential games as conflict situations with an infinite set of alternatives that can be described by differential equations. In the book, the main attention is paid to fundamental mathematical questions, the solution methods are illustrated by a large number of interesting examples, which are important in themselves. A number of unsolved problems are proposed.

The technical report [15] provides a detailed description of the implementation of a clean pursuit curve tracking algorithm. Given the general success of the algorithm in recent years, it is likely that it will be used again in land navigation tasks. The report also includes a geometric version of the method and provides some information on the performance of the algorithm as a function of its parameters.

The article [16] describes a third-order chase, a game of evasion in which both players have the same speed and minimum turning radius. A idiosyncratic game is first resolved for a barrier or shell of states that can be captured. When capture is possible, the game at some level is decided for optimal control of the two players as a function of relative position. It is found that the solution of the problem includes a universal surface for the pursuer and an acceleration surface for the evaders.

The article [17] considers the coplanar problem of evading pursuit, in which two pursuers P_1, P_2 and one fugitive participate E. The fugitive E evades with a constant speed $w > 1$ greater than that of the pursuers, and must pass between the two pursuers P_1 and P_2 with a single speed, the gain is the distance of closest approach to any of the pursuers. This is a typical two-player zero-sum game theory problem. Velocity directions P_1, P_2 and E are chosen as controlling variables. A closed-loop solution is obtained through elliptic functions of the first and second kind. The closed-loop solution is shown graphically in several diagrams for different values of w.

Research [18] uses differential equations with fractional variable coefficients that can describe our task, but the simplicity and the accuracy of the proposed numerical scheme not optimal for work on the surface.

In [19, 20] collective described predicting rate of divorce tendency and based on it rate of human happiness in Nigeria. Differential equations and limits closely describe same task, but the correlation in results cannot describe binary result.

Economical bonuses by routing optimization [21] makes research task very attractive. Terminal nodes limitations for author's solutions were chosen. This limit type cannot be use in our task without weapon usage in the model.

Surface geometry tasks in [22] describes slow dynamical models with strong (stochastic) noise components. This type of prediction too hard for constant speed tasks.

A review of literature sources showed that the problem of pursuit with a formulation in which the maximization of the fugitive's movement towards a certain "saving" limit due to the selection of the angular coefficient of the escape line was chosen as the quality criterion was not considered before.

2 Problem Statement

In some cases, with the passage of time, the old problems of finding the pursuit curves "one pursuer-one fugitive" on the plane are revised in new formulations. This may be related to the possibility of changing different strategies for avoiding interception by the fugitive due to the choice of one or another quality criterion. For example, as a criterion of quality in such tasks, the maximum advance of the fugitive towards a certain border, which is considered "life" line for the fugitive, can be considered. Therefore, from a practical point of view, solving the problem of pursuit "one pursuer-one fugitive" in a production with a new quality criterion will be relevant in such areas as transport, logistics, military affairs, sports events, computer games, etc.

The object of research is the process of chasing a fugitive, who tries to move as far as possible in the direction of a certain given limit before the moment of interception, moving along an inclined straight line.

The subject of the research is the properties and characteristics of the pursuit curve, which influence the choice of a fugitive's successful escape strategy.

The research objective. The task is: to find 1) the trajectory $L : \{y = y(x)\}$ of the pursuit boat; 2) the point on the plane at which it will catch up with the boat-fugitive; 3) the angular coefficient that ensures the maximum displacement of the boat-fugitive along the horizontal axis in the direction of the border between the two countries. The flow rate of the reservoir is not taken into account.

Consider the starting position of two moving points, one of which is chasing the other (Fig. 1). On one shore (marked as the OY axis in Fig. 1) of the reservoir, two boats stand at a distance $y_0 = 1$ from each other – a coast guard boat U (hereinafter the pursuit-boat) and a border violator boat V (hereinafter the boat-fugitive).

The reservoir separates the two countries, as a natural border between them. At the same time, the runaway boat V is trying to cross the body of water as quickly as possible in the direction of the border ("lifeline")AB between the two countries. We believe that this border is parallel to the shoreline of the start of the two boats ($AB \parallel OY$). The chasing boat U tries to overtake the boat-fugitive V before it reaches a certain point, which is on the border AB of two countries.

So, at the initial time $t = 0$, the violator boat V begins to move along a straight line $\eta = k\xi$ (Fig. 1) with a speed of v. The boat U starts at the same time as the boat V and in the process of pursuit chooses the direction of movement to the current location of the boat V on the plane OXY. At the same time, the pursuit boat U has a constant speed

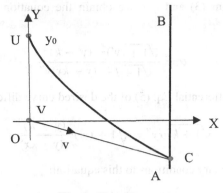

Fig. 1. Pursuit curve

u, which is α times greater than the speed of the boat V: $u = \alpha v$. The fugitive's task is to move as far as possible in the direction of the border AB between the two countries, choosing for escape the optimal angular coefficient k of the inclined line with a constant α coefficient.

3 Construction of the Differential Equation of the Trajectory of the Pursuit

We will call the L trajectory of the boat U the **pursuit curve**. Let us identify the location of the boats on the plane OXY with the points U and V. We denote by (x, y) the coordinates of the point U and by (ξ, η) are the coordinates of the point V at the current time t. Let's draw a tangent at a point $U(x, y)$ to the curve L connecting the points U and V and establish the relation for the derivative function $y(x)$:

$$y' = \frac{\eta - y}{\xi - x} \Rightarrow y' = \frac{k\xi - y}{\xi - x}. \tag{1}$$

Let's express the variable ξ from Eq. (1):

$$\xi = \frac{xy' - y}{y' - k}. \tag{2}$$

Let's find the time derivative of the variable ξ:

$$\dot{\xi} = \frac{y - kx}{(y' - k)^2} y'' \dot{x}. \tag{3}$$

Now let's determine the ratio α for the speed modules of the boats, taking into account that

$$u = \frac{ds}{dt} = \sqrt{1 + (y')^2} \dot{x} \text{ and } v = \sqrt{1 + k^2} \dot{\xi} \tag{4}$$

Taking into account (3) and (4), we obtain the equation of the trajectory L : $\{y = y(x)\}$:

$$\alpha = \frac{\sqrt{1 + (y')^2}}{\sqrt{1 + k^2}} \frac{(y' - k)^2}{(y - kx)y''} \tag{5}$$

Let's rewrite the differential Eq. (5) of the desired curve differently:

$$\sqrt{1 + k^2}\alpha y'' = \sqrt{1 + (y')^2} \frac{(y' - k)^2}{(y - kx)}. \tag{6}$$

Let's add two boundary conditions to this equation:

$$y(0) = 1; \quad y'(0) = -\infty. \tag{7}$$

So, to find the pursuit curve, we have a boundary value problem with differential Eq. (6) and boundary conditions (7).

4 Solving the Boundary Value Problem of Finding the Trajectory Equation

In Eq. (6), we will replace the variable:

$$z = y - kx \Rightarrow z' = y' - k; z'' = y''$$

As a result, we obtain a nonlinear differential equation of the second order:

$$\alpha\sqrt{1 + k^2} \cdot z'' \cdot z = (z')^2 \cdot \sqrt{1 + (z' + k)^2} \tag{8}$$

Equation (8) allows the order to be reduced by one if the following substitution is made:

$$\frac{dz}{dx} = p(z); \frac{d^2z}{dx^2} = p(z) \cdot \frac{dp}{dz}. \tag{9}$$

In the new variables, Eq. (9) will take the following form:

$$\alpha\sqrt{1 + k^2} \cdot p'_z \cdot p \cdot z = p^2 \cdot \sqrt{1 + (p + k)^2}. \tag{10}$$

In Eq. (10), we reduce by a factor $p(z) \neq 0$ and get the equation with separated variables:

$$\alpha\sqrt{1 + k^2} \cdot \frac{dp}{p\sqrt{1 + (p + k)^2}} = \frac{dz}{z}. \tag{11}$$

After integrating the Eq. (11), we get the equation of the first order:

$$\text{tg}^{\alpha}\left(\frac{1}{2} \cdot \text{arctg}\left(\frac{p}{1 + k(p + k)}\right)\right) = C_1 \cdot z. \tag{12}$$

To determine the first constant in Eq. (12), we use the relation $z = y - kx$ and boundary conditions (7). As result of the limit transition, we obtain the first constant:

$$C_1 = \text{tg}^\alpha \left(\frac{1}{2} \cdot \text{arctg} \left(\frac{1}{k} \right) \right). \tag{13}$$

Let's express the function $p = z'$ from (12) and (13) in terms of z:

$$\frac{p}{1 + k(p + k)} = \text{tg} \left\{ 2\text{arctg} \left[z^{\frac{1}{\alpha}} \text{tg} \left(\frac{1}{2} \text{arctg} \left(\frac{1}{k} \right) \right) \right] \right\}. \tag{14}$$

We denote the right-hand side of relation (14) by $f(z, k, \alpha)$. As a result, we got equations of the first order (here $p = z'$):

$$\frac{dz}{dx} = \frac{f(z, k, \alpha) \cdot (k^2 + 1)}{1 - f(z, k, \alpha) \cdot k}. \tag{15}$$

Let's write the second integral using (15):

$$x(z, k, \alpha) = \int_1^z \frac{1 - f(z, k, \alpha) \cdot k}{f(z, k, \alpha) \cdot (k^2 + 1)} dz. \tag{16}$$

Let's supplement expression (16) with an expression for $y(z, k, \alpha)$, which together describe the pursuit trajectory $y(x)$ in a parametric form of a pair of functions, where the variable z acts as a parameter, namely:

$$\begin{cases} x = x(z, k, \alpha); \\ y = z + k \cdot x(z, k, \alpha) \end{cases} \tag{17}$$

5 Research Results for Numerical Analysis of the Pursuit Trajectory and Its Characteristics

So, two Eqs. (17) describe the trajectory $L : \{y = y(x)\}$ of the pursuing boat U. We present a graph (Fig. 2) constructed for the case of the pursuit process, in which the following movement parameters are selected: $k = -2$ and $\alpha = 2$.

On the graph in Fig. 2, the blue curve represents the trajectory of the fleeing boat V, and the red curve represents the trajectory of the boat U.

The point of the plane XOY in which the escape boat is apprehended has coordinates $(0{,}431; -0{,}863)$. Geometrically, the fact of arrest means that the red curve is tangent to the blue line.

The time T required to detain a fugitive boat V in the process of pursuit is equal to: $T = \sqrt{0{,}431^2 + (-0{,}863)^2}/v = 0{,}965/v$. Let's calculate the distance S traveled by the pursuit boat U during the pursuit: $S = T \cdot u = 0{,}965 \cdot \alpha = 1{,}93$.

Figure 3 presents a curve that reflects the dependence of the change in the distance $x(0, k, \alpha)$ (on the horizontal axis OX) of the arrest of the runaway boat depending on the angular coefficient k of its direct movement at a fixed coefficient α.

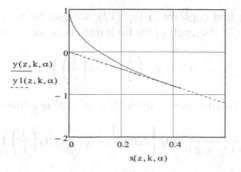

$$\frac{y(z,k,\alpha)}{y1(z,k,\alpha)}$$

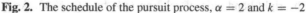

$$x(z,k,\alpha)$$

Fig. 2. The schedule of the pursuit process, $\alpha = 2$ and $k = -2$

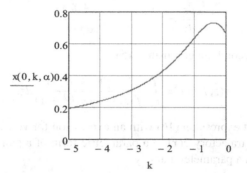

$$\frac{x(0,k,\alpha)}{}$$

k

Fig. 3. Graph of the dependence of $x(0, k, \alpha)$ on the coefficient k

The point of the plane XOY in which the escape boat is apprehended has coordinates $(0,431; -0,863)$. Geometrically, the fact of arrest means that the red curve is tangent to the blue line. The time T required to detain a fugitive boat V in the process of pursuit is equal to: $T = \sqrt{0,431^2 + (-0,863)^2}/v = 0,965/v$.

Let's calculate the distance S traveled by the pursuit boat U during the pursuit:$S = T \cdot u = 0,965 \cdot \alpha = 1,93$. Figure 3 presents a curve that reflects the dependence of the change in the distance $x(0,k,\alpha)$ (on the horizontal axis OX) of the arrest of the runaway boat depending on the angular coefficient k of its direct movement at a fixed coefficient α. It has a well-defined local maximum, which indicates that there is a certain non-zero angle of inclination of the direct movement of the escape boat, which gives it the opportunity to achieve the maximum movement towards the border between the two countries.

So, we observe a certain paradox: the curve in Fig. 3 establishes that the strategy of choosing the angle of inclination of a straight line equal to zero (that is, moving along the shortest segment connecting two parallel lines - which would be natural) is not correct for a successful escape. In this version of the calculation $\alpha = 2$, $k = -0, 4$), this movement is as follows:$x(0; -0,4; 2) = 0,734$.

Considering the indicated in Fig. 3, the paradoxical property of the pursuit process using (17), we establish the dependence $y = y(x)$ (Fig. 4), where $x = x_{max}(0,k,2)$ is the maximum value of the fugitive's V movement along the OX axis with a variable coefficient k and a fixed coefficient $\alpha = 2$. The graph of the curve (Fig. 4) also shows the point of maximum movement of the fugitive V along the OX axis with a non-zero angular coefficient $k = -0, 4$, which acts as a parameter of this curve.

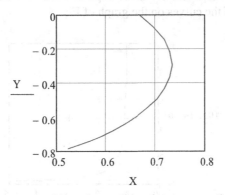

Fig. 4. Graph of dependence $y = y(x)$ on the condition that $x = x_{max}(0, k, 2)$ and the coefficient k changes

Let's find out the behavior of the quantity $x(0, k, \alpha)$ under the condition that the coefficient α goes either to infinity or to unity. Figure 5 shows the graph of the dependence of the value $x(0, k, \alpha)$ on the coefficient α for a fixed value of the angular coefficient $k = -0.45$.

From the behavior of the curve in Fig. 5, it follows that if the coefficient $\alpha \to \infty$, then the distance $x(0, k, \alpha)$ at which the latter will be detained tends to zero.

On the other hand, if $\alpha \to 1$(that is, the speeds of the boats are equal), then the indicated distance goes to infinity, that is, the arrest of the runaway boat will not take place.

The Table 1 summarizes the results of numerical experiments to determine the movement of the fugitive $x = x(0, k, \alpha)$ along the axis OX and $y = y(0, k, \alpha)$- along the axis OY before its interception, the time spent $T = T(0, k, \alpha)$ by the pursuer on delaying the fugitive, and the path $S = S(0, k, \alpha)$ traveled by the pursuer before the delay of the fugitive.

Table 1. Research result comparison

-62172563900k	0.00	0.1	0.2	0.3	0.4	0.5	0.6	0.7	0.8	0.9	1.0
x	0.667	0.696	0.718	0.730	0.734	0.730	0.719	0.703	0.683	0.661	0.638
y	0.00	0.07	0.144	0.218	0.294	0.365	0.431	0.492	0.547	0.595	0.638
T	0.667	0.700	0.732	0.762	0.790	0.816	0.838	0.858	0.875	0.890	0.902
S	1.334	1.400	1.464	1.525	1.581	1.632	1.676	1.716	1.750	1.779	1.805

In this case, the parameter is fixed, the speed of the fugitive is equal to one, and the parameter varies from zero to one. From the given table it follows that the best escape strategy of the fugitive is to choose a straight line with the tangent of the slope angle of the straight line of escape, equal to about. At the same time, the pursuer will spend units of time and travel units of the path to delay him.

Figure 6 shows the corresponding graph of the dependence of the value on the coefficient, which is constructed for those angular coefficients of the escape line that provide the maxima of the curves on the graph of Fig. 3.

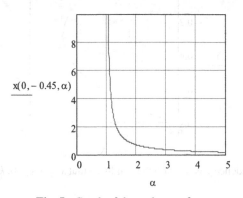

Fig. 5. Graph of dependence of on

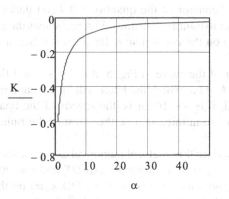

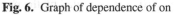

Fig. 6. Graph of dependence of on

6 Summary and Conclusions

The urgent problem of building a mathematical model of the process of "flat" pursuit with the selection of the optimal angle of inclination of the line of escape with the aim of avoiding the capture of the fugitive by the longest movement has been solved.

The fugitive's goal is to advance as far as possible to the lifeline on the border between the two countries, choosing the new optimal angular ratio of the straight line to escape.

The equation of the pursuit optimized curve is obtained in a parametric form. Its numerical analysis was carried out, the influence of the parameters of the pursuit process on the form of the curve was investigated.

The best case, the parameter is about 0.4, the time units and travel in result comparison show optimal path.

First time the dependence of the change in the distance (on the horizontal axis) to the point of capture of the fugitive, depending on the magnitude of the angular coefficient of his movement along an inclined line at a fixed angular coefficient, was established. It is shown that this dependence has a well-defined local maximum, which indicates the presence of a certain non-zero angle of inclination of the fugitive's movement line, which provides him with the opportunity to reach the maximum movement towards the lifeline on the border between countries.

New paradoxical phenomenon was founded: for a successful escape, the strategy of choosing the angle of inclination of a straight line equal to zero (that is, moving along the shortest segment connecting two parallel straight lines) is not correct.

The future development of this research can be use in military tasks in the Southeast of Ukraine in torpedoes usage. The mathematical results of this research may also be useful for computer game industry that needs movement engines for much realistic in game physic.

References

1. Patsko, V., Kumkov, S., Turova, V.: Pursuit-Evasion Games. In: Başar, T., Zaccour, G. (eds.) Handbook of Dynamic Game Theory, pp. 951–1038. Springer, Cham (2018). https://doi.org/10.1007/978-3-319-44374-4_30
2. Xu, C., Zhang, Y., Wang, W., Dong, L.: Pursuit and evasion strategy of a differential game based on deep reinforcement learning. Frontiers in Bioengineering and Biotechnology (March 2022). https://doi.org/10.3389/fbioe.2022.827408
3. Lopez, V.G., Lewis, F.L., Wan, Y., Sanchez, E.N., Fan, L.: Solutions for multiagent pursuit-evasion games on communication graphs: finite-time capture and asymptotic behaviors. IEEE Trans. Autom. Control. **65**, 1911–1923 (2019). https://doi.org/10.1109/TAC.2019.2926554
4. Zhou, Z., Xu, H.: Mean field game and decentralized intelligent adaptive pursuit evasion strategy for massive multi-agent system under uncertain environment with detailed proof. Proceedings of the Artificial Intelligence and Machine Learning for Multi-Domain Operations Applications II; Online. (27 April–8 May 2020)
5. Wan, K., Wu, D., Zhai, Y., Li, B., Gao, X., Hu, Z.: An improved approach towards multi-agent pursuit-evasion game decision-making using deep reinforcement learning. Entropy (Basel). **23**(11), 1433 (2021). https://doi.org/10.3390/e23111433. Oct 29
6. Malacká, Z.: Pursuit Curves and Ordinary Differential Equations. Komunikacie **14**(1), 66-68. University of Žilina (March 2012). https://doi.org/10.26552/com.C.2012.1.66-68
7. Nahin, P.J.: Chases and Escapes: The Mathematics of Pursuit and Evasion. Princeton-New Jersey: (Princeton Puzzlers) 272 (2012)
8. Bruckstein, A.M., Cohen, N., Efrat, A.: Ants, crickets and frogs in cyclic pursuit, Preprint, Technion CIS Report, Haifa #9105 (July 1991)
9. Bruckstein, A.M.: Why the ant trails look so straight and nice. The Mathematical Intelligencer **15**(2), 59–62 (1993). https://doi.org/10.1007/BF03024195
10. Behroozi, F., Gagnon, R.: Cyclic pursuit in a plane. J. Math. Phys. **20**, 2212–2216 (1979). https://doi.org/10.1080/00029890.1975.11993941. November

11. Klamkin, M.S., Newman, D.J.: Cyclic pursuit or "The Three Bugs Problem". Amer. Math. Monthly 631–639 (June 1971). https://www.jstor.org/stable/2316570
12. Hajek, O.: Pursuit Games: An Introduction to the Theory and Applications of Differential Games of Pursuit and Evasion. Academic Press, New York (1975)
13. Kuchkarov, A.S., Rikhsiev, B.B.: A pursuit problem under phase constraints. Autom. Remote. Control. **62**(8), 1259–1262 (2001)
14. Isaacs, R.: Differential Games, p. 260. John Wiley and Sons, New York (1967)
15. Craig Conlter, R.: Implementation of the pure pursuit path tracking algorithm. CMU-RI-TR-92-01. The Robotics Institute Carnegie Mellon University Pittsburgh, Pennsylvania 15213 (January 1992)
16. Merz, A.: The game of two identical cars. Journal of Optimization Theory and Applications **9**, 324–343 (1972). https://doi.org/10.1007/BF00932932
17. Hagedorn, P., Breakwell, J.V.: A differential game with two pursuers and one evader. Multicriteria Decision Making and Differential Games 443–457 (1976). https://doi.org/10.1007/978-1-4615-8768-2_27
18. Falade, K.I., Tiamiyu, A.T.: Numerical solution of partial differential equations with fractional variable coefficients using new iterative method (NIM). Int. J. Mathem. Sci. Comp. (IJMSC) **6**(3), 12–21 (2020). https://doi.org/10.5815/ijmsc.2020.03.02
19. David, O.O., et al.: A mathematical model for predicting rate of divorce tendency in nigeria: a study of taraba state, nigeria. Int. J. Mathem. Sci. Comp. (IJMSC) **6**(5), 15–28 (2020). https://doi.org/10.5815/IJMSC.2020.05.02
20. David, O.O., et al.: Mathematical model for predicting the rate of human happiness: a study of federal university Wukari community of Nigeria. Int. J. Mathem. Sci. Comp. (IJMSC) **6**(6), 30–41 (2020). https://doi.org/10.5815/IJMSC.2020.06.05
21. Gnanapragasam, S.R., Daundasekera, W.B.: Optimal solution to the capacitated vehicle routing problem with moving shipment at the cross-docking terminal. Int. J. Mathem. Sci. Comp. (IJMSC) **8**(4), 60–71 (2022). https://doi.org/10.5815/ijmsc.2022.04.06
22. Nazimuddin, A.K.M., Ali, S.: Application of differential geometry on a chemical dynamical model via flow curvature method. Int. J. Mathem. Sci. Comp. (IJMSC) **8**(1), 18–27 (2022). https://doi.org/10.5815/ijmsc.2022.01.02

Oscillations at Nonlinear Parametric Excitation, Limited Power-Supply and Delay

Alishir A. Alifov[1]([✉]) and M. G. Farzaliev[2]

[1] Mechanical Engineering Research Institute of the Russian Academy of Sciences, Moscow 101990, Russia
a.alifov@yandex.ru
[2] Azerbaijan State University of Economics, Baku, Azerbaijan

Abstract. The work is devoted to the development of the theory of oscillatory systems with limited power-supply, the relevance of which has increased due to energy and environmental problems. A system with an energy source of limited power and a time delay in the elastic force under the action of nonlinear parametric excitation is considered. On the basis of the direct linearization method, the linearization of the nonlinear elastic force is carried out, the equations of unsteady motions are obtained. Stationary modes and their stability are considered. Using the Routh-Hurwitz criteria, the stability conditions of stationary oscillations are obtained. To obtain information about the effect of lag on stationary modes of motion, calculations were performed. The obtained graphical results clearly show the influence of nonlinear parametric excitation on the dynamics of the system in the presence of a delay in the elastic force. Depending on the value of the delay, different features appear in the dynamics of oscillations and their stability. The curves shift in amplitude and frequency, and a rather significant shift appears in the case of a nonlinear elastic force.

Keywords: Oscillations · Nonlinearity · Parametric excitation · Energy source · Limited power · Delay · Method · Direct linearization

1 Introduction

A machine in the modern definition is a device that performs some of the functions assigned to it and is able to adapt to environmental changes with the help of artificial intelligence. Any device uses an energy source to perform its assigned functions. As an object of control in an artificial intelligence system (robots, automatic machines, manipulators, etc.), there can be time-varying displacement, speed, electrical voltage, etc. They are included in the design schemes and models of the description of the device when it is created. Differential equations, including nonlinear ones, are very often used as an analysis of dynamic models of devices.

To solve nonlinear differential equations, a number of approximate methods of nonlinear mechanics have been developed [1–8, etc.], which are characterized by high time and labor costs. In these methods, the form of the solution is first specified, and then

Z. Hu et al. (Eds.): ICCSEEA 2023, LNDECT 181, pp. 682–689, 2023.
https://doi.org/10.1007/978-3-031-36118-0_61

rather laborious calculations are performed. The complexity and laboriousness of mathematical relations increases sharply with an increase in the order or number of the approximation, which causes the problem of their applicability in practice. The same growth takes place with an increase in the degree of nonlinearity, which in most cases leads to the construction of only the first approximation. With reference to [9–12], in [13] this problem is noted in the study of coupled oscillatory networks that play an important role in electronics, neural networks, physics, etc.

The *method of direct linearization* (MPL) described in [14–17, etc.] is fundamentally different from the commonly used known methods of nonlinear mechanics. In MPL, the nonlinear function is replaced by a linear one, the form of solving the original nonlinear equation is not required, and there are no approximations of different orders. It is quite simple to use, reduces labor and time costs by several orders of magnitude. Simplicity and low labor intensity of using any method are very important when calculating real devices (machines, equipment, electrical systems, etc.).

In devices of various kinds (electronics, tracking systems, regulators, etc.), systems with a delay are widespread. The source of lag in autonomous and remote-controlled systems (mobile robots, manipulators, etc.) is the data transmission channel. For example, when managing a group of robots, neglecting them can lead to significant errors [18]. A large number of papers [18–22, etc.] are devoted to the problems of oscillations in systems with a delay without taking into account the properties of the energy source. And the work in which there is such a record is incomparably less. In this article, oscillations are considered under nonlinear parametric excitation, taking into account an energy source of limited power and a retarded elastic force. The purpose of the work is to development the theory of oscillatory systems with limited excitation to use the results obtained in the calculation of real objects and the selection of energy sources with its minimum consumption. The relevance of this direction of the theory of oscillations has increased due to the growing energy and environmental problems in the world. The ecological crisis is a crisis of human society. The total energy consumption on Earth is growing in proportion to the square of the population, humanity has very significantly exceeded the limit of its energy capacity and has gone beyond the threshold of self-destruction of the biosphere.

2 Model and Equations of the System

Figure 1 shows the model of the system described in [23]. It represents a rod with a spring, which is connected to a crank driven by a limited-power motor.

The system is nonlinear and, taking into account the nonlinearity of parametric excitation and the delay in the elastic force, is described by the equations

$$\ddot{y} + \beta_1 \dot{y} + \omega^2 y + by^3 \sin \varphi = -m^{-1} f(y) - c_\tau y_\tau$$
$$J\ddot{\varphi} = M(\dot{\varphi}) - 0.5 c_2 y^2 \cos \varphi - 0.5 c_3 \sin 2\varphi - c_4 \cos \varphi$$

(1)

where

$$\omega^2 = \frac{c}{m}, \ c = \frac{\pi^4 EI_x}{2l^3}\left(1 - \frac{P_0}{P_1}\right), \ b = \frac{c_2}{m}, \ c_2 = -\frac{\pi^2 r_1 c_1}{2l}, \ m = \frac{\rho l}{2}$$
$$\beta_1 = \frac{\beta}{m}, \ P_0 = f_0 c_1, \ P_1 = \frac{\pi^2 EI_x}{l^2}, \ c_3 = c_1 r_1^2, \ c_4 = f_0 r_1 c_1$$

Here $c_1 = const$ and $\beta = const$ are the coefficients of stiffness and resistance, $b = const$, $f(y)$ is the nonlinear part of elasticity, $c_\tau = const$, $y_\tau = y(t - \tau)$, $\tau = const$ is time delay, $J = const$ is the moment of inertia of the motor rotor rotating the crank with radius r_1, $\dot{\varphi}$ is the speed of rotation of the motor rotor, $M(\dot{\varphi})$ is the driving torque of the motor taking into account the forces of resistance to rotation.

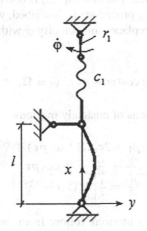

Fig. 1. System model

Let us represent the nonlinear function $f(y)$ as a polynomial $f(y) = \sum_s \gamma_s y^s$, $s = 2,3,4,\ldots$. Using DLM [14–17], we replace it with a linear function

$$f_*(y) = B_f + c_f y \tag{2}$$

where B_f and k_f are the *linearization coefficients* defined by the expressions

$$B_f = \sum_s N_s \gamma_s a^s, \quad s = 2, 4, 6, \ldots (s \text{ is even number})$$

$$k_f = \sum_s \overline{N}_s \gamma_s a^{s-1}, \quad s = 3, 5, 7, \ldots (s \text{ is odd number})$$

Here $a = \max|x|$, $N_s = \frac{(2r+1)}{(2r+1+s)}$, $\overline{N}_s = \frac{(2r+3)}{(2r+2+s)}$, r is the *linearization accuracy parameter* which affects the accuracy of the method and is introduced into the functional for determining the linearization coefficients.

Taking into account (2), Eqs. (1) take the form:

$$\ddot{y} + \beta_1 \dot{y} + \omega_0^2 y + by^3 \sin\varphi = -m^{-1}B_f - c_\tau y_\tau \tag{3}$$

$$J\ddot{\varphi} = M(\dot{\varphi}) - 0.5c_2 y^2 \cos\varphi - 0.5c_3 \sin 2\varphi - c_4 \cos\varphi$$

where

$$\omega_0^2 = \omega^2 + c_f m^{-1}$$

3 Equation Solutions

Solutions of Eqs. (3) can be constructed by two methods [14], of which the *method of replacing variables with averaging* allows us to study stationary and non-stationary processes. In [14] there is a *standard form of the relation* for a *linearized nonlinear equation of a fairly general form*. The first Eq. (3) is one of the special cases of such an equation. In [16], the averaging procedure is described, which is applicable to solving the second equation in (3). It replaces the velocity $\dot{\varphi}$ with its averaged value Ω. Based on the above, using

$$y = a\cos\psi, \ y_\tau = a\cos(\psi - p\tau), \ \dot{\varphi} = \Omega, \ \psi = pt + \xi, \ p = \frac{\Omega}{2}$$

we obtain the following equations of unsteady motions

$$\begin{array}{c} \frac{da}{dt} = -\frac{a}{2}(\beta_1 - 2c_\tau\Omega^{-1}\sin p\tau) + \frac{ba^3}{4\Omega}\cos 2\xi \\ \frac{d\xi}{dt} = \frac{4\omega^2-\Omega^2}{4\Omega} + \frac{c_f}{4\Omega} + \frac{c_\tau}{\Omega}\cos p\tau - \frac{ba^2}{2}\sin 2\xi \\ \frac{d\Omega}{dt} = \frac{1}{J}\left[M(\Omega) - \frac{c_2 a^2}{8}\right] \end{array} \tag{4}$$

The equations of stationary motions follow from the conditions $\dot{a} = 0$, $\dot{\xi} = 0$, $\dot{\Omega} = 0$:

$$4m^2B^2 + E^2 = 4m^2b^2a^4, \ \ tg2\xi = -\frac{E}{2mB}$$
$$M(\Omega) - S(a) = 0$$

where

$$B = 2\Omega\beta_1 - 4c_\tau\sin p\tau, \ E = m(4\omega^2 - \Omega^2 + 4c_f + 4mc_\tau\cos p\tau$$
$$S(a) = \frac{c_2 a^2}{8}$$

The expression $S(a)$ defines the load on the energy source and the intersection point of the curves $S(a)$, $M(\Omega)$ gives a stationary velocity value Ω.

4 Stability of Stationary Motions

It is necessary to check the stability of stationary modes of movement. By composing the equations in variations for (4) and using the Routh-Hurwitz criteria

$$D_1 > 0, \ \ D_3 > 0, \ \ D_1D_2 - D_3 > 0 \tag{5}$$

where

$$D_1 = -(b_{11} + b_{22} + b_{33}), D_2 = b_{11}b_{33} + b_{11}b_{22} + b_{22}b_{33} - b_{23}b_{32} - b_{12}b_{21} - b_{13}b_{31}$$
$$D_3 = b_{11}b_{23}b_{32} + b_{12}b_{21}b_{33} - b_{11}b_{22}b_{33} - b_{12}b_{23}b_{31} - b_{13}b_{21}b_{32}$$

we have

$$b_{11} = \tfrac{1}{J}Q, \quad b_{12} = -\tfrac{c_2 a}{4J}, \quad b_{13} = 0$$

$$b_{21} = -\tfrac{a}{\Omega^2}c_\tau \sin p\tau - \tfrac{ba^3}{4\Omega^2}\cos 2\xi, \quad b_{22} = -\tfrac{1}{2}(\beta_1 - 2c_\tau\Omega^{-1}\sin p\tau) + \tfrac{3ba^2}{4\Omega}\cos 2\xi$$

$$b_{23} = -\tfrac{ba^3}{2\Omega}\sin 2\xi, \quad b_{31} = -0.25 - \tfrac{\omega^2}{\Omega^2} - \tfrac{c_f}{m\Omega^2} - \tfrac{c_\tau}{\Omega^2}\cos p\tau + \tfrac{ba^2}{2\Omega^2}\sin 2\xi$$

$$b_{32} = \tfrac{1}{m\Omega}\tfrac{\partial c_f}{\partial a} - \tfrac{ba}{\Omega}\sin 2\xi, \quad b_{33} = -\tfrac{ba^2}{\Omega}\cos 2\xi$$

$$Q = \tfrac{d}{d\Omega}M(\Omega)$$

When calculating $\tfrac{\partial B_f}{\partial \Omega}, \tfrac{\partial B_f}{\partial a}$, only even powers of s are taken into account, and when calculating $\tfrac{\partial c_f}{\partial \Omega}, \tfrac{\partial c_f}{\partial a}$ odd powers of s are taken into account.

5 Calculations

To obtain information about the influence of delay on the dynamics of the system, calculations were carried out. The nonlinearity $f(y)$ was chosen as γy^3 and the following parameters were used: $\omega_0 = 1\ s^{-1}$, $m = 1\ \text{kgf·s}^2\text{·cm}^{-1}$, $c_2 = 0.07\ \text{kgf·cm}^{-1}$, $\beta = 0.02\ \text{kgf·s·cm}^{-1}$, $c_\tau = 0.05\ \text{kgf·cm}^{-1}$, $\gamma = \pm 0.2\ \text{kgf·cm}^{-3}$ The linearization coefficient $\overline{N}_3$ is calculated with the linearization accuracy parameter $r = 1.5$ and has the value $\overline{N}_3 = \tfrac{3}{4}$ (for $r = 1.5$, the results of the direct linearization method completely coincide with the results based on the asymptotic method of Bogolyubov-Mitropolsky [2]).

Figure 2 shows the amplitude-frequency dependences $a(\Omega)$, where the double-dashed curves correspond to $p\tau = \pi/2$, dashed curves to $p\tau = \pi$, dash-dotted curves to $p\tau = 3\pi/2$, and solid curves to the absence of delay. Stationary solutions do not take place at $\gamma = -0.2$ and $p\tau = 3\pi/2$. The steepness of the characteristic of the energy source $M(\Omega)$, which is within the shaded sector, corresponds to stable oscillations. These sectors should be indicated on the $S(a)$ load curve, but for brevity they are shown on the amplitude-frequency curves. Criteria (criterion) of stability (5) in the black-filled parts of the sectors are performed in the form $0.000X > 0$, where $X \le 9$, i.e. there is a rather weak stability. At $p\tau = \pi/2$, the curves are unstable for both linear ($\gamma = 0$) and non-linear ($\gamma = \pm 0.2$) elasticity. There is also the following interesting feature. For $p\tau = \pi$, in the case of $\gamma = 0.2$, small amplitude values are stable, and in the case of $\gamma = -0.2$, large ones. Moreover, in both cases, stability takes place in small sections of the amplitude curves.

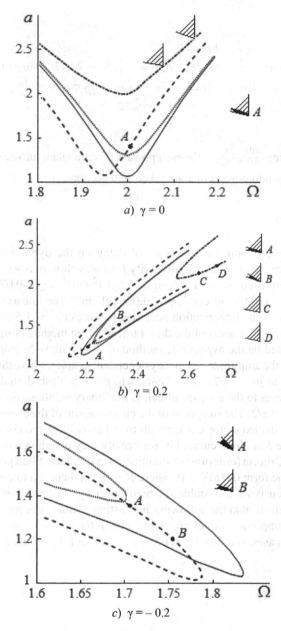

Fig. 2. Amplitude-frequency curves

6 Discussion and Conclusion

We have considered the effect of the delay in the elastic force on the dynamics of a system with nonlinear parametric excitation at a limited power of the energy source supporting the oscillations. It turned out that depending on the magnitude of the delay, different features appear in the system. These features relate both to the location of the amplitude-frequency curves of stationary oscillations in the frequency domain, and their stability. The curves shift in amplitude and frequency, and a rather significant shift appears in the case of a nonlinear elastic force. At some value of delay, stationary oscillations turn out to be unstable both for linear and nonlinear elasticity. At other value of delay and nonlinear elasticity, stability takes place in small sections of the amplitude curves, and in the case of "hard" nonlinearity, small values of the amplitude are stable, and in the case of "soft" – large.

References

1. Butenin, N.V., Neymark, Y., Fufaev, N.A.: Introduction to the Theory of Nonlinear Oscillations. Nauka, Moscow (1976). (in Russian)
2. Bogolyubov, N.N., Mitropolskii, Y.: Asymptotic Methods in Theory of Nonlinear Oscillations. Nauka, Moscow (1974). (in Russian)
3. Moiseev, N.N.: Asymptotic Methods of Nonlinear Mechanics. Nauka, Moscow (1981). (in Russian)
4. Tondl, A.: On the interaction between self-exited and parametric vibrations. National Research Institute for Machine Design Bechovice. Series: Monographs and Memoranda, p. 25 (1978)
5. Hayashi, C.: Nonlinear Oscillations in Physical Systems. Princeton University Press, New Jersey (2014)
6. He, J.H.: Some asymptotic methods for strongly nonlinear equations. Int. J. Modern Phys. B 20(10), 1141–1199 (2006)
7. Karabutov, N.: Structural Identification of Nonlinear Dynamic Systems. Int. J. Intell. Syst. Appl. 9, 1–11 (2015)
8. Qi, W., Fu, F.: Variational iteration method for solving differential equations with piecewise constant arguments. Int. J. Eng. Manuf. 2(2), 36–43 (2012). https://doi.org/10.5815/ijem.2012.02.06
9. Acebrón, J.A., et al.: The Kuramoto model: a simple paradigm for synchronization phenomena. Rev. Mod. Phys. 77(1), 137–185 (2005)
10. Bhansali, P., Roychowdhury, J.: Injection Locking Analysis and Simulation of Weakly Coupled Oscillator Networks. Simulation and Verification of Electronic and Biological Systems. Springer Science+Business Media B.V., Dordrecht pp. 71–93 (2011)
11. Ashwin, P., Coombes, S., Nicks, R.J.: Mathematical frameworks for oscillatory network dynamics in neuroscience. J. Math. Neurosci. 6(2), 1–92 (2016)
12. Masoud Taleb Ziabari, Ali Reza Sahab, Seyedeh Negin Seyed Fakhari: Synchronization New 3D Chaotic System Using Brain Emotional Learning Based Intelligent Controller[J]. International Journal of Information Technology and Computer Science (IJITCS), 2015, 7(2): 80–87
13. Gourary, M.M., Rusakov, S.G.: Analysis of oscillator ensemble with dynamic couplings. In: AIMEE 2018. The Second International Conference of Artificial Intelligence, Medical Engineering, Education, pp. 150–160 (2018)
14. Alifov, A.A.: Methods of Direct Linearization for Calculation of Nonlinear Systems. M.-Izhevsk: Regular and Chaotic Dynamics (2015) (in Russian)
15. Alifov, A.A.: Method of the direct linearization of mixed nonlinearities. J. Mach. Manuf. Reliab. 46(2), 128–131 (2017)
16. Alifov, A.A.: About some methods of calculation nonlinear oscillations in machines. In: Proc. Inter. Symp. of Mechanism and Machine Science, Izmir, pp. 378–381 (2010)

17. Alifov, A.A.: About Direct Linearization Methods for Nonlinearity. In: Advances in Artificial Systems for Medicine and Education III. Advances in Intelligent Systems and Computing, vol. 1126, pp.105–114. Springer, Cham (2020)
18. Zolotukhin, Y., Kotov, K., Maltsev, A.S., Nesterov, A.A., Filippov, M.N., Yan, A.P.: Correction of transport lag in the control system of a mobile robot. Autometry **2**(47), 46–57 (2011)
19. Than, V.Z., Dementiev, Y., Goncharov, V.I.: Improving the accuracy of calculating automatic control systems with a delay. Softw. Products Syst. **31**(3), 521–526 (2018)
20. Garkina, I.A., Danilov, A.M., Nashivochnikov, V.V.: Simulation modeling of dynamical systems with delay. Mod. Prob. Sci. Educ. **1–1**, 285 (2015). (in Russian)
21. Kashchenko, S.A.: Dynamics of a logistic equation with delay and diffusion and with coefficients rapidly oscillating over a spatial variable. Rep. Acad. Sci. **482**(5), 508–512 (2018)
22. Chun Hua Feng: Oscillation for a mechanical controlled system with time delay. Adv. Mater. Res. **875–877**, 2214–2218 (2014)
23. Kononenko, V.O.: Vibrating Systems with Limited Power-Supply. Iliffe, London (1969)

Fairness Audit and Compositional Analysis in Trusted AI Program

Sergiy Gnatyuk, Pylyp Prystavka, and Serge Dolgikh[✉]

National Aviation University, 1 Lubomyra Huzara Ave, Kyiv, Ukraine
sdolgikh@nau.edu.ua

Abstract. In this work, definitions and approaches in the analysis of fairness and bias in "black box" learning systems were examined from the perspective of unsupervised generative learning. Challenges in direct analysis of fairness and bias in "black box" systems that cannot provide suitable explanations for their decisions have several facets, including: sufficiency and representativity of marked trusted data sets for confident correlation analysis; dimensionality and processing challenges with complex real-world data; and trustworthiness of the trusted data itself. We investigated approaches to fairness audit from the perspective of generative compositional analysis that does not depend on significant amounts of prior data and for that reason can provide additional perspective to the analysis. With generative structure of informative low-dimensional representations that can be obtained with different methods of unsupervised learning, certain forms of bias analysis can be performed without prior data. As demonstrated on a practical example, the significance of the proposed method is that it can provide additional directions and valuable insights in the analysis of bias and fairness of learning systems.

Keywords: Fairness and bias · Black box learning systems · Explainable AI · Trusted AI

1 Introduction

While AI technology has been developing at an outstanding pace in the past decades, finding applications in many areas and domains of research, industry and societal functions, the progress has not been entirely and unconditionally positive. One area of concern and persistent research and discussions is understanding the reasoning of AI systems and ability to provide explanations or audit of their decisions ("black box" learning systems, explainable AI [1, 2]). Another, though closely related area of research is developing conceptual and ontological frameworks describing the fairness, trustworthiness and bias of AI systems [3, 4].

Many studies examined and described examples of bias in different functional applications of AI, including criminal justice, health care, human resources, social networks and others [5–7]. It was noted that the problems of explainable and trusted AI can be closely interrelated [8]: understanding the reasons why learned systems make certain decisions can be a key factor in determining whether it can be trusted; on the other hand,

Z. Hu et al. (Eds.): ICCSEEA 2023, LNDECT 181, pp. 690–699, 2023.
https://doi.org/10.1007/978-3-031-36118-0_62

it is not obvious to imagine a mechanism or a process of confident and reliable determination of a trusted AI without some insight into the reasons of its decisions, though directions for exploration of this question exist and some of them are discussed in this paper. A number of different approaches in verification of learned intelligent systems and extraction of knowledge were developed in the recent years [9, 10].

To address the limitation of unknown, a priori, reasoning in decisions made by black box systems and limitations on available trusted prior data is some tasks and domains, in this work we attempted to outline a possible approach to these essential and actual questions that does not involve massive prior trusted data for examination, and in some cases, confident determination of trustworthiness and bias. This approach allows to untie the problem of "chicken and egg" in determination of trustworthiness, for example: could the system that produced prior decisions be itself trusted? while suggesting sound and practical methods of analysis of fairness and bias based on the natural conceptual structure of the data in the problem. In the authors' view, methods of unsupervised generative concept learning that have been actively developed [11–13] can provide a both insightful and useful perspective for such analysis.

Assuming that generative structure of the data can be identified by one or combination of known methods of dimensionality reduction preserving essential informative characteristics of the input distribution, an analysis of distributions of decisions produced by the audited learning systems across general types of input data can provide valuable insights about the system, including whether it can be trusted or produces biased decisions. Such analysis can directly address the objective formulated for this work, that is, examine and introduce methods of analysis of decisions of black-box learning systems that do not depend significantly and essentially, on large prior trusted knowledge about the system.

2 Methodology

2.1 Black-Box Learning Systems

A black-box learning system L_v, $v = \{v_1,.. v_k\}$, k: 1.. L: internal parameters of the system can be characterized by the condition that for a given set of inputs, $I_x = \{i_1,.. i_k\}$, k: 1.. N described by parameters x in the input parameter space X, decisions $D_y(I_x, v) = \{i_1,.. i_k\}$, k: 1.. N described by parameters y in the output (decision) space are produced based entirely from the parameters of the inputs and internal parameters of the system, without any additional context ("explanations"). In this definition, inputs and decisions are the only variables that can always be observed externally, while the internal parameters, nor the decision logic of the system that were essential for producing certain decision may not be known to the external observer.

2.1.1 Decisions and Correctness

Following a process of learning, the parameters of a black-box system are "trained" to produce decisions that are considered to be correct or optimal. In other words, a system is trained to maximize the correctness, and correspondingly, minimize the error of the produced decisions, that can be measured by some metric ("cost" or error functional) in the output space, Y.

The accuracy, and error are numerical measures of correctness of the outputs or decisions produced by a black-box system L on a certain set of inputs:

$$A, E(I, D, w) : D \to M \tag{1}$$

where M: a numerical space in which the accuracy, error take values; w: additional parameters of the accuracy functional. For example, in a common practice, the set of decisions produced by the learning system on the known, "training" inputs can be compared to prior known correct decisions (standard decisions, $w = \{ D_S \}$) with the resulting deviation of the produced decisions from the standard ones related to the error, and accuracy of the system.

$$E(I_s, D, D_s) \cong dist(D, D_s) \tag{2}$$

where $dist$: a distance metrics in the output (decision) space Y.

The following characteristics of the accuracy and error functions are essential for the subsequence analysis:

- The measure of error (and accuracy) can be calculated for a single input i_x, as long as the correct decision $d_c(i_x)$ is available or can be produced (for example, in the process of audit): $E(d(i_x), d_c(i_x)), A(d(i_x), d_c(i_x))$.
- The measure of error (accuracy) is valid within the statistical margins of the standard set of known (*input, decision*) pairs.
- While the measure of accuracy based on standard decisions (2) can be relatively easy to calculate, and is a standard practice in machine learning, its relevance for arbitrary inputs depends on a number of assumptions. Primarily, the standard (training) set has to be sufficiently representative to describe a wide range of realistic inputs, and the input on which the error is being evaluated, sufficiently similar to some of the inputs in the standard set. If the distance between certain real input i_x and the nearest standard inputs i_S is sufficiently large, estimation of the accuracy of the decision $d(i_x)$ produced by the system can be more challenging.

In conclusion of this section, it needs to be noted that while the error and accuracy represent a measure of correctness of the decisions produced by a black-box system that can be estimated by standard statistical methods [14], it cannot always produce essential conclusions about a) internal context or "motivation" of the system and b) ethical factors or fairness of the produced decisions. This essential aspect will be addressed in the next section.

2.1.2 Accuracy, Fairness and Bias

An ideal system that after being trained produces a correct decision on any input by definition, has perfect accuracy and is perfectly fair, in the sense that no input is more or less preferred by the system than another.

Such a system though, even if it existed, could not be recognized by any practical means because it would require testing candidate systems with every possible input in the input space. Therefore, one has to consider more practical black-box learning systems that produce the output on a certain standard set of decisions D_S with the accuracy $a = (A, E)$. A question can be examined then, is or could such a system be fair or biased, and would the accuracy measures be sufficient for such determination? To advance in addressing it, one needs to define fairness formally.

A learning system L that is a black-box system can be said to be "fair", if its decisions satisfy, or are compatible with, certain fairness objectives (fairness goal, criteria, etc.), O_f:

$$f(L, I, O_f) = F(I, D, O_f) \tag{3}$$

It is presumed that a fair system has to satisfy the objective with any (realistic) set of its decisions. So, if a system fails the objective on a certain set of inputs I_f, a determination of a "bias" is made:

$$\neg f(L, I_f, O_f) : B(L) = True \tag{4}$$

where $F(I, D, O_f)$, $B(I, D, O_f)$: fairness, bias functionals.

Some natural or basic fairness requirements can be defined:

- Consistency (no arbitrary decisions): the system has to produce similar decisions on similar inputs:

$$i_{x1} \cong i_{x2} \rightarrow d(i_{x1}) = d(i_{x2});$$

- An extension of this rule is "group fairness": if a subset of inputs belongs to the same general type G, characterized by essential similarity of inputs within the group, then the decisions produced on the inputs in the group have to be similar as well:

$$d(i_x \in G) \sim const$$

- Implications for the correctness: if a decision on an input is correct / wrong, the decision on a similar input has to be the same (correct or wrong, respectively):

Now one can note that fairness objective is not immediately equivalent to the accuracy characteristics and has to constitute an additional input for the learning system. For clarification of this point, let us consider an example.

2.1.3 Example: Correctness vs. Fairness

Consider a trained black-box learning system with the accuracy characteristics $a = (A, E)$ measured on a set of inputs D_S. Would such a system be fair or biased, relative to additionally defined fairness objective, O_f? Can the answer to this question be determined based only on characteristics of accuracy?

Suppose the objective is defined as "egalitarian": the decisions of the system should be distributed similarly, within statistical margins, for all inputs based on clearly defined criteria. Further, we consider a simpler case where the space of possible inputs is comprised of similarity domains containing inputs of the same general type, with significantly higher similarity within the type than across the entire space of possible inputs. Accuracy characteristics above can then be compiled by the general types of inputs, G_S:

$$a = \{a_s\} = \{(A_s, E_s)\}, G_s \in D_s \tag{5}$$

Then, based on the general fairness principle and the fairness objective one has to conclude that the distributions of decisions based on the set criteria have to be the same (or similar, within statistical margins) for all known groups of inputs. That is, if the criteria for a certain decision are met, the resulting decision has to be the same, regardless of the general type of the input.

Not in the least, verification of this requirement implies that all characteristic types of inputs have to be represented in the verification set D_S, importantly, all inputs that can be encountered by the system in its operation, not only those seen by the system in training.

Clearly, the conclusion that can be derived from this example imposes serious constraints on the accuracy results for determination of fairness that may not be met for all previously trained learning systems. Based on these considerations one has to conclude that correctness measurements alone, without additional information on composition of the observable input space, may not be sufficient for determination of fairness of a black-box learning system.

2.2 Unsupervised Concept Learning

Methods of unsupervised concept learning, where successful, can identify a structure of characteristic patterns, general types or natural concepts in the input distributions by stimulating learning models to improve quality of generation from informative latent distributions, often of significantly reduced complexity.

Unlike methods of conventional supervised learning, these approaches do not depend on specific assumptions about the distribution of the data, large sets of prior data and can be used in a general process with data of different types, origin and domains of application and for that reason, can be applied in the analysis of different types of data and learning models, tasks, domains and applications.

Obtaining such structures of natural groups or types, with entirely unsupervised methods that do not depend on prior trusted data, can be instrumental, as demonstrated in the sections that follow, in the analysis of trust and fairness produced by a learning system.

2.3 Fairness Objective and Criteria

Following the arguments presented earlier, designing learning systems, in particular those of machine intelligence for fairness and trust requires explicit and upfront definition of the fairness objective (FO). As the range of possible tasks and applications of learning systems can be very broad, formalization of all possible objectives can be a challenging, though a worthy in the authors view, task. Here, only some straightforward examples will be provided.

– "Egalitarian" objective: any input satisfying clearly set decision criteria has to produce the same decision.

– "Protective" or non-discrimination: distributions of decisions for specific, identified groups of inputs (focus groups) have to be similar, withing statistical margins, to that of the general group, and certainly, many others.

Thus, it can be essential for clear definition and interpretation of fairness audit to agree upon and formulate:

- Clear, sufficient and robust definition of the Fairness Objective(s) explicitly and upfront, for each area or case of application of learning system
- Definition of the criteria of passing the objective test and determination of fairness.

Whereas defining design practices and guidelines can improve chances of the systems being developed being fair (in some implicitly understood sense), only explicit definition of the objective and criteria (of passing the objective test) will allow to perform fairness audit and make a formal determination of the fairness and trustworthiness of the resulting system.

2.4 Unsupervised Compositional Analysis in Fairness Audit

Combining the definitions and analysis of the preceding sections one can propose an approach in the analysis of fairness, trust and bias based on the structure of general types (concepts) that in some / many cases can be resolved with methods of unsupervised learning, dimensionality reduction and similar. One can argue that many, if not most of practical cases and applications of learning systems satisfy the assumption of generally typed data, (5):

The space of observable inputs is composed of a finite number of general types, with significantly higher similarity of inputs within the type subgroups, plus possible non-typed inputs and / or random noise.

The challenge in the case of a general application of a learning system is that all characteristic types of inputs may not be known prior to, or in the process of development and training of the learning system. Yet, based on the defined fairness objective and criteria, a determination of fairness and trust would have to be made for such systems, within the program of Trusted AI.

A possible direction to address this challenge was outlined in [15], based on the methods of unsupervised composition analysis of general data developed in the recent years. The aim of the analysis is to identify similarity groups or "general types" in the input data with entirely unsupervised means that do not require additional prior knowledge. The approach includes the following general stages of processing the data that can be used with different types of data in a range of tasks and applications:

- Preprocessing observable data and producing a representative set of inputs described by observable or input parameters, that have to be sufficiently sensitive and informative to describe essential variation of observable inputs.
- Application of methods of unsupervised dimensionality reduction to produce informative representations of the representative set of inputs. The resulting, commonly low-dimensional latent space describing the distribution of inputs is defined by the coordinates of informative features found by the methods. A sufficiently large array of methods that can be used to this end, including: linear dimensionality reduction

(PCA); non-linear generative learning, including autoencoder-type neural networks; non-linear dimensionality reduction, manifold learning and others.

- Application of clustering methods, specifically, non-parametric density clustering, in the resulting informative representation space to identify characteristic clusters, groups, or general types of inputs.

An essential distinct feature of the described process and is that it is entirely unsupervised, and at no point requires additional, prior knowledge about the content, composition, assumptions about distributions or other. Consequently, the approach of Unsupervised Composition Analysis (UCA) can be applied to general types of data in a wide range of tasks and applications (Fig. 1), *additionally and in parallel* with conventional methods of development of learning system, with no dependence on trusted prior training data.

For these reasons, there is a natural case for application of the methods of Unsupervised Composition Analysis in the area of fairness and trust audit of learning systems as they can provide critical information and insight into the composition of inputs, including description of general types of inputs.

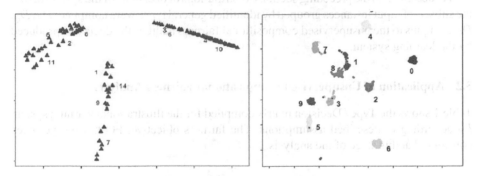

Fig. 1. Unsupervised Composition Analysis, image data (left: geometrical shapes, generative neural model, [16]; right: printed digits, manifold learning, [17])

Combining the case for clearly defined fairness objective, FO and unsupervised composition analysis, UCA, one can propose an approach for auditing operational learning systems with respect to defined fairness and trustworthiness criteria. Importantly, the proposed approach does not depend on, nor require additional inputs or prior trusted data.

As the input to the analysis, a matrix of distributions of decisions by general types of inputs in the axes of: decision criteria and general types of inputs in the data is compiled based on:

- The recorded decisions of the system on provided inputs.

- The distribution of inputs across the general types, or clusters of the inputs identified with UCA methods.

Based on compiled distributions of decisions across characteristic types of inputs, and the defined fairness objective a determination of compliance (i.e., a fair trusted system) or bias can be made.

3 Results

3.1 Unsupervised Compositional Fairness Audit: a Practical Case

For an illustration of the proposed approach to fairness audit of pre-trained and operational black-box learning system consider an illustrational example. Assume that the system, L_P has to produce distinct decisions in three different ranges of some observable parameter of significance, P_S: R_1, R_2, R_3. According to the process of unsupervised composition analysis, suppose application of UCA produced three distinct groups or general types of inputs: A, B, C.

As described in the preceding sections, distributions of decisions of the system across the subsets of input instances grouped by identified general types were compiled: D_A, D_B, D_C as inputs to the unsupervised compositional fairness audit of the decisions produced by the learning system.

3.2 Application of Unsupervised Compositional Fairness Analysis

Table 1 shows the Type / Decision matrix compiled for the illustrational learning system L_P according to described assumptions. The fairness objective, FO has not yet been considered at this stage of the analysis.

Table 1. Type / Decision matrix, practical learning system

General type	Decision criteria 	R_1 (P_S)	R_2	R_3
A		$D_A^{(1)}$	$D_A^{(2)}$	$D_A^{(3)}$
B		$D_B^{(1)}$	$D_B^{(2)}$	$D_B^{(3)}$
C		$D_C^{(1)}$	$D_C^{(2)}$	$D_C^{(3)}$
Other		-	-	-

As discussed earlier, the next essential input in the fairness audit is the Fairness Objective (FO). We will assume that in the case under consideration, the FO for the system L_P has been defined as "Egalitarian", as described in Sect. 3.1, requiring that outside of defined decision criteria, other factors including general type, must have no influence on the decision of the system.

Based on the clearly stated Fairness Objective and distributions of group decisions obtained with UCA methods, one can compare the distributions of decisions in the same criteria range, for example, R_2 across different general types (A – C, Table 1) to make a substantiated, including statistically significant, determination of the fairness or bias in the decisions produced by the audited system, by calculating statistical variances between decision sets of different groups X, Y in the same criteria range, $V_R(X,Y)$.

The logical sequence of the audit is shown in Table 2.

Table 2. Fairness audit: determination of fairness / bias, practical learning system

Learning system	FO	Observed variance, general types [*]	Determination
L_1	Egalitarian	No (($V_R(X,Y)$ ~ 0)  0)	Fair
L_2	--	Yes	Biased

[*] Statistically significant variance of distributions of decisions in general type groups

In the first case (learning system L_1), a determination of fairness was made because distribution of decisions across characteristic types of inputs satisfied the stated fairness objective within the margin of statistical analysis. In the second case (L_2), the criterion of the FO was failed, resulting in the determination of bias.

As can be concluded from this practical example, application of methods of Unsupervised Compositional Analysis in combination with explicitly defined fairness objective allowed to obtain a substantiated determination of fairness / bias in a practical learning system.

4 Summary and Conclusion

Due to variety of methods, architectures, domains, tasks and applications, confident determination of trustworthiness or bias in artificial learning systems can be a challenging and time-consuming task. Principal insights proposed in this paper are first, explicit formulation of the Fairness Objective (FO) that is relevant to the task or application domain and definition of clear and functional criteria of compliance of the decisions produced by the system with the defined fairness objective.

The second essential contribution proposed here is application of methods of Unsupervised Compositional Analysis in the audit of decisions produced by a learning system within the framework of the defined Fairness Objective. Methods of unsupervised composition analysis of general data developed in the recent years have proven both effective and efficient in identifying characteristic informative structure in the data such as characteristic clusters or general types, natural concepts, etc. These methods do not depend on nor require additional prior knowledge about the data and can be used additionally and in parallel to conventional methods of development of artificial learning systems.

It was shown that a combination of methods of conventional statistical analysis with the methods of unsupervised composition analysis proposed in this work can be effective in producing substantiated and reasoned determination of fairness or bias in black-box learning systems, based on the formulated Fairness Objective and the structure of general types characteristic for the data and can significantly improve the outcomes of detection of bias and confidence in the trustworthiness of intelligent systems compared to the status quo. Due to demonstrated wide and successful application of unsupervised composition analysis, it is expected that the approach proposed here can have general applicability in a wide range of tasks, domains and applications of artificial learning systems.

References

1. Longo, L., Goebel, R., Lecue, F., Kieseberg, P., Holzinger, A.: Explainable Artificial Intelligence: concepts, applications, research, challenges and visions. Proceedings of CD-MAKE, LNCS 12279, pp. 1–16 (2020)
2. Rai, A.: Explainable AI: from black box to glass box. J. Acad. Mark. Sci. **48**(1), 137–141 (2019). https://doi.org/10.1007/s11747-019-00710-5
3. Schwartz, R., Vassilev, G.K., Perine, L., Bart, A.: Towards a standard for identifying and managing bias in Artificial Intelligence. National Institute of Standards and Technology, USA, Special Publication 1270 (2022)
4. Bartneck, C., Lütge, C., Wagner, A., Welsh, S.: Trust and fairness in AI systems. In: An Introduction to Ethics in Robotics and AI. Springer Briefs in Ethics. Springer, Cham (2020)
5. Bogen, M.: All the ways hiring algorithms can introduce bias, Harvard Business Review, 06.05 (2019)
6. Gianfrancesco, M.A., Tamang, S., Yazdany, J., Schmajuk, G.: Potential biases in Machine Learning algorithms using electronic health record data. JAMA Intern. Med. **178**(11), 1544–1547 (2018)
7. European Union Agency for Fundamental Rights (FRA). Bias in algorithms - Artificial intelligence and discrimination. European Union (2022)
8. Zhou, J., Verma, S., Mittal, M., Chen, F.: Understanding relations between perception of Fairness and Trust in algorithmic decision making. arXiv (2021). 2109.14345 [cs.CY]
9. Derakhshan, M.A., Maihami, V.B.: A review of methods of instance-based automatic image annotaton. Int. J. Intell. Sys. Appli. **8**(12), 26–36 (2016)
10. Fotsoh, A., Sallaberry, C., Lacayrelle, A.LP.: Retrieval of complex named entities on the Web: proposals for similarity computation. Int. J. Info. Technol. Comp. Sci. **11**(11), 1–14 (2019)
11. Benjio, Y., Courville, A., Vincent, P.: Representation Learning: a review and new perspectives. arXiv (2014). 1206.5538 [cs.LG]
12. Jing, L., Tian, Y.: Self-supervised visual feature learning with deep neural networks: a survey. IEEE Trans. Pattern Anal. Mach. Intell. **43**(11), 4037–4058 (2021)
13. Liu, W., Wang, Z., Liu, X., et al.: A survey of deep neural network architectures and their applications. Neurocomputing **234**, 11–26 (2017)
14. Li, J., Gao, M., D'Agostino, R.: Evaluating classification accuracy for modern learning approaches. Stat. Med. **38**(13), 2477–2503 (2019)
15. Dolgikh, S.: Fairness and bias in Learning Systems: a generative perspective. In: BEWARE-2022 Workshop, 21st International Conference of the Italian Association for Artificial Intelligence, Udine, Italy, CEUR Workshop Proceedings 3319, pp. 61–66 (2022)
16. Dolgikh, S.: Topology of conceptual representations in unsupervised generative models. In: 26th International Conference Information Society and University Studies, Kaunas, Lithuania, 2021, CEUR Workshop Proceedings, pp. 150–157 (2915)
17. Uniform Manifold Approximation and Projection (UMAP) (2018)

Determination of Brachistochronous Trajectories of Movement of a Material Point in a One-Dimensional Vector Field

Viktor Legeza[✉], Oleksandr Neshchadym, and Lyubov Drozdenko

National Technical University of Ukraine "KPI Named After Igor Sikorsky", Kyiv 03056, Ukraine
Viktor.legeza@gmail.com

Abstract. A new approach to solving the Zermelo navigation problem using variational methods is proposed, which differs from methods based on the Pontryagin maximum principle. In this work, the motion of the ship is presented as the motion of a material point in a flat horizontal vector field of a moving fluid, for which the classic variational problem of finding brachistochronous trajectories is formulated. The purpose of the work is to establish the equations of brachistochronous trajectories of the movement of a material point, which ensure the minimum time of its movement between two given points The objective function of the time of movement of a material point between given points is constructed, which is then minimized. With its use, Euler's equation was obtained, which allows us to directly write the differential equations of extreme trajectories. Algebraic equations of the brachistochronous motion of a material point have been established for a given version of the boundary conditions using numerical experiments. A comparative analysis of the time spent both on extreme trajectories and on an alternative straight path was carried out. It is established that the given variational problem generates two solutions of opposite sign, and only one of them delivers the minimum of the time functional. A comparative analysis of the calculation results showed that the brachistochronous trajectory of the movement of a material point is not a straight line connecting two given points, but is an oscillating curve.

Keywords: Zermelo problem · variational problem · brachistochronous motion · objective time functional · euler equations · extremal trajectory

1 Instruction

Let us briefly outline the review of scientific research carried out within the framework of the problem under consideration, related to the search for equations of extremal trajectories, in various problem formulations. I.Bernoulli was the first to set and solve the problem of finding an equation that describes the shape of the brachistochron curve. This is a curve along which a material point moves from a position of rest and without initial speed and resistance in a vertical gravitational field from one known point to another in the minimum time. As a result of the study of this problem in such a statement, it turned out that its solution is a cycloid [1]. In the classic textbook [2], I. Bernoulli's problem about the brachystochrone was solved by the methods of calculus of variations. The formulation of the problem of finding a brachistochrone within the framework of geometric optics (the problem of light beam propagation) was implemented in work [3]. In works [4, 5], the "brachystochronous" movement of a material point in a vertical gravitational field was considered by the methods of variational calculus, taking into account the forces of dry Coulomb friction. In [6], the same problem was solved using the methods of optimal control theory. The author chose the time derivative of the angle, which determines the direction of the brachistochronous movement of the point along the curve, as the controlling parameter in this formulation of the problem. The paper [7] provides a solution to the problem in which a geometric constraint in the form of a cylindrical surface is imposed on the brachistochronous motion of a material point, and dry friction acts between the point and this surface. The problems of finding brachistochronous curves in the process of movement of a material point in non-conservative force fields were considered in articles [8, 9]. In works [10, 11], the search for brachistochrone curves was considered in uniform force fields, and in [10] the movement of a material point was limited by a bond in the form of a cylindrical surface, and in [11] in the form of a spherical surface. The search for a brachistochronous curve during the movement of a material point in non-homogeneous force fields was implemented in works [12–14], and in [14] this problem was solved for the case when a material point moved in a linear radial force field. In works [13, 15, 16], solutions to the brachistochrone problem were found for the case when a point moves in a radial field with a force dependence that is inversely proportional to the distance between the interacting points. In works [17–19], the classical problem about the brachistochrone was generalized to the case of relativistic motion of a material point. In addition, work [19] also solved problems in which the brachistochronous motion of a point was considered on algebraic surfaces of rotation without energy loss. For this, Euler's equations and a special approach developed for solving a number of geometrical optics problems were applied. The article [20] presents the following generalization of the problem of the brachistochronous movement of a material point without friction along a curve that connects a given point and a given curve or two given curves. This formulation of problems leads to a class of variational problems with free boundary conditions. In work [20], such problems were solved using variations at different end points. The next step in the generalization of brachistochrone problems was made in work [21] and related to the search for a brachistochrone curve for a body of finite dimensions that rolls on a cylindrical surface without friction. However, the author of the article [21] did not provide a mathematically correct solution to the brachistochrone problem in such a formulation. In the article [22], the brachistochronous motion of a thin homogeneous disk rolling on a horizontal plane without slipping under the action of three moments was considered. This task was set and solved within the framework of the theory of optimal control. The possibility of realizing the brachistochronous movement of the disc due to the forces of dry friction has been established. The influence of the value of the dry friction coefficient on the structure of extreme trajectories was also analyzed. The article [23] gives a generalization of the brachistochrone problem to the case of a cylindrical container filled with a viscous liquid, which plays the role of a material point in the classical Bernoulli problem. A brachistochronous curve is found that connects two given points between which a container with liquid moves in the minimum time. It is proved that the brachistochron obtained for such a case differs from the cycloid. This effect is explained by the fact that the increase in the rate of change of the kinetic energy of the liquid container entails an increase in energy dissipation due to the movement of the viscous liquid. In articles [24, 25], the brachistochrone equation was found for a heavy cylinder that rolls along a concave cylindrical notch without slipping under the influence of the forces of a vertical gravitational field. The classical calculus of variations was used to find the brachistochrone equation. It is shown that the center of mass of the cylinder traces out a cycloid in the vertical plane, and the oscillations of the cylinder itself in the recess are isochronous. In a number of scientific studies, the tasks of finding extreme trajectories, which are related to various formulations of the Zermelo navigation problem, were considered. In works [26–29], the specified problem was solved within the framework of the theory of optimal control. In these studies, the task was set as follows: it is necessary to find such an optimal control that would ensure the ship's arrival at a given point in the minimum time, i.e., optimal speed control. The time of the ship's movement when closing these problems was determined from transcendental equations, which were solved by numerical methods. As examples, the results of specific solutions of problems for different forms of the profile of the velocity field are given. There are cases when the maximum principle does not give an optimal solution. Since in a number of the above-mentioned works, the problem of finding brachistochronous trajectories leads to certain boundary value problems, we present related works in which new approaches to their numerical solution are considered [30, 31].

The review of literature sources in the field of finding extreme curves showed that previously the Zermelo navigation problem was not solved by variational methods. In addition, other authors [26–29] noted that in some cases this problem cannot be solved using the Pontryagin maximum principle, since it does not provide an optimal solution. Another advantage of the direct variational approach is that it provides the possibility to directly obtain the differential equation of the brachistochrone curve from the target functional. Therefore, in this report, it is proposed for consideration the solution of the problem of finding brachistochronous trajectories of the movement of a material point in a horizontal vector field using the methods of variational calculus. First, we construct the target functional, then, based on the Euler equations, we obtain the brachistochrone differential equation, which is solved by numerical methods.

1.1 The Purpose of the Research

The purpose of the research is 1). obtaining the equations of extreme (brachystochronous) trajectories of motion by variational calculus methods, along which a material point moves from a given starting point to a given final point in a flat vector field of a moving fluid in the shortest time; 2). numerical analysis of the features of brachistochronous trajectories.

1.2 Formulation of the Problem

Let a material point move with variable velocity $\vec{V} = (u, v)$ in a flat one-dimensional vector field along a certain curve $y = y(x)$ connecting two given points $O(0, 0)$ and

$M(L, y(L))$ in the minimum time. At the same time, the modulus of the point velocity vector is constant, i.e. $\left|\vec{V}\right| = C$:

$$\sqrt{u^2 + v^2} = C, \tag{1}$$

where $u-$ is the projection of the speed $\vec{V}$ on the horizontal axis OX; $v-$ projection of speed $\vec{V}$ on the vertical axis $OY u$ is the horizontal projection of the velocity; v vertical projection of speed. The vector $\vec{V}$ is always directed along the tangent drawn at the current point of the sought trajectory $y = y(x)$ of the moving material point. The vector field considered in this problem is formed by a flat one-dimensional river flow. The speed of the river is variable depending on the horizontal coordinate and is determined by some known function $f(x)$. The direction of the river velocity vector is opposite to the positive direction of the vertical axis OY.

2 The Method of Constructing the Time Functional and Determining the Equation of the Desired Trajectory

Let us write expressions for the projections of the speed of a point on the axes OX and OY, taking into account the speed of the river flow:

$$\begin{cases} \dfrac{dx}{dt} = u; \\ \dfrac{dy}{dt} = v - f(x). \end{cases} \tag{2}$$

Let us define the objective time functional to be minimized:

$$T = \int_0^L \frac{dx}{u} \underset{y(x)}{\to} \min. \tag{3}$$

We use the system of Eqs. (2) to construct functional (3). After some transformations, we get:

$$T = \int_0^L \frac{-f(x)y' \pm \sqrt{(Cy')^2 - (f^2(x) - C^2)}}{f^2(x) - C^2} dx. \tag{4}$$

Let us denote by $F(x, y')$ the integrable function in expression (4). We use the Euler equation [2] to find the differential equation for the desired point trajectory:

$$F_y' - \frac{d}{dx}\left(F_{y'}'\right) = 0. \tag{5}$$

The integrand in (4) does not depend on the variable y, so $F_y' = 0$ and Eq. (5) are reduced.
   As a result, we obtain a first-order equation:

$$F_{y'}' = C_1, \tag{6}$$

where C_1 is the first arbitrary constant.

After identical transformations from expression (6) we find the differential equation of the trajectory of a material point:

$$y' = \pm h(x, C, C_1), \tag{7}$$

where

$$h(x, C, C_1) = \frac{1}{C} \sqrt{\frac{f^2(x) - C^2}{1 - \dfrac{C^2}{[C_1 \cdot (f^2(x) - C^2) + f(x)]^2}}}.$$

After integrating (7), we obtain the final equation of the extremal curve:

$$y(x) = \pm \int h(x, C, C_1)dx + C_2, \tag{8}$$

where C_2 is the second arbitrary constant.

Formula (8) establishes the final form of the extremal equation $y(x)$, to which two boundary conditions should be added:

$$y(0) = 0; y(L) = y_L. \tag{9}$$

As we can see, the differential Eq. (7) and its solution (8) for the brachistochronous trajectory of a material point can be immediately obtained from the constructed time functional (4) without constructing other intermediate mathematical constructions, as it happens in works [27–29].

3   The Method of Solving the Boundary Value Problem

Therefore, the differential Eq. (7) obtained above and the boundary conditions (9) form a boundary value problem for finding the brachistochron curve. Let the speed of the river flow be represented by a trigonometric function:

$$f(x) = \sin\left(\frac{\pi x}{L}\right). \tag{10}$$

In formula (8), we replace the function $f(x)$ with the trigonometric function (10) and perform all the necessary identical calculations. After transformations similar to those carried out above, we obtain the function $h(x, C, C_1)$ corresponding to expression (10). Note that from the general form of the integral function in the integral (8), it can be stated that it does not have a primitive in the elementary functions. Hence, we conclude about the need to integrate this integral by numerical methods. To this end, let's represent the function $h(x, C, C_1)$ using Taylor's formula, taking into account the first six terms in order to achieve sufficient accuracy. The development of the function $h(x, C, C_1)$ was carried out by the variable x around the point $x = 0$. Based on the results of the calculations, we will first present all six derivative functions $h(x, C, C_1)$ to construct this development:

$$h^{(0)}(0, C, C_1) = -\frac{C \cdot C_1}{K(C,C_1)}, \quad h^{(1)}(0, C, C_1) = \frac{\pi}{L \cdot C \cdot K^3(C,C_1)},$$

$$h^{(2)}(0, C, C_1) = \frac{\pi^2 \cdot C \cdot C_1^3[(C \cdot C_1)^2 - 4]}{L^2 \cdot K^5(C,C_1)},$$

$$h^{(3)}(0, C, C_1) = -\frac{\pi^3 \cdot [(C \cdot C_1)^4 - 15 \cdot C^2 \cdot C_1^4 - 2 \cdot (C \cdot C_1)^2 + 1]}{L^3 \cdot C \cdot K^7(C,C_1)},$$

$$h^{(4)}(0, C, C_1) = -\frac{\pi^4 \cdot C \cdot C_1^3[4(C \cdot C_1)^6 - 3C^4 \cdot C_1^6 - 24(C \cdot C_1)^4 + 36C^2C_1^4 + 36C^2 \cdot C_1^2 + 72C_1^2 - 16]}{L^4 \cdot K^9(C,C_1)},$$

$$h^{(5)}(0, C, C_1) = \frac{\pi^5[(C \cdot C_1)^8 - 150C^6 \cdot C_1^8 - 4(C \cdot C_1)^6 + 525C^4C_1^8 + 300C^4 \cdot C_1^6 + 6(C \cdot C_1)^4}{C \cdot L^5 \cdot K^{11}(C,C_1)}$$
$$+ \frac{420C^2 \cdot C_1^4 - 150C^2C_1^4 - 4(C \cdot C_1)^2 + 1]}{C \cdot L^5 \cdot K^{11}(C,C_1)},$$

where

$$K(C, C_1) = \sqrt{[1 - (C \cdot C_1)^2]}.$$

Let's represent the function $h(x, C, C_1)$ using Taylor's formula in compressed form:

$$h(x, C, C_1) = \sum_{n=0}^{5} h^{(n)}(0, C, C_1)\frac{x^n}{n!}. \tag{11}$$

Let's substitute the expression (11) in the integral (8) and integrate it. As a result, we get the final brachistochrone equation:

$$y(x, C, C_1, C_2) = \pm \sum_{n=0}^{5} h^{(n)}(0, C, C_1)\frac{x^{n+1}}{(n+1)!} + C_2, \tag{12}$$

where C_1, C_2 are unknown constants, which should be determined from the boundary conditions (9).

4 Numerical Implementation and Analysis of Results

Numerical analysis was performed for three cases. The first two cases concerned two extreme solutions of the variational problem, and the third case was chosen as an alternative one, which represents the motion of a material point along a straight line connecting two given limit points. The measure of the optimality of the chosen path will be the time spent on moving a material point between two given points on the plane. The following boundary conditions were chosen for all three cases:

$$L = 1, \quad C = 1, 1, \quad y(0) = 0, \quad y(L) = 0.$$

1. The first case. In expression (12), we choose a positive sign and find the unknown constants C_1 and C_2 using numerical methods and using boundary conditions. The result of calculating the unknown constants is as follows: $C_1 = 0,385$ and $C_2 = 0$. After that, the final expression for the desired function $y(x)$ for the first case can be obtained. To determine the time of movement of a material point along the brachistochrone, we use the functional (4). After calculating the functional, we get: $T = 2,46$ (time units). Figure 1 shows the graph of the brachistochronic trajectory $y(x)$ for the first case.

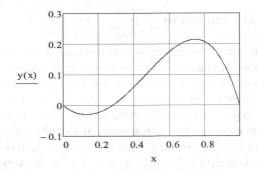

Fig. 1. Brachistochronic trajectory for the first case

2. The second case. In expression (12), we choose a negative sign and again find the unknown constants C_1 and C_2 by numerical methods and using boundary conditions. The results of the calculation of unknown constants coincide with the previous case: $C_1 = 0, 385$ and $C_2 = 0$. Next, we obtain the final expression for the desired function $y(x)$ for the second case. To determine the time of movement of a material point along the brachistochrone, we will again use the functional (4). The calculation of the functional gives less value than in the first case: $T = 1, 140$ (time units). Figure 2 shows the graph of the brachistochronic trajectory $y(x)$ for the second case. Note that the ordinates of the points of this graph on the same abscissa have opposite signs in relation to the graph in Fig. 1.

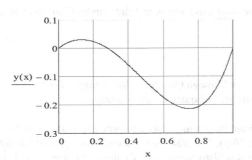

Fig. 2. Brachistochronic trajectory for the second case

3. The third case. Here we consider the movement of a point along a straight line connecting two given points: $O(0,0)$ and $M(L,0)$. So, the trajectory of the point is described by the function $y(x) = 0$. It would seem that this trajectory should be optimal in terms of time, but the results of the numerical experiment refute this assumption. The time functional (4) for the third case gives a greater result than for the second case: when substituting the function $y(x) = 0$ in (4), we have the following result: $T \approx 1, 344$ (units of time).

5 Summary and Conclusion

In this report, the classical Zermelo problem is formulated and solved as a variational problem of determining the brachistochronous trajectory of a material point in a horizontal one-dimensional vector field of a moving fluid. The velocity module of the moving fluid is given as a function of the transverse coordinate (relative to the velocity vector of the moving fluid). A target functional was built, which determines the time of movement of a material point along the brachistochron curve. With its help, the Euler equation was found, which leads to the differential equation of extreme trajectories. Next, the boundary conditions of the boundary value problem are added to this equation, which allows obtaining a solution in quadratures. Based on the expansion of the integral function into a power series, an approximate equation of the extreme trajectories of the point's motion is obtained, taking into account the given boundary conditions.

Within the framework of the numerical experiment, specific examples of the implementation of the proposed approach are given. For the selected version of the boundary conditions, the equations of extreme (brachystochronous) trajectories were determined, along which the material point passes from the start to the finish in the minimum time. An estimate of the time of movement of a material point was made both by brachistochronic trajectories and by an alternative rectilinear path. It is shown that this variational problem has two different solutions that differ only in sign. A comparative analysis of the point movement time on two extreme trajectories was carried out. At the same time, only one of them provides a minimum of the target time function.

An interesting scientific fact has been established: the brachistochronous trajectories of the point's movement are not rectilinear, but have an oscillatory character. Thus, the "shortest path" of the ship's motion is not the "fastest" in this variational problem.

References

1. Dunham, W.: Journey Through Genius, p. 304. Penguin Books, New York (1991)
2. Eltsgolts, L.P.: Differential Equations and Variational Calculus, p. 432. Nauka, Moscow, SU (1974)
3. Erlichson, H.: Johann Bernoulli's brachistochrone solution using Fermat's principle of least time. Eur. J. Phys. **20**(5), 299–304 (1999). https://doi.org/10.1088/0143-0807/20/5/301
4. Ashby, N., et al.: Brachistochrone with Coulomb friction. Am. J. Phys. **43**(10), 902–906 (1975). https://doi.org/10.1119/1.9976
5. van der Heijden, A.M.A., Diepstraten, J.D.: On the brachistochrone with dry friction. Int. J. Non-Linear Mech. **10**(2), 97–112 (1975). https://doi.org/10.1016/0020-7462(75)90017-7
6. Lipp, S.: Brachistochrone with Coulomb friction. SIAM J. Control Optim. **35**(2), 562–584 (1997). https://doi.org/10.1137/S0363012995287957
7. Covic, V., Veskovic, M.: Brachistochrone on a surface with Coulomb friction. Int. J. Non-Linear Mechanics **43**(5), 437–450 (2008). https://doi.org/10.1016/j.ijnonlinmec.2008.02.004
8. Hayen, J.C.: Brachistochrone with Coulomb friction. Int. J. Non-Linear Mech. **40**(8), 1057–1075 (2005). https://doi.org/10.1016/j.ijnonlinmec.2005.02.004
9. Vratanar, B., Saje, M.: On analytical solution of the brachistochrone problem in a non-conservative field. Int. J. NonLinear Mech. **33**(3), 489–505 (1998). https://doi.org/10.1016/S0020-7462(97)00026-7

10. Yamani, H.A., Mulhem, A.A.: A cylindrical variation on the brachistochrone problem. Am. J. Phys. **56**(5), 467–469 (1988). https://doi.org/10.1119/1.15755

11. Palmieri, D.: The brachistochrone problem, a new twist to an old problem. Undergraduate Honors Thesis, Millersville University of PA (1996)

12. Aravind, P.K.: Simplified approach to brachistochrone problem. Am. J. Phys. **49**(9), 884–886 (1981). https://doi.org/10.1119/1.12389

13. Denman, H.H.: Remarks on brachistochrone-tautochrone problem. Am. J. Phys. **53**(3), 224–227 (1985). https://doi.org/10.1119/1.14125

14. Venezian, G.: Terrestrial brachistochrone. Am. J. Phys. **34**(8), 701 (1966). https://doi.org/10.1119/1.1973207

15. Parnovsky, A.S.: Some generalisations of the brachistochrone problem. Acta Physica. Pol. Suppl. A **93**, 5–55 (1998)

16. Tee, G.: "Isochrones and brachistochrones", Neural. Parallel Sci. Comput. **7**, 311–342 (1999)

17. Goldstein, H.F., Bender, C.M.: Relativistic brachistochrone. J. Math. Phys. **27**(2), 507–511 (1986)

18. Scarpello, G.M., Ritelli, D.: Relativistic brachistochrone under electric or gravitational uniform field. Z. Angew. Math. Mech. **86**(9), 736–743 (2006). https://doi.org/10.1002/zamm.200510279

19. Gemmer, J., et al.: Generalizations of the brachistochrone problem. Pi Mu Epsilon J. **13**(4), 207–218 (2011)

20. Mertens, S., Mingramm, S.: Brachistochrones with loose ends. Eur. J. Phys. **29**, 1191–1199 (2008). https://doi.org/10.1088/0143-0807/29/6/008

21. Rodgers, E.: Brachistochrone and tautochrone curves for rolling bodies. Am. J. Phys. **14**, 249–252 (1946). https://doi.org/10.1119/1.1990827

22. Obradovic, A., et al.: The brachistochronic motion of a vertical disk rolling on a horizontal plane without slip. Theor. Appl. Mech. **44**(2), 237–254 (2017). https://doi.org/10.2298/TAM1710020150

23. Gurram, S.S., et al.: On the brachistochrone of a fluid-filled cylinder. J. Fluid Mech. **865**, 775–789 (2019). https://doi.org/10.1017/jfm.2019.70

24. Legeza, V.P.: Cycloidal pendulum with a rolling cylinder. Mech. Solids **47**(4), 380–384 (2012). https://doi.org/10.3103/S0025654412040024

25. Legeza, V.P.: Brachistochrone for a rolling cylinder. Mech. Solids **45**(1), 27–33 (2010). https://doi.org/10.3103/s002565441001005x

26. Bryson, A.E. Jr., Ho, Y.-C. Applied Optimal Control: Optimization, Estimation and Control, pp. 496. Taylor&Francis Group, New York (1975). https://doi.org/10.1201/9781315137667

27. Pashentsev, S.V.: Zermelo navigation problem: an analytical solution for a vortex field. Bull. MSTU **3**(1), 17–22 (2000). (in Russ.)

28. Pashentsev, S.V.: Zermelo problem: analytical solution for a linear velocity field. Vestnik MSTU **5**(2), 187–192 (2002). (in Russ.)

29. Pashentsev, S.V.: Solution of the zermelo navigation problem for an arbitrary axial velocity field. Vestnik MSTU **13**(3), 587–591 (2010). (in Russ.)

30. Belhocine, A.: Exact analytical solution of boundary value problem in a form of an infinite hypergeometric series. Int. J. Math. Sci. Comput. (IJMSC) **3**(1), 28–37 (2017). https://doi.org/10.5815/ijmsc.2017.01.03

31. Falade, K.I.: A Numerical approach for solving high-order boundary value problems. Int. J. Math. Sci. Comput. (IJMSC) **5**(3), 1–16 (2019). https://doi.org/10.5815/ijmsc.2019.03.01

A Fuzzy Based Predictive Approach for Soil Classification of Agricultural Land for the Efficient Cultivation and Harvesting

Rajalaxmi Hegde and Sandeep Kumar Hegde[✉]

Nitte (Deemed to Be University), Department of Computer Science and Engineering, NMAM Institute of Technology, Nitte, Karkala Taluk, Udupi District 574110, India
laxmi123.prabhu@gmail.com, sandeep.hegdey@gmail.com

Abstract. Soil classification is a crucial component for geotechnical engineering purposes. When compared to conventional methods, smart and soft computing technology can quickly and accurately classify various soil types with high precision. Based on earlier studies, a database of soil characteristics is compiled and created, and it is used to train and test machine learning classifier algorithms like Naive Bayes and decision tree induction. Depending on the attitude place of origin, or inherent qualities soils can be grouped in a variety of ways. The goal is to forecast soil classification based on soil characteristics and information gathered about soil mechanics parameters. In the recommended paper, the fuzzy C-Means classifiers are used to classify the soil. Several data mining techniques such as data preprocessing, data analysis techniques are used to analyze and preprocess the data. Soil classification techniques influences several properties such as use and management of land. Soil texture is an important part of soil classification it influences fertility tillage, holding capacity of water, drainage etc. The algorithm's performance is contrasted with that of other conventional machine learning methods. The experiment's findings indicate that the recommended technique outperformed the alternatives in terms of accuracy as well as efficiency.

Keywords: Fuzzy C Means · Soil Morphology · Machine Learning · Agriculture · Decision Tree

1 Introduction

India is primarily a farming nation. India is the second-largest country in terms of total agricultural land with over 60% of its land area being used for agriculture. Farmers still use conventional methods like manual field [1] monitoring, which involves frequently watering crops and pesticides to control pests, whether or not they are aware of the appropriate dosage. These conventional methods have drawbacks like flooded fields, overuse of pesticides and fungicides, and ignorance of the crops that should be developed for a specific type of soil. Low quality yield, decreased soil fertility, and decreased expected yield from agricultural land are the impacts of these drawbacks.

Z. Hu et al. (Eds.): ICCSEEA 2023, LNDECT 181, pp. 709–720, 2023.
https://doi.org/10.1007/978-3-031-36118-0_64

Soils can be categorised into different groups based on their behaviour, place of origin, or inherent characteristics [1]. The classification approach may be impacted by differences in the importance of physical characteristics to various land uses and disease theories. Contrarily, a technical system approach to classifying soils divides soils into groups based on their suitability for particular purposes and microclimatic characteristics. Geologists changes in soil structure, texture, and properties may take tens of thousands of years to occur [2]. The soil data must be analyzed based on the pH for effective cultivation and harvesting. Applications of machine learning algorithms have been utilized to categorize soil by both commercial and academic institutions [3]. These methods have been applied for commercial, scientific, and industrial objectives. For instance, machine learning techniques have been applied to evaluate enormous data sets and identify insightful categorizations and patterns. Research investigations in agriculture and biology have used a variety of data analysis approaches, including decision trees, statistical machine learning, and other analysis methods [4]. This study explores the potential of machine learning techniques to enhance the analysis and pattern detection of large experimental soil profile datasets.

The goal of this effort is to comprehend potential processes underlying the aging effects in sands. This was achieved through a review of the available data from the literature and the creation of a lab testing program to categorize them according to the attributes [5]. The laboratory testing program was created so that the effects of several factors, including soil type, relative density, temperature, pH, and geographic locations, could be researched [6]. Finding a collection of fuzzy rules to address a particular classification problem from training data with uncertainty is the most crucial task in the design of fuzzy classification systems [7].

One of the key tasks in data mining investigations and statistical data analysis, which is widely applied in many fields, is clustering. Datasets are grouped into groups called clusters using the relationships between them as the basis for clustering. The relationship here between data in datasets and the effectiveness of the cluster formation determine the different clustering algorithms. The most common types of relationships between data are those defined by their distance from one another, their connectivity, their density in the data space, their interval or distribution, and other factors. Clustering is viewed as a multi-purpose optimization problem from the perspective of how to accomplish efficient cluster formation.

In the proposed approach the soil data were classified using the fuzzy C-Means technique through an unsupervised learning approach. The complete training data set is transformed into fuzzy rules, and the relationships between the input variables are used to calculate the weights for each input variable that appears in the resulting fuzzy rules. The test dataset is fed into the proposed model, and the accuracy obtained with the test dataset is examined. In the proposed paper, C language was utilized to implement the fuzzy rules by first establishing the membership functions for the input attributes of the soil data, and then generating the initial fuzzy rules for the training data. The proposed approach's accuracy is compared to that of conventional machine learning methods like SVM, random forest, naive Bayes classifier, and decision tree techniques. The results of the experiments show that the proposed approach achieved better accuracy when compared to conventional machine learning techniques.

2 Literature Review

The following section summarizes the various research studies that have been proposed in the field of soil classification using various machine-learning approaches. In [8] Support vector machines were proposed for recognizing, mapping, and categorizing different types of soil. Hyper spectral data with a high spectral resolution is used in this method. When working with a tiny sample size of data, a support vector machine classifier produces accurate results. It has been found that high-dimensional datasets with few training samples benefit from the support vector machine technique. The investigated soil datasets were categorized using a technique that is proposed in [9]. By using the Naive Bayes and k-Nearest Neighbor algorithms, soils are divided into three categories: low, medium, and high. The Soil Testing Laboratory in Jabalpur, Madhya Pradesh, is where the data is gathered. The tuples in the dataset describe the number of nutrients and micronutrients present in the soil. By dividing these nutrients and micronutrients into two groups, it is possible to determine the soil's capacity for yield. The medium group of soils shows good yielding capabilities, according to the JNKW Jabalpur department of soil science. The category high (H) and very high soils exhibit modern yielding capability, while the category low (L) and very low soils do not. A categorization approach for soil texture based on hyper spectral data was presented in [10]. A random forest classifier's performance is compared to that of the CNN techniques. The soil dataset from the Land/Use and Cover Area Frame Statistical Survey (LUCAS) is used in this study. Hyper spectral and soil texture data are included. One drawback of the proposed technique is that atmospheric error, which is typically present in hyper spectral data, must be corrected. When compared to traditional statistical methods, in [11] the proposed methodology performs better which uses a limited amount of attributes from the dataset to analyze. The agricultural soil profiles were chosen to be comprehensive and to make soil classification easier. In [12] author proposed a convolutional neural network-based empirical model to investigate the daily soil moisture retrieval for passive microwave remote sensing. The soil moisture retrieval model based on deep learning can learn detailed features from a vast amount of remote sensing data. In [13] author developed a support vector machine-based model for the classification of soil samples with the help of numerous scientific factors. Several algorithms, characteristics, and filters are used to collect and process color photos of the soil samples. These algorithms extract numerous characteristics, including color, texture, and others. Support Vector Machine only uses a small subset of the training samples that are situated at the boundaries of the class distribution in feature space to fit an ideal hyper plane between the classes. The supervised categorization's accuracy is influenced by the training data that was used. In [14] authors proposed an improved model with more features than the current method, such as crop lists, urea levels, and soil nutrients. To categorize soils where color, energy, and HSV may be visible, image processing techniques can be employed. In [15] authors applied J48, JRip algorithm, decision tree, and Naive Bayes classification techniques that are used on the soil data sets for the soil classification. The Naive Bayes algorithms obtained higher accuracy compared to other machine learning approaches when imposed on soil datasets. This paper introduces the fuzzy Sequence - to - sequence neural network clustering technique along with a mediated activation function to address combinatorial optimization problems. The fuzzy Hopfield neural network therefore uses a particular numerical process. It can reduce an energy function to locate the membership grade. [16].The process of clustering entails dividing or classifying a collection of objects into distinct subsets or clusters. The assumption that there are particular predefined c clusters in the data sets is predicated on the clustering algorithm clustering. It is possible to locate the ideal group of clusters by minimizing an objective function.

3 Methodology

The architecture of the proposed approach is shown in Fig. 1 below. As shown in the figure, initially soil data is given as input to the proposed model. The data are passed as input and undergo pre-processing stage where noise and irrelevant features are filtered from the given dataset.

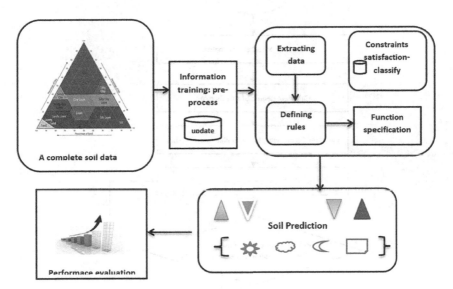

Fig. 1. Proposed Architecture Diagram

3.1 Data Pre-processing

To increase performance accuracy when classifying or forecasting time series data, data pre-processing is a crucial step that must be completed first. In order to transform large datasets into more complete, organized, and noise-free data, incomplete words and noise must be reduced or eliminated [18, 19]. This pre-processing stage aims to get the findings of extracting the features that is not widely used and minimize errors induced by bias from every dataset [19].This study's pre-processing stage included a number of procedures for a wide range of data types, beginning with the steps for removing noise, cleaning incomplete data, and finishing incomplete data. The default value, 0, may be utilized in a mathematical operation if a value data is absent or incomplete [19].

Crop harvest and rainfall data were among the different types of data included in the dataset used in this study. Data on plant commodities obtained from the primary sources, specifically from the Selman Agriculture Service, are presented as a series of graphs for the last five years, starting with the volume of production from different plant varieties, pests that attack, and pest management data from 2014 to 2018. Data on rainfall were also gathered concurrently from secondary sources, specifically from the BPS (Statistics Indonesia) state website in Selman from 2005 to 2018.The data is still obtained in raw and unstructured form, so preprocessing is required to clean and normalize the data before further processing can be done with it.

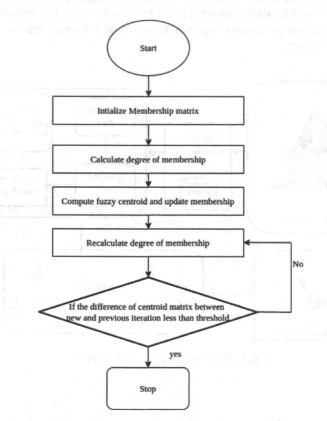

Fig. 2. Flowchart of Fuzzy C means Algorithm

The process of forecasting rain is carried out using rainfall data from previous years after that the data is deemed normal. The classification of horticultural crops that are suitable for plants in particular months is based on the outcomes of rainfall forecasting and a number of parameters or dependent variables. The expectation is that the words or text produced by the normalization process can retrieve into features that impact their respective classes because normalization is one of the data transformation texts that works to process raw and disjointed data from the data mining process [17]. In this study, the normalization of the data is accomplished by grouping the data according to the pertinent attributes and will be used as dependent variables. The next step is to remove the unnecessary information while keeping the variables that influence how plant varieties are determined untouched. The normalization of statistics in a database schema differs from the normalization of data in general. Once the dataset is pre-processed by removing noise and outliers, a fuzzy rule set is defined on the extracted dataset as per functional specification from which the classification of soil will be derived. In the end, the performance of the predictive model will be evaluated.

The dataset required to experiment was gathered from the Kaggle website. The dataset is normalized using a min max normalizer and outliers are removed using the interquartile range method. The insignificant features are removed from the dataset using the recursive feature elimination technique. Figure 2 above depicts the flowchart of the proposed Fuzzy C means algorithm. Initially, a membership matrix is generated for the given dataset. From the generated membership matrix degree of the membership is generated using Eq. (1)

$$\mu_{jk} = \frac{1}{\sum_{l=1}^{d}\left(\frac{e_{jk}}{e_{kl}}\right)^{\left(\frac{2}{n-1}\right)}} \tag{1}$$

where μ_{jk} indicates the membership of the jth and kth cluster centroid, 'd' represents the centers of the cluster, e_{jk} indicates the Euclidean distance between the jth and kth cluster centroid, and 'n' is the index of fuzziness. For each of the memberships, a fuzzy centroid is derived and membership has been upgraded using Eq. (2) below.

$$u_k = \frac{\left(\sum_{j=1}^{u}(\mu_{jk})^n S_j\right)}{\left(\sum_{j=1}^{u}(\mu_{jk})^n\right), \forall_k} = 1, 2..d \tag{2}$$

u_k represents the kth fuzzy center, X_j represents the data points and 'k' is the objective function. The fuzzy centroid computations will be performed until the difference in the centroid matrix between the new and previous iterations will be less than the threshold.

The pseudocode of the proposed fuzzy c means algorithm is as below.

Let S = {s1, s2, s3 ..., sn} are the data points from the considered dataset and U = {u1, u2, u3 ..., uc} are the cluster centre.

1) Randomly choose the centers of the cluster 'c'.
2) Compute the fuzzy membership 'μ_{jk}' as

$$\mu_{jk} = \frac{1}{\sum_{l=1}^{d}\left(\frac{e_{jk}}{e_{kl}}\right)^{\left(\frac{2}{n-1}\right)}}$$

3) Determine the fuzzy centroid 'u_k' as

$$u_k = \frac{\left(\sum_{j=1}^{u}(\mu_{jk})^n S_j\right)}{\left(\sum_{j=1}^{u}(\mu_{jk})^n\right), \forall_k} = 1, 2..d$$

4) Repeat stages 2 and 3 till the minimum value for 'k' is achieved.

One of the advantages of the proposed algorithm is that it obtains better results with the overlapped dataset. The fuzzy c means algorithms are applied to the considered dataset and the result obtained with the test dataset is analyzed. The section below illustrates a detailed discussion of the result obtained with the proposed approach.

4 Results and Discussions

The enormous amounts of information that are currently practically harvested alongside the crops need to be analyzed and used to the fullest extent possible. The analysis of a soil input data using data mining methods is the goal of this study. It focuses on classifying soil using the various available algorithms. Regression analysis is used to predict untested attributes, and automated sample soil classification is put into practice. A soil classification process groups soils with comparable characteristics or attributes together. The data from the soil series can be classified using machine learning techniques. To forecast the crop production that are best suited for a specific region's soil series and climatic conditions, the results of this classification are merged with crop datasets. Finding a collection of fuzzy rules to address a particular classification problem from training data with uncertainty is the most crucial task in the design of fuzzy classification systems. The Values can be predictably determined by grouping soil forms. After going through the classification procedure, the database for future references has successfully been updated with soil data. The figure below indicates the result obtained with the proposed approach when it is applied to the soil dataset.

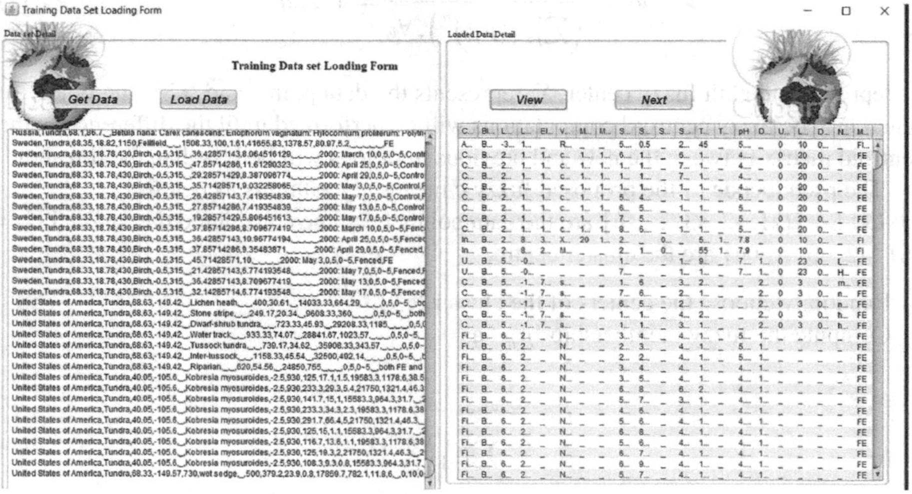

Fig. 3. Snapshot of Training Dataset Loading form

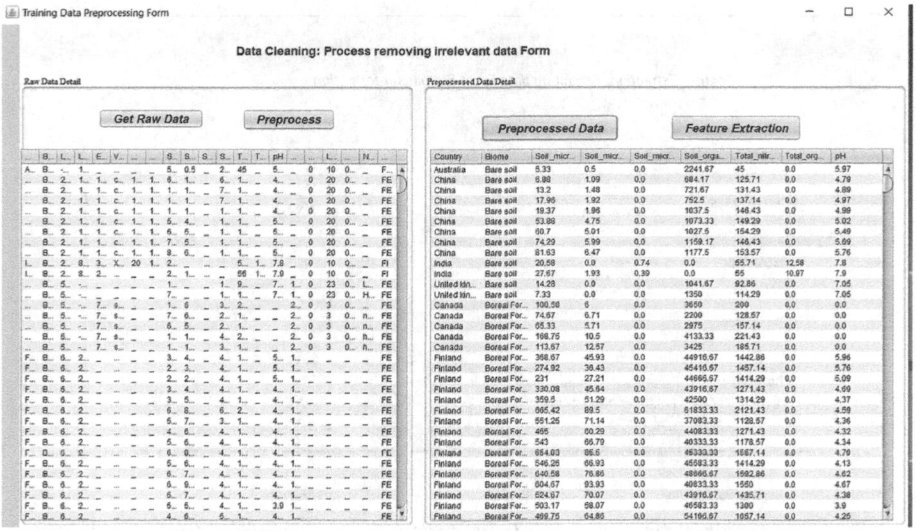

Fig. 4. Snapshot of Data Pre-processing Stage

As shown in Figs. 3 and 4, initially the collected soil dataset is loaded for the further visualization process. The dataset in its original form may contain noise and outliers which has to be pre-processed to obtain optimal accuracy.

As shown in Fig. 5 feature extraction process is carried out to extract the data from the Biomass information. The dataset is classified based on different parameters such as cultivated area, Forest area, and Grass area. The implemented interface also has the option to display different information such as the number of countries, number of lands, and number of forests from which the soil dataset is taken from.

As shown in Fig. 6, the forest area is classified using different parameters such as Temperate Coniferous Forest, Temperate Broadleaf Forest, and Tropical/Subtropical forest by applying the proposed algorithm.

As shown in Fig. 7 above Cultivated lands are classified from the biome by applying Fuzzy rules. The grasslands are classified from the biome using different parameters such as tundra, glacier, wetland, Shrub, and Pasture.

Figure 8 above illustrates the validation rate achieved using various machine learning algorithms. From the graph, it can be evident that the proposed fuzzy c means algorithms obtained higher accuracy and precision rate compared to the traditional machine learning approaches.

Feature Extraction: Extracting data From Biomass information

	Biome	
MICRO BIOMASS	Bare soil	☑ No of Countries 47
	Boreal Forest	
	Cropland	
	Desert	
Extraction	Glacier	☑ No of Forests 14
	Grassland	
	Natural Wetland	
	Not mentioned	
Filtering lands	Pasture	
	Savanna	☑ No of Lands 15
	Shrub	
Fuzzy Logic	Temperate Broadleaf Forest	
	Temperate Coniferous Forest	
	Tropical/Subtropical Forest	
	Tundra	

Classification 1: Click Here   Classification 2   Classification 3

◎ Cultivated area:Classification   ◎ Forest Area: classification   ◎ Grass Area: Classification

Fig. 5. Snapshot of Feature Extraction Process

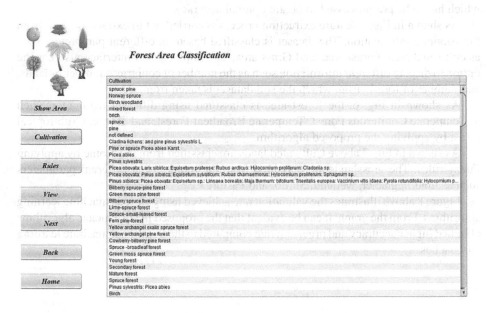

Forest Area Classification

	Cultivation
	spruce: pine
	Norway spruce
Show Area	Birch woodland
	mixed forest
	brich
	spruce
Cultivation	pine
	not defined
	Cladina lichens: and pine pinus sylvestris L.
Rules	Pine or spruce Picea abies Karst.
	Picea abies
	Pinus sylvestris
	Picea obovata: Larix sibirica: Equisetum pratense: Rubus arcticus: Hylocomium proliferum: Cladonia sp.
View	Picea obovata: Pinus sibirica: Equisetum sylvaticum: Rubus chamaemorus: Hylocomium proliferum: Sphagnum sp.
	Pinus sibirica: Picea obovata: Equisetum sp.: Linnaea borealis: Maja themum: bifolium: Trientalis europea: Vaccinium vitis idaea: Pyrola rotundifolia: Hylocomium p...
	Bilberry spruce-pine forest
	Green moss pine forest
Next	Bilberry spruce forest
	Lime-spruce forest
	Spruce-small-leaved forest
	Fern pine-forest
Back	Yellow archangel exalis spruce forest
	Yellow archangel pine forest
	Cowberry-bilberry pine forest
	Spruce -broadleaf forest
	Green moss spruce forest
Home	Young forest
	Secondary forest
	Mature forest
	Spruce forest
	Pinus sylvestris: Picea abies
	Birch

Fig. 6. Snapshot of Forest Area Classification

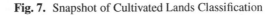

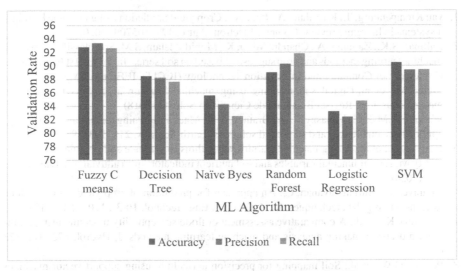

Fig. 7. Snapshot of Cultivated Lands Classification

Fig. 8. Accuracy analysis of machine learning algorithms

5   Conclusion

In the proposed paper new method to generate fuzzy rules from training data to deal with the soil classification problem was presented. The entire training data set is converted into fuzzy rules and then the weight of each input variable appearing in the generated fuzzy rules is derived by determining the relationships between input variables. Accordingly, the test dataset is passed to the proposed model, and the accuracy obtained with the test dataset is analyzed. The collected dataset contains four parameters including soil mechanics properties such as cohesion and friction angle, and the physical characteristics such as water and dry density. The training data contain 70 data (70%) and the testing data includes the remaining values of 34 data (30%). The 30% remaining data were also entered into the modeling section without being used in the design section. Also, both sections contain 7 types of soil to test the flexibility of the models. The results showed that only 6 samples were not correctly identified among the total testing data (34 samples) in Naïve Bayes model and 7 samples were not correctly identified among the total testing data (34 samples) in other ML model. Thus, these models provide the most reliable results and can help the automatic classification of different types of soil. The performance of the proposed approach is compared with the traditional machine learning algorithms such as Accuracy, Precision, and Recall. The experimental results illustrate that the proposed weighted fuzzy rules generation approach obtained a higher classification accuracy compared to the traditional technique. Hence it can be concluded that fuzzy approaches are more suitable for soil classification through which the efficiency of agricultural cultivation and harvesting can be improved.

References

1. van Klompenburg, T., Kassahun, A., Catal, C.: Crop yield prediction using machine learning: a systematic literature review. Comput. Electron. Agric. **177**, 105709 (2020)
2. Rahman, S.K., Zaminur, A., Chandra Mitra, K., Mohidul Islam, S.M: Soil classification using machine learning methods and crop suggestion based on soil series. In: 2018 21st International Conference of Computer and Information Technology (ICCIT), IEEE (2018)
3. Meier, M., et al.:Digital soil mapping using machine learning algorithms in a tropical mountainous area. Revista Brasileira de Ciência do Solo **42** (2018)
4. Wadoux, A.M.J.-C., Minasny, B., McBratney, A.B.: Machine learning for digital soil mapping: applications, challenges and suggested solutions. Earth-Sci. Rev. **210**, 103359 (2020)
5. Kingsley, J., et al.: Using machine learning algorithms to estimate soil organic carbon variability with environmental variables and soil nutrient indicators in an alluvial soil. Land **9**(12), 487 (2020)
6. Palanivel, K., Surianarayanan, C.: An approach for prediction of crop yield using machine learning and big data techniques. Int. J. Comput. Eng. Technol. **10**(3), 110–118 (2019)
7. Khosravi, K., et al.: A comparative assessment of flood susceptibility modeling using multi-criteria decision-making analysis and machine learning methods. J. Hydrol. **573**, 311–323 (2019)
8. Pereira, G.W., et al.: Soil mapping for precision agriculture using support vector machines combined with inverse distance weighting. Precision Agric. **23**(4), 1189–1204 (2022)
9. Taher, K.I., Abdulazeez, A.M., Zebari, D.A.: Data mining classification algorithms for analyzing soil data. Asian J. Res. Comput. Sci. **8**, 17–28 (2021)
10. Pandith, V., et al.: Performance evaluation of machine learning techniques for mustard crop yield prediction from soil analysis. J. Sci. Res. **64**(02), 394–398 (2020)
11. Zhigang, X., et al.: Multisource earth observation data for land-cover classification using random forest. IEEE Geosci. Remote Sens. Lett. **15**(5), 789–793 (2018)
12. Xuebin, X., et al.: Applying convolutional neural networks (CNN) for end-to-end soil analysis based on laser-induced breakdown spectroscopy (LIBS) with less spectral preprocessing. Comput. Electron. Agric. **199**, 107171 (2022)
13. Zhu, Q., et al.: Drought prediction using in situ and remote sensing products with SVM over the Xiang River Basin, China. Nat. Hazards **105**(2), 2161–2185 (2020)

14. Abraham, S., Huynh, C., Huy, V.: Classification of soils into hydrologic groups using machine learning. Data **5**(1), 2 (2019)
15. Wankhede, Disha S.: Analysis and prediction of soil nutrients pH, N, P, K for crop using machine learning classifier: a review. In: International Conference on Mobile Computing and Sustainable Informatics, Springer, Cham (2020)
16. Alzaeemi, S.A.S., Sathasivam, S., Velavan, M.: Agent-based Modeling in doing logic programming in fuzzy hopfield neural network. Int. J. Mod. Educ. Comput. Sci. **13**(2), 23–32 (2021)
17. Putra, A.B.W., Gaffar, A.F.O., Suprapty, B.: A performance of the scattered averaging technique based on the dataset for the cluster center initialization. Int. J. Mod. Educ. Comput. Sci. **13**(2), 40–50 (2021)
18. Kajol, R., Akshay Kashyap, K.: Automated agricultural fieldanalysis and monitoring system using IOT. Int. J. Inform. Eng. Electron. Bus. **12**(2), 17 (2018)
19. Yudianto, M.R.A., et al.: Rainfall forecasting to recommend crops varieties using moving average and naive bayes methods. Int. J. Mod. Educ. Comput. Sci. **13**(3), 23–33 (2021)

Sector-Independent Integrated System Architecture for Profiling Hazardous Industrial Wastes

Wilson Nwankwo[1(✉)], Charles O. Adetunji[3], Kingsley E. Ukhurebor[4], Acheme I. David[2], Samuel Makinde[2], Chukwuemeka P. Nwankwo[2], and Chinecherem Umezuruike[5]

[1] Department of Cyber Security, Delta State University of Science and Technology, Ozoro, Nigeria
wnwankwo@dsust.edu.ng
[2] Department of Computer Science, Edo State University, Uzairue, Nigeria
[3] Department of Microbiology, Edo State University, Uzairue, Nigeria
[4] Department of Physics, Edo State University, Uzairue, Nigeria
[5] College of Computing, Bowen University, Iwo, Nigeria

Abstract. The accummulation of waste in urban societies has been christened the "industrialization and urbanization aftermath" in some quarters. In recent times, the climate change phenomenon is attributed to the generation and release of massive wastes from humanity's economic activities globally. The move to restore the ecological integrity of the earth's habitat is paramount and many programmes and policies are currently being designed to solve this global challenge. This paper reviews the current industrial waste management approaches and practices including legislation, machineries, and policies in Lagos Nigeria, a major urban area and commercial centre in sub-Saharan Africa. The study adopts a narrative-expository posture with enquiries made in respect of all operations and machineries of waste management in the urban and surrounding areas. Findings show that while machineries and policies are substantially crafted to meet the changing socio-economic developments, there is a gap in the deployment of modern information technologies in fostering the policy provisions in such a manner that complements the efforts of the relevant government machineries while ensuring sustainable development. The current waste management practices do not reflect what is considered the minimum global best practices. Accordingly, this study proposes a technology-driven architecture that employs an intelligent integrated resource planning system to support collective and collaborative management of industrial wastes particularly in the area of risk management, profiling, and resource allocation. The proposed architecture is significant in that it provides a realistic coordinating single window to harness all relevant existing practices through the introduction of a computerized platform that includes risk and hazard profiling of various industry subsectors, industry operators, location of operation, citizen reporting, environmental impact, feedback, etc. Thus, this architecture promotes collective and citizen-driven waste management ecosystem which in turn enhance the operations of all environmental management stakeholders through data provisioning, messaging, real-time data analytics, and forecasting.

Z. Hu et al. (Eds.): ICCSEEA 2023, LNDECT 181, pp. 721–747, 2023.
https://doi.org/10.1007/978-3-031-36118-0_65

Keywords: Hazard · Industrial Wastes · Computerized Risk Management · Sustainable development · Targeting

1 Introduction

Waste management (WM) poses challenges to developing and developed countries. It has been observed that the daily increase in human population currently experienced globally has led to increase in waste generation. This could also be linked to high level technological advancement most especially in the industrial sector, consumption, lifestyle choices, which has reinforced the necessity to address this environmental concern [1]. WM is an essential part of industrialization programs directed at minimizing the impact of large quantities of wastes generated from the industrial activities. WM policy is imperative in every clime as it is designed to ensure sustainable economic, environment, social relations that promote sustainable development (SD). The following are the major stages involves in the process of waste management: generation, accumulation, collection, movement, processing and the eventual disposal of these wastes. All these stages involve numerous techniques that could be utilized at each stage in order to ensure adequate management of wastes [2].

Experience has shown that one of the most crucial strategies that could be applied for effective waste management is to ensure their eventual minimization thorough an effective developmental design and plan. In addition, it has been shown that if minimal waste were generated at the initial stage, definitely there would be minimal pressure on other later stages involved in the process of managing wastes [3, 4]. The later stages may factor in some evaluation techniques especially on the efficacy and efficiency of disposal plants in the WM process. Such evaluation techniques could reveal the points where due diligence or care was not adhered to by some persons, organizations or agencies, for instance, where unselective and poor waste disposal had led to numerous environmental maneuverings and complications.

In many societies, it has become mandatory to carefully chose and adopt sustainable disposal approaches that are ecofriendly for effective WM. Such choices might be linked to some merits contemplated and some degree of certainty to which such measures could be adjudged to yield useful resources in addition to their primary intendment, for instance, where a chosen WM approach could aid translate wastes to necessary raw materials for other industries [5, 6].

Various researches had stated that waste disposal management includes the following: pyrolysis, burning, recycling, incineration, shredding, composting, reuse, open dumping, and landfilling [7–11]. Recycling entails the careful utilization of materials that had attained their end-of-life state for the creation, production, manufacturing, and/or rebranding of new products. The products so produced could afford users further uninterrupted uses [12]. The process of recycling involves the reduction of wastes materials most especially the finite resources while the significant resources are preserved [13–15].

Some forms of recycling may involve some other waste disposal methods: incineration, composting, landfilling, open dumping, burning, reuse, shredding, and pyrolysis landfilling [7–11]. However, recycling has its negatives, which include decline of product's value and effectivenessp [16] and the tendency to recover inappropriate energy [17,

18]. Incineration involves combustion at very high temperatures [19], on the site. The merits of incineration include decrease of waste, generation of electricity and heat energy [8], while the demerits entail generation of acid rain and fabrication of ash. Composting entails the application of beneficial microorganisms in the decomposition of organic waste [8]. This might lead to the incidence of global warming through generation of carbon dioxide, plastic and glass like residue [20]. Landfill reflects an allocated space on the earth's surface where waste materials are buried. Landfills are usually made specifically for a certain period and there are two types which include natural and sanitary attenuation [10]. Some merits of landfills include: opportunities for scavengers to perform their role, cost effectiveness, and release of gases from which reasonable energy could be generated; while some of the demerits include high level of air and ground water pollution as well as air borne diseases and fire hazards [21, 22]. Burning is recognized to have similar features to open dumpsites where wastes are set on fire. The risks related to this include air pollution, explosion and fire. Burning produces huge amounts of greenhouse gases that readily predisposes the ozone layer to severe depletion. Reuse involves any operation through which components or products which are not wastes (in the true sense of the term but are embedded in wastes) are utilized again for the purpose they are designed for. Reuse is a disposal option that guarantees that supposed wastes are reused in an effective manner through indirect or direct application [10]. Some of the benefits of reuse include ecofriendiness, reduction of burden of disposal, and utility. Pyrolysis is adjudged an alternative to incineration. Pyrolysis guarantees the thermal disintegration of materials when subjected to very high temperatures in the presence of oxygen [11]. Pyrolysis has the potential to minimize the volume of waste, destroy toxic constituents, and produce gases that could be utilized as fuel. Nevertheless, pyrolysis generate ashes that could pollute the air, leading to pollution. One of the greatest disadvantages of pyrolysis is that it is not economical [11]. It follows that apart from recycling and reuse some other techniques are not ecofriendly, hence, are unsustainable. Unfortunately, not all categories of wastes could be recycled or reused. In recent times, there is a high demand for new efficient and integrative approaches to WM in developing nations including Nigeria where the production of material wastes is high owing to deficiencies in the existing waste disposal techniques. Research has affirmed that the high volumes of wastes generated in Nigeria is linked to absence of modern inexpensive disposal techniques which has placed a high demand on government to enforce organizational regulatory policy [23–25]. Similarly, it is submitted that the management of waste in Nigeria is generally poorly handled [26]. In this study, we focus on Lagos as a case study. Lagos is one of the largest cities in Africa, and it is widely believed that what works in Lagos would work in other parts of Africa especially in large African cities. In Lagos, the Lagos State Waste Management Authority (LAWMA) is the public agency overseeing all WM operations. According to LAWMA, it operates five landfill sites with two as temporary and three major sites in the state where different types of waste are disposed [27]. Similar postures are maintained by other waste management authorities across the 36 States in Nigeria.

It is observed that the rate of recycling of waste in Lagos State is very low [28], while huge volumes of waste are generated from households, factories, and markets, an insignificant volume particularly papers and plastics are subjected to the process of recycling [29]. [28] notes that burying and burning are the major practices of WM in

Lagos State. Furthermore, [30] states that the reuse of some waste material is still feasible but [31] observed that these processes are under-utilized. The use of incineration for the management of waste is not commonly carried out in Lagos and in most cities in Nigeria [32].

It could be said that there is no functional machinery that targets the management of industrial wastes in the State. Wastes from industrial processes are often in three forms: solid, liquid water waste or effluents, gaseous. It is observed that most of these gaseous emissions leaches; and it is common knowledge that some liquid effluents pose health risks to the workers within and around the industries [33]. Majority of such industrial wastes (IW) contain hazardous wastes that pose threat and a source of environmental and health hazards when not properly managed. Prior to further discourse on hazardous wastes it would be ideal to recourse to necessary legislative provisions in determining what constitutes hazardous or harmful wastes. Accordingly, Section 15 of Nigeria's Harmful Wastes (Special Criminal Provisions etc.) Act, rightly provides that "harmful wastes means any injurious poisonous, toxic or noxious substance and, in particular, nuclear wastes emitting any radioactive substance ... as to subject a person to the risk of death, fatal injury or incurable impairment of physical or mental health".

The effect of this provision is that any waste whether or not industrial that could cause some harm is a harmful waste.

Hazardous wastes could pollute the air, water, and soil. The contamination of these three environmental layers is akin to humankind intentionally creating a hazardous ecosystem antithetical to sustainable development and growth. Hazardous waste materials could route toxic chemicals through the soil, air, water thereby providing the platform for pollution, which predisposes man to various health anomalies such as infections, respiratory stress, visual impairment, allergic reactions, poisoning, blood-related problems, etc. Workers are exposed to numerous risks of varying magnitude depending on the jobs and work environments. A typical scenario is an asbestos plant where workers may have regular exposures to air pollutants such as benzene and toluene [34]. The result of the above scenario is very consistent with the position of [35] wherein he observed that approximately 2 million working hours is wasted during work – associated illness while about two million people testified that they are confronted with illness that could be associated with their workplace.

In view of the aforestated gaps and challenges, this review proposes a technology-driven architecture that would employ a resource planning system to support collective and collaborative management of risks associated with industry wastes. The proposed architecture extends the existing practices through the introduction of computerized approaches that includes risk and hazard profiling of various industry subsectors, industry operators, location of operation, environmental impact etc.

2 Categorization of Industrial Wastes by Industry/Economic Sector

Categorically, the main sources of poisonous waste are grouped into two: human and natural sources. Regardless of the category to which industrial wastes belongs, there is a common concern which is 'harm' to humans and the environment [36–38]. We shall discuss some of the categories briefly.

2.1 Human Sources

Mostly, poisonous wastes arise from industrial plants, firms, companies, refineries or industries. The quantity of wastes produced by such entities may be alarming owing to the cumulative degenerative effects they cause on the environment. The poisonous wastes produced by such organizations are often in solid or fluid (liquid or gases). Some culpable organizations [36–38] include:

a. Chemical manufacturing industries and plants that produce chemical wastes such as spent solvents, strong acids and bases as well as other chemical reactive wastes, etc.
b. Cosmetic/cleaning agents manufacturing industries that produce strong acids and bases, heavy metal dust, combustible waste and solvents, etc.
c. Printing industries that produce heavy metal, wastes ink and spent electroplating, solvents, etc.
d. Furniture and wood manufacturing and refinishing industries, which generate spent solvents, combustible wastes, etc.
e. Metal manufacturing industries that generate heavy metals, cyanide, strong chemical (acids and bases), etc.
f. Leather products manufacturing industries that release wastes such as toluene and benzene, etc.
g. Paper manufacturing industries, which release heavy metal wastes, combustible solvents, strong chemicals (acid and bases), etc.
h. Vehicle manufacturing and maintenance companies and plants that release huge amounts of heavy metal and flammable wastes, as well as spent solvents and used lead acid cells, etc.
i. Petroleum exploration industries also generate enormous toxic organic and biochemical wastes.

Presently, Nigeria does not have any nuclear power plant, firm, company, refinery or industry but the country is greatly exposed to high technology wastes (poisonous effluents, apprehensions and emissions) imported from the industrialized societies, some of which exhibit radioactive tendencies.

2.2 Natural Sources

The major apparent natural sources of toxic wastes are volcanic eruptions (which produce several tons of poisonous gases and undesirable emissions as well as harmful lava), and some agricultural and chemical raw materials containing phytoxins, poisonous dust, lead, etc. which may be released during processing [36–38]. It is sad to note that more than 80% of the high technology wastes in the world are dumped directly or indirectly in landfills, rivers, seas or oceans in Africa and Asia [36–40] which when discovered many years after may appear as if such wastes are the result of nature's activity. Nigeria is noted as one of the emerging dumping grounds for these toxic, organic, biochemical and electronic waste from the industrialized world [37, 39–43]. In addition to the aforementioned sources, oil spills had become a serious concern for decades now in the Niger delta area of Nigeria.

2.3 Industrial Health Hazards (IHH)

Industrial infections and poisoning are a potential and possible risk to the health of persons that are unprotected in an insalubrious environment [36, 37, 44–46].

According to Occupational Health and Safety (OHS) provisions, occupational hazards and/or infection (industrial health hazard) occur following exposures to toxic and/or infectious biological, physical or chemical substances in the workplace, which could negatively affect the person's normal physiochemical mechanisms leading to some impairments to their general wellbeing [44–47].

IHH could result to accidents or infections after such exposures in the workplace. Exposure could be to the immediate locality or to substances from a remote site. Some health challenges manifest after extensive exposure to working tools, equipment, apparatus and other devices, as well as infection-causing microorganisms, chemicals, waste products or even dust over a considerable period due to insanitary and other unhealthy situations at our various places of work [46–48], whereas some health problems may be immediate e.g. allergic reactions following inhalation of poisonous lead substances from a mining site.

Some known diseases and/or anomalies arising from IHH include musculoskeletal difficulties, tuberculosis, blood borne infections, latex allergy, fierceness and some other work-associated trauma and stress [46–48]. IHH are not given appropriate attention and are mostly pronounced as public health problems amongst healthcare practitioners in most developing economies [38, 46, 47]. Complications amongst healthcare practitioners in health facilities is also within the contemplation of IHH and is one of the common issues that occur following exposure to substances and fluids from the body of patients, accidents and inadequate personal and other protective working devices.

2.4 Formal and Policy-Based IWM Practices

In discussing formal policy-based practices reference must be made to industrialization and civilization in Nigeria. These two constitute the cornerstone on which the rise of several industries involved in production of goods and services to meet the demands of the growing national population, surely rests. Research has documented the continual release of poisonous effluents, and emissions to the surrounding environment by these industries [36, 38–40] regardless of the subsisting legislations and policies on industrial WM. It is interesting to submit that a conglomerate of socioeconomic concerns that tend to mar the effective waste management practices is contributing to the aforementioned pitiable state. These concerns are:

a. Uncontrolled consistent spread of disorderly and disorganized settlements
b. Traffic congestion in major cities
c. Insecurity and terrorism
d. Ignorance and illiteracy

Given the astronomical increase in the country's population, industrialization, and ongoing developmental projects, it would be interesting to submit that contemporary and operative waste management practices need be deployed to foster socioeconomic wellbeing of the general public [36, 38, 45]. Such practices and strategies should encompass all stakeholders, including the industries (that produce the wastes), the private sector, the public sector as well as the government regulatory agencies.

That most societies are regularly confronted with some environmental problems resulting from industrial wastes as well as other human and natural phenomena is well-documented [36–40]. It is trite that majority of the risks associated with these environmental dilapidations do flow from human activities and progression [50–52]. Thus, it may be safely held that the effective design and implementation of effective but efficient strategies is what distinguished a progressive society from a retrogressive sociopolitical system.

While customary waste management practices do offer some reliefs there is no substitution to formal policies, legislations, procedures, guidelines and machineries directed at WM. Accordingly, it may be said that Nigeria has rich environmental laws, policies and regulations which provide valuable frameworks for the protection of the environment from dilapidations [39, 40, 53]. However, it is apt to state that these environmental laws and regulations have not yielded their intended results as there is little or no adherence to the provisions of these laws, policies and regulations [36–40, 49]. Notwithstanding the aforementioned shortfalls, it is instructive to submit that environmental and waste management practices in Nigeria are driven by legislations and policies.

IWM practices in Nigeria are tightly connected to the environmental management (EM) laws, policies and agencies. Accordingly, our discussion would be dealing on both WM and EM and in some cases; the terms would be used interchangeably. Industrial wastes are the major cause of disorderliness in the environment's ecosystem hence EM presupposes all practices directed at maintaining the integrity of the air, water, and land. EM therefore encompasses the management of emission to air and water, environmental impact assessments, waste and pollutions, land and environmental issues connected to wastes, etc. It also includes the management and mitigation of cross-border dealings [39, 40].

IWM practices in Nigeria in a broad sense, is chiefly based on the National Policy on Environment (NPE) and several legislative provisions (which are briefly discussed in Sect. 2.6 of this paper). The NPE came into effect in 1989 and has undergone reviews in 1999, 2007 and 2009 respectively. The NPE and other legal frameworks grant legitimacy to the various machineries of government including agencies, ministries and departments, and their operational guidelines and procedures all of which are made to guarantee the conservation and management of natural resources, safety of the environment and the wellbeing of the citizenry and other members of the ecosystem [37, 39, 40]. These machineries are discussed further in Sect. 2.5.

It is vital to note that the operations of these machineries of Government are influenced by political, cultural and social mantra of Nigeria, which include political instability, policy somersaults, nepotism, and religio-political affiliations, etc. all of which are quite bewildering so to say. Such immense influence does pose some threats to the efficiency of these agencies. Consequently, compliance to guidelines, procedures and environmental laws are greatly marred [37, 39, 40].

2.5 Environmental Management Agencies

Environmental agencies or environmental protection agencies are commissions, establishments or organizations which are established within the government or separate from the government agencies whose function is to protect the environment as well to manage and preserve the nation's ecological, environmental, and biological resources [36, 38–40]. There are relevant laws to prevent or control any form of intervention that may hamper the integrity of the collective environmental resources. These laws empower these commissions, establishments, agencies or organizations. These agencies are responsible for protecting and managing the environment from dilapidation arising from climate change, environmental pollution, etc. Such agencies may have international, national, regional, state or local council parastatals or representatives as well as some non-governmental or private commissions, establishments, agencies or organizations [36, 38–40, 48].

It should be noted that Nigeria as a federal entity runs three tiers of government i.e. federal, state, and local hence environmental protection responsibilities is not the sole responsibility of any single tier but are shared across the various tiers. This accounts for the side-by-side existence of federal legislations and policies with state laws and environmental policies. Thus, federal ministries, agencies and departments perform environmental protection functions at the federal level whereas in the states, the various state ministries and agencies also perform their responsibilities. The same is true for local councils. However, there is a hierarchy in the complexity of responsibilities undertaken at each tier. The principal legislation on environmental standards, regulation and control is a federal legislation hence all other laws at the state level and bye-laws at the local council tier derive their legality and legitimacy from this principal legislation save where the principal legislation or other federal legislations in this context do not 'cover the field' and in this case the respective states may invoke the doctrine of not covering the field wherein they are constitutionally empowered to make laws that must not be inconsistent with any federal legislation. Essentially, the agencies, commissions and units at the State and Local council levels complement the federal government's effort in the management and preservation of the integrity of the environment. The key environmental establishments in Nigeria are presented in Table 1.

With respect to the regulation of major environmental protection activities including waste management, and enforcement of various environmental legislations in Nigeria, NESREA, a federal agency is the primary agency at the centre. NESREA operates the NESREA Act with the 33 subsidiary regulations attached to it as well as other vital environmental policies, legislations (e.g. Environment Impact Assessment Act, etc.), international conventions, treaties, and protocols to which Nigeria is a party. NESREA does not work in isolation; it collaborates with other federal agencies such as the NNRA, NEMA, etc. NESREA also collaborates with the following establishments:

a. State ministries and agencies
b. International community and donor agencies
c. Community-based organisations.
d. Faith based organisations.
e. Non-governmental organisations
f. Civil society organisations.

Table 1. Various Public Environmental Establishments

No.	Establishment	Type
1	Forestry Research Institute of Nigeria (FRIN)	Agency
2	Federal Ministry of Health	Ministry
3	National Oil Spill Detection and Response Agency (NOSDRA)	Agency
4	National Emergency Management Agency (NEMA)	Agency
5	Nigeria Hydrological Services Agency (NIHSA)	Agency
6	River Basin Authority (RBA)	Agency
7	Environmental Health Officers Registration Council of Nigeria (EHORCN)	Agency
8	Nigerian Conservation Foundation (NCF)	Agency
9	National Environmental Standards and Regulations Enforcement Agency (NESREA)	Agency
10	Directorat of Petroleum Resources (DPR)	Department
11	Nigerian Nuclear Regulatory Authority (NNRA)	Agency
12	National Biosafety Management Agency (NBMA)	Agency
13	Energy Commission of Nigeria (ECN)	Agency
14	Drought and Desertification Agency (DDA)	Agency
16	Federal Ministry of Environment	Ministry
17	Federal Ministry of Water Resources	Ministry
18	Department of Climate Change	Department
19	Department of Planning, Research and Statistics	Department
20	Federal Ministry of Science and Technology	Ministry
21	Federal Ministry of Water Resources	Ministry
22	Erosion, Floods and Coastal Zone Management	Department
23	State and Local Council commissions and departments	Commissions and Departments

g. Other law enforcement agencies e.g. Nigeria Police, Nigeria Security and Civil Defence Corps, etc.

The operations of NESREA are captured in Fig. 1. NESREA generally implements pre-emptive mechanisms in ensuring compliance with appropriate legislative necessities and licensing requirements [54]. However, in a situation where voluntary compliance is not imminent, NESREA would use its enforcement machineries. The main elements in NESREA's enforcement policies include [54]:

a. Inspection.
b. Compliance monitoring.
c. Negotiation.
d. Legal action.
e. Prosecution.

The procedures of enforcement by NESREA are:

a. Issue of permits and licences.
b. Issue of prohibition and enforcement notices.
c. Variation of license conditions.
d. Implementing the "polluter pays" principle.
e. Suspension and/or revocation of permits and licences.
f. Injunction and carrying out of remedial works.

2.6 Environmental Laws in Nigeria

Environmental laws in Nigeria have their roots in the Constitution of the Federal Republic of Nigeria [as amended]. Environmental remediation provision could be adduced from Section 33(1) of the Constitution. According to the said section, "every person has a right to life and no one shall be deprived intentionally of his life". This provision presupposes a healthy environment in that healthy life is dependent on a healthy environment. It follows that the citizenry ought to be protected against hazards that may exist in degrading environments. Another provision that lends credence to constitutionality of environmental protection is in Section 2 of the Environmental Impact Assessment Act. This said provision requires that:

"every company whether public or private must prior to embarking on any industrial project evaluate the environmental effects of such proposed projects".

The said assessment of proposed project impact on the environment is to be evaluated, approved/disapproved by a legitimate authority. This supervisory role is the mandate of NESREA. The content of environmental legislations in Nigeria is quite comprehensive [37–40, 42]. It covers the issues bordering on "land use, soil conservation, wildlife, forestry, protected natural area, water and marine resources, coastal areas, wastes and sanitation, environment/occupational health and safety, air quality, hazardous substances, pollution/pollutants land degradation and planning, afforestation, deforestation, desertification among others" [37–40, 42]. Apart from the constitution, other sources of environmental laws in Nigeria [37, 39, 40, 42] are: Nigerian judicial precedents/Nigerian case law; Subsidiary legislations; the received English law; Customary law, and the Islamic law.

In the present dispensation not all these sources listed above are applicable in the strictest sense to the environmental management in Nigeria [37–42].

The major environmental legislations in Nigeria are listed in Table 2.

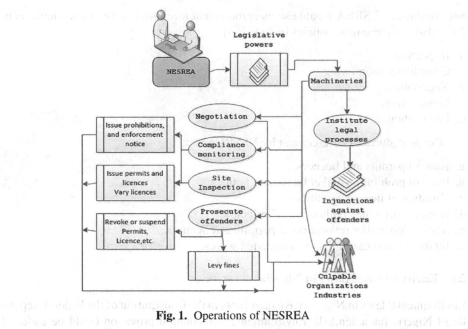

Fig. 1. Operations of NESREA

Table 2. Major Environmental Legislations

No	Legislation	In force	Main object
1	Constitution of the FRN 1999 (as amended)	Yes	Grundnorm and law of the land
2	Environmental Impact Assessment Act 2004	Yes	Techniques and rules of environmental impact assessment in different economic sectors
3	NESREA Act 2007	Yes	Principal environmental regulatory agency, enforces environmental laws; prosecutes offenders
4	Harmful Waste (Special Criminal Provisions etc.) Act 2004	Yes	Prohibits dropping, carrying, and dumping of injurious waste on land and in territorial waters
5	Oil in Navigable Waters Act	Yes	Regulates petroleum discharges from ships
6	Associated Gas re-injection Act	Yes	Controls gas flaring by petroleum firms. Bans, unlawful flaring by oil/gas companies
7	Water Resources Act	Yes	promote development, utilization and protection of water resources

(continued)

Table 2. (*continued*)

No	Legislation	In force	Main object
8	Endangered Species (Control of International Trade and Traffic) Act 2016	Yes	conservation and management of wildlife and the protection of endangered species
9	Nuclear Safety and Radiation Protection Act	Yes	Controls use of radioactive substances and apparatus that emit or produce ionizing radiation
10	National Oil Spill, Detection and Response Agency Act	Yes	Establish machinery to co-ordinate and implement National Oil Spill Contingency Plan to ensure safe, timely, effective and appropriate response to major or disastrous oil pollution
12	Hydrocarbon Oil Refineries Act	Yes	licensing and control of refining activities
13	National Park Services Act	Yes	conservation and protection of natural resources and plants in national parks
15	The Land Use Act	Yes	Use of land
16	Sea fisheries Act	Yes	Control, regulate, protect sea fisheries in Nigeria's territorial waters
17	Nigerian Mining Corporation Act	Yes	Regulation of exploration of solid minerals
18	Environmental laws in 36 states	Yes	Environmental protection and waste management

2.7 Challenges of IWM Operations

The loopholes in the operation of the policies and legislations on WM are not only documented by researchers but routine observation or surveillance in urban areas such as Lagos and Port Harcourt is enough to provide a convincing ground that the said legislative frameworks and machineries are not yielding the right results. Several pitfalls connected to the operations of these frameworks have been noted ranging from poor and crude infrastructure, poor funding, incoherent initiatives of government operatives, poor maintenance culture, corruption, nepotism, social stratification with the emergence of the 'untouchables' that include individuals and organizations which are covertly above the law, poor enforcement, to negligence and abdication of responsibilities on the part of the staff manning these agencies. One of the greatest blows to the principal legislation (NESREA Act) and its guidelines noncompliance and weak enforcement [50, 55, 56]. Noncompliance and weak enforcement had been shown to be connected to some sociopolitical and economic problems [37, 41, 52]. These problems are:

a. Weak environmental awareness campaigns
b. Deficiency of competent workforce, instruments and technology
c. Corruption and embezzlement of environmental fund.

d. Levity on the side of the Government
e. Inadequate databases
f. Poor funding of environmental operations
g. Poor maintenance culture
h. Weak or absence of environmental pressure groups
i. Weak enforcement strategies

These problems have negated the true objects of environmental laws, policies, programs and regulations in driving the safeguarding and conservation of the environment [36, 37, 41]. We submit that these challenges could be addressed through the use of a versatile technology framework or single window that provides a transparent platform for all stakeholders in the waste management ecosystem. The single window will provide planning/scheduling and monitoring for all necessary environmental awareness campaigns, workforce, budgets (allocations and expenditure), waste management databases, profiles, etc.

3 Why Technology-Based Waste Management?

Coordinated and organized WM campaign is neither about collation and disposal of wastes nor about enforcement or prosecution of offenders. Efficiency, effectiveness, sustainability, and better approaches are factors that must be considered continually. Accordingly, the concept of technology-based WM is an inevitable part of modern IWM programs and campaigns. In recent times, the waste stream size and composition serve as determinants for the type of waste management system to be designed and/or implemented. Modern WM ideologies advocate integrated technology-based waste management approaches with consideration on resource recycling, recovery and energy generation facilities from industrial waste [57]. In fact, waste-to-energy conversion is now considered as one of the best ways of addressing the problem of waste management in a sustainable manner [58, 59]. Therefore, various thermochemical and biological waste-to-energy technologies are now being used to treat industrial wastes including harmful wastes. Thus, new approaches that could lead to efficient industrial WM have now extended the horizon of WM. Some of these are included in Fig. 2. Some of these approaches could provide a platform for transforming waste into useful energy [60].

Like other socio-economic sectors, waste management requires decision making at the tactical, operational and strategic levels. To achieve the objectives of such decision-making needed in the management of hazardous industrial wastes, information technologies (ITs) provide the right platform and resources for the 21st century waste management service delivery. Considering the growing need for large-scale complicated data storage, communication, analysis, and applications in connection with fast and parallel computational capabilities, IT is becoming more critical for IWM. In terms of monitoring, collection, transport, treatment and management, the scope of IT for IWM can be divided

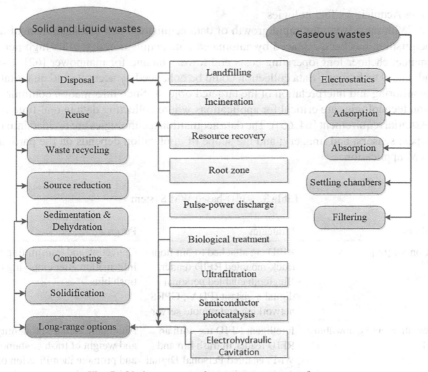

Fig. 2. Various approaches to management of waste

into spatial technologies, identification technologies, data acquisition technologies and data communication technologies.

1. Spatial Technologies

   Spatial technologies are the most commonly used technologies in environmental modelling, and in many environmental studies, spatial analysis is of great importance. Such technologies are useful for managing complex spatial information and for providing mechanisms for the integration of various models, interfaces and subsystems—for instance, the integration of incineration and landfill. Spatial technologies may be categorized into three major groups: Geographic Information Systems (GIS), Global Positioning System (GPS) and Remote Sensing (RS) [61]. The essential functions of such technologies include the collection, storage, analysis and visualization of spatial data. A spatial dataset's content can include attribute data, spatial topology, raster, features, and even network datasets.

2. Identification Technologies

   Researchers and organizations involved in WM have been studying different kinds of technology in recent decades to improve WM performance and automate bins collection. Several researchers have researched the possibility of implementing advanced WM systems that are based on identification technologies [62, 63]. WM systems are being improved through the use of identification technologies, such as barcode and RFID technologies as described in Tables 3 and 4 respectively.

3. Data Acquisition Technologies

With the advent and rapid growth of data acquisition technologies, manual data acquisition has been replaced by automated data acquisition due to its high performance, cheaper long-operating costs and lower demand for manpower [62]. Using advanced techniques, data collection could be enhanced by accurate and quantitative monitoring and interpretation of the targeted objects. Such information communication technologies are critical for applications where collecting data in real-time is an essential requirement [64, 65]). The data acquisition technologies are divided into two types: sensors, and imagery; and the scope of application depends on the particular WM application.

Table 3. RFID-based WM System

Authors	Strategies	Purpose
O'Connor [66]	RFID tag attached to bin floor, truck-mounted RFID reader, Bluetooth enabled personal digital assistant (PDA), GPRS networking and remote server	Collecting and transmitting tag information after emptying of trash bins
Chowdhury and Chowdhury [67]	Intelligent RFID tag with an RFID reader in trash bin and Wi-Fi enabled Personal Digital Assistant (PDA)	Automatic capture of identity and weight of trash containers and promote identification of stolen containers
Pratheep and Hannan [68]	RFID tag built on bins and RFID reader with a vehicle-mounted antenna	Provides better visibility of the pick-up activities of drivers and track large trash receptacles
Gnoni et al. [69]	Trash bins and collection vehicles were fitted with an RFID tag and vehicle RFID reader	Facilitate growth of waste management services Pay as You Throw Billing
Nielsen et al. [70]	An RFID reader in a vehicle with a GPS receiver, tag in garbage bins and backend decision support device	Optimisation of waste disposal equipment and calculation of weight for billing purposes in some situations
Ali et al. [71]	RFID tag mounted in bin, black box mounted in vehicles containing RFID receiver, camera, GPS receiver, GSM/GPRS module and remote server	Vehicle collection and solid waste bin control to simplify waste management system
Glouche et al. [72]	Smart bins fitted with an RFID reader, RFID tag with information about recyclable materials for each waste object	Bin level effective waste object handling for selective sorting and recycling

4. Data Communication Technologies

Rapid growth of networking systems and universal internet connectivity opens the possibilities for instant data transfer from remote locations. Wireless communication technologies that are used in WM applications include Global System for Mobile communications (GSM), General Packet Radio Service (GPRS), wireless fidelity (WiFi), Very High-Frequency Radio (VHFR for long-range communication) and Bluetooth (for short-range communication).

Table 4. Barcode-based WM System

Authors	Strategies	Purpose
Li et al. [73]	Keeps track of materials and their status and combines with incentives to reduce waste	Reducing and handling the building of waste
Saar et al. [74]	Relates current waste material barcodes to websites via mobile phones to get information about disassembly	Gives the recyclers exact details about dismantling for automatic recycling
Greengard [75]	Barcodes are placed on waste items	Real-time monitoring of waste trajectory
Maruca et al. [76]	Scanning data from a waste item identifier and comparison with reference data to meet predetermined safety requirements	Determining and sorting banned waste products

3.1 Application of Artificial Intelligence (AI) in Managing Risks Associated with Industry Wastes

The goal of AI in the management of risks associated with industrial wastes is to provide intelligent control by supporting the planning, scheduling, status monitoring, change initiation in components like sensors, analysis and reporting. These are done with the aid of the AI tools, which may include expert system, fuzzy logic, neural networks, decision support system and a hybrid expert network system [77–90]. The key opportunities for AI applications are in the area of multi-source data integration, including input data and timely and defensible recommendations. The use of AI in the management of risks with industrial wastes can generally be grouped into two areas: (a) decision support and (b) monitoring and reporting. The concept of AI in the decision support field will address the tools for burn plans and assist with the incinerator activities decision-making process. Whereas the monitoring and reporting system is aimed at providing monitoring status information as well as historical or predictive details and patterns. These components may be integrated into waste repositories, incinerators, scheduling and performance evaluation routines. The AI component could monitor pre-burn waste from inventories to incineration and secondary post-burn waste. Following are some of the AI applications in regards to risk management of industrial wastes:

a. Maintenance schedules
b. Development and scheduling blends; waste feedstocks, etc.
c. Approval/rejection of requests for incineration;
d. Inventory control
e. Scheduling industrial waste shipments;
f. Evaluation of data processes and recommendations
g. Data quality control
h. Integrated performance assessments

3.2 Risk Management and Profiling in IWM

Risk is the likelihood of obtaining unexpected or uncontrolled loss of value: this could be money, social status, environment variables, and a person's physical health. [91]. Certain actions of man such as: rapid industrialization and urbanizations, are a trade-off for gaining or losing some of these values. Risks tend to exists in almost every area of human endeavor, hence the need for effective methods of risk management. Of substantial value to us in this study is health risk. Health risk reflects unforeseen chances of illnesses following direct and/or remote exposure to toxic environmental wastes. Closely related to health risks are safety and environmental risks which are consequences of hazardous biological and physico-chemical substances that negatively impacts the environment, consequently, environmental risk analysis (ERA) which seeks to study and understand actions and events that introduce risks to human health through the environment is a requirement during establishment of factories. Risk management involves the identification, evaluation, and prioritization of risks [92]. When every risk involved in a project has been identified and evaluated, then coordinated actions and wise utilization of resources can be applied to minimize and control the impact of unforeseen and undesirable occurrences. Effective risk management encompasses all strategies employed to manage all threats, which are uncertainties with negative consequences. The methods applied to risk management vary widely depending on the context of the risk as the strategies that would be employed in managing financial risk will be different from the industrial risks, however, generally, risk management involves avoidance of identified threats, reduction of the negative effect of the threat, reduction in the probability of the threat as well as transferring the majority of the threat to another party. Industrial and manufacturing activities result in the production of waste materials. Among several others are chemical solvents, industrial by-products, metals, and radioactive waste. Research has shown the direct relationship between population size and waste generation as a larger population tends to be associated with bigger industries and factories leading to the generation of larger amounts of industrial wastes. [93]. Solid wastes generated in the big manufacturing industries like lead and zinc batteries, detergent containers, and PVC contain cadmium which derives its toxicological properties from its chemical similarity to zinc is the most reported. These wastes especially those with cadmium content accumulate in the human body and causes harm to all the vital organs including the liver, kidneys, lungs, bones, the placenta, brain, and the central nervous system. Damage to human reproductive systems has also been reported as well as hematological and immunological damages [94].

These dire consequences and the threat to human lives and the environment have led to the need for the development of proper profiling and management of industrial wastes for efficient risk management as this will safeguard the environment and human lives by reducing the risk associated with industrial wastes [95]. In Nigeria, several legislations contain the objects of regulation and risk management for the protection of human lives through the reduction of pollution to air, water, land, and food [96].

Table 5. Risk Profiles of Industrial Wastes In Nigeria

Industrial Sector	Waste Generated	Associated risks/hazards
Nigerian Oil and Gas Production	Drill mud, Glass, Metals, Plastic drums, Oil spills, Sludge, Chemicals	Inherently hazardous materials Water and Land Pollution Risk to Aquatic life Cancer causing substances from plant pollution
Nigerian Plastic and Polymer Producers	Accumulation of plastic objects, Polyethylene, Terephthalate (PET), High-Density Polyethylene (HDPE), Polyvinyl chloride (PVC), Polypropylene (PP)	Accumulation of plastic objects leading to land and water pollution Non-biodegradable ones causing drain blockages
Cement and Mining	Ore processing Slag, Phosphogypsum(from phosphoric acid production), Gasifier ash, Massive air pollution (from dust/sludge), Process wastewater, Calcium sulfate, Chemical vapors	Air Pollution Respiratory diseases Lead Poisoning *Ionizing radiation* *Increased Heat*
Beverage Bottling Factories	Wastewater from, Left-over and discarded bottles, straws, etc	Land and water pollution
Pharmaceuticals	Cytotoxic waste materials, Cytostatic waste materials, Expired pharmaceutical stock Kits for destroying controlled drugs, Recalled pharmaceutical stock	Land and water pollution
Paper and Pulp industry	Sludge from wastewater treatment plants Air emissions	Air Pollution Respiratory diseases

3.3 Profiling Industries and Waste Generated

Manufacturing and exploration industries at the point of manufacturing or delivering services accumulates wastes. The majority of these wastes pose adverse consequences [97–103]. Government policies and regulations have made it mandatory for these organizations to focus on the minimization of wastes in their production processes. Failures on part of these organization and the regulatory agencies have been clearly documented. It is submitted that without adequate mechanisms to ensure that every stakeholder contributes towards the objectives of such regulations would avail nothing. Hence, profiling risks/hazards prevalent in the environment and providing an "all-in-one" platform through which concerns may be communicated and addressed, is considered a worthy venture.

Going forward, a sector-by-sector profiling of some industries in Nigeria as regards the wastes generated and the associated risks is presented in Table 5.

4 Integrated System for Risk and Hazard Profiling

Having noted the various gaps in the workings of the various agencies, legislative and policy frameworks, we propose an integrated system for managing the risks associated with harmful wastes from industrial sources. This proposition is borne out of recent developments in AI, Big Data, and Internet of Things (IoT). These technologies are revolutionizing every facet of the socioeconomic sector.

As have been noted, decision-making, efficiency, eco-friendliness, and sustainability are the pillars on which any system whether political or economic should anchor in these modern times.

Authorities like NESREA and other agencies and organizations it collaborates with should be able to relate through a single window, which affords active collaboration unlike the existing passive system. Risk or hazard management is a collaborative effort and we submit that it is only a smart ICT-based system that can drive such a collaborative effort. All stakeholders must be recognized and taken into consideration while evolving the architecture for managing these hazards.

The aforestated is entirely is true for risk management especially in the public sector where the majority of actions taken are on behalf and for the betterment of everyone. Transparency, accountability, and fairness must reflect in the architecture.

Unlike the existing system wherein impact assessment is conducted by NESREA and the company/industry in question and approval granted without seeking inputs from the residents who live and do business around the area where such industry/company plans to implement a project, the proposed architecture is to be all-encompassing with all stakeholders involved and an unbounded platform provided for such stakeholders to relate directly with the operators on any development of mutual concern.

The proposed architecture will require the documentation of all industries/companies registered with the corporate affairs commission whose businesses revolve around exploration, production, mining, fabrication, manufacturing, parts importation. As presented in Table 4 above, the risk profiles associated with the activities of the said industries would also be documented, as this would help during the modeling of a risk assessment model that could be embedded in a more robust national integrated hazard management and planning system.

Figure 3 presents the block structure of the proposed architecture. Four layers are recognized:

a. The Social layer
b. Enterprise layer
c. Logic and analytics layer
d. Messaging layer

The Social layer would address the concerns of the ordinary people. For instance, this layer would present a platform for stakeholders (members of the public, experts, industry practitioners, etc.) to interact with the enterprise layer (government operatives). Inquiries, submissions, opinions, etc. could be made and tracked on this layer. The Enterprise layer would not exercise any powers as to deleting any event that occurs on this layer.

The Enterprise layer incorporates the operations of the government operatives regardless of their placements in hierarchy. In other words, all collaborating agencies, ministries, departments would be incorporated into this layer with their identities maintained

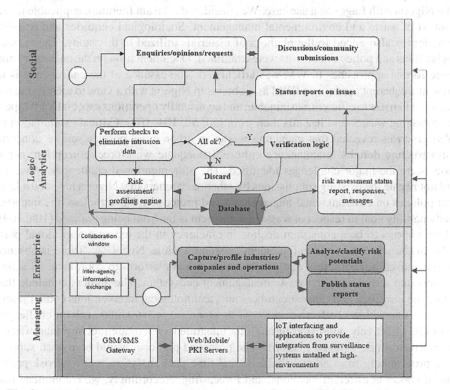

Fig. 3. Proposed architecture of an Integrated System for Hazardous Wastes Management

independently. This layer would afford real-time collaborations among these establishments. The supervising regulatory agency would be able to document and classify industries based on their risk potentials.

The Logic and analytics layer would incorporate necessary risk management and profiling engine, robust database (of all companies as defined earlier). This layer provides intelligence to the entire system. It would receive inputs from stakeholders such as residents in areas where industries are located, expert opinions, etc. and would use such real-time information to re-compute the risk profiles of the industries involved and present such reports to all stakeholders.

The Messaging layer would support seamless electronic data exchange between collaborating agencies. It would provide room for IoT applications wherein information from surveillance systems installed by the government agencies on various locations around high-risk industries could automatically relay such information to the central system over the Internet.

5 Conclusion

In this paper, we present succinct details on IWM as it applies to a developing country like Nigeria with Lagos as a use case. We considered relevant literature applicable to the context of waste and environmental management. Sociological enquiries and real-life experiences also added to the plethora of materials utilized in this study. The various legislations and policy documents were examined. Documentation on the customary and more formal approaches to WM was articulated. The essence of the entire study is to present a coherent account on IWM as it obtains in Nigeria with a view to identifying the gaps and barriers to efficient, sustainable, and ecofriendly operations especially in respect of managing wastes and hazards therefrom [50, 52, 104–107]. Critical examination of WM practices revealed that managing wastes in urban cities poses a serious concern. Infrastructure deficits, maintenance culture, inadequate workforce, corruption, nepotism, were the major challenges identified. The aforementioned challenges, however, did not negate the existing submissions by other researchers that Nigeria has robust laws and policies on environmental management and protection. Nevertheless, the intended sustainability goal in respect of waste management is far from being realized [108, 109]. There appears to be a consistent decline in efficiency in the existing operational portfolio of the regulatory and enforcement agencies such as NESREA. There is absence of specialized infrastructure that provides a common platform to coordinate all stakeholders and activities in the waste management ecosystem. The authors conclude that while the identified gaps appear cumbersome, technological interventions could present a more sustainable and effective option. Going forward, we submit that all the challenges could be effectively addressed through an integrated system architecture that offers a data-driven single window platform. The proposed platform would be robust, support data provisioning from all stakeholders, real-time data analytics, real-time risk profiling of wastes at different locations, and forecasting. Accordingly, we recommend that the proposed architecture is necessary to drive the modernization of the EM and WM agencies in line with the sustainability development goals. More so, it will eliminate the existing skewness in decision-making (where every decision is almost taken by the respective authority alone), present appropriate platform for community participation, and engender fairness in handling issues of collective concerns.

Acknowledgment. We thank the assistance provided to us by officers of the lagos state waste management authority especially waste management reforms.

References

1. Asase, M., Yanful, E.K., Mensah, M., Stanford, J., Amponsah, S.: Comparison of municipal solid waste management systems in Canada and Ghana: a case study of the cities of London, Ontario, and Kumasi, Ghana. Waste Manage. **29**, 2779–2786 (2009)
2. Rodgers, M.: Fundamentals of Development Administration. S.K. Publishers, London, United Kingdom (2011)
3. Osmani, M., Glass, J., Price, A.D.F.: Architects' perspective on construction waste reduction by design. Waste Manage. **28**, 1147–1158 (2008)
4. Ekanayake, L.L., Ofori, G.: Building waste assessments score: design based tool. Build. Environ. **39**, 851–861 (2004)
5. Geng, Y., Fu, J., Sarkis, J., Xue, B.: Towards a national circular economy indicator system in China: an evaluation and critical analysis. J. Clean. Prod. **23**, 216–224 (2012)
6. Preston, F.A.: Global Redesign: Shaping the Circular Economy. Chatham House, London, UK (2012)
7. Ghisellini, P., Cialani, C., Ulgiati, S.A.: Review on circular economy: the expected transition to a balanced interplay of environmental and economic systems. J. Clean. Prod. **114**, 11–32 (2016)
8. Seo, S., Aramaki, T., Hwang, Y., Hanaki, K.: Environmental impact of solid waste treatment methods in Korea. J. Environ. Engineering **130**, 81–89 (2004)
9. Kamalan, H., Sabour, M., Shariatmadari, N.A.: Review on available landfill gas models. J. Environ. Sci. Technol. **4**, 79–92 (2011)
10. Den Hollander, M.C., Bakker, C.A., Hultink, E.J.: Product design in a circular economy: development of a typology of key concepts and terms. J. Ind. Ecol. **21**, 517–525 (2017)
11. Ajie, U.E., Dienye, A.: Spatial data analysis of solid waste management system in Port Harcourt metropolis after 100 years of its existence. In: Proceedings of the FIG Congress (2014)
12. Van den Berg, M.R., Bakker, C.A.: A product design framework for a circular economy. Product. Lifetimes Env. **17**, 365–379 (2015)
13. Shi, L., Xing, L., Bi, J., Zhang, B.: Circular economy: a new development strategy for sustainable development in China. In: 3rd World Congress of Environmental and Resource Economists, pp. 3–7 (2006)
14. Su, B., Heshmati, A., Geng, Y., Yu, X.: A review of the circular economy in China: moving from rhetoric to implementation. J. Clean. Prod. **1**(42), 215–227 (2013)
15. Jun, H., Xiang, H.: Development of circular economy is a fundamental way to achieve agriculture sustainable development in China. Energy Procedia **1**(5), 1530–1534 (2011)
16. Bartl, A.: Withdrawal of the circular economy package: a wasted opportunity or a new challenge? Waste Manag. (New York, NY). **44**, 1–2 (2015)
17. Moreno, M., Braithwaite, N., Cooper, T.: Moving beyond the circular economy, p. 10 (2014)
18. Commission, E.: Directive 2008/98/EC of the European parliament and of the council on waste and repealing certain directives. Off. J. Eur. Union **312**, 3–10 (2008)

19. Wiles, C.C., Kosson, D.S., Holmes, T.: United States Environmental Protection Agency Municipal-Waste Combustion Residue Solidification/Stabilization Evaluation Program. Environmental Protection Agency, Cincinnati, OH (United States). Risk Reduction Engineering Lab. (1990)
20. Ezeigwe, C.: Appropriate solid waste disposal methods for developing countries. NSE Tech. Trans. **32**(2), 33–34 (1995)
21. Ajani, O.I.: Determinants of an effective solid waste management in Ibadan metropolis, Oyo State, Nigeria. Int. J. Food Agric. Env. **6**, 152–157 (2008)
22. Igoni, A.H., Ayotamuno, M.J., Ogaji, S.O., Probert, S.D.: Municipal solid-waste in Port Harcourt, Nigeria. Appl. Energy **84**(6), 664–670 (2007)
23. Akanni, P.O.: An empirical survey of the effect of materials wastage on contractors' profit level in construction projects. The Professional Builders. J. Niger. Instit. Build. 35–46 (2007)
24. Adewuyi, T.O., Idoro, G.I., Ikpo, I.J.: Empirical evaluation of construction material waste generated on sites in Nigeria. Civ. Eng. Dimension **16**(2), 96–103 (2014)
25. Kareem, W.A., Asa, O.A., Lawal, M.O.: Resources conservation and waste management practices in construction industry. Oman Chapter Arab. J. Bus. Manag. Rev. **34**(2603), 1–2 (2015)
26. Dania, A.A., Kehinde, J.O., Bala, K.: A study of construction material waste management practices by construction firms in Nigeria. In: Proceedings of the 3rd Scottish Conference for Postgraduate Researchers of the Built and Natural Environment, Glasgow, pp. 121–129 (2007)
27. LAWMA: Landfill https://www.lawma.gov.ng/inside-lawma/departments/landfill/ (2019)
28. Ogunmakinde, O.E., Sher, W., Maund, K.: An assessment of material waste disposal methods in the Nigerian construction industry. Recycling **4**(1), 13 (2019)
29. LAWMA: Landfill—Monthly Dump Reports. http://www.lawma.gov.ng/insidelawma/departments/landfill/2014. Accessed on 20 May 2022
30. Obadina, A.: Solid Waste Management Livelihood on Lagos Dumpsite: Analysis of Gender and Social Difference. Ph.D. Thesis, Loughborough University, Loughborough, UK (2016)
31. Odewumi, S.G.: Appraisal of storage and collection strategies of municipal solid waste in Lagos State. J. Humanit. Soc. Sci. **10**, 61–67 (2013)
32. Kofoworola, O.F.: Comparative assessment of the environmental implication of management options for municipal solid waste in Nigeria. Int. J. Waste Resour. **7**(1), 1–5 (2016)
33. García, V., Pongrácz, E., Keiski, R.: Waste minimization in the chemical industry: from theory to practice. In: Proceedings of the Waste Minimization and Resources Use Optimization Conference, pp. 93–106, 10 Jun 2004
34. Ana, G.R., Sridhar, M.K., Olawuyi, J.F.: Air pollution in a chemical fertilizer complex in Nigeria: the impact on the health of the workers. J. Env. Health Res. **4**(2), 57–62 (2005)
35. Armstrong, M.: A Handbook of Human Resource Management Practice. Kogan Page Publishers (2006)
36. Mallak, S.K., Elfghi, F.M., Rajagopal, P., Vaezzadeh, V., Fallah, M.: Overview of waste management performance of industrial sectors by selected Asian countries: current practices and issues. Int. Proc. Chem. Biol. Env. Eng. **99**, 66–75 (2016)
37. Gana, A.J., Ngoro, D.: An investigation into waste management practices in Nigeria (a case study of lagos environmental protection board and Abuja environmental protection board). West Afr. J. Ind. Acad. Res. **12**(1), 112–126 (2014)
38. Guerrero, L.A., Maas, G., Hogland, W.: Solid waste management challenges for cities in developing countries. Waste Manage. **33**(1), 220–232 (2013)
39. Emejuru, C.T.: Human rights and environment: whither Nigeria. JL Pol'y Globalization **35**, 108 (2015)

40. Gadzama, N.M., Ayuba, H.K.: On major environmental problem of desertification in Northern Nigeria with sustainable efforts to managing it. World J. Sci. Technol. Sustainable Dev. **13**(1), 18–30 (2016)
41. Ijaiya, H., Joseph, O.T.: Rethinking environmental law enforcement in Nigeria. Beijing Law Rev. **05**(04), 306–321 (2014)
42. Kankara, A.I., Adamu, G.K., Tukur, R., Ibrahim, A.: Examining environmental policies and laws in Nigeria. Int. J. Env. Eng. Manage. **4**(3), 165–170 (2013)
43. Ladan, M.T.: Trend in Environmental Law and Access to Justice in Nigeria (2012)
44. World Health Organization: International Health Regulations, 3rd edn. Switzerland, Geneva (2016)
45. Etemire, U.: Insights on the UNEP Bali guidelines and the development of environmental democratic rights. J. Env. Law **28**(3), 393–413 (2016)
46. Patwardhan, A.D.: Industrial solid wastes. The Energy and Resources Institute (TERI) (2013)
47. Pariatamby, A., Fauziah, S.H.: Sustainable 3R practice in the Asia and Pacific Regions: the challenges and issues. In: Municipal Solid Waste Management in Asia and the Pacific Islands: Challenges and Strategic Solutions, pp. 15–40 (2014). https://doi.org/10.1007/978-981-4451-73-4_2
48. Zhang, D.Q., Tan, S.K., Gersberg, R.M.: Municipal solid waste management in China: status, problems and challenges. J. Environ. Manage. **91**(8), 1623–1633 (2010)
49. May, J.R., Daly, E.: Global Environmental Constitutionalism. Cambridge University Press (2014). https://doi.org/10.1017/CBO9781139135559
50. Nwankwo, W., Olayinka, S.A., Ukhurebor, K.E.: Green computing policies and regulations: a necessity. Int. J. Sci. Technol. Res. **9**(1), 4378–4383 (2020)
51. Ukhurebor, K.E., Aigbe, U.O., Olayinka, A.S., Nwankwo, W., Emegha, J.O.: Climatic change and pesticides usage: a brief review of their implicative relationship. AU eJ. Interdisc. Res. **5**(1), 44–49 (2020)
52. Nwankwo, W., Ukhurebor, K.E.: An x-ray of connectivity between climate change and particulate pollutions. J. Adv. Res. Dyn. Control Syst. **11**(8), 3002–3011 (2019)
53. Sengar, D.S.: Environmental Law Practices. Hall, New Delhi, India (2007)
54. NESREA: Establishment Act, CAP 164 Laws of the Federation of Nigeria (2007)
55. Naibbi, A.I., Mustapha, A.B.: Streamlining sustainability: environmental regulations in Nigeria: a mini review. Int. J. Env. Sci. Nat. Resour. **1**(5), 142–144 (2017)
56. Akamabe, U.B., Kpae, G.: A critique on Nigeria national policy on environment: Reasons for policy review. IIARD Int. J. Geogr. Env. Manage. **3**(3), 22–36 (2017)
57. Oh, J., Hettiarachchi, H.: Collective action in waste management: a comparative study of recycling and recovery initiatives from Brazil, Indonesia, and Nigeria using the institutional analysis and development framework. Recycling **5**(1), 4 (2020)
58. Galeno, G., Minutillo, M., Perna, A.: From waste to electricity through integrated plasma gasification/fuel cell (IPGFC) system. Int. J. Hydrogen Energy **36**(2), 1692–1701 (2011)
59. Zaman, A.U.: Comparative study of municipal solid waste treatment technologies using life cycle assessment method. Int. J. Environ. Sci. Technol. **7**(2), 225–234 (2010)
60. Czajczyńska, D., et al.: Potential of pyrolysis processes in the waste management sector. Thermal science and engineering progress. **1**(3), 171–197 (2017)
61. Milla, K.A., Lorenzo, A., Brown, C.: GIS, GPS, and remote sensing technologies in extension services: where to start, what to know. J. Ext. **43**(3), 1–8 (2005)
62. Palczynski, R.J., Scotia, W.N.: Study on Solid Waste Management Options for Africa. Project Report. Final Draft Version. Prepared for African Development Bank Sustainable Development and Poverty Reduction Unit, Abidjan (2002)
63. Nadaf, R.A., Katnur, F.A., Naik, S.P.: Android application based solid waste management. In: Pasumpon Pandian, A., Palanisamy, R., Ntalianis, K. (eds.) Proceeding of the International Conference on Computer Networks, Big Data and IoT (ICCBI – 2019), pp. 555–562.

Springer International Publishing, Cham (2020). https://doi.org/10.1007/978-3-030-43192-1_63

64. Faccio, M., Persona, A., Zanin, G.: Waste collection multi objective model with real time traceability data. Waste Manage. **31**(12), 2391–2405 (2011)
65. Ratnasabapathy, S., Perera, S., Alashwal, A.: A review of construction waste data and reporting systems used in Australia. 43RD AUBEA. 2019 Nov 6:396
66. O'Connor, M.: Greek RFID pilot collects garbage. RFID Journal, http://www.rfidjournal.com/article/view/2973. Accessed. 14:22 Jan 2007
67. Chowdhury, B., Chowdhury, M.U.: RFID-based real-time smart waste management system. In: 2007 Australasian Telecommunication Networks and Applications Conference, pp. 175–180. IEEE, 2 Dec 2007
68. Pratheep, P., Hannan, M.A.: Solid waste bins monitoring system using RFID technologies. J. Appl. Sci. Res. **7**(7), 1093–1101 (2011)
69. Gnoni, M.G., Lettera, G., Rollo, A.: A feasibility study of a RFID traceability system in municipal solid waste management. Int. J. Inf. Technol. Manage. **12**(1–2), 27–38 (2013)
70. Nielsen, I., Lim, M., Nielsen, P.: Optimizing supply chain waste management through the use of RFID technology. In: 2010 IEEE International Conference on RFID-Technology and Applications, pp. 296–301. IEEE, 17 Jun 2010
71. Ali, M.L., Alam, M., Rahaman, M.A.: RFID based e-monitoring system for municipal solid waste management. In: 2012 7th International Conference on Electrical and Computer Engineering, pp. 474–477. IEEE 20 Dec 2012
72. Glouche, Y., Sinha, A., Couderc, P.: A smart waste management with self-describing complex objects. Int. J. Adv. Intell. Syst. **8**, 63–70 (2015)
73. Li, H., Chen, Z., Wong, C.T.: Barcode technology for an incentive reward program to reduce construction wastes. Comput.-Aided Civ. Infrastruct. Eng. **18**(4), 313–324 (2003)
74. Nabegu, A.B.: An analysis of municipal solid waste in Kano metropolis, Nigeria. J. Human Ecol. **31**(2), 111–119 (2010)
75. Iyer, M., et al.: Environmental survival of SARS-CoV-2–a solid waste perspective. Environ. Res. **1**(197), 111015 (2021)
76. Maruca, D., Leone, J., Cronin, N.E., inventors; Casella Waste Systems Inc, assignee. Systems and methods for identifying banned waste in a municipal solid waste environment. United States patent US 7,870,042. 11 Jan 2011
77. Nwankwo, W., et al.: Integrated fintech solutions in learning environments in the post-Covid-19 era. IUP J. Knowl. Manage. **20**(3), 22–43 (2022)
78. Nwankwo, W., Nwankwo, C.P., Wilfred, A.: Leveraging on artificial intelligence to accelerate sustainable bioeconomy. IUP J. Knowl. Manage. **20**(2), 38–59 (2022)
79. Acheme, I.D., Makinde, A.S., Udinn, O., Nwankwo, W.: An intelligent agent-based stock market decision support system using fuzzy logic. IUP J. Inform. Technol. **16**(4), 7–25 (2020)
80. Nwankwo, W., Adetunji, C.O., Olayinka, A.S.: IoT-driven bayesian learning: a case study of reducing road accidents of commercial vehicles on highways. In: Artificial Intelligence-based Internet of Things Systems, pp. 391–418 (2022). https://doi.org/10.1007/978-3-030-87059-1_15
81. Nwankwo, W., Ukhurebor, K.E.: Big data analytics: A single window IoT-enabled climate variability system for all-year-round vegetable cultivation. IOP Conf. Ser.: Earth Env. Sci. **655**(1), 012030 (2021)
82. Nwankwo, W., Ukhurebor, K., Ukaoha, K.: Knowledge discovery and analytics in process reengineering: a study of port clearance processes. In: 2020 International Conference in Mathematics, Computer Engineering and Computer Science (ICMCECS), pp. 1–7. IEEE, 18 Mar 2020

83. Chinedu, P.U., Nwankwo, W., Masajuwa, F.U., Imoisi, S.: Cybercrime detection and prevention efforts in the last decade: an overview of the possibilities of machine learning models. Rev. Int. Geo. Educ. Online **11**(7), 956–974 (2021)
84. Nwankwo, W., et al.: The adoption of AI and IoT technologies: socio-psychological implications in the production environment. IUP J. Knowl. Manag. **19**(1), 50–75 (2021)
85. Olayinka, A.S., Adetunji, C.O., Nwankwo, W., Olugbemi, O.T., Olayinka, T.C.: A study on the application of bayesian learning and decision trees IoT-enabled system in postharvest storage. In: Pal, S., De, D., Buyya, R. (eds.) Artificial Intelligence-based Internet of Things Systems, pp. 467–491. Springer International Publishing, Cham (2022). https://doi.org/10.1007/978-3-030-87059-1_18
86. Anani, O.A., Adetunji, C.O., Olugbemi, O.T., Hefft, D.I., Wilson, N., Olayinka, A.S.: IoT-based monitoring system for freshwater fish farming: analysis and design. In: AI, Edge and IoT-based Smart Agriculture, pp. 505–515. Elsevier (2022). https://doi.org/10.1016/B978-0-12-823694-9.00026-8
87. Adetunji, C.O., Anani, O.A., Olugbemi, O.T., Hefft, D.I., Wilson, N., Olayinka, A.S.: Toward the design of an intelligent system for enhancing salt water shrimp production using fuzzy logic. In: AI, Edge and IoT-based Smart Agriculture, pp. 533–541. Elsevier (2022). https://doi.org/10.1016/B978-0-12-823694-9.00005-0
88. Adetunji, C.O., et al.: Machine learning and behaviour modification for COVID-19. In: Inuwa, H.M., et al. (eds.) Medical Biotechnology, Biopharmaceutics, Forensic Science and Bioinformatics, pp. 271–287. CRC Press, Boca Raton (2022). https://doi.org/10.1201/9781003178903-17
89. Nwankwo, W., Ukhurebor, K.E.: Web forum and social media: A model for automatic removal of fake media using multilayered neural networks. Int. J. Sci. Technol. Res. **9**(1), 4371–4377 (2020)
90. Makinde, A.S., Agbeyangi, A.O., Nwankwo, W.: Predicting mobile portability across telecommunication networks using the integrated-KLR. Int. J. Intell. Inform. Technol. (IJIIT) **17**(3), 50–62 (2021). https://doi.org/10.4018/IJIIT.2021070104
91. Kiran, S., Kumar, K.P., Sreejith, B., Muralidharan, M.: Reliability evaluation and risk based maintenance in a process plant. Procedia Technol. **1**(24), 576–583 (2016)
92. Jerie, S.: Occupational risks associated with solid waste management in the informal sector of Gweru, Zimbabwe. J. Environ. Public Health **21**, 2016 (2016)
93. Sanders, A.P., Flood, K., Chiang, S., Herring, A.H., Wolf, L., Fry, R.C.: Towards prenatal biomonitoring in North Carolina: assessing arsenic, cadmium, mercury, and lead levels in pregnant women. PLoS ONE **7**(3), e31354 (2012)
94. Aluko, O.O., Adebayo, A.E., Adebisi, T.F., Ewegbemi, M.K., Abidoye, A.T., Popoola, B.F.: Knowledge, attitudes and perceptions of occupational hazards and safety practices in Nigerian healthcare workers. BMC. Res. Notes **9**(1), 1–4 (2016)
95. Nwufo. C.: Legal framework for the regulation of waste in Nigeria. Afr. Res. Rev. **4**(2), 491-501 (2010)
96. Yunusa, U., Lawal, U.B., Idris, A., Garba, S.N.: Occupational health hazards among commercial motorcyclists in Ahmadu Bello University, Zaria. IOSR-JNHS. **3**(17), 46–52 (2014)
97. Afube, G.C., Nwaogazie, I.L., Ugbebor, J.N.: Identification of industrial hazards and assessment of safety measures in the chemical industry, Nigeria using proportional importance index. Arch. Current Res. Int. **19**(1), 1–5 (2019)
98. Agagu, O.: Threats to the Nigerian environment: a call for positive action. Nigerian Conservation Foundation (2009)
99. Kalu, C., Modugu, W.W., Ubochi, I.: Evaluation of solid waste management policy in Benin metropolis, Edo State, Nigeria. Afr. Sci. **10**(1), 117–125 (2009)

100. Oyeniyi, B.A.: Waste management in contemporary Nigeria: the Abuja example. Int. J. Politics Good Governance **2**(2.2), 1–8 (2011)
101. Lu, J.W., Chang, N.B., Liao, L.: Environmental informatics for solid and hazardous waste management: advances, challenges, and perspectives. Crit. Rev. Environ. Sci. Technol. **43**(15), 1557–1656 (2013)
102. Manu, P., Poghosyan, A., Mshelia, I.M., Iwo, S.T., Mahamadu, A.M., Dziekonski, K.: Design for occupational safety and health of workers in construction in developing countries: a study of architects in Nigeria. Int. J. Occup. Saf. Ergon. **25**(1), 99–109 (2019)
103. Maphosa, V.: Sustainable e-waste management at higher education institutions' data centres in Zimbabwe. Int. J. Inform. Eng. Electron. Bus. (IJIEEB) **14**(5), 15–23 (2022)
104. Latif, S.D.: Technical improvement of air pollution through fossil power plant waste management. Int. J. Eng. Manuf. **10**(4), 43–53 (2020). https://doi.org/10.5815/ijem.2020.04.04
105. Nwankwo, W., Ukhurebor, K.E.: Data Centres: a prescriptive model for green and eco-friendly environment in the cement industry in Nigeria. Int. J. Sci. Technol. Res. **9**(5), 239–244 (2020)
106. Nwankwo, W., Olayinka, A.S., Ukhurebor, K.E.: The urban traffic congestion problem in Benin city and the search for an ICT-improved solution. Int. J. Sci. Technol. **8**(12), 65–72 (2019)
107. Nwankwo, W., Olayinka, A.S., Etu, M.: Management of poultry waste in Edo state: a case study of Etsako district. Niger. Res. J. Eng. Env. Sci. **4**(2), 893–902 (2019)
108. Nwankwo, W., Kifordu, A.: Strengthening private sector participation in public infrastructure projects through concession policies and legislations in Nigeria: a review. J. Adv. Res. Dyn. Control Syst. **11**(8), 360–1370 (2019). special issue
109. Nwankwo, W., Chinedu, P.U.: Green computing: a machinery for sustainable development in the post-covid era. In: Sabban, A. (ed.) Green Computing Technologies and Computing Industry in 2021. IntechOpen (2021). https://doi.org/10.5772/intechopen.95420

Design and Development of a Hybrid Eye and Mobile Controlled Wheelchair Prototype using Haar cascade Classifier: A Proof of Concept

Yusuf Kola Ahmed[1,2(✉)], Nasir Ayuba Danmusa[1], Taofik Ahmed Suleiman[3], Kafilat Atinuke Salahudeen[1], Sani Saminu[1], Abdul Rasak Zubair[4], and Abdulwasiu Bolakale Adelodun[5]

[1] Department of Biomedical Engineering, University of Ilorin, Ilorin, Nigeria
[2] Department of Occupational Therapy, University of Alberta, Edmonton, Canada
ykahmed@ualberta.ca
[3] Department of Medical Imaging and Computing, University of Girona, Girona, Spain
[4] Department of Electrical and Electronic Engineering, University of Ibadan, Ibadan, Nigeria
ar.zubair@ui.edu.ng
[5] ECWA College of Nursing and Midwivery, Egbe, Kogi State, Nigeria

Abstract. According to the wheelchair foundation, about 1.86% of the world's population requires a functional wheelchair. Most of these wheelchairs have manual control systems which puts millions of people with total paralyzes (total loss of muscle control including the head) at a disadvantage. However, the majority of those who suffer from muscular and neurological disorders still retain the ability to move their eyes. Hence the concept of eye-controlled wheelchair. This paper focused on the design and development of a hybrid control system (eye and mobile interface) for a wheelchair prototype as a proof of concept. The system was implemented using the pre-trained Haar cascade ML classifier in open CV. Focus was shifted from high accuracy common to lab-based studies to deployment and power consumption which are critical to usability. The system consists of a motor chassis that takes the place of a wheelchair, a raspberry pi4 module which acts as a mini-computer for image and information processing, and a laser sensor to achieve obstacle avoidance. The Bluetooth module enables serial communication between the motor chassis and the raspberry pi, while the power supply feeds the raspberry pi and the camera. The system performance evaluation was carried out using obstacle avoidance and navigation tests. An accuracy of 100% and 89% were achieved for obstacle avoidance and navigation, respectively, which shows that the system would be helpful for wheelchair users facing autonomous mobility issues.

Keywords: Paralysis · Wheelchair · Eye control · Obstacles avoidance · Camera

© The Author(s), under exclusive license to Springer Nature Switzerland AG 2023
Z. Hu et al. (Eds.): ICCSEEA 2023, LNDECT 181, pp. 748–758, 2023.
https://doi.org/10.1007/978-3-031-36118-0_66

1 Introduction

Paralysis is a total or partial loss of physiological functions in some part of the human body [1]. Paralysis is often characterized by either motor or sensory loss or both in the affected area. Common causes of paralysis include Stroke, Multiple sclerosis, Spinal cord injury, Cerebral palsy among others [1, 2]. An age-adjusted comparison of results from the 1980 and 1990, National Health Interview Survey (NHIS) found a 26% increase in the use of canes and a 57% and 65% increase in the use of walkers and wheelchairs, respectively, among all ages which increased to 83% by 1994 [3]. Besides, the latest study has reported that 1.86% of the world population need functional wheelchair [4]. The design of wheelchairs for paralyzed patients has witnessed advances from conventional manually powered to electrical wheelchairs [5]. The population of elderly citizens in the world is also increasing, with this increase, there is a need to design welfare robotic devices for such people. A conventional wheelchair tends to focus exclusively on manual use which assumes users are still able to use their hands and excludes those unable to do so. Many of those suffering from complete paralysis usually still retain the ability to control eye movements, which inspired the development of an eye-controlled electric wheelchair. There are in existence different modes of eye-controlled wheelchair such as EOG – control, gaze estimation control and video based control systems. However, most of them only have one control system (the eye) and few who incorporated an auxiliary control system employed manual gear steering (joy stick) which requires some form of energy and skill to use.

In order to increase versatility and ease of use, a dual-controlled eye tracking system for wheelchair navigation is proposed, which uses eye gaze as the primary control system and mobile interface as an auxiliary. The proposed system uses an off-the-shelf webcam for eye image acquisition using Open CV/CV2 library to process the signal in a raspberry pi 4 device which has improved functions and uses Linux based operating system. Going by the ubiquity of mobile phone users, the auxiliary control module is implemented on a mobile phone interface such that instead of a joystick, the users' phone is connected by Bluetooth to the Arduino Uno control board to steer the motors of the wheelchair. Hence, increasing the system's usability for patients with either complete or partial paralysis and the elderly. A wheelchair prototype is employed in the form of a robotic car, and the obstacle avoidance mechanism is achieved using a laser sensor

1.1 Literature Review

Over the years, research works have been using different control applications on wheelchair systems aside from manual control approaches. The common physiological signals often adopted include Electrooculography (EOG) signal, Electromyography (EMG) signal, and Electroencephalography (EEG) signal [2, 4]. The EOG signal and video-based control systems have been extensively studied for eye control applications in a wheelchair [6]. When the eye moves, the EOG signal is generated from the changing potential difference between the cornea and the retina. This change is registered as a potential difference useful in detecting the location or gaze of the eye. This system often requires the direct attachment of electrodes to the face/forehead of the user, as in Fahim *et al.* [1]. Some EOG-based studies developed better eye movement tracking algorithms and wheelchair control [7–9]. [1, 9, 10] adopted EOG and EMG signals in the control system design, while Naeem and Qadar [11] used voice control. The major drawback of EOG-based design is that it often constitutes a physical burden on the user besides the heavy noise attributed to the signal [4] and the high level of cognition required. Voice control is often plagued with interference in a sound wave from the environment coupled with its high voice energy demands [11].

A few works on video-based eye control include the work of Mani et al. [12]. His work processed the image of the eye captured by a head-mounted camera to activate the direction of motion in 3 dimensions (left, right, and forward). However, the performance evaluation of the system was not wholly reported.

Another design by Patel et al. [13] was built around Raspberry pi using an open CV library for image processing of the acquired face image for eye location. The system activates and deactivates in 3 seconds; eye closure and directional motion are initiated with the eye's blinking. The procedure was adequately explained, but the system's performance was not assessed using accuracy and response latency which are vital tools for such systems evaluation.

Lately, machine learning algorithms/models [14] have been employed in the image processing of eye-controlled wheelchairs with a significant increase in accuracy [15, 16]. George et al. [16] implemented a convolution neural network (CNN) model using a webcam image for real-time gaze classification in seven directions: right, left, center, forward, upright, up left, down left, and down right. Localization of facial regions was carried out by a modified Viola-Jones algorithm. The CNN model for combined eyes (left and right eye) was observed to yield better accuracy than training each eye separately. It gave good accuracy, but it's computationally complex. Krafka et al. [15] also proposed an eye tracking system using deep end-to-end CNN with a larger data set obtained from 1450 participants. To check computational time and high complexity, dark knowledge was used with a simpler network, however, the frame rate is less than the standard 20-25 FPS needed for smooth real-time application.

2   Methodology

2.1   Material

Material selection was a key part of this project. Before the development of the project, certain factors were considered for the design to meet technical specifications while minimizing cost and increasing easy movement [17, 18]. The important materials selected are the Web camera, Raspberry pi board, Arduino Uno, H-bridge Motor driver (L298), Bluetooth module, and Motor chassis.

2.2   System Design and Description

The system is divided into three modules: the image data acquisition module, the wheelchair control/steering module, and the wheelchair. Fig. 1 shows the block diagram of the eye-controlled wheelchair system. The web camera captures the eye image data which is sent to the Raspberry-pi. The Raspberry-pi serves as the mini processor for synthesizing the pupil location [19]. The location identified by the raspberry–pi is fed to the Arduino Uno board where different eye positions are programmed into left, right, and forward movement commands. These commands are fed to the motor driver of the wheelchair for navigation in the desired direction. The auxiliary control module involves the mobile App with a control interface connected by Bluetooth to the Arduino board [20]. This can also be used to navigate the wheelchair aside the eye control. The laser sensor feedback mechanism for obstacle avoidance takes care of system safety.

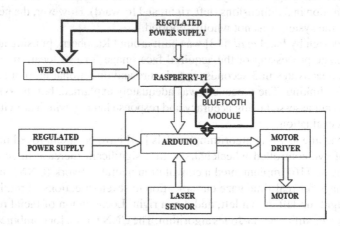

Fig. 1. System block diagram

2.2.1 Image Data Acquisition System

An off-the-shelf raspberry pi camera is wired to the raspberry pi board with an extended cable. The camera is placed at a distance of 5cm from the face and at eye level such that it can capture the facial images from which eye/pupil location would be extracted.

The dynamics of navigation proposed for the wheelchair includes forward, left, right, and backward movement with a start and stop process initiated by the blinking of the eye. The first left eye winking starts the image acquisition and subsequent movement of the wheelchair, while any second left eye winking would automatically stop data acquisition and, subsequently, the wheelchair stops, as implemented in [4]. Hence, four (4) eye states are employed in the navigation process.

Haar cascade was used in the gaze tracking algorithm for pupil detection because of its simplicity and fast processing time, making it useful in real-time applications. This method uses the concept of features and classifiers in machine learning (ML) for its detection scheme. The fundamental idea behind Haar cascades is that they employ a collection of salient "features" that can be computed from an image, including the intensity, difference between neighboring pixels or tiny areas of the image [21–24].

The Haar-like features in Haar cascade are calculated using the following equation:

$$f(x, y) = \sum_{i=x}^{x+w-1} \sum_{j=y}^{y+h-1} I(i, j) \tag{1}$$

where I(i,j) represents the intensity of the pixel located at position (i,j). The Haar-like feature is calculated as the difference between the sum of intensity of pixels in two rectangular regions, defined as:

$$f(x, y) = \sum_{i=x}^{x+w/2-1} \sum_{j=y}^{y+h-1} I(i,j) - \sum_{i=x}^{x+w/2} \sum_{j=y}^{y+h-1} I(i,j) \qquad (2)$$

The Haar cascade uses an Adaboost classifier, which is a type of machine learning classifier that combines multiple simple classifiers to form a strong classifier. The Adaboost classifier makes its decision using the following equation:

$$h(x) = \begin{cases} 1 & \sum_{t=1}^{T} \alpha_t h_t(x) \geq \frac{1}{2} \sum_{t=1}^{T} \alpha_t \\ 0 & otherwise \end{cases} \qquad (3)$$

where T is the number of weak classifiers, α_t represents the weight associated with each weak classifier, $h_t(x)$ represents the decision made by the tth weak classifier, and sign(x) is the sign function, which returns 1 if x is positive and -1 if x is negative

A classifier is therefore trained using these characteristics to differentiate between the pupil and the background. The trained classifier is used by the Haar cascade to assess the characteristics of several eye regions, including the eye lid, iris, and cornea, and determine whether or not the eye frame is likely to contain the object of interest. This procedure is repeated for several eye regions and points most likely to contain the pupil is marked "detection". However, the frame illumination often changes with the nature of the environment; hence Histogram Equalization is used to change the probability distribution of the image to a uniform distribution for better image contrast [25].

2.2.2 Wheelchair Control/Steering System

The decision/output of the pupil data acquisition system is used to navigate the wheelchair. These decisions are coded in form of left, right, forward, backward, and start/stop modes such that when fed to the H-bridge motor driver, it steers according to the code in the received signal.

2.2.2.1 Mobile Control Interface

The primary control system for the wheelchair prototype is the eye gaze detection system. However, a secondary control system is incorporated in an RC-controlled mobile interface to improve the versatility of the device for both tetraplegic and non-tetraplegic patients, such as the elderly. This uses a Bluetooth module to communicate between the app interface and the H-bridge Arduino motor driver to achieve motion control. The H-bridge motor driver is responsible for steering the wheels' directions. An open-access RC controller program was adapted to achieve forward, backward, left, and right movements and other sub-directions (Fig.3). Two DC motors were installed in the prototype and linked with the front wheels.

2.2.3 Safety Subsystem: Laser Sensors

For safety precautions, the wheelchair is designed to have an automated stopping mechanism that will disconnect the gaze-based controller and stop the chair as soon as it approaches any nearby object. A proximity sensor quantitively measures how close an object is to the sensor. It has a range, and a flag is raised whenever an object enters this threshold region. The sensor sends out electromagnetic waves, which are picked up after being reflected off nearby objects. The procedure is quite similar to how radar works. A laser proximity sensor is adopted in this work.

2.2.4   Mobile Application

The Bluetooth RC controller is a mobile application that connects with the Bluetooth attached to the wheelchair (prototype) and is used to send commands serially to the Bluetooth, which in turn controls the motors, as an alternate method of control for partially paralyzed subjects (lower limbs) to aid easy control of the wheelchair. Fig. 2 (c) shows the Bluetooth RC controller mobile application interface.

2.2.5   Wheelchair

A motor chassis serves as a prototype for a wheelchair and necessary components and modules are attached to it.

2.3   Overall Design Outlook

The various modules are interconnected to form the complete system. Fig. 2 describe the complete system in more details. The device operational flow chart is shown in Fig. 3. The web camera picks the face and eye image frames and the algorithm checks if the eye is open or closed. If closed, the loop returns and check again for eye status. If open, the gaze direction is identified by template matching and accurate direction is detected before obstacle avoidance is checked. The obstacle avoidance is set at a threshold (stopping

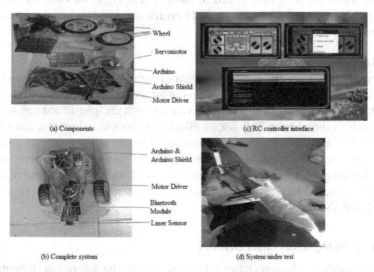

(a) Components
Wheel
Servomotor
Arduino
Arduino Shield
Motor Driver

(c) RC controller interface

(b) Complete system
Arduino &
Arduino Shield
Motor Driver
Bluetooth
Module
Laser Sensor

(d) System under test

Fig. 2. Pictorial description of the system

distance) of 1m as used in [4]. If an obstacle is detected at a distance <=1 m, the device stops abruptly or would not move if detected at the point of start; else, it moves in the determined direction.

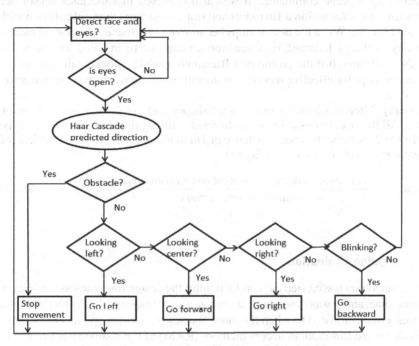

Fig. 3. Flowchart of the navigation control algorithm

3   Results and Discussion

3.1   Performance Evaluation

Testing the whole design is carried out to certify that the constructed design is in its proper working condition and ensure that operating expectations are achieved. Thus, tests carried out include the following: Test for the output voltage of the power supply and charging unit using a digital multimeter, test for the right connection of components with Arduino, components like the motor driver, Bluetooth module, laser sensor, servo motor, testing the functionality of the serial connection between the Bluetooth module and the Raspberry pi, testing for accuracy in the detection of various eye gaze by the camera as seen in Fig. 2(d), and testing for the effectiveness of the detected eye gaze's condition transmission and what is received by the motors.

From the tests carried out, it was confirmed that the camera detects certain eye gazes, including: blinking, looking center, looking left, and looking right, and by clicking buttons right, left, up, and down on the mobile application, the motors carried out the respective appropriate commands. It was also observed that the laser sensor detects obstructing obstacles within a 1m radius and stops movement to avoid collision with the detected obstacle. When the device stops because of an obstacle, a backward movement (given by blinking), leftward, rightward movements can be initiated for a new path. It was also confirmed that the prototype's Bluetooth module automatically connects with the Raspberry pi for effective receiving/transmitting of information upon powering ON the device.

Twenty different obstacles of different shapes and sizes were put in front of the device. All the twenty obstacles were detected, causing the wheelchair (prototype) to stop forward movements when a distance of 1m to the obstacle was reached. Therefore, the accuracy is 100% as given by Eq. (4).

$$Accuracy = \frac{\text{No of obstacles detected} - -\text{No of obstacles undetected}}{\text{Total number of tests carried out}} = \frac{20-0}{20} \times 100 = 100\%$$

$$(4)$$

3.2 Navigation Evaluation

Another important test carried out was to monitor the design for seamless navigation and response time, and it was observed that there is a slow response time after the command has been given compared to when the motor carries out its appropriate commands, and this is due to the low random access memory (RAM) of the raspberry pi4 model being used in this design. Therefore, obtained average navigation/direction accuracy was 89.16 %. This accuracy was calculated using equation 5.

$$accuracy = \frac{No\ of\ correct\ predictions}{Total\ no\ of\ predictions} \times 100 \qquad (5)$$

In turn, the speed of the motors was reduced to compensate for this setback so that more distance is not covered before another command is received. Table 1 shows the average accuracy (%) from the movement of the wheelchair for four (4) different directions after five (5) trials for each subject experimented with a total of three (3) subjects. Two different kind of average accuracies were calculated: the average navigation/directional accuracy and the average accuracy/person. The average accuracy per person showed how each participants performed on the overall task. Subject 1 had the best score of 90, suggesting that he was slightly able to direct is eye better than others. Table 2 showed the score from the mobile interface test and overall score was 100%. Hence, navigation was smooth with the mobile interface.

Table 1. Average accuracy (%) from the movement of the wheelchair for four different directions after five trials for each subject experimented with a total of 3 subjects using eye control.

Direction of movement	Subject 1	Subject 2	Subject 3	Average accuracy per direction (%)
Right	85	90	90	88.3
Left	90	85	85	86.67
Forward	85	85	85	85
Backward	100	95	95	96.67
Average accuracy per subject (%)	90	88.75	88.75	89.16 89.16

Table 2. Average accuracy (%) from the movement of the wheelchair for four different directions after five trials for each subject experimented with a total of 3 subjects using the mobile interface.

Direction of movement	Subject 1	Subject 2	Subject 3	Average accuracy per direction (%)
Right	100	100	100	100
Left	100	100	100	100
Forward	100	100	100	100
Backward	100	100	100	100
Average accuracy per subject (%)	100	100	100	100 100

4 Summary and Conclusion

The paper is centered on the design and development of a prototype electric-powered wheelchair that has an eye-controlled navigation system with obstacle avoidance technology for safety. The system provides navigation autonomy for people with total paralysis by using the eye to drive the wheelchair and even extended the research by incorporating a mobile control interface instead of the common joystick auxiliary control to extend the system's usability to the elderly and people with hemiplegia. Haar cascade ML algorithm with Histogram Equalization (for illumination control) was adopted in the design for its simplicity, high perception rate and fast processing time. Four movement directions were achieved from the eye image frames, while more directional movements were obtainable from the auxiliary mobile control interface. Hence, the system is versatile enough to accommodate persons with total and partial paralysis. The average navigation accuracy of the device/person is 89.16%; this is slightly low because of the low computation power of Raspberry Pi4, and better accuracy could be achieved when the eye detection algorithm is fine-tuned, or deep learning classification is introduced. However, using CNN often require GPU which makes implementation and deployment on edge devices like raspberry pi impractical. The obstacle avoidance score was 100%. The paper provided an extensive review of state-of-the-art technology in this domain, identifying that most studies are lab-based focusing more on accuracy rather than deployment and power consumption which are critical to usability. This work also put into consideration implementation in low-resource settings. Future recommendations to enable optimum satisfaction of this design should include: the implementation of the project on a typical wheelchair and utilization of the necessary components for the effective running of the wheelchair, the usage of the latest model of Raspberry pi with higher random access memory(RAM) to ensure swift navigation of the wheelchair, a tracking device using a motion sensor can be incorporated in the design for emergency tracking of patient's whereabouts, the utilization of an infrared webcam for night gaze detection. Finally, more laser sensors can be incorporated to cater to the whole range of motion of the device.

References

1. Bhuyain, Md.F., Kabir Shawon, Md.A.-U., Sakib, N., et al.: Design and development of an EOG-based system to control electric wheelchair for people suffering from quadriplegia or quadriparesis. In: 2019 International Conference on Robotics, Electrical and Signal Processing Techniques (ICREST). Pp. 460–465 (2019)
2. Subramanian, M., Park, S., Orlov, P., et al.: Gaze-contingent decoding of human navigation intention on an autonomous wheelchair platform. IEEE/EMBS Neural Eng. (2021). https://arxiv.org/abs/2103.03072.
3. LaPlante, M.P., Kaye, H.S.: Demographics and trends in wheeled mobility equipment use and accessibility in the community. Assistive Technol. 22, 3–17 (2010). https://doi.org/10.1080/10400430903501413
4. Dahmani, M., Chowdhury, M.E.H., Khandakar, A., et al.: An intelligent and low-cost eye-tracking system for motorized wheelchair control. Sensors 20, 3936 (2020). https://doi.org/10.3390/s20143936
5. Simpson, R.C.: Smart wheelchairs: a literature review. J Rehabil Res Dev 42, 423–436 (2005). https://doi.org/10.1682/jrrd.2004.08.0101
6. Eid, M.A., Giakoumidis, N., El Saddik, A.: A novel eye-gaze-controlled wheelchair system for navigating unknown environments: case study with a person with ALS. IEEE Access 4, 558–573 (2016). https://doi.org/10.1109/ACCESS.2016.2520093
7. Djeha, M., Sbargoud, F., Guiatni, M., et al.: A combined EEG and EOG signals based wheelchair control in virtual environment. In: 2017 5th International Conference on Electrical Engineering – Boumerdes (ICEE-B), pp. 1–6 . IEEE, Boumerdes (2017)
8. Pingali, T.R., Dubey, S., Shivaprasad, A., et al.: Eye-gesture controlled intelligent wheelchair using Electro-Oculography. In: 2014 IEEE International Symposium on Circuits and Systems (ISCAS), pp. 2065–2068 (2014)
9. Hashimoto, M., Takahashi, K., Shimada, M.: Wheelchair control using an EOG- and EMG-based gesture interface. In: 2009 IEEE/ASME International Conference on Advanced Intelligent Mechatronics, pp 1212–1217. IEEE, Singapore (2009)
10. Tsui, C.S.L., Jia, P., Gan, J.Q., et al.: EMG-based hands-free wheelchair control with EOG attention shift detection. In: 2007 IEEE International Conference on Robotics and Biomimetics (ROBIO), pp 1266–1271 (2007)
11. Naeem, A., Qadar, A.: Voice Controlled Intelligent Wheelchair using Raspberry Pi. Int. J. Technol. Res. 2(2), 65–69 (2014)
12. Mani, N., Sebastian, A., Paul, A.M., et al.: Eye controlled electric wheel chair. IJAREEIE 4, 3494–3497 (2015). https://doi.org/10.15662/ijareeie.2015.0404105
13. Patel, S.N., Prakash, V.: Autonomous camera based eye controlled wheelchair system using raspberry-pi. In: 2015 International Conference on Innovations in Information, Embedded and Communication Systems (ICIIECS), pp. 1–6. IEEE, Coimbatore, India (2015)

14. Gupta, P., Saxena, N., et al.: Deep neural network for human face recognition. IJEM **8**, 63–71 (2018). https://doi.org/10.5815/ijem.2018.01.06

15. Krafka, K., Khosla, A., Kellnhofer, P., et al.: Eye tracking for everyone. In: 2016 IEEE Conference on Computer Vision and Pattern Recognition (CVPR), pp. 2176–2184. IEEE, Las Vegas, NV, USA, (2016)

16. George, A., Routray, A.: Real-time eye gaze direction classification using convolutional neural Network. In: 2016 International Conference on Signal Processing and Communications (SPCOM), Bangalore, India, pp. 1–5 (2016). https://doi.org/10.1109/SPCOM.2016.7746701

17. Ahmed, Y.K., Zubair, A.R., Sani, S., Akande, K.A., Afolayan, M.A., Afonja, A.A.: Design and construction of a portable electronic sleep inducer for low resource settings. FUOYE J. Eng. Technol. **5**(2), 84–88 (2020). https://doi.org/10.46792/fuoyejet.v5i2.475

18. Ahmed, Y.K., Ibitoye, M.O., Zubair, A.R., et al.: Low-cost biofuel-powered autoclaving machine for use in rural health care centres. J. Med. Eng. Technol. **44**, 489–497 (2020). https://doi.org/10.1080/03091902.2020.1825847

19. Andrea, R., Ikhsan, N., Sudirman, Z.: Face recognition using histogram of oriented gradients with tensorflow in surveillance camera on raspberry Pi. IJIEEB **14**, 46–52 (2022). https://doi.org/10.5815/ijieeb.2022.01.05

20. Mahadevaswamy, U.B.: Wireless wearable smart healthcare monitoring using android. Int. J. Comput. Netw. Inform. Secur. **10**(2), 12–19 (2018). https://doi.org/10.5815/ijcnis.2018.02.02

21. Cuimei, L., Zhiliang, Q., Nan, J., Jianhua, W.: Human face detection algorithm via Haar cascade classifier combined with three additional classifiers. In: 2017 13th IEEE International Conference on Electronic Measurement & Instruments (ICEMI), pp. 483–487 (2017)

22. Ahmed, Y.K., Zubair, A.R.: Performance evaluation of wavelet de-noising schemes for suppression of power line noise in electrocardiogram signals. Nig. J. Technol. Dev. **18**, 144–151 (2021)

23. Zubair, A.R., Ahmed, Y.: Engineering education: computer-aided engineering with MATLAB; discrete wavelet transform as a case study. Int. J. Comput. Appl. **182**, 6–17 (2019). https://doi.org/10.5120/ijca2019918598

24. Zubair, A.R., Alo, O.A.: Grey level co-occurrence matrix (GLCM) based second order statistics for image texture analysis. Int. J. Sci. Eng. Invest. **8**(93), 64–73 (2019)

25. Zubair, A.R.: Comparison of image enhancement techniques. Int. J. Res. Commer. IT Manage. **2**(5), 44–52 (2012)

A Case-Based Approach in Astrological Prediction of Profession Government Officer or Celebrity Using Machine Learning

Snehlata Barde[✉], Vijayant Verma, and Apurva Verma

MSEIT, MATS University, Raipur, Chhattisgarh 493441, India
v.snehabarde@gmail.com, {drvijyantverma,
apurv}@matsuniversity.ac.in

Abstract. Astrology is regarded as a method of foreseeing the future. Everyone who consults an astrologer, from ancient times to the present, has knowledge of the future. Astrologers use planetary positions to discuss the many stages in a person's life. Understanding the interplay of planetary positions can reveal critical life stages. Astrologers utilize a person's horoscope to make predictions. But the accuracy of horoscopes depends only on the knowledge of "Jyotish acharya" especially when it comes to employment forecasts. Machine learning offers the greatest solution for the analysis and prediction of such types of applications. The paper's primary goal is to find the scientific validity and rules for astrological prediction using a case-based reasoning method that mitigates the shortcomings of the conventional method, clarifies the fundamental principles of astrological prediction, and establishes the reliability of astrology using machine learning classification approaches Naive Bayes, Logistic-R, and J48. Experiment applied to the horoscope to identify whether the person becomes a government officer or a celebrity or not. 300 people's data were gathered in order to conduct these studies. 100 were celebrities, 100 were Govt.-officer and the remaining people were unemployed or students. The information gathered included the person's time, place, and date of birth.

Keywords: Astrology · Classification techniques · Decision table · Logistic -R · Naïve Bayes J48 · Horoscope

1 Introduction

Numerous astrological research programs and pieces of software are now being created. Astrology is the subject of many businesses, publications, journals, articles, courses, and studies. Additionally, astrologers promise that they can predict someone's future using their astrological expertise and intuition. All of this has increased our curiosity about the accuracy and validity of astrology. There aren't many clear-cut, standardized rules for astrological prediction. Each astrologer has their own set of guidelines and techniques for making predictions. The astrologer uses the diagnosis and prognostication of the prior horoscopes of people in related industries to calculate the prognosis of the individual in

© The Author(s), under exclusive license to Springer Nature Switzerland AG 2023
Z. Hu et al. (Eds.): ICCSEEA 2023, LNDECT 181, pp. 759–773, 2023.
https://doi.org/10.1007/978-3-031-36118-0_67

front of him/her. It is predicated on a system of thinking that shows how interconnected the planets are and helps identify human personalities, as opposed to biometrics, which use features like a person's face, ear, iris, palm, and footprint to identify them [1].

In order to determine the scientific validity and guidelines for astrological prediction, this study discusses astrological experimentation and use the Case Base Reasoning method.

To learn from a vast amount of previously known data and gain knowledge from those facts, case base machine learning algorithms and other classification methods can be employed. It is predicated on the idea that an issue can be easily solved using the answer offered to a situation that is similar to it in the past. As a result, the system can benefit from prior cases [2]. By utilizing a classification technique that is appropriate for the specific issue area, learning can be accomplished. The birth chart and associated status of the person as Government officer, celebrity or not are saved in the case base storage, which can be used to learn from past instances. Therefore, the above method can offer a solid foundation for doing research in the area of astrology and assist in giving astrology scientific validity [3].

A fact-based structure that accumulates algorithmic knowledge over time is the basic notion underpinning machine learning, a subset of artificial intelligence. With little hominoid touch, decision-making and forecasting are made easier because of this set of information, known as training data. Data filtering and analytics are two areas where machine learning techniques are used [4, 5]. Only a few parameters were used that were directly available to us but as in astrology many other parameters are used to do predictions similarly, we can use parameters such as the environment in which a person lives, the person's background persons education, parent's education, parents background, money status, the strength of planets and so on for prediction.

2   Related Work

There aren't many significant investigations on the scientific validity of astrology that have been published, from both a research and an applied perspective. To develop practical applications of astrological prediction and to demonstrate the accuracy of astrology, academic scholars and astrologers must collaborate. Although astrology is regarded as a pseudo-science, there is a wealth of literature on the subject. Studies were conducted after the 15th century to establish astrology's scientific validity. Some scientists believe that astrology is not scientific [6, 7], while others believe that a thorough investigation is necessary in this area to demonstrate its scientific veracity [8]. Some people think that astrology is divided into two categories: predictive and non-predictive. Researchers believe that predictive astrology is the best method for determining whether astrology may be used to foretell a person's future occurrences [9]. In a research investigation on the angularity of the moon, Irving reported encouraging findings. Morin hypothesized that significant angular planets in a person's birth chart had an impact on their lives at the same time [10].They are looking for the existence of astrology as well as universal rules using a variety of scientific approaches. To ascertain the person's occupation, they will apply the Xero, Simple Cart, and Decision Table categorization algorithms. 24 Singer recordings, 24 Player records, and 10 Scientist recordings made up the data set used to learn classification. To do analysis and prediction tasks, the Weka tool, which is open-source and free, is used [11]. The Astrological Prediction System about Job to discover the relationship between a person's profession and the planetary alignment of the stars at the time of his or her birth. This relationship is then used by the system to predict how new cases will be handled [12]. by carrying out a methodical scientific experiment on the birth charts of people with cancer and people without cancer, one can learn some fundamental principles of Vedic astrology. Without a trustworthy set of parameters for cancer prediction, but there was no statistically significant difference between the two data sets in terms of the astrological positivity or negativity of any of the entities we looked into [13]. Public health policymakers need a reliable forecast of verified cases in the future to establish medical facilities. Algorithms that use machine learning can estimate the future based on historical data [14].

Artificial Neural Network and Decision Tree algorithms, along with meteorological data collected between 2000 and 2009 in Ibadan, Nigeria, were used to predict weather reports such as rainfall, temperature, and wind speeds. The classifier algorithms were trained using a data model that was created for the meteorological data. Using standard performance metrics, the algorithms' effectiveness was assessed, and the best algorithm was used to produce classification criteria for the mean weather variables [15]. The description of the behaviour of the k-NN rule in different situations. In the current study, class overlapping, feature space dimensionality, and class density are defined along with their links to the practical accuracy of this classifier using a variety of data complexity measures [16].

The models are produced by recursively partitioning the data space and fitting a simple prediction model within each division. As a result, the partitioning may be visually represented as a decision tree. Classification trees are employed for dependent variables that have a finite number of unordered values, and the prediction error is measured in terms of misclassification cost [17]. CBR is gaining popularity since it needs to gather case histories and find the characteristics that significantly describe a case in order to solve problems using this approach. The development of case-based reasoning has received a great deal of effort and focus, from establishing the principles [18] to implementing methodologies [19] to the creation of many applications. The fields of industrial implementation [20], medical practice, weather forecasting [21], stock market forecasting[16], design and planning[22], geographic analysis, and natural language processing [23] have all made major contributions to the subject of case-based reasoning.

3   Conventional Approach in Astrology

Astrology is a traditional Indian science that makes calculations based on factors such as geography, algebra, astronomy, and cultural influences. Planets are the subject of astrology. A person's horoscope is made by gathering information about that individual, such as their birth date, exact time of birth, and the location of their birth, and then figuring out every aspect of their life, from conception to passing away. In the conventional approach, it is challenging to create a horoscope for a stranger because nobody is an expert at calculating planet positions or rashies. Only "Jyotishaarcharya" is aware of the kundalini and is capable of making predictions about a person's family, finances, education, career, and overall health. Prediction accuracy, though, is confusing and difficult to understand as providing information that may or may not be trustworthy [24]. It is a process of learning. The satisfaction of the client is the primary goal of astrologers. Some reluctance to tell the truth. The conventional approach is shown in Fig. 1.

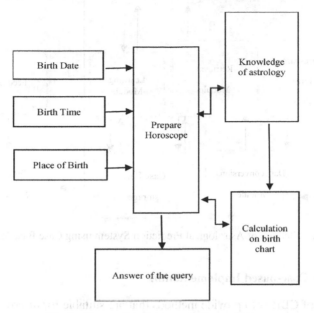

Fig. 1. Conventional Approach for prediction

4 The Architecture of the Case-based Approach in Astrology

According to what we know, astrology is not a branch of science, and the scientific community has rejected it since there is no reliable explanation for how the entire world is presented. Numerous modifications take place as astrology theories' scenarios evolve from conventional to scientific [25]. Figure 2 provides the overall layout of the system for astrological predictions. Through an input interface, the user communicates with the system and provides data for prediction. With the aid of the Data Conversion Module, the data supplied through the input interface are transformed into a format compatible with the data in the case base storage. Based on the data that the Input Interface to the Data Conversion Module provides, the Retrieval Module retrieves the cases from Case-Based Storage. Based on the cases that are updated and saved in the case-based storage, the learning module continuously spreads knowledge. The Inference Engine makes an astrological prediction of the data based on the findings from the Retrieval Module and Learning Module.

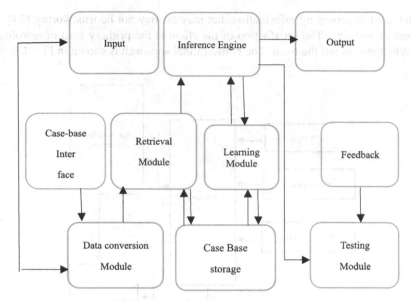

Fig. 2. Architecture of Astrological Prediction System using Case Base Reasoning

4.1 Steps for Case-based Implementation

The objective of CBR is to provide methods that are suitable for discovering information and resolving issues in certain application domains. The primary topics that CBR research concentrates on can be categorized into five groups.

Knowledge Representation
We expressed cases for the astrological prediction system using the Category & Exemplar Model. Here, we provide several categories for famous persons and less well-known individuals, such as celebrities and non-celebrity. It is feasible to find a case in the case base that matches an input description by combining the input features of a problem case into a pointer to the case or category that shares the majority of the features. The information that needs to be entered into the astrological prediction system's case interface is saved in the case base storage. This information is converted into a format that is appropriate for the cases saved after data conversion [26, 27].

Retrieval Techniques
Nearest Neighbor, Induction, Knowledge Guided Induction, and Template retrieval are a few of the several forms of retrieval. Knowledge-guided induction is used in this situation. The knowledge is built by choosing the features that will be utilized to retrieve cases from the case base storage, and these techniques include Logistic, Naive Bayes,

J48, and Decision Table. The cases from the case base are retrieved using the retrieval module.

Reuse Strategies
The two fundamental strategies for making use of prior examples are copying and adapting. The transformational reuse of adapt approach is used when information in the form of a transformational operator is applied to a solution for new circumstances.

Revision Strategies
A person's forecast is made by an astrological prediction system, and the result is then cross-verified with the user using a feedback interface. No revision is required if the results are accurate; however, if the results are incorrect, revision is required.

Retention Strategies
When a case is forecasted, the forecast is evaluated to determine whether it is accurate. If it is, the case is retained in the case base storage without any changes, but if it is inaccurate, the case is destroyed and not kept.

5 Experimental Analysis

A key component of data-driven system development, machine learning is a branch of artificial intelligence that focuses on gradually improving the computational abilities of machine algorithms. The training data set facilitates decision-making and forecasting with little to no human involvement [28].

5.1 Raw Data Set

Data on 300 people were gathered, 200 for the training dataset of which 100 were celebrities and 100 were Govt.-officer. Astro Data Bank was used to gather the records of notable people from throughout the world. It is a division of Astrodienst AG and a non-profit community endeavor with no financial gain. The final 100 testing data points were gathered on paper and via emails from trustworthy locals. These people were not in government job or not celebrities. The information that was gathered included the person's status, place of employment, date of birth, location of birth, and time of birth. The person's birth chart was then made using this information.

5.2 Preprocessing Data

After compiling the necessary information, we created the individual's horoscope chart, which depicts the twelve zodiac signs and a planet's position.

5.3 Feature Extraction

The twelve houses in a horoscope define the placements of nine planets, each house has a zodiac sign, several combinations of planets and their yogic relationships, the strength of the houses, and other astrological terms. Segment charts that examine a person's various characteristics aid in obtaining a highly precise conclusion.

6 Classification Techniques

Artificial intelligence has a variety of supervised and unsupervised categorization techniques that can be used to produce general theories from data sets that can be utilized to make future predictions. It can be difficult to determine whether this approach is suitable for data collection. Due to their simplicity and complexity, three AI classification algorithms were selected [29].

6.1 Logistic Regression

A widely used supervised classification method in machine learning is logistic-R. For the purpose of making predictions from the collection of independent factors, various class label dependent variables were used. Discrete outputs from logistic-R include true/false, yes/no, on/off, and 0/1. Provides output between 0 and 1, rather than the precise values of 0 and 1.

The logistic regression was produced by the linear regression. Equation for a straight line written as:

$$Z = B0 + B1X1 + B2X2 + B3X3 + \ldots\ldots + BnXn \tag{3}$$

The equation of logistic-R is in the form
Z = 1.

$$log\left(\frac{p}{1-p}\right) = B0 + B1X \tag{4}$$

Projected probability value of X is

$$P = \frac{e^{(B0+B1X)}}{1 + e^{(B0+B1X)}} \tag{5}$$

6.2 J48

Using the idea of information entropy, this method, like the ID3 algorithm, builds decision trees from a collection of training data. The training data consists of a collection of previously identified objects [29]. a1, a2,... Samples Each sample, ai, is represented by a p-dimensional vector, (v1, I v2, I vp, I), where the vj denotes the class to which the sample belongs as well as the attribute values or qualities that apply to that sample. The ideal property to divide on for the best classification accuracy is the feature with the most data.

6.3 Naïve Bayes Classification

Another popular and effective probability-based classification method is the naive Bayes approach. When the class has more open-ended problems, it is appropriate to anticipate whether the class's characteristics are present or absent, connected to or disconnected from other components.

Despite the fact that Bayesian approaches are more accurate, many used naive Bayes classification applications rely on them.

Let R1, R2, R3......Rn be 'n' classes

All Naïve Class independent conditionally

$$P(X/Rk) = \prod_{i=1}^{n} P\left(Xi/Rk\right) \tag{1}$$

$$P(X|Rk) = P(X1|Rk)*P(X2|Rk)*\ldots*P(Xn|Rk) \tag{2}$$

7 Result and Analysis

T horoscope chart was used to construct a tabular record of 300 pertinent data that contained the person's birth details. After gathering the data, we used 'Astrosage' Kundli software to generate a horoscope chart for each individual to view the positions of the planets, their combinations, and zodiac sign in the 12 houses depicted in Fig. 3.

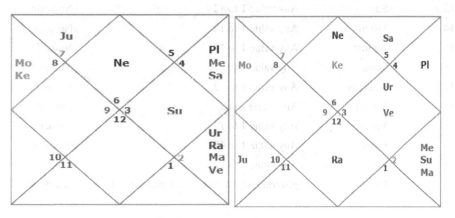

Fig. 3. Horoscope of a person

We then created a table to take the attributes such as all planets, zodiac signs with values ranging from 1 to 12, and class with the values of government officers, celebrities or not in Table 1.

The prepared dataset is loaded into the WEKA tool with CSV (comma, delimited) extension format different classes of government officers or celebrity or not any one shown in Fig. 4.

Logistic regression, J48, and naive Bayes were the three classification strategies we worked on. As a result, Table 2 displays the value in terms of correctly classified instances (CCI), incorrectly classified instances (ICI), mean absolute error (MAE), and root mean squared error (RMSE).

According to Fig. 4, three classifiers Logistic-R, Naïve Bayes, and J48, show the results of correctly classified instances (CCI) and incorrectly classified instances (ICI)

Table 1. Basic details of the horoscope of a person in.csv file

S. No.	Attribute	Value	Type
1	Aries	Any value 1 to 12	Numeric
2	Taurus	Any value 1 to 12	Numeric
3	Gemini	Any value 1 to 12	Numeric
4	Cancer	Any value 1 to 12	Numeric
5	Leo	Any value 1 to 12	Numeric
6	Virgo	Any value 1 to 12	Numeric
7	Libra	Any value 1 to 12	Numeric
8	Scorpio	Any value 1 to 12	Numeric
9	Sagittarius	Any value 1 to 12	Numeric
10	Capricorn	Any value 1 to 12	Numeric
11	Aquarius	Any value 1 to 12	Numeric
12	Pisces	Any value 1 to 12	Numeric
13	Sun	Any value 1 to 12	Numeric
14	Moon	Any value 1 to 12	Numeric
15	Jupiter	Any value 1 to 12	Numeric
16	Mars	Any value 1 to 12	Numeric
17	Marcury	Any value 1 to 12	Numeric
18	Saturn	Any value 1 to 12	Numeric
19	Venus	Any value 1 to 12	Numeric
20	Rahu	Any value 1 to 12	Numeric
21	Ketu	Any value 1 to 12	Numeric
22	Class	government officers or Celebrity or not	Nominal

Table 2. Three classifiers results in terms of CCI, ICI, MAE and RMSE

Classifier	Class	CCI	ICI	MAE	RMSE
Naïve Bayes	G	504	96	0.2422	0.3558
Naïve Bayes	C	405	195	0.3294	0.4625
Logistic -R	G	519	81	0.2137	0.3239
Logistic -R	C	512	88	0.2414	0.3449
J48	G	574	26	0.0817	0.2021
J48	C	579	21	0.0651	0.1791

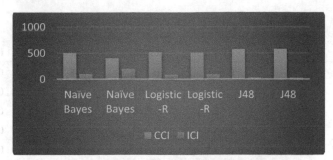

Fig. 4. Results of correct and incorrect classifications

of the two classes of government officers, and celebrities, 579 is the highest CCI and 21 is the lowest ICI generated by J48 for the class celebrity.

Figure 5 shows the mean absolute error (MAE) and root mean squared error (RMSE) values of three classifiers Logistic-R, Naïve Bayes, and J48, for the two classes of government officers and celebrities. 0.3294 maximum and 0.0651 minimum MAE value for celebrity class and 0.4625 maximum and 0.1791 minimum RMSE value generated by Naïve Bayes and J48.

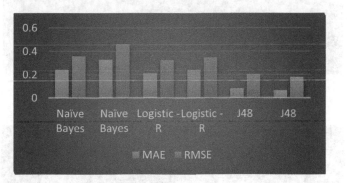

Fig. 5. MAE and RMSE rate for various Classification Techniques

Table 3 displays the value in terms of true positive rate (TPR), False positive rate (FPR), precision and Recall.

Figure 6 shows the true positive rate (TPR) and false positive rate (FPR) values of two classes of government officers and celebrities by the three classifiers Logistic-R, Naïve Bayes, J48. J48 gave 0.965 maximum TPR and 0.119 minimum FPR for celebrity.

According to Fig. 7 the precision and recall values of three classifiers Logistic-R, Naïve Bayes, and J48, for the two classes of government officers and celebrities. The highest precision and recall value 0.964 and 0.965 generated by the J48 for the celebrity class.

Table 4 indicates the F-measure, MCC, ROC, and PRC area values for the two-class labels government officer and celebrity.

Table 3. TPR, FPR, precision, and recall for the classes

Classification	Class	TP Rate	FP Rate	precision	Recall
Naïve Bayes	G	0.840	0.552	0.820	0.840
	C	0.675	0.497	0.765	0.675
Logistic -R	G	0.865	0.595	0.852	0.865
	C	0.853	0.717	0.856	0.853
J48	G	0.957	0.217	0.959	0.957
	C	0.965	0.119	0.964	0.965

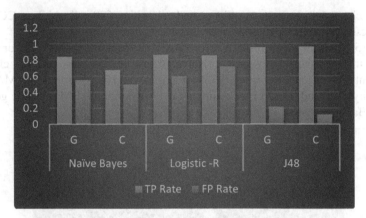

Fig. 6. TP and FR rate for various classification Techniques

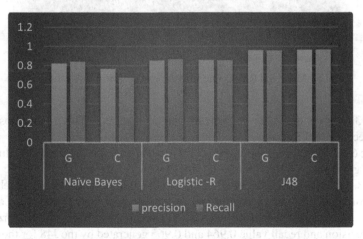

Fig. 7. Precision and Recall for various Classification Techniques

Table 4. F-Measure, MCC, ROC, and PRC area for the classes

Classification	Class	F-Measure	MCC	ROC area	PRC area
Naïve Bayes	G	0.826	0.343	0.732	0.844
	C	0.709	0.143	0.663	0.816
Logistic -R	G	0.839	0.408	0.797	0.877
	C	0.806	0.315	0.730	0.850
J48	G	0.954	0.839	0.896	0.940
	C	0.964	0.871	0.965	0.973

According to Fig. 8 the F-measure and MCC values of three classifiers Logistic-R, Naïve Bayes, and J48, for the two classes of government officers and celebrities. The highest F-measure and MCC values 0.964 and 0.871 generated by the J48 for the celebrity class.

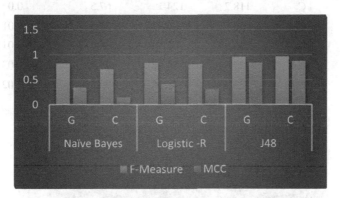

Fig. 8. F-Measure, MCC for three Classification Techniques

Figure 9 shows the ROC and PRC values for the two classes of government officers and celebrities by the three classifiers Logistic-R, Naïve Bayes, and J48. The highest ROC area of 0.965 and PRC area of 0.973 was generated by the J48 for the celebrity class.

Table 5 indicates the two class labels Govt. Officer (G), and celebrity (C), calculation in terms of RAE, RRSE, Accuracy, Training, and Testing Model time in seconds.

Figure 10 shows the accuracy of three classifiers Logistic-R, Naïve Bayes, and J48, for the two classes of government officers and celebrities. The highest accuracy 96.5% measure by the J48 for the celebrity class.

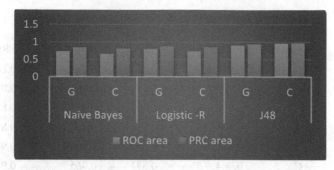

Fig. 9. ROC and PRC area for three Classification Techniques

Table 5. Indicate the RAE, RRSE, Accuracy and Training, and Testing Model Time (in sec.)

Classification	Class	RAE %	RRSE %	Accuracy %	Time in sec
Naïve Bayes	G	86.94	95.46	84	0.02
	C	118.2	124.1	67.5	0.0
Logistic -R	G	76.71	86.91	86.5	0.01
	C	86.68	92.54	85.33	0.01
J48	G	29.32	54.22	95.66	0.03
	C	23.35	48.05	96.5	0.02

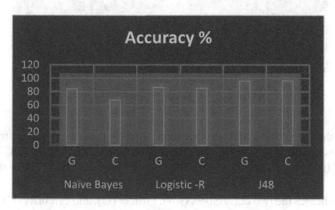

Fig. 10. Accuracy for three Classification Techniques

8   Conclusion

For the same reason, good astrological software is available nowadays, but it does not calculate any data for any event. We used restricted data in this study article; we gathered information from 300 people to calculate their career area, which is government officer or celebrity, based on their birth information, in order to forecast their future. We computed the result and found that the high percentage of accuracy by J48 is 96.5% and the low percentage of accuracy by naive Bayes is 67.5%, respectively, using three classification techniques: nave bays, logistic-R, and J48. We discovered that the outcome is not precisely what we had hoped for, but it is encouraging. Some strategies produce better outcomes than others, necessitating the use of the hybrid methodology to enhance. We're experimenting with different classifiers and evaluating their strengths and limitation.

9 Limitation of the Study

A prognosis for a person who has decided whether or not to pursue the career of Professor, Businessman, or Doctor. However, it is dependent on a variety of factors, such as a person's surroundings. His qualifications are other important factors in determining the type of work he gets, and money is another factor to consider.

References

1. Iqbal, A., Aftab, S.: Prediction of defect prone software modules using MLP based ensemble techniques. Int. J. Inform. Technol. Comput. Sci. 12(3), 26–31 (2020)
2. Barde, S., Zadgaonkar, A.S., Sinha, G.R.: Multimodal biometrics using face, ear and iris modalities. Int. J. Comput. Appl. NCRAIT (2), 9–15 (2014)
3. Aamodt, A., Plaza, E.: Case-based reasoning foundational issues, methodological variations, and system approaches. AICom-Artif. Intell. Commun. 7, 39–59 (1995)
4. Holt, A., Benivell, G.L.: Case Based Reasoning & Spatial Analysis, pp. 95–99. University of Otago Dept. of Information Science (1995)
5. Tiwari, S., Barde, S.: Machine learning make possible to astrological prediction for government job using classification techniques. In: 2021 Asian Conference on Innovation in Technology (ASIANCON), pp. 1–4 (2021). https://doi.org/10.1109/ASIANCON51346.2021.9544586
6. Barde, S., Tiwari, S.: Job prediction astrology for using classification techniques in machine learning. In: Patnaik, S., Kountchev, R., Jain, V. (eds.) Smart and Sustainable Technologies: Rural and Tribal Development Using IoT and Cloud Computing: Proceedings of ICSST 2021, pp. 317–324. Springer Nature Singapore, Singapore (2022). https://doi.org/10.1007/978-981-19-2277-0_29
7. Kelly, I.W.: A Concept of Modern Astrology a Critique. Psychological Reports, 81 (1997)
8. Mcgrew, J.H., Mcfall, R.M.: A scientific inquiry into the validity of astrology. J. Sci. Explor. 4(1), 75–83 (1990)
9. McRitchie, K.: Support for astrology from the carlson double-blind experiment. ISAR Int. Astrologer 40(2), 34–39 (2011)
10. Barde, S., Zadgaonkar, A.S., Sinha, G.R.: PCA based multimodal biometrics using ear and face modalities. Int. J. Inform. Technol. Comput. Sci. 6(5), 43–49 (2014)
11. Irving, K.: Willie Sutton's Rules of Research. NCGR Inc. Research symposium compendium (2007)
12. Morin, J.B.: trans. Holden, J.H., Tempe, A.Z. Astrologia Gallica, Book XXII: American Federation of Astrologers, Inc., 2nd ed (1994)
13. Chaplot, N., Dhyani, P., Rishi, O.P.: Astrological prediction for profession using classification techniques of artificial intelligence. Int. Conf. Comput. Commun. Autom. 2015, 233–236 (2015). https://doi.org/10.1109/CCAA.2015.7148378

14. Rishi, O.P., Chaplot, N.: Predictive role of case based reasoning for astrological predictions about profession: System modeling approach. In: 2010 International Conference on Communication and Computational Intelligence (INCOCCI), pp. 313–317 (2010)
15. Ahmad, A., Garhwal, S., Ray, S.K., Kumar, G., Malebary, S.J., Barukab, O.M.: The number of confirmed cases of covid-19 by using machine learning: methods and challenges. Arch. Comput. Methods Eng. **28**(4), 2645–2653 (2020). https://doi.org/10.1007/s11831-020-09472-8
16. Bhattacharjee, P.: Scientific astrological prediction of human life. Int. J. Jyotish Res. **5**(1), 12–16 (2020)
17. Barde, S.: PCA based multimodal biometrics using ear and face modalities. Int. J. Inf. Technol. Comput. Sci. (IJITCS) **6**(5), 43–49 (2014)
18. Olaiya, F., Adeyemo, A.B.: Application of data mining techniques in weather prediction and climate change studies. Int. J. Inform. Eng. Electr. Bus. **4**(1), 51–59 (2012)
19. Sánchez, J.S., Mollineda, R.A., Sotoca, J.M.: An analysis of how training data complexity affects the nearest neighbor classifiers. Pattern Anal. Applic. **10**, 189–201 (2007)
20. Loh, W.: Classification and regression trees. Wiley Interdisc. Rev.: Data Min. Know. Discovery **1**(1), 14–23 (2011)
21. Kolodner, J.L.: An Introduction to case based Reasoning. Artif. Intell. Rev. **6**(1), 3–34 (1992)
22. Leake, D.B., Plaza, E. (eds.): Case-Based Reasoning Research and Development: Second International Conference on Case-Based Reasoning, ICCBR-97 Providence, RI, USA, July 25–27, 1997 Proceedings. Springer Berlin Heidelberg, Berlin, Heidelberg (1997)
23. Bergmann, R., et al.: Developing industrial case based reasoning applications. In: Inreca Methology, 2nd edn. Springer-Verlag, Berlin (1999)
24. Bichindaritz, I., Kansu, E., Sullivan, K.M.: Case-based reasoning in CARE-PARTNER: gathering evidence for evidence-based medical practice. In: Advances in Case-Based Reasoning Proceedings 4th European Workshop on Case-Based Reasoning, pp. 334–345. Springer-Verlag (1998). https://doi.org/10.1007/BFb0056345
25. Lenz, M., Burkhard, H.-D., Bartsch-Spörl, B., Wess, S. (eds.): Case-Based Reasoning Technology: From Foundations to Applications. Springer, Berlin, Heidelberg (1998)
26. Chakkour, F., Toussaint, Y.: Sentence Analysis by Case-Based Reasoning. In: Monostori, L., Váncza, J., Ali, M. (eds.) IEA/AIE 2001. LNCS (LNAI), vol. 2070, pp. 546–551. Springer, Heidelberg (2001). https://doi.org/10.1007/3-540-45517-5_61
27. Faccini, D., Maggioni, F., Potra, F.A.: Robust and distributionally robust optimization models for linear support vector machine. Comput. Oper. Res. **147**, 105930 (2021)
28. Singla, M., Ghosh, D., Shukla, K.K.: A survey of robust optimization-based machine learning with special reference to support vector machines. Int. J. Mach. Learn. Cyber **11**, 1359–1385 (2020)
29. Dey, P.P., Nahar, N., Mainul Hossain, B.M.: Forecasting stock market trend using machine learning algorithms with technical indicators. Int. J. Inform. Technol. Comput. Sci. **12**(3), 32–38 (2020)

Method of Automatic Depersonalization of Databases for Application in Machine Learning Problems

Artem Volokyta[✉] and Vlada Lipska

Igor Sikorsky Kyiv Polytechnic Institute, Kyiv 03056, Ukraine
artem.volokita@kpi.ua

Abstract. The paper proposes a method of deanonymization of data based on the approach of building Bayesian models in combination with the synthetic creation of new instances of data by generative models built using directed graphs with weights for each vertex that guarantee correspondence to the initially calculated probabilities on the interval [0, 1], and on this basis, they construct synthesized artificial data with the boundaries of the projection or the curvature of the curve, according to the given functions. The importance of this task is related to the lack of possibility of using real personal data for building models and at the same time the need to preserve the nature of the data, taking into account the range of probabilities and the distortion/shift of the value curve in the industries in connection with the protection of personal data. The effectiveness and efficiency of using the proposed method was empirically confirmed, and the results are highlighted in this article with the provided metrics.

Keywords: Depersonalization · Anonymization · Data

1 Introduction

The amount of data collected by various services increases asymptotically every day. Sometimes this data is used to train models and predict threshold values to improve user performance and related metrics (churn time, segments, LTV, ARPU, ARPPU, prices). It depends on the specific task but the most popular algorithms for this field are logistic/linear regression, decision tree/random forest, Naive Bayes, KNN algorithm, K-means.

As a result, the problems of working with data are confidentiality and their possible leakage. Because of these issues, the processing and storage of data is governed by various agreements and conventions (e. g. GDPR [3]).

For ChatGPT [15] and similar AI-powered solutions, the same thing with the incorrect use of raw data is becoming more and more relevant every day [33, 34].

A logical solution to this issue is to remove and replace important parameters according to the characteristics of the original data. It is important not to distort the original data picture by changing the layers of information, otherwise the work will have a high probability of an unreliable final result.

© The Author(s), under exclusive license to Springer Nature Switzerland AG 2023
Z. Hu et al. (Eds.): ICCSEEA 2023, LNDECT 181, pp. 774–787, 2023.
https://doi.org/10.1007/978-3-031-36118-0_68

As mentioned, to solve this problem, it is necessary to find a way to change and reproduce the data in such a way that the nature of the data is not distorted. Currently, there are already research and development [10–13] on this topic. In the mentioned works, the extrapolation of one set of values to the interval of another set, taking into account certain deviations, is most often encountered. This method is suitable for changing data or comparing data at different scales, but from a security point of view we get the so-called Caesar cipher. The Caesar cipher [16] is known to be vulnerable to a frequency analysis attack. Also, these studies do not take into account the element of randomness and interdependence of values.

This paper proposes a combination of known static replacement methods, but adds processing based on Bayesian model building combined with synthetic creation of new instances of the data using generative models built using directed graphs with weights for each vertex guaranteed to match the initial computed probabilities per interval [0, 1], and build on this basis synthesized artificial data with filled projection boundaries or curve curvature, according to the given functions. In this way, an element of randomness will be taken into account according to the nature of the data and interdependencies, and artificial data will be added, which will make it impossible to decipher the results.

2 Methodology

2.1 Static Substitutions

The first step suggested by the system is the simplest and, depending on the quality of the input data, may be more or less necessary. The bottom line is that sometimes the data contains incorrect indicators that significantly distort the mathematical characteristics, which are then taken as a basis for building solutions. Most often, this is a small category of points in the submitted set, which are also called noises (Fig. 1).

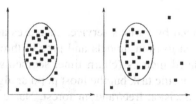

Fig. 1. Out-of-range values

The person working on the data can determine the boundaries of acceptable parameters and understand which records are irrelevant for a particular sample. Based on this, the system proposes to do primary data processing: replace the indicated values, replace the minimum and maximum, change certain data using a regular expression pattern, change the values by a certain delta, change one range to another, taking into account the distribution and the step of values. After clearing the data from noise, it is possible to proceed to the next step - building the network.

2.2 Network Construction and New Data Synthesis

The initial data have their own characteristics and connections. From the fundamentals of machine learning theory, we know that should not be simultaneously included fully dependent parameters in the processed set if it does not carry additional semantic load. This is due to the fact that this characteristic will be used twice (or more) and its weight will increase, thereby reducing the weight of other parameters due to the frequency of use only. For example, it should not be included the number of years and the year of birth at the same time, because there is a linear relationship f (y) = kx + b (Fig. 2).

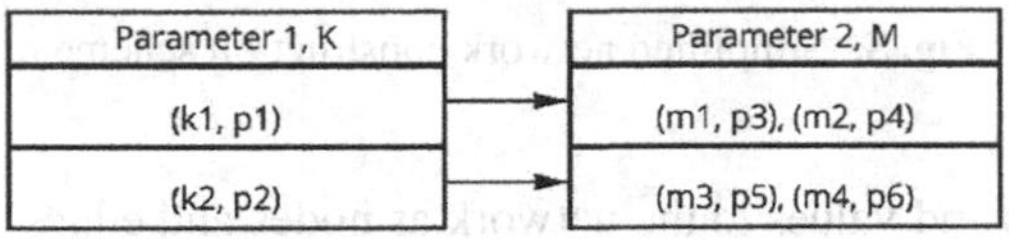

Parameter 1, K		Parameter 2, M
(k1, p1)	→	(m1, p3), (m2, p4)
(k2, p2)	→	(m3, p5), (m4, p6)

Fig. 2. Relationships between features

According to Bayes' theorem, we know the formulas for the probabilities of events A and B, given that event B is known to have occurred:

$$P(A|B) = (P(B|A)P(A))/(P(B)), P(B) \text{ is not equal to } 0.$$

Which is also true for continuous values, not just discrete ones:

$$f(X|Y) = \left(f_{Y|X=x}(y)f_x(x)\right)/f_y(y), f_y(y) \text{ is not equal to } 0.$$

The current research takes into account that if an event occurs that excludes other events, then codependent events, if any, are also excluded. Using these theories, it is possible to build a Bayesian network by first computing probabilities from the raw data and understanding the relationship between the features. A Bayesian network is a representation of the joint probability distribution of a set random variables with a possible mutual causal relationship. [22] The network will look like a directed graph with many possible values and their probabilities. A simplified network construction scheme for one of the features during the research (Fig. 3):

This scheme is extrapolated for each parameter. This means that the complexity of the network is $O(n*n)$, $n =$ the number of values, which means that it takes a long time for large data sets, so this should be taken into account when using this approach.

The algorithm for constructing the network is as follows:

1) Remove noises
2) Update values with static replaces (indicated values, minimum/maximum, changes using a regular expression pattern/certain delta, scale ranges, take into account the distribution and the step of values)
3) Select feature X
4) Calculate probabilities for every parameter of X-feature
5) Select feature Y (Y ! = X)
6) Calculate P(Y|X) for each value of Y-feature

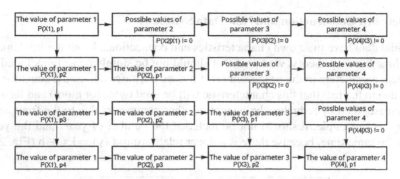

Fig. 3. Simplified network construction scheme

7) Add probabilities and values to the network as nodes and edges
8) Return to step 5 and repeat for each feature except processed ones
9) Return to step 3 and repeat for each feature except processed ones

Thus, we have a Bayesian network that captures the nature of the data and its patterns. Before direct data synthesis, final preparations must be made. We use one of the typical approaches used in machine learning problems when a new value needs to be determined. Namely, the approach consists in the fact that the probability is distributed in the interval from 0 to 1, and the possible values occupy the corresponding space in the sector along the length of its probability. A random value from 0 to 1 is pseudo-randomly generated, the point of the generated value is marked on the sector, and depending on whether it falls into a gap, the value of the parameter is determined. An example is offered for consideration. There is a value with probability $p1 = 0.25$, b with probability $p2 = 0.25$, c with probability $p3 = 0.5$. Then values from 0 to 0.25 in the interval $[0, 1]$ correspond to a, from 0.25 to 0.5 to b, and from 0.5 to 1 to c. Let's assume that the value 0.78 has fallen out by chance, then we choose the value c. Therefore, according to the described approach, it is necessary to go around (BFS-like) the graph and calculate such sectors on the interval for child nodes, keeping in the mind that the probability of all events is equal to 1. The sum of the probabilities of child nodes of a common node cannot exceed 1. Also, to speed up calculations, the preparation of intervals can be carried out before the direct selection of the value of a specific feature. Acceleration occurs due to the absence of the need to calculate intervals for mutually exclusive events.

$$(P(A|B) = 0 \text{ if } P(B) = 1, \text{ and } P(B|A) = 0 \text{ if } P(A) = 1)$$

At this stage, we have a network that reflects the nature of the processed data and corresponds to the laws of their formation. Any generated sets will now have the same characteristics and metrics as the original data.

3 Results

3.1 Terminology

Classification problems are problems in which an object must be assigned to one of n classes according to the degree of similarity of its features to the features of each class. [23] Classes are a set of similar objects. Below is a graphical representation of the classification problem, where squares are classified objects, circles are not yet classified objects (Fig. 4).

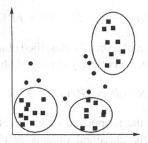

Fig. 4. Graphic representation of the classification problem

Regression problems concern the determination of continuous values depending on the correlation between other variables. (Typical tasks: determining the price of houses, shares) (Fig. 5).

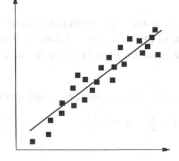

Fig. 5. Graphic representation of the regression problem

There are also certain metrics that determine how well the problem is solved. For classification problems, TP, TN, FP, FN concepts are often used.

True positive (TP) is a result when the model correctly predicts a positive class. An object was classified as a representative of class A and it is true.

True negative (TN) is the result when the model correctly predicts the negative class. An object was classified as a not representative of class A and it is true.

False positive (FP) is when a model incorrectly predicts a positive class. An object was classified as a representative of class A and it is false.

False negative (FN) is when a model incorrectly predicts a negative class. An object was classified as a not representative of class A and it is false.

Accuracy reflects the proportion of correctly classified objects relative to all objects.

$$Accuracy = (TP + TN)/(TP + TN + FP + FN)$$

Precision reflects the proportion of correctly classified objects as class members relative to correctly and incorrectly classified objects as class members.

$$Precision = TP/(TP + FP)$$

Recall reflects the proportion of correctly classified objects as class members relative to correctly classified objects and incorrectly classified objects as not class members.

$$Recall = TP/(TP + FN)$$

False positive rate reflects the proportion of incorrectly classified objects as class members relative to incorrectly classified objects as class members and correctly classified objects as not class members.

$$False\,positive\,rate = FP/(FP + TN)$$

F1 score is a composite measure used to measure the performance of both correct and incorrect classifications of machine learning classification models.

$$F1score = 2TP/(2TP + FP + FN) = 2(Precision * Recall)/(Precision + Recall)$$

Since regression problems work with continuous values, the article also discusses the metrics used for this type of problem. One of them is R-squared.

R-squared is a statistical measure of how close the data is to the fitted regression line.

$$R2 = 1 - sum\,squared\,regression(SSR)/total\,sum\,of\,squares(SST)$$
$$= 1 - \sum (y_i - f_i)^2 / \sum (y_i - y_a)^2,$$

y_i – actual value,
f_i – predicted value,
y_a – average of actual values.

Explained variance score (EVS) is a metric for calculating the ratio between the variance of the error and the variance of the true values.

$$EVS = 1 - \sum (E_i - E_a)^2 / \sum (y_i - y_a)^2,$$
$$E_i = y_i - f_i,$$
$$E_a = \left(\sum E_i \right) / n$$

y_i – actual value,
f_i – predicted value,
y_a – average of actual values,
n – number of elements.

Mean absolute percentage error (MAPE) is a measure of the accuracy of the forecasting method. [30]

$$MAPE = 100\% * \left(\sum |(y_i - f_i)/y_i| \right) / n,$$

y_i – actual value,
f_i – predicted value,
n – number of elements.

Machine learning models always use two sets of data - training and testing.

Training set is a set of data samples that are used to tune the parameters of a machine learning model to train it on an example.

Test set is a set of data to test a model after it has been trained on an training set.

One of the most common data structures when working with machine learning models is a decision tree. A decision tree is a structure that includes a root node, branches, and leaf nodes. Each internal node represents an attribute validation, each branch represents the result of a validation, and each terminal node contains a class label. The topmost node in the tree is the root node. [9] A set of decision trees forms a forest of decision trees. Based on this theory, there are methods RandomForestClassifier and RandomForestRegressor. A random forest is a meta estimator that fits a number of decision tree classifiers on various sub-samples of the dataset and uses averaging to improve the predictive accuracy and control over-fitting [31, 32].

3.2 Gender Classification Problem

During the research, the method was tested on the basis of the problem of classifying a person by gender based on his preferences. The dataset was taken from Kaggle [9]. According to the methodology, new data are generated based on Bayesian models that preserve the nature of the data. These steps should solve the problem of depersonalization and preserving the nature of the data. The problem of depersonalization is solved by the generation of artificial data by definition, so it remains to check the consistency of the prediction results of the same models on the original and generated data. The data structure used in the research is shown below.

To test the approach, 3 data sets were generated. Each feature was graphed for a visual representation of the comparison with the original data. Generated1, Generated2, Generated3 correspond to the first, second and third generated sets of data. Original corresponds to the initial data. On the ordinate axis are the percentages corresponding to the number (percentage) of the corresponding values. The similarity of the curves on the graphs once again confirms that the nature of the data remained unchanged and the distribution among the values took place taking into account the frequency of occurrence of the values in the sample (Figs. 6 and 7).

	1	2	3	4	5
1	Favourite Colour (Category)	Favourite Music Genre	Favourite Pet	Favourite Soft Drink	Gender
2	Cool	Pop	Parrot	Coca Cola/Pepsi	M
3	Cool	Rock	Dog	Coca Cola/Pepsi	M
4	Neutral	Rock	Rabbit	Coca Cola/Pepsi	F
5	Cool	Rock	Parrot	Coca Cola/Pepsi	F
6	Warm	Pop	Doesn't like	7UP/Sprite	M
7	Cool	Rock	Cat	Coca Cola/Pepsi	M
8	Cool	Pop	Parrot	Fanta	M
9	Cool	Pop	Rabbit	Coca Cola/Pepsi	F
10	Cool	Pop	Cat	Fanta	M
11	Cool	Pop	Rabbit	Coca Cola/Pepsi	F
12	Warm	Hip hop	Cat	Coca Cola/Pepsi	M
13	Cool	Pop	Cat	Coca Cola/Pepsi	F
14	Cool	Rock	Cat	7UP/Sprite	M
15	Warm	Rock	Dog	Coca Cola/Pepsi	M
16	Neutral	Rock	Dog	Coca Cola/Pepsi	M
17	Warm	Jazz/Blues	Cat	Fanta	M
18	Warm	Folk/Traditional	Parrot	Fanta	F
19	Cool	Pop	Dog	Coca Cola/Pepsi	M
20	Warm	Pop	Parrot	Fanta	F
21	Warm	R&B and soul	Dog	Coca Cola/Pepsi	M
22	Cool	R&B and soul	Dog	Coca Cola/Pepsi	F
23	Cool	Folk/Traditional	Other	Fanta	M
24	Cool	Pop	Doesn't like	7UP/Sprite	F
25	Cool	Hip hop	Cat	Coca Cola/Pepsi	M

Fig. 6. The task data structure

So, we have a different amount of new data that preserves the properties of the original, so we can start testing hypotheses. An important point is that the sets for testing should be taken from the initial ones, because even if the properties are preserved, the generated data still remains artificial.

To solve the problem, it is necessary:

1) Choose a training set
2) Choose a test set from the source data
3) Create a machine learning model
4) Train the model on the training set
5) Make predictions on the test set
6) Calculate the metrics of the results
7) Draw conclusions to what extent the predicted values correspond to reality

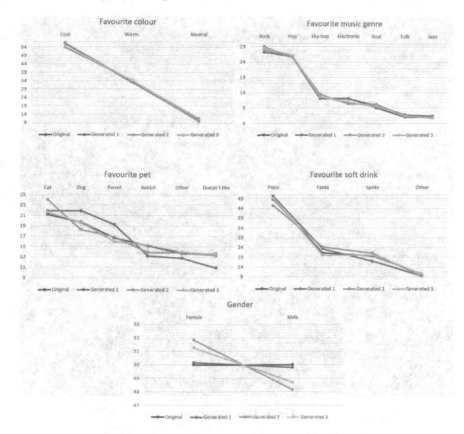

Fig. 7. Data sets values occurrences

Below is the code of the described actions, namely how the data and the model were prepared. RandomForestClassifier method was used for this task (Fig. 8).

Next, in the next image, the final steps are performed - training and obtaining predictions, after which the metrics are calculated (Fig. 9).

Metrics are recorded in a table and can be compared (Table 1).

It can be seen from the table that the indicators are quite successful on all data sets. By definition, for all indicators except FPR, the success of the model is directly proportional to the approach of the metric values to 1. For FPR, on the contrary, it is better when the value approaches 0 by definition. The most important thing to focus on is F1 score, as this metric is comprehensive and balanced to determine the success of the entire model. For some datasets (Generated 1 and Generated 2), F1 score is better than for the original dataset. This is possible because the artificial data is generated from probabilities of the original data, which ensures data uniformity and the absence of noise and outliers in the artificial data, while isolated noise or corrupted data in the original data may be missed during preparation for use. The similarity of the metric values once again supports the hypothesis that the nature of the data is preserved and the predictions are performed quite well (> ~ 90%) according to the metric results.

```
30   # Reading data from test_path and train_path variables
31   data = pd.read_csv(test_path)
32   data_train = pd.read_csv(train_path)
33
34   # Preparing data for classification
35   data['Favourite Colour'] = pd.factorize(data['Favourite Colour'])[0]
36   data['Favourite Music Genre'] = pd.factorize(data['Favourite Music Genre'])[0]
37   data['Favourite Pet'] = pd.factorize(data['Favourite Pet'])[0]
38   data['Favourite Soft Drink'] = pd.factorize(data['Favourite Soft Drink'])[0]
39   data['Gender'] = pd.factorize(data['Gender'])[0]
40
41   data_train['Favourite Colour'] = pd.factorize(data_train['Favourite Colour'])[0]
42   data_train['Favourite Music Genre'] = pd.factorize(data_train['Favourite Music Genre'])[0]
43   data_train['Favourite Pet'] = pd.factorize(data_train['Favourite Pet'])[0]
44   data_train['Favourite Soft Drink'] = pd.factorize(data_train['Favourite Soft Drink'])[0]
45   data_train['Gender'] = pd.factorize(data_train['Gender'])[0]
46
47   # Create target object and call it y (to train and to verify)
48   y_train = data_train.Gender
49   y_valid = data.Gender
50
51   # Remove values that should be predicted from set to test
52   features = data.columns.drop(['Gender'])
53
54   # Define features to train with
55   X_train = data_train[features]
56   X_valid = data[features]
57
58   # Normalize data
59   scaler=StandardScaler()
60   X_valid = scaler.fit_transform(X_valid)
61   X_train = scaler.fit_transform(X_train)
62
63   # Define the model
64   model_1 = RandomForestClassifier(max_features=4, random_state=1)
```

Fig. 8. Data preparation

```
# Function to train and predict, get a graph and metrics
def score_model(model, X_t, X_v, y_t, y_v):

    # Fit the model
    model.fit(X_t, y_t)

    # Get predictions
    preds = model.predict(X_v)

    # Metrics calculation
    tn, fp, fn, tp = confusion_matrix(y_v, preds, labels=[1, 0]).reshape(-1)
    print("Model tn: %f fp: %f, fn: %f tp: %f" % (tn, fp, fn, tp))
    fpr = fp / (fp + tn)

    # Print for information purposes
    return accuracy_score(y_v, preds), precision_score(y_v, preds), recall_score(y_v, preds), f1_score(y_v, preds), fpr
```

Fig. 9. Training and obtaining predictions

3.3 Determining the Price of Health Insurance (Regression Problem)

The next task is to predict the price of health insurances. The dataset was taken from Kaggle [9]. This time we are dealing with continuous values since this is a regression problem. The data structure is presented below (Fig. 10).

Table 1. Gender classification problem results

Data (number)	Accuracy	Precision	Recall	F1 score	False positive rate
Original	0.940279	0.933133	0.881493	0.906578	0.118507
Generated 1 (+200)	0.978135	0.976032	0.890736	0.931435	0.109264
Generated 2 (+500)	0.968941	0.963491	0.871173	0.915009	0.128826
Generated 3 (+2000)	0.959172	0.958912	0.847345	0.899682	0.152655

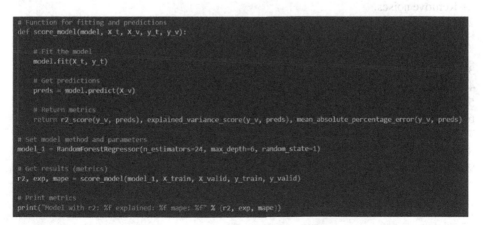

Fig. 10. The task data structure

The data preparation algorithm is similar to the previous problem, therefore, the method, prediction of values and metrics are considered in more detail in the next image (because they differ from the previous example) (Fig. 11).

```
# Function for fitting and predictions
def score_model(model, X_t, X_v, y_t, y_v):

    # Fit the model
    model.fit(X_t, y_t)

    # Get predictions
    preds = model.predict(X_v)

    # Return metrics
    return r2_score(y_v, preds), explained_variance_score(y_v, preds), mean_absolute_percentage_error(y_v, preds)

# Set model method and parameters
model_1 = RandomForestRegressor(n_estimators=24, max_depth=6, random_state=1)

# Get results (metrics)
r2, exp, mape = score_model(model_1, X_train, X_valid, y_train, y_valid)

# Print metrics
print("Model with r2: %f explained: %f mape: %f" % (r2, exp, mape))
```

Fig. 11. Training and obtaining predictions

The obtained results are recorded in a table and can be analyzed.

Table 2. Results of the algorithm usage for regression problem

Data	R2 score	Explained variance score	Mean absolute percentage error
Original	0.854321	0.865893	0.284498
Generated 1 (+500)	0.873433	0.873496	0.263907
Generated 2 (+800)	0.865672	0.865871	0.267938
Generated 3 (+2000)	0.839087	0.841002	0.309629

Table 2 lists the metrics used for regression problems. By definition, the more successful the model is, the closer R2 and EVS are to 1 and MAPE to 0. The similarity of the indicators confirms the preservation of the nature of the data and their homogeneity. All datasets showed good metric values. Better performance of artificial data sets can occur due to the absence of exceptions or damaged objects as determined by the artificial data set construction algorithm. Thus, artificial sets constructed by the proposed method can be used for regression problems, thereby allowing depersonalization and data synthesis.

4 Summary and Conclusion

This research proposes, in addition to known static extrapolations and data substitutions, to generate artificial data sets based on Bayesian models. It allows to automatically determine the nature of the data and generate artificial depersonalized data, the effectiveness of which has been practically confirmed. The steps in the algorithms were described as follows:

1) Remove noises.
2) Update values with static replaces (indicated values, minimum/maximum, changes using a regular expression pattern/certain delta, scale ranges, take into account the distribution and the step of values).
3) Select feature X.
4) Calculate probabilities for every parameter of X-feature.
5) Select feature Y (Y ! = X).
6) Calculate P(Y|X) for each value of Y-feature.
7) Add probabilities and values to the network as nodes and edges.
8) Return to step 5 and repeat for each feature except processed ones.
9) Return to step 3 and repeat for each feature except processed ones.

References

1. Big & Personal: data and models behind Netflix recommendations. URL: https://xamat.git hub.io/pubs/BigAndPersonal.pdf
2. Geron, A.: Hands-On Machine Learning with Scikit-Learn, Keras, and TensorFlow: Concepts, Tools, and Techniques to Build Intelligent Systems, 2nd Edition, p. 57 (2019)
3. GDPR. URL: https://en.wikipedia.org/wiki/General_Data_Protection_Regulation
4. If an app asks to track your activity. URL: https://support.apple.com/en-us/HT212025
5. Shai, S.-S., Shai, B.-D.: Understanding Machine Learning: From Theory to Algorithms (2014). URL: https://www.cs.huji.ac.il/~shais/UnderstandingMachineLearning/und erstanding-machine-learning-theory-algorithms.pdf
6. Ng, A.: Machine Learning by Stanford University, Coursera. URL: https://www.coursera.org/ learn/machine-learning
7. Deep Learning – An MIT Press Book. URL: https://www.deeplearningbook.org/
8. Beyond Accuracy: Precision and Recall. URL: https://towardsdatascience.com/beyond-acc uracy-precision-and-recall-3da06bea9f6c
9. Kaggle (Gender Classification Dataset). URL: https://www.kaggle.com/hb20007/gender-cla ssification
10. Scale Synthetic. URL: https://scale.com/synthetic
11. Depersonalizing Your Test Data. URL: https://github.com/CleverComponents/Data-Depers onalizer
12. Synth. URL: https://www.getsynth.com/docs/getting_started/synth
13. Twinify. URL: https://github.com/DPBayes/twinify
14. What Is Synthetic Data? URL: https://blogs.nvidia.com/blog/2021/06/08/what-is-synthetic-data/
15. OpenAI (ChatGPT). URL: https://openai.com/research/overview
16. ScienceDirect. URL: https://www.sciencedirect.com/topics/computer-scicnce/caesar-cipher
17. Malani, R., Putra, A.B.W., Rifani, M.: Implementation of the naive bayes classifier method for potential network port selection. Int. J. Comp. Netw. Info. Secu. (IJCNIS) 12(2), 32–40 (2020)
18. Pandey, A., Jain, A.: Comparative analysis of KNN algorithm using various normalization techniques. Int. J. Comp. Netw. Info. Secu. (IJCNIS) 9(11), 36–42 (2017)
19. Zimba, A.: A bayesian attack-network modeling approach to mitigating malware-based banking cyberattacks. Int. J. Comp. Netw. Info. Secu. (IJCNIS) 14(1), 25–39 (2022)
20. Dasari, K.B., Devarakonda, N.: Detection of DDoSAttacks Using Machine Learning Classification Algorithms. Int. J. Comp. Netw. Info. Secu. (IJCNIS) 14(6), 89-97 (2022)
21. Bayes Server Learning Center. URL: https://www.bayesserver.com/docs/introduction/bay esian-networks/
22. Horny, M.: Bayesian Networks N.5 (2014). https://www.bu.edu/sph/files/2014/05/bayesian-networks-final.pdf
23. What are classification problems? URL: https://www.educative.io/answers/what-are-classific ation-problems
24. Oracle AI & Data Science Blog. URL: https://blogs.oracle.com/ai-and-datascience/post/a-simple-guide-to-building-a-confusion-matrix
25. What is a good F1 score and how do I interpret it? URL: https://stephenallwright.com/good-f1-score/
26. Lipton, Z.C., Elkan, C., Naryanaswamy, B.: Thresholding Classifiers to Maximize F1 Score. URL: https://arxiv.org/pdf/1402.1892.pdf
27. Metrics for evaluating regression models. URL: http://www.enlistq.com/top-5-metrics-eva luating-regression-models/

28. Mean Absolute Percentage Error. URL: https://www.statisticshowto.com/mean-absolute-per
    centage-error-mape/
29. de Myttenaerea, A., Goldena, B., Le Grand, B., Rossi, F.: Mean Absolute Percentage Error
    for Regression Models. URL: https://arxiv.org/pdf/1605.02541.pdf
30. What is MAPE? URL: https://www.indeed.com/career-advice/career-development/what-is-
    mape
31. RandomForestClassifier. URL: https://scikit-learn.org/stable/modules/generated/sklearn.ens
    emble.RandomForestClassifier.html
32. RandomForestRegressor. URL: https://scikit-learn.org/stable/modules/generated/sklearn.ens
    emble.RandomForestRegressor.html
33. ChatGPT is a data privacy nightmare. URL: https://theconversation.com/chatgpt-is-a-data-
    privacy-nightmare-if-youve-ever-posted-online-you-ought-to-be-concerned-199283
34. ChatGPT vs GDPR – what AI chatbots mean for data privacy. URL: https://www.inform
    ation-age.com/chatgpt-vs-gdpr-what-ai-chatbots-mean-for-data-privacy-123501570/

Mathematical Modeling of States of a Complex Landscape System with a Wind Farm

Taras Boyko[1], Mariia Ruda[2], Elvira Dzhumelia[3], Orest Kochan[4,5(✉)], and Ivan Salamon[6]

[1] Department of Intellectual Mechatronics and Robotics, Lviv Polytechnic National University, Lviv 79013, Ukraine
[2] Department of Ecological Safety and Nature Protection Activity, Lviv Polytechnic National University, Lviv 79013, Ukraine
[3] Department of Software, Lviv Polytechnic National University, Lviv 79013, Ukraine
[4] Department of Measuring Information Technologies, Lviv Polytechnic National University, Lviv 79013, Ukraine
orest.v.kochan@lpnu.ua
[5] Department of Automation, Lublin University of Technology, 20618 Lublin, Poland
[6] Department of Ecology, Faculty of Humanities and Natural Sciences, University of Presov, Presov 080 01, Slovakia

Abstract. The aim of this study is to model the processes of mutual influence of the wind turbine and the environment. An elementary structural element of an ecosystem is a compartment of a complex landscape system in which a wind turbine is considered during its life cycle. The task of creating a 'cyber-twin' of a wind turbine, as a technogenic object of the compartment of the complex landscape system, was realized by mathematical modeling of the states of the layers and subsystems of the compartment of the complex landscape system under integrated impact. Impacts are pollutants and harmful factors, primarily of anthropogenic origin, the values of which were estimated by simulation modeling in the experimental part of this study. As a result of mathematical modeling a system of differential equations was obtained, the input data for which were the values of environmental impacts, expressed by the specified indicators. The resulting model will act as ideal for a real system 'wind turbine-environment' and will allow predicting the consequences of harmful impact of a wind turbine on a complex landscape system.

Keywords: environment · landscape complex · renewable energy sources · ecological indicators · harmful ecological influence · cyber-physical system

1 Introduction

In recent times, there has been a surge in the adoption of innovative energy technologies that are aimed at developing efficient power grids (Smart Grid), advanced accounting and settlement systems (Smart Metering), managing energy demands (Demand Response), and energy storage solutions. These technologies are designed to address the increasing

© The Author(s), under exclusive license to Springer Nature Switzerland AG 2023
Z. Hu et al. (Eds.): ICCSEEA 2023, LNDECT 181, pp. 788–802, 2023.
https://doi.org/10.1007/978-3-031-36118-0_69

demand for reliable and high-quality electricity, promote energy efficiency, and enhance the sustainability of the environment [1, 2]. However, despite the extensive research conducted by ecologists, the impact of wind energy facilities on the environment is not adequately explored and remains largely unaddressed [3–7].

When we consider the impact of human activities on the environment, we can view each component as a cyber-physical system. This perspective is possible when we use computer algorithms to monitor the integration mechanisms. The development of industrial cyber-physical systems typically involves five standard processes, one of which is the 'cyber-implementation' process. This process involves creating a 'cyber-twin' to describe the system's state and comparing it with the real device before synthesizing the entire system.

The purpose of the study is to create a "cyber-twin" for a wind turbine to simulate and predict the states of an individual complex landscape system (CLS), in which a wind farm operates. The object of the study is the process of modeling the man-made load of the wind turbine on the environment during its life cycle.

The subject of the study is a model of a system that includes a complex landscape system – a biological system that combines biotic and abiotic components. For example, by definition the Eastern flysch Carpathians (the Borzhava Polonyna) belong to a CLS [8–19].

2 Literature Review

There is a shortage of life cycle assessment (LCA) studies for wind turbines with a high-rated power of 600 kW. Although some studies are available [20, 21], they have varying scopes, but they consistently highlight the significant impact of the manufacturing and usage of various materials on the ecological metrics of wind farms. These factors must be taken into account when assessing the adverse effects of wind turbines on the ecosystem. Some estimates also point to a large amount of indirectly generated waste [22], which must also be considered. In the works [8, 23] it is shown that the interaction between wind turbines and the CLS is advisable to consider based on the compartment concept of the CLS.

Similar results are presented in the works [24]. The authors conclude that the construction of the corresponding model involves the interpretation of the CLS as a system characterized by a certain composition and structure which is a natural combination of natural components – landscape systems of different taxonomic rank [25, 26]. The CLS, or complex landscape system, in which wind turbines operate is characterized by the structural and functional unity of the interconnected components and the integrity of both biotic and abiotic elements [27–34].

The authors have demonstrated that the biotic elements are organized into compartments comprising hierarchically interconnected subsystems of varying levels of organization, with numerous distinct layers that have close hierarchical, material, and energetic relationships between them. This approach to the CLS allows identifying the purpose and objectives of the study as well as the initial conditions and limits of modeling.

In work [32], the authors 'proposed a new way to assessing the impact of the multicomponent composition of the power battery on the state of the environment, which consists in determining the stability of ecosystems, and makes it possible to obtain quantitative indicators of the stability and loss of natural ecosystems, which can be used as indicators of the environment state, and therefore assessment of the environmental component, which is important for determining the real impact of the polyelement composition of power batteries'. The paper [34] proposes to use the SimaPro software to assess the impact of the wind power station on the landscape system.

3 Methodology

This study presents a methodology for modeling the interaction between man-made and ecological systems in various regions. To accomplish this, an integrated approach is required that simulates the interaction of a wind turbine (WT) and the complex landscape system (CLS) throughout its life cycle, such as a cyber-physical system (CPhS). The CPhS is a complex, multi-component mechanism in which computer algorithms monitor and control processes. These algorithms must adequately describe the processes. It is widely known that the development and deployment of production CPhS can be implemented through five processes, which include communication, conversion, cyber-implementation, cognition, and configuration [35].

The process of 'cyber-implementation' involves creating a 'cyber-twin' to describe the system's state, comparing it with a real device, and subsequently synthesizing the system. Creating a 'cyber-twin' for a wind turbine that interacts with the environment is a multi-stage process. This process can be based on the application of the compartment concept of the complex landscape system (CLS) for an ecosystem and the product life cycle concept for wind turbines [36].

This study utilized the 5C architecture, which is a common design for cyber-physical-human systems (CPhS) and was originally proposed by [35]. The International Organization for Standardization's (ISO) 14040/44 standard defines LCA as a comprehensive assessment of a product's environmental characteristics. The result of the LCA process is a quantitative measure of the product's environmental friendliness [37–45]. In work [34], M. Ruda used an assumption for study the influence of elemental composition of batteries on the sustainability of ecosystems, that would be valid for assessing the impact of wind power plants.

4 Basic Theory of the Proposed Method

As a result of interaction of the listed natural components and anthropogenic factors, the specific system of CLS is formed. In the presented system all values acquire mainly three indices: i_1 – to denote the area of space (CLS compartment); i_2 – to denote the geophysical environment; and i_3 – to indicate the impact. Then the expression $Q_{i_1 i_2 i_3}$ will mean the influence of wind turbines at all stages of the life cycle $-i_3$, in the layers or subsystems $-i_2$, CLS compartment $-i_1$. According to, such indices are called multi-indices and are denoted by one letter i $= (i_1 i_2 i_3)$, and the set of multi-indices in the model – by the letter Ω.

If the CLS consists of n compartments and m impacts are considered, then.

$$1 \leq i_1 \leq n; 1 \leq i_2 \leq f; 1 \leq i_3 \leq m \tag{1}$$

Let Q_i be the impact of i_3 in the layers or subsystems i_2 of the compartment growth area i_1 that arising at time t. Denote by $k_{i,j}(t)$ the value of this impact, which after time τ will be in the layers or subsystems j_2 of the growth region j_1, having carried out in the process of transforming the flow of matter f into the impact of j_3. Since the impacts in the process of transfer from the layers to the subsystems of the compartment and from the compartment to another compartment do not arise and do not disappear, then.

$$\sum_{j=\Omega} k_{i,j} = \sum_{j_1=1}^{n} \sum_{j_2=1}^{f} \sum_{j_3=1}^{m} k_{i_1 i_2 i_3 j_1 j_2 j_3} = 1. \tag{2}$$

The coefficients $k_{i,j}(t)$ are the probabilities of the states $P_{i,j}(t)$ of the corresponding matrix of states P and the transition intensity $\lambda_{i,j}(t)$ of the corresponding matrix of transition intensities Λ of the concentration of harmful substances or compartments from state i to state j.

Formalization of graphs of the interaction of the CLS compartment and wind turbine is carried out by means of the Kolmogorov system of differential equations. If, for example, we restrict to $m, l = 1, 2, 3, 4$ and $m \neq l$, then the system of differential equations will take the form (3), and solving them with the help of computer technology, it is possible to study the development of the CLS in dynamic and stationary modes:

$$\begin{cases} \frac{dP_{i,j,S_1}}{dt} = -\lambda_{i,j,S_1,S_2} \cdot P_{i,j,S_1} + \lambda_{i,j,S_2,S_2} \cdot P_{i,j,S_2} \\ \frac{dP_{i,j,S_2}}{dt} = \lambda_{i,j,S_1,S_2} \cdot P_{i,j,S_1} - \left(\lambda_{i,j,S_2,S_1} + \lambda_{i,j,S_2,S_3}\right) \cdot P_{i,j,S_2} + \lambda_{i,j,S_3,S_2} \cdot P_{i,j,S_3} \\ \frac{dP_{i,j,S_3}}{dt} = \lambda_{i,j,S_2,S_3} \cdot P_{i,j,S_2} - \left(\lambda_{i,j,S_3,S_2} - \lambda_{i,j,S_3,S_4}\right) \cdot P_{i,j,S_3} + \lambda_{i,j,S_4,S_3} \cdot P_{i,j,S_4} \\ \frac{dP_{i,j,S_4}}{dt} = \lambda_{i,j,S_3,S_4} \cdot P_{i,j,S_3} - \lambda_{i,j,S_4,S_3} \cdot P_{i,j,S_4} \end{cases} \tag{3}$$

where $i = 1, 2, \ldots, N$ and $j = 1, 2, \ldots, M$ are ordinal numbers of multiindexes i and j, respectively; $S_1, \ldots, S_l, S_m, \ldots, S_n$ are states of CLS compartments; P_{i,j,S_i} is the probability of location the CLS compartment or the concentration of harmful substances in the state S_l or the state S_m; λ_{i,j,S_i,S_m} is the intensity of the transitions of the CLS compartments or the concentration of harmful substances from the state S_l to the state S_m for each multiindex.

5 Research Results

The intensity values of state transitions for each element of the hierarchical structure, including layers and subsystems of the compartment, are statistical information derived from studying the functioning of the complex landscape system (CLS).

To evaluate and predict the states of these CLSs and their compartments, it is recommended to gather information about the impact of wind turbines at various stages of their life cycle and at different time intervals during the functioning of the CLS. At the same time such information can be obtained through simulation.

Simulation computer modeling of the process of obtaining the results of experimental studies on wind turbines during their life cycle was carried out. The results of modeling the effects of harmful substances, considering the scenario of waste management, are presented in Table 1.

Table 1. Positive Impact of Waste Management Scenarios (WMS) on Reducing the Impact of Harmful Substances for each Category

N	Impact categories	Harmful substances (effects)	Indicators	Value of impact	Reduction due to WMS	Value of impact after WMS
1	Carcinogenic effects	electricity from coal for steel and copper production	DALY	0.41	−100%	−
		electricity from coal for reinforcing steel production	DALY	0.38	−100%	−
		arsenic in water and air	DALY	0.81 + 0.27	−	1.08
		unidentified metals in water and air	DALY	0.028 + 0.342	−	0.37
		cadmium in the air	DALY	0.06	−	0.06
2	Effects of respiratory inorganics	electricity from coal for concrete production	DALY		−15.4%	5.91
		– dust		2.33		
		– nitrogen dioxide		1.65		
		– sulfur dioxide		1.51		
3	Effects of respiratory organics	non-methane volatile organic compounds	DALY	25.2E-4	+ 5.93%	33.2E-4
		methane and unidentified aromatic hydrocarbons	DALY	7.45E-4		
4	Climate change – global warming	electricity from coal for reinforcing steel production	DALY	1.58	−	1.58
		impact of the use of wind turbines	DALY	2.14	−41%	1.257
5	radiation effects	radon-222	DALY	2.55E-3	−16.5%	3,79E-3
		carbon-14	DALY	1.22E-3		

(continued)

Table 1. (*continued*)

N	Impact categories	Harmful substances (effects)	Indicators	Value of impact	Reduction due to WMS	Value of impact after WMS
6	ozone layer depletion	Bromotrifluoromethane	DALY	5.02E-4	+ 7.34%	5.02E-4
7	ecotoxicity	unidentified metals	PAF × m² × year	1.25E6	−47.8%	2.32E6
		nickel and zinc in the air	PAF × m² × year	6,33E5		
		lead in the air	PAF × m² × year	9.64E4		
8	acidification and eutrophication	nitrogen oxides	PAF × m² × year	1.06E5	−4.07%	1.38E5
		sulfur oxides	PAF × m² × year	2.88E4		
9	Land use	transformation of industrial zones	PDF × m² × year	9.18E3	−32.3%	3.1E4
		occupation of industrial zones	PDF × m² × year	1.3E4		
		occupation of landfills	PDF × m² × year	1.46E4		
10	Depletion of minerals	nickel in silicates − 1.98%	MJ	6.47E4	−79.1%	−1.01E5
		nickel in raw ore − 1.04%	MJ			
		copper in sulfide − 0.99%	MJ	3.68E4		
		pure copper − 0.36%	MJ			
		copper in raw ore − 8.2E-3%	MJ			
11	Depletion of fossil fuels	coal, oil, gas	–	–	–	–

The typical ecological models are represents the impacts. In particular, the impact on human health is expressed by the indicator DALY (Disability Adjusted Life Years).

The impact on the ecosystem quality (ecotoxicity) is expressed as an indicator of affected species of living organisms PAF (Potentially Affected Fraction) or an indicator of loss of living organisms PDF (Potentially Disappeared Fraction).

PAF (Potentially Affected Fraction) and PDF (Potentially Disappeared Fraction) are determined by evaluating the toxicity of a substance to various organisms such as microorganisms, fish, crustaceans, mollusks, amphibians, algae, worms, and plants in both terrestrial and aquatic environments.

The outcome of modeling the effects of hazardous substances is a compilation of potential impacts on the environment. This record of effect scores, designated for each category, is referred to as the environmental profile of the product or service.

The Fig. 1 the environmental profile of entire wind turbine life cycle is shown. It shows the environmental load of the impact categories which were chosen for the analysis. It displays the sets of four scores, each one representing the impact on one of the four categories: 1. Resource depletion; 2. Global warming; 3. Acidification; 4. Ozone depletion.

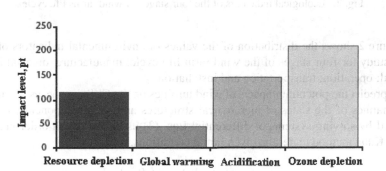

Fig. 1. Entire wind turbine life cycle environmental profile

During the next step, the load exposure is calculated using the European average data. There are three categories of damage: 1. Human health; 2. Ecosystem quality; 3. Resource depletion.

The first impact category is related to human health and includes radiation, eco-toxicity, carcinogens, respiratory substances, ozone depletion, and climate change. The second category covers acidification or eutrophication and land use. The third category, known as 'resource depletion,' encompasses natural resources, minerals, and fossil fuels. The values for each impact category are combined to provide a total impact score. The effect on human health is expressed as DALY, which stands for disability-adjusted life years.

The impact on the ecosystem quality is expressed as $PAF \times m^2 \times year$ and $PDF \times m^2 \times year$. Resource depletion is expressed as the energy surplus required for the subsequent extraction of minerals and petroleum (fossil fuels).

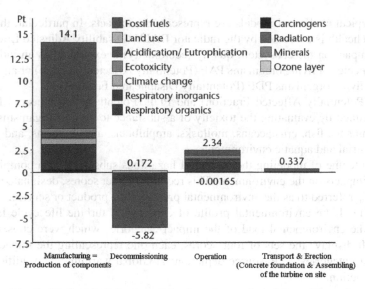

Fig. 2. Ecological indicators of the four stages of wind farms life cycle

Figure 2 shows the distribution of the values of environmental indicators obtained in the study for four stages of the wind farm life cycle: manufacture; dismantling and disposal; operation; transportation and installation.

To predict the potential impacts of wind turbines on the CLS, it is necessary to study the dynamics of the states of hierarchical structures and their components. It can be achieved by solving systems of differential Eqs. (3) using the fourth-order numerical Runge-Kutta method under the given initial conditions.

$$P(0) = \begin{pmatrix} P_0(0) \\ P_1(0) \\ P_2(0) \\ .... \\ P_n(0) \end{pmatrix} = \begin{pmatrix} P_{00} \\ P_{10} \\ P_{20} \\ ... \\ P_{n0} \end{pmatrix}, \tag{4}$$

Using the methods of approximate calculations and computer technology, the solution of the system of Eqs. (3) can be written as

$$P_{i,k+1} = P_{ik} + \frac{H}{6} \cdot (K_{1i} + 2 \cdot K_{2i} + 2 \cdot K_{3i} + K_{4i}), \tag{5}$$

where

$$K_{1i} = \sum_{j=0}^{n} a_{ij} \cdot P_{jk}, i = 0, 1, 2, \dots, n;$$

$$K_{2i} = \sum_{j=0}^{n} a_{ij} \cdot \left(P_{jk} + K_{1i} \cdot \frac{H}{2}\right), i = 0, 1, 2, \dots, n;$$

$$K_{3i} = \sum_{j=0}^{n} a_{ij} \cdot \left(P_{jk} + K_{2i} \cdot \frac{H}{2}\right), i = 0, 1, 2, \dots, n; \tag{6}$$

$$K_{4i} = \sum_{j=0}^{n} a_{ij} \cdot \left(P_{jk} + K_{3i} \cdot H\right), i = 0, 1, 2, \dots, n;$$

$$H = \Delta t.$$

The study of the CLS as an object with a hierarchical structure in the stationary mode, when $t \to \infty$ and $\frac{dP}{dt} = 0$, is carried out using numerical methods based on the solution of a system of algebraic equations, whose matrix notation takes the form

$$\Lambda \cdot P = 0 \tag{7}$$

where Λ is the matrix of intensities of transitions from state to state and P is the matrix of states probabilities.

Using the Khaletsky scheme of the Gauss numerical method [9], we find $P_n, P_{n-1}, \dots, P_0$, using recurrent formulas

$$\gamma_{i0} = a_{i0}, (i = 0, 1, 2, \dots, n),$$

$$\alpha_{0k} = \frac{\alpha_{0k}}{\alpha_{00}}, (k = 1, 2, \dots, n),$$

$$\gamma_{ik} = a_{ik} - \sum_{j=0}^{k-1} \gamma_{ij} \cdot \alpha_{jk}, (i, k = 1, 2, \dots, n; n \geq k), \tag{8}$$

$$\alpha_{ik} = \frac{1}{\gamma_{ii}} \cdot \left(a_{ik} - \sum_{j=k}^{i=1} \gamma_{ij} \cdot \alpha_{jk}\right), (i, k = 1, 2, \dots, n; i < k).$$

in the following order

$$P_0 + a_{01} \cdot P_1 + a_{02} \cdot P_2 + \dots + a_{0n} \cdot P_n = 0,$$
$$P_1 + a_{12} \cdot P_2 + \dots + a_{1n} \cdot P_n = 0,$$
$$\dots\dots\dots\dots\dots\dots\dots\dots\dots\dots\dots \tag{9}$$
$$P_n = 0$$

The methods proposed in this study for collecting input data, building, and solving differential and algebraic equation systems could serve as a fundamental framework for mathematically analyzing the CPhS. This approach can be used to investigate the state of individual compartments within the CLS and predict the potential environmental impacts of wind turbines. Below there are the results of the probable wind turbines' impact on the simulated wind farms during the four stages of their life cycle (see Fig. 2, Table 2) on the subsystems and layers of the Complex Landscape System compartments, which is confirmed by the report on the research work 'Study of the presence of plants and animals listed in the Red Book of Ukraine and plant groups that are protected by the legislation of Ukraine in the territory of the planned activities'.

To ensure the reliability and accuracy of the theoretical results, we plan to conduct experiments in our future research.

Table 2. The impact of wind turbine LC on the subsystems and compartments of the Complex Landscape System

Stage of WT life cycle	Area & Substance	Compartment	Unit	Total	Formalized description of subsystems and layers of the compartment affected by the impact of wind turbine LC
Transport	Occupied area as industrial area	Grass-shrub layer & soil subsystem of the forest compartment	ha	0.2	Soil compaction
	Cetane		%	47	Pollution
	Alfamel-naphthalene		%	53	
	CO	Air	kg	18	
	CO_2		kg	17.5	
	C_xH_y		kg	14.3	
	Occupied area as industrial area	Fauna	dBa	80.0	Changing migration routes
			X	X	Vehicles crash. Intrusion of invasive species
	Acid as H^+	Wastewater	g	21	Pollution
	BOD		g	412	
	Calcium ions		g	25	
	Cl		mg	17	
	COD		g	14	
	C_xH_y		g	36	

(continued)

Table 2. (*continued*)

Stage of WT life cycle	Area & Substance	Compartment	Unit	Total	Formalized description of subsystems and layers of the compartment affected by the impact of wind turbine LC
Transport & Erection (Concrete foundation & Assembling) of the turbine on site	Occupied area as industrial area	Grass-shrub layer of the compartment	ha	~ 30.6	Extinction or damage to especially valuable species
		aboveground phytomass of the Grass-shrub layer of the compartment	X	X	Destruction of or damage to a variety of herbaceous communities – *Hypochoeridi uniflorae-Nardetum* and *Hieracio vulgati-Nardetum*, according to the classification of biotopes Eur 28 (EUNIS 6230) – species-rich *Nardus* grasslands and fragments of relict phytocoenoses with dominance of arcto-alpine species *Juncus trifidus* L., which can be considered within the association *Oreochloetum (distichae) juncosum (trifidi)*, listed in the Green Book of Ukraine
	Occupied area as industrial area	(land use)	X	X	12 types of settlements that are subject to protection (namely: 'C2.12', 'C2.18', 'C2.19', 'C2.25', 'D2.226', 'D2.3', 'E1.71', 'E4.11', 'E4.3', 'E5.5', 'F4.2', 'G1.6')
		soil subsystem of the forest compartment	ha	~ 30.6	Removal of the top fertile layer of soil. Wind and water erosion Changes in the soil profile
		(land use)	X	X	Change of hydrological regime
		Fauna	X	X	Strengthening the edge effect Loss of natural habitats, nesting sites, food for birds and bats
	CO	Air	kg	27	Pollution
	CO_2		kg	17	
	C_xH_y		kg	14	

(*continued*)

Table 2. (*continued*)

Stage of WT life cycle	Area & Substance	Compartment	Unit	Total	Formalized description of subsystems and layers of the compartment affected by the impact of wind turbine LC
Operation & Maintenance	Occupied area as industrial area	soil subsystem of the forest compartment	ha	~ 30.6	Soil compaction
		Fauna	X	X	Damage to bats and birds during a collision with wind turbines, power lines
		(land use)	m	150	Influence on the visual perception of the environment
	Fuel and lubricants	Grass-shrub layer of the compartment	ha	0.003	Possible spills when replacing component parts
	BOD	Wastewater	g	456	Surface run-off
Decommissioning	Occupied area as industrial area	grass-shrub layer & soil subsystem of the forest compartment	ha	~ 30,6	Extinction of or damage to especially valuable species
	Acid as H^+	Wastewater	g	48	Pollution
	BOD		g	850	
	Calcium ions		g	47	
	Cl		mg	37	
	COD		g	76	
	C_xH_y		g	26	
	Detergent/oil		mg	49	
	Occupied area as industrial area	Fauna	X	X	Hindering the free migration of animal species
	CO	Air	kg	23	Pollution
	CO_2		kg	20	
	C_xH_y		kg	19	

6 Summary and Conclusion

It is obvious that any man-made energy-generating facility, even if it is classified as the so-called 'green energy', has a negative impact on the environment.

A wind turbine, as an object of this kind, should be considered throughout all stages of its life cycle.

It is assumed that the process of spreading and transformation of impact or pollutant in the layers and subsystems for a certain period of time can be described in simplified form by a linear operator that connects the state of the system at the previous and the next point of time, and the process itself is investigated using the theory of the Markov chains. Then the state of the CLS compartments can be represented as a graph with vertices identifying these states.

Formal representation of such graphs is proposed to be carried out by Kolmogorov's differential equations which link the time change in the probability of finding a compartment with the intensity of the transitions of the concentration of harmful substances from the previous state to the next one.

It was obtained an integrated system of categories and indicators of wind farm impacts on the layers and subsystems of the CLS compartment.

The use of WMS made it possible to assess the potential reduction of harmful effects, but it was found that for the categories of 'fossil fuels' they cannot be used and for the categories of 'ozone layer depletion' and 'respiratory non-organic effects', the use of WMS has the opposite effect and increases the harmful impacts.

According to the research, the dismantling and disposal stages of the wind turbine's life cycle show the most potential for significant reductions in harmful environmental impacts when using WMS.

Ecological profiles were developed, which made it possible to present the impacts during the life cycle of wind turbine system in the form of eco-indicators. An example is the system of eco-indicators, distributed over four stages of the life cycle of the wind farm: manufacture; operation; transportation and installation; dismantling and disposal. The greatest harmful environmental impact with an eco-indicator value of 14.1 occurs during the manufacturing stage of wind turbines, which is associated with the type of electricity used.

The study of the dynamics of states of hierarchical structures and their elements is carried out on the basis of solving systems of differential equations using the fourth order Runge-Kutta numerical method.

References

1. Parsons, D.: The environmental impact of disposable versus re-chargeable batteries for consumer use. The International Journal of Life Cycle **12**, 197 (2007). https://doi.org/10.1065/lca2006.08.270
2. Sunil, K.K., Muniamuthu, S., Tharanisrisakthi, B.T.: An investigation to estimate the maximum yielding capability of power for mini venturi wind turbine. Ecological Engineering and Environmental Technology **23**(3), 72–78 (2022)
3. Ginevicius, R., Trishch, R., Remeikiene, R., et al.: Complex evaluation of the negative variations in the development of lithuanian municipalities. Transform. Bus. Econ. **20**(2), 635–653 (2021)
4. Chaudhary, A.S., Chaturvedi, D.K.: Efficient Thermal Image Segmentation for Heat Visualization in Solar Panels and Batteries using Watershed Transform. I.J. Image, Graphics and Signal Processing **11**, 10–17 (2017)
5. He, Y., Qian, X.: Contemporary Development and Trend of Jiangsu Province Wind Power Generation Technology. I.J. Education and Management Engineering **2**, 46–51 (2013)
6. Hui, Z., Jinhua, Z.: Landscape Pattern Evolvement in Mining Area: a Case of Liyuan Town in China. I.J. Education and Management Engineering **2**, 48–56 (2011)

7. Strušnik, D., Brandl, D., Schober, H., Ferčec, J., Avsec, J.: A simulation model of the application of the solar STAF panel heat transfer and noise reduction with and without a transparent plate: A renewable energy review. Renew. Sustain. Energy Rev. **134**, 110149 (2020)
8. Bojko, T., Paslavskyi, M., Ruda, M.: Stability of composite landscape complexes: model formalization. Scientific Bulletin of UNFU **29**(3), 108–113 (2019)
9. Korn, G., Korn, T.: Mathematical handbook for scientists and engineers: definitions, theorems, and formulas for reference and review. Dover Publications (2013)
10. Javorskyj, I., Yuzefovych, R., Lychak, O., Kurapov, P.: Hilbert Transform for analysis of daily changes of the earth magnetic field. Proc. of the 2021 IEEE 12th International Conference on Electronics and Information Technologies, ELIT pp. 181–185 (2021)
11. Javorskyj, I., Yuzefovych, R., Kurapov, P., Semenov, P.: The covariance and spectral properties of the quadrature components of narrowband periodically non-stationary random signals. Proc. of the International Scientific and Technical Conference on Computer Sciences and Information Technologies **2**, 95–98 (2020). 9321964
12. Chen, J., Su, J., Kochan, O., Levkiv, M.: Metrological software test for simulating the method of determining the thermocouple error in situ during operation. Measurement Science Review **18**(2), 52–58 (2018)
13. Hu, Z., Gnatyuk, S., Okhrimenko, T., Tynymbayev, S., Iavich, M.: Computer network and information security. Int. J. Comp. Netw. Info. Secu. **12**(3), 1–10 (2020)
14. Jun, S., Kochan, O., Kochan, R.: Thermocouples with built-in self-testing. Int. J. Thermophys. **37**(4), 37 (2016)
15. Melnyk, R., Hatsosh, D., Levus, Y.: Contacts detection in PCB image by thinning, clustering and flood-filling. Proc. of the International Scientific and Technical Conference on Computer Sciences and Information Technologies **1**, 370–374 (2021)
16. Harrington, Jr., E.C.: The desirability function. Industrial Quality Control **21**, 494-498 (1965)
17. Shkrabak, V., Kalugin, A., Averyanov, Y.: Assessing effectiveness of technical measures for improving working conditions of wheeled vehicle operators. Proc. of the 5th International Conference on Industrial Engineering (ICIE 2019), 1, pp. 29–39 (2020)
18. Schulte, P.A., Delclos, G., Felknor, S.A., Chosewood, L.C.: Toward an expanded focus for occupational safety and health: a commentary. Int. J. Environ. Res. Public Health **16**, 4946 (2019)
19. Aurbach, D., et al.: On the surface chemical aspects of very high energy density, rechargeable li–sulfur batteries. J. Electrochem. Soc. **156**, A694–A702 (2009)
20. Andersen, P.D., Bjerregaard, E.: Prospective life cycle assessment on wind power technology 2020. TA-Datenbank-Nachrichten **4**, 10 (2001)
21. Elsam. Livscyklusvurdering af hav- og landplacerede vindmølleparker. 02–170261, Elsam Engineering A/S, Kraftværksvej 53, Fredericia, DK (2004)
22. Pombo, O., Allacker, K., Rivela, B., Neila, J.: Sustainability assessment of energy saving measures: a multi-criteria approach for residential buildings retrofitting. A case study of the Spanish housing stock. Energy and Buildings **116**, 384–394 (2016). https://doi.org/10.1016/j.enbuild.2016.01.019
23. Trishch, R., Cherniak, O., Kupriyanov, O., Luniachek, V., Tsykhanovska, I.: Methodology for multi-criteria assessment of working conditions as an object of qualimetry. Engineering Management in Production and Services **13**(2), 107–114 (2021)
24. Matsuda, H., Nishioka, K., Nakanishi: High-throughput combinatorial screening of multi-component electrolyte additives to improve the performance of Li metal secondary batteries. Scientific Reports **9**, 6211 (2019)
25. Lu, Y., et al.: Ecosystem health towards sustainability. Ecosystem Health and Sustainability **1**(1), 2 (2015)
26. Trishch, R., Nechuiviter, O., Dyadyura, K., Tsykhanovska, I., Yakovlev, M.: Qualimetric method of assessing risks of low quality products. MM Science Journal 4769–4774 (2021)

27. Martinez, E., Sanz, F., Pellegrini, S., et al.: Life cycle assessment of a multi-megawatt wind turbine. Renewable Energy **34**(3), 667–673 (2009)
28. Dzhumelia, E., Pohrebennyk, V.: Methods of soils pollution spread analysis: case study of mining and chemical enterprise in Lviv region (Ukraine). Ecolog. Eng. Environ. Technol. **22**(4), 39–44 (2021)
29. Hu, Z., Tereikovskyi, I., Chernyshev, D., Tereikovska, L., Tereikovskyi, O., Wang, D.: Procedure for processing biometric parameters based on wavelet transformations. Int. J. Modern Edu. Comp. Sci. (IJMECS) **13**(2), 11–22 (2021)
30. Pohrebennyk, V., Koszelnik, P., Mitryasova, O., Dzhumelia, E., Zdeb, M.: Environmental monitoring of soils of post-industrial mining areas. J. Ecolog. Eng. **20**(9), 53–61 (2019)
31. Kupriyanov, O., Trishch, R., Dichev, D., Bondarenko, T.: Mathematic model of the general approach to tolerance control in quality assessment. Lecture Notes in Mechanical Engineering 415–423 (2021)
32. Ruda, M., Bubela, T., Kiselychnyk, M., Boyko, O., Trishch, R.: Assessment of the impact of the elemental composition of batteries on the sustainability of ecosystems. 2022 IEEE 16th International Conference on Advanced Trends in Radioelectronics, Telecommunications and Computer Engineering (TCSET) 390–393 (2022)
33. Hu, Z., Tereykovskiy, I., Tereykovska, L., Pogorelov, V.: Determination of Structural parameters of multilayer perceptron designed to estimate parameters of technical systems. Int. J. Intell. Sys. Appli. (IJISA) **9**(10), 57–62 (2017)
34. Ruda, M., Boyko, T., Chayka, O., Mikhalieva, M., Holodovska, O.: Simulation of the influence of wind power plants on the compartments of the complex landscape system. Journal of Water and Land Development **2**(I–III), 156–165 (2022)
35. Lee, J., Bagheri, B., Kao, H.A.: A cyber-physical systems architecture for Industry 4.0-based manufacturing systems. Manufacturing Letters **3**, 18–23 (2015)
36. Hamadea, R., Al, A.R., Bou, G.M., et al.: Life cycle analysis of aa alkaline batteries. Procedia Manufacturing **43**, 415–422 (2020)
37. DSTU ISO 14040:2004. Ekologhichne keruvannja. Ocinjuvannja zhyttjevogho cyklu. Pryncypy ta struktura [Environmental management. Life cycle assessment. Principles and structure]. Derzhstandart Ukrajiny, Kyiv (2007). (In Ukrainian)
38. Muralikrishna, I.V.: Valli Manickam. Chapter Five - Life Cycle Assessment. Environmental Management, pp. 57–75 (2017)
39. Simensen, T., Horvath, P., Vollering, J., Erikstad, L., Halvorsen, R., Bryn, A.: Composite landscape predictors improve distribution models of ecosystem types. Biodiversity Research **26**, 928–943 (2020)
40. Küçükkaraca, B., Barutcu, B.: Life cycle assessment of wind turbine in Turkey. Balkan Journal of Electrical & Computer Engineering **10**(3), 230–236 (2022)
41. Chipindula, J., Sai, V., Botlaguduru, V.: Life cycle environmental impact of onshore and offshore wind farms in Texas. Sustainability **10**(6), 2022 (2018)
42. Bonou, A.L., Olsen, S.I.: Life cycle assessment of onshore and offshore wind energy: from theory to application. Applied Energy **180**, 327–337 (2016)
43. Hu, Z., Ivashchenko, M., Lyushenko, L., Klyushnyk, D.: Artificial neural network training criterion formulation using error continuous domain. Int. J. Modern Edu. Comp. Sci. (IJMECS) **13**(3), 13–22 (2021)
44. Kent-Dobias, J., Kurchan, J.: Complex complex landscapes. Physical Review Research **3**, 023064 (2021)
45. Cushman, S.: Calculation of configurational entropy in complex landscapes. Entropy **20**(4), 298 (2018)

Development of Distributed System for Electric Personal Transporters Charging

Roman Kochan[1,2](✉), Nataliia Kochan[3], Nataliya Hots[2], Uliana Kohut[2], and Volodymyr Kochan[4]

[1] University of Bielsko-Biala, 43-300 Bielsko-Biala, Poland
[2] Lviv Plitechnic National University, Lviv 79013, Ukraine
roman.v.kochan@lpnu.ua
[3] Ivan Franko National University of Lviv, Lviv 79001, Ukraine
[4] West Ukrainian National University, Ternopil, Ukraine

Abstract. The analysis of the market of Electric Personal Transporters (EPT) and charging infrastructure is presented in this paper. This analysis shows the positive trend of the EPT market and the lack of charging possibilities for EPT of existed charging infrastructure. This paper presents the development of the distributive systems that provide EPTs' battery charging during their staying in public parking lots. It provides an operating range of EPT increasing. The presented system utilizes the Internet of Things methodology and provides charging process control via the mobile application. Its implementation provides increasing of mobility level EPT users.

Keywords: Battery Charging · Distributive System · Structural Synthesis · IEEE 1451 Standard

1 Introduction

The development of mechatronics and electrochemistry has provided the ability to build vehicles based on electric engine for different application, including land, air and voter transport [1]. Limitations of public transport functionality due to the COVID-19 pandemic have led to an avalanche increasing the number of users of light personal electric transporters (EPT) [2]. The forecast of the global EPT market obtained from open sources [3–18] is presented in Table 1. It shows a steady trend towards an increase in the number of EPT in the future.

The existing system of specialized charging stations, which can be found using the query "charging points, <city>, <country>" in the search engine of Google Maps for some largest cities of Poland and Ukraine is presented in Table 2. The total population of presented cities, according to the search query "population <city>, <country>" in Poland cities is approximately 4.1 million, they use 458 charging stations. It means that an average number of users is approximately 9 thousand inhabitants per charging station. The total population of presented Ukrainian cities is approximately 4, 5 million persons. They use 381 charging stations. Therefore, the average number of users is approximately

Z. Hu et al. (Eds.): ICCSEEA 2023, LNDECT 181, pp. 803–814, 2023.
https://doi.org/10.1007/978-3-031-36118-0_70

12 thousand inhabitants per charging station. It is necessary to mention that indicated population is based on Wikipedia and do not account the migration caused by the current war in Ukraine. Anyway, all these charging stations are classified in Google Maps as car charging stations. Approximately half of them are marked by electric power of the charging source. Most of them provide charging using source with a power 11… 50 kW. The number of channels is usually 2…4. It means that these charging stations can be classified as the level "SF4: DC high power charging infrastructure" of the model of electric vehicle support infrastructure presented in [19]. These charging stations are focused on high-capacity batteries that are installed in the car and are incompatible with EPT. Some of the charging stations listed in the Table 1 provide charging channels with a power less than 5 kW, but the total number of such stations does not exceed 20% of the total number of charging channels. These channels can be classified as the level "SF3: Public slow charging infrastructure" [19]. And only these charging channels are compatible with EPT. Therefore, the level SF3 is poorly developed, and the number of charging channels does not correspond to the number of available EPT.

Table 1. Compound annual growth rate (CAGR) of the global EPT market

№	Type of EPT	Years	CAGR	Source
1.	Bikes	2020–2030	10,5	[3]
2.	Bikes	2020–2027	7,9	[4]
3.	Motorcycles	2019–2025	10,35	[5]
4.	Motorcycles	2020–2025	10,35	[6]
5.	Trike	2018–2025	13,5	[7]
6.	Three wheeler	2020–2027	1,7	[8]
7.	Three wheeler	2020–2025	9	[9]
8.	Scooter & Bike & Skateboard	2019–2025	8,4	[10]
9.	Kick scooter	2021–2028	10,3	[11]
10.	Kick scooter	2020–2026	12,3	[12]
11.	Skateboard	2019–2025	3,1	[13]
12.	Skateboard	2019–2023	3,1	[14]
13.	Unicycle	2017–2021	26,9	[15]
14.	Hoverboard	2017–2022	6,1	[16]
15.	Onewheel	2019–2027	5	[17]
16.	Onewheel	2020–2027	5	[18]

One of the main factors that limits the development of EPT is the underdeveloped battery charging infrastructure, which, at least in Poland and Ukraine, is based on "Home charging infrastructure" – the level of "SF2" model of electric vehicle support infrastructure presented in [19].

Table 2. Number of charging stations

№	City	Country	Population, thousands[*]	Number of charging stations
1.	Warsaw	Poland	1765	121
2.	Krakow	Poland	766	89
3.	Lodz	Poland	696	50
4.	Gdansk	Poland	582	104
5.	Katowice	Poland	320	94
6.	Kyiv	Ukraine	2884	177
7.	Dnipro	Ukraine	966	98
8.	Lviv	Ukraine	721	106

[*] based on Wikipedia and do not account the migration caused by the current war in Ukraine

Solving the EPT charging problem by replacing separately pre-charged batteries has a few disadvantages, including the following:

- Increasing the load on home electrical power networks, that is not oriented for such operation mode;
- Complicates the construction of EPT;
- Increasing the electrical losses on the connectors;
- Decreasing the reliability of EPT;
- The support of sharing services becomes more complicated.

These disadvantages of existing approaches to solving the problem of battery charging limits the range of EPT effective using [20], and therefore inhibit the migration of users from traditional modes of transport to the EPT. Despite this, the forecast of the CAGR parameter of the global EPT market is optimistic with the average value 7.8% for the period 2021–2025 with a standard deviation 3.5%.

The objective of this work is in development the system of charging stations that provides batteries of EPT recharging during mooring in public parking lots. It promises to increase the operating range of EPT. This solution will ensure the implementation of the level "SF3: Public slow charging infrastructure", and probably the level "SF4: DC high power charging infrastructure" [19] for EPT. It provides maximization the distance of available trips using EPT and potentially will help to solve the transport and air pollutions problems in the cities by user migration from cars to EPT. The analysis of the charging station market, as well as modeling of the decision-making process by buyers [21] allows predicting that such a service will be in demand.

2   Concept of Distributed System for EPT Battery Charging

The conception of proposed system of EPT batteries charging is based on Internet of Things technology [22, 23]. The developed system is distributed, and it provides the EPT battery charging by connection charger to electric power lines. It is oriented on batteries with relatively small capacity. This system is oriented for implementation in public parking lots, for example near apartments, offices, malls, etc. The proposed system has a modular structure and open architecture, which ensures its scalability and high modernization potential. This system supports the set of following functions:

– Personalized user accounting;
– Access and control using a mobile application;
– On-line monitoring of charging;
– Remote switch on/off of charging;
– Measurement of charge current and value of electric charge;
– Load balancing for all phases.

The developed distributed system has a hierarchical structure that corresponds to the structure presented in the IEEE 1451 family of standards [24] for smart distributed systems. Structure of this system is presented on Fig. 1. It consists of three levels:

– Server;
– Controllers of local area network – NC;
– Charging Controllers – CC.

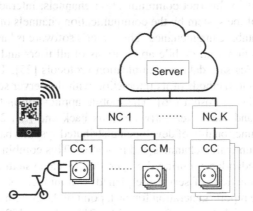

Fig. 1. Structure of developed distributed charging system

Server of this system provides connection the set of controllers of local area networks – NC. Each controller of local area network provides connection the set of Charging controllers – CC. Each charging controller provides the set of charging channel. Drivers of EPT that want to charge the battery of EPT are users of this system. They connect own charger of EPT battery to allowable charging channel of charging controller and run special mobile application. This mobile application connects to the server, provides user identification and personal account access. This mobile application provides scanning the ID tag of selected charging channel than user order electric charge value, that he wishes to charge or time of charger connection to electric power network. Then user prepays the charging service. After payment the Server sends the command of switching on the selected charging channel. This command is addressed to Charging controller that controls the selected charging channel. The parameters sent in this command are: charging time or charge value. Then charging controller switch on the selected charging channel and measure connection time as well as transmitted electric charge. The charging controller provides finishing the charging process by switching off the selected charging channel after achievement the predefined value of charging time or electric charge. Then the charging controller sends to the server the report about successful finishing the charging process. If EPT battery charger of some charging channel will be disconnected before finishing the charging process charging controller sends to the server special urgent query. Server informs user about charging process. This information consists of following fields: coordinates of charger, time of charging starting, time of charging finishing, charging time, value of transmitted electric charge, e.t.c. This information is presented to user by messages of mobile application or selected messenger.

3   Components of the Distributed Charging System

3.1   Server

Server is the component that manages the system. It provides interaction with mobile applications of users by Internet communication channels, interaction with the lower hierarchical levels of the system by the communication channels of controllers of local area network and database maintenance. The server's software is based on a web server, and provides interaction with mobile applications of all users and the set of charging controllers. It provides safe data communication protocols [25]. The prototype of the mobile application and methods of its interaction with the server software is the parcel monitoring system Nova Poshta [26]. The mobile application is the front end of the developed system, and the server software is its back end [27, 28]. Server software provides real-time functionality of developed distributed systems basing on the non-real time Internet communication channels and protocols. This combination is provided by some information redundancy during the operation of the system [29]. The following additional service functions of server are planed: communication with electric power distributor including reports generation for each point of the system connection to power grid [30] and testing of the system functionality [31] and error [32].

3.2   Controller of Local Area Network

The controller of local area network provides the division of the system into segments [33]. Its basic functionality corresponds to the network capable application processor of the IEEE 1451 standard [34, 35]. The specific functions of controller of local area network that are adapted to the target system are:

- routing of messages between server and charging controllers;
- accounting of consumed electric energy;
- measurement the voltage values that is connected to chargers;
- measurement the charging current value;
- emergency shutdown of systems;
- charging controllers remote control;

– testing of modules of the system.

The UML model of this controller is presented in [36]. Its connection with the server is provided using the stack of Internet protocols. Each controller of local area network provides connection to the set of charging controllers. This connection is provided by electric power lines using high frequency modulation of data signals. The structure of controller of local area network is presented on Fig. 2. This controller consists of four main subsystems that are cooperated between each other. These subsystems are presented lower, and they are marked by dotted lines:

– Measuring;
– Control;
– Communication;
– User interface.

The measuring subsystem provides measurement of the values of consumed electric energy, and also rms values of voltage and current of each phase of three phase electric power network. The measuring subsystem consists of:

– Counter of of consumed electric energy – Energy Counter
– Instrumentation for voltage rms value measurement of all phases – Voltmeter;
– Instrumentation for current rms value measurement of all phases (Ammeter);

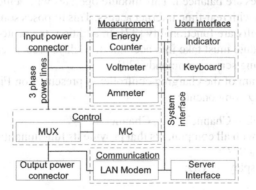

Fig. 2. Structure of controller of local area network

The control subsystem provides the interaction of all components with each other, as well as the shutdown of the system in the event of an emergency, for example, when the parameters of the system exceed the permissible limits. It includes: microcontroller (MC) and protection switch (MUX).

The communication subsystem consists of two interface controllers that provide data transmission between system hierarchical levels, particularly between the charging controllers and the controller of local area network (LAN Modem), and between the server and the controller of local area network (Server Interface).

The controller of local area network is an invisible component of the system for users, and it provides automatic operation, therefore the user interface subsystem provides interaction with the controller of local area network only during system installation and maintenance. Users are then specialists of the relevant profile, and this subsystem displays the current operation parameters of the segment of charging system, which is controlled by the appropriate controller of local area network.

3.3 Charging Controller

Charging controllers provides connection/disconnection of chargers of EPT batteries to electric power lines 230 VAC by the command of server. These modules provide:

- Multichannel charging process;
- Detection of the connection&disconnection of chargers;
- Switching on/off the charging current;
- Measurement the charging current of all channels;
- Measurement of charging time and accounting the electric charge transmitted to each EPT battery;
- Load balancing of all phases;
- Remote control and monitoring of the charging process;
- Indication the parameters of charging process.

The charging controller measures the charging current of all connected EPT batteries and switches them by such combination to three phase electric power network so that the load on all phases are balanced. This module operates with remote control from the server of the system with real-time Internet access. This imposes some requirements and limitations on the software functionality [37]. In addition, the protocols of server interaction with this module must take into account the delays associated with the response time for implementing server commands [29].

The block diagram of the charging controller is presented on Fig. 3. It contains the following interacting components:

- Charging channels – Channel 1, ..., Channel n;
- Microkontroller with all components that provide its functionality – Microcontroller;
- Real time clock – RTC;
- Liquid crystal display – LCD;
- Modem – MOD.

Charging channels provide:

- detection of chargers connection to the connector of the charging channel;
- connection of chargers to the electric power network;
- indication of the charging process;
- measurement the charging current & electric charge;
- fully automatic operation under control of server queries;
- scaling the number of channels.

Each charging channel consists from:

- Connector for charger connection – CONN;
- Multiplexer of charger to the selected phase A, B or C of three phase electric power network – MUX A, MUX B, MUX C;

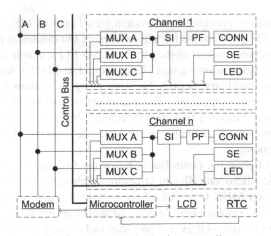

Fig. 3. Structure of charging controller

- Pulse distortion filter – PF;
- Sensor of electric current – SI;
- Sensor of charger connection – SE;
- LED indicator – LED.

Microcontroller provides control of all components of charging module, server command execution and messages exchanging with server of this system.

Real time clock provides measurement of the connected time of EPT batteries.

LCD provides current status and parameters indication of charging controller.

Modem provides connection to controller of local area network. We propose to use radio modules of radio transmitter/receivers with frequency 433 MHz, with electric power lines using as external antenna. All transmitters use the same phase as external antenna. All receivers also use the same phase as external antenna but this phase is another than transmitter's phase. This solution provides trusty connection because the large antennas dimension and small distance between these antennas. Also it provides increasing the distance of connection in comparison with radio modules using and does not demand additional communication lines.

4 Prototype of the Developed System

The prototype of the system is based on the Arduino microprocessor kit and Raspberry Pi single-board computers and is designed to debug algorithms of system components and data transmission protocols.

The prototype of the controller of local area network is based on a module with remote reprogramming [38] connected to single-board computer Raspberry Pi 3 Model B. It partially implements the control and communication subsystems. The control subsystem additionally includes a three-phase solid state relay, which is controlled by general-purpose input/output lines. The communication subsystem assumes that communication with the server is supported by the built-in WiFi adapter. Communication with the charging controler is provided by a modem based on 433 MHz wireless modules of the Arduino kit [39], which use the phase lines of the electrical power network as antennas. These modules implement the physical layer of bit transmission of the seven-level model of open systems interaction [40]. Channel and higher levels of this model are implemented by Raspberry Pi [41]. The measuring subsystem is connected to the serial ports of the general purpose bus. The user interface subsystem implements a standard USB keyboard and monitor that are connected only during system installation and maintenance. Archives with system parameters are stored on the SD-card that provides operating system loading.

The prototype of charging controller is based on Arduino Mega 2560 module. It implements module Microcontroller of structure of charging controller. Multiplexers MUX A, MUX B and MUX C are based on 4-channel relay multiplexer of Arduino shields. It provides fast implementation of 4-channel charging system. Current sensors are based on current transformators that provides isolation of power lines from control lines. Sensor SE is based on reed switch. LED is based on two colors LED. It is installed together with sensor SE inside the socket box of charger connector. LCD is based on 16*2 positions module of LCD. RTC is based on chip DS1307. Modem MOD is based on wireless radiofrequency modules with frequency 433 MHz from Arduino set. The modem of Controllers of local area network uses the same approach. The distortion rejection is provided by connection of all transceivers' antennas to one phase, and all receivers' antennas to another.

5 Conclusions

The market analysis of EPT and charging infrastructure show the positive trend of EPT market and lack of charging possibilities compatible with EPT. This paper presents synthesized the distributive systems that provides batteries f EPT charging. It is oriented on implementation on public parking lots of EPT. The developed distributive charging system is based on the network of charging controllers connected to electric power network and controlled by server. It provides on-line control of charging process using mobile application. The developed system is modular and based on distributive approach. It provides its easy installation and multiplication. The developed system implementation provides increasing the distance of EPT operation. This increasing is up to doubling in the condition of waiting enough time for charging the batteries (for example, during lessons in Universities or Schools, concerts, cinema, etc.). Generally proposed system implementation bring intensification of EPT implementation, promote solving transport and ecological problems of megalopolises.

References

1. Williamson, S.S., Rathore, A.K., Musavi, F.: Industrial electronics for electric transportation: Current state-of-the-art and future challenges. IEEE Transa. Ind. Electron. **62**(5), 3021–3032 (2015). https://doi.org/10.1109/TIE.2015.2409052
2. 11 types of electric personal transportation vehicles: a comparison.https://www.hobbr.com/types-of-electric-personal-transport-devices/

3. Electric Bike Market by Product (Pedelecs, Speed Pedelecs, Throttle on Demand, and Scooter & Motorcycle), Drive Mechanism (Hub Motor, Mid-Drive, and Others), and Battery Type (Lead Acid, Lithium-Ion (Li-ion), and Others): Global Opportunity Analysis and Industry Forecast, 2020–2030. https://www.alliedmarketresearch.com/electric-bikes-market

4. E-bike Market Growth, Industry Trends, and Statistics by 2027. https://www.marketsandma rkets.com/Market-Reports/electric-bike-market-110827400.html

5. Electric Motorcycles Market to Exhibit 10.35% CAGR by 2025 | Market Research Future. https://www.globenewswire.com/fr/news-release/2021/02/11/2173763/0/en/Electric-Mot orcycles-Market-to-Exhibit-10-35-CAGR-by-2025-Market-Research-Future.html

6. Electric Motorcycle Market Size, Share, Growth, Report, 2027. https://www.marketresearchf uture.com/reports/electric-motorcycles-market-8136

7. Electric Trike Market Showcase Approximately vigorous CAGR of Around 13.5% In Global Market Forecast Till 2025. https://www.mynewsdesk.com/us/news-world-center/pressrele ases/electric-trike-market-showcase-approximately-vigorous-cagr-of-around-13-dot-5-per cent-in-global-market-forecast-till-2025-2848684

8. Electric Three Wheeler Market Size, Share, Global Report, 2027. https://www.fortunebusin essinsights.com/electric-three-wheeler-market-105028

9. Electric Three-Wheeler Market – Growth, Trends, and Forecasts (2020–2025). https://www. globenewswire.com/news-release/2020/08/27/2084586/0/en/Electric-Three-Wheeler-Mar ket-Growth-Trends-and-Forecasts-2020-2025.html

10. Electric Transporters Market Size, Share & Trends Analysis Report By Vehicle (Bike, Scooter, Skateboards), By Battery (Sealed Lead Acid, NiMH, Li-Ion), By Voltage, And Segment Forecasts, 2019–2025. https://www.grandviewresearch.com/industry-analysis/electric-transporters-market

11. Electric Kick Scooters Market Size Worth $5.2 Billion By 2028 | CAGR: 10.3%: Grand View Research, Inc. https://www.prnewswire.com/news-releases/electric-kick-scooters-mar ket-size-worth-5-2-billion-by-2028--cagr-10-3-grand-view-research-inc-301210598.html

12. Global Electric Kick Scooters Market By Battery, By Voltage, By Region, Industry Analysis and Forecast, 2020–2026. https://www.globenewswire.com/news-release/2021/02/25/218 2331/0/en/Global-Electric-Kick-Scooters-Market-By-Battery-By-Voltage-By-Region-Ind ustry-Analysis-and-Forecast-2020-2026.html

13. Skateboard Market Size Worth $2.4 Billion By 2025 | CAGR: 3.1%. https://www.grandview research.com/press-release/global-skateboard-market

14. Electric Skateboard Market. https://askwonder.com/research/electric-skateboard-market-bud 11r96e

15. Electric Unicycle 2017 Global Market Expected to Grow at CAGR of 26.90% and Forecast to 2021. https://www.einpresswire.com/article/414374748/electric-unicycle-2017-glo bal-market-expected-to-grow-at-cagr-of-26-90-and-forecast-to-2021

16. Hoverboard Market Forecast, Trend Analysis & Competition Tracking – Global Market insights 2017 to 2022. https://www.factmr.com/report/219/hoverboard-market

17. One Wheel Electric Scooter Market by Product Type (Electric Unicycle, and Electric One wheel Hoverboard), Application (Off-road Activities, and Daily Commute), Sales Channel (Online Sales, and Offline Sales), and Speed Limit (Kmh) (20 Kmh – 30 Kmh, 30 Kmh – 50 Kmh, and More than 50 Kmh): Global Opportunity Analysis and Industry Forecast, 2020–2027. https://www.alliedmarketresearch.com/one-wheel-electric-scooter-market-A08744

18. One Wheel Electric Scooter Market to Garner $146.7 Million by 2027: Allied Market Research, http://www.globenewswire.com/en/news-release/2021/01/18/2160005/0/en/ One-Wheel-Electric-Scooter-Market-to-Garner-146-7-Million-by-2027-Allied-Market-Res earch.html

19. Funke, S.Á., Sprei, F., Gnann, T., Plötz, P.: How much charging infrastructure do electric vehicles need? A review of the evidence and international comparison. Transp. Res. Part D: Transp. Environ. **77**, 224–242 (2019). https://doi.org/10.1016/j.trd.2019.10.024
20. Zagorskas, J., Burinskienė, M.: Challenges caused by increased use of e-powered personal mobility vehicles in European cities. Sustainability **12**(1), 273 (2019). https://doi.org/10.3390/su12010273
21. Kowalska-Styczeń, A., Sznajd-Weron, K.: From consumer decision to market share – unanimity of majority? J. Artif. Soc. Soc. Simul. **19**(4), 10 (2016). https://doi.org/10.18564/jasss.3156
22. Greengard, S.: The Internet of Things. The MIT Press (2015). https://doi.org/10.7551/mitpress/10277.001.0001
23. Jun, S., Kochan, O., Kochan, V., Wang, C.: Development and investigation of the method for compensating thermoelectric inhomogeneity error. Int. J. Thermophys. **37**, 1–14 (2016)
24. Song, E., Lee, K.: Understanding IEEE 1451-Networked smart transducer interface standard – What is a smart transducer? IEEE Instrum. Meas. Mag. **11**(2), 11–17 (2008). https://doi.org/10.1109/MIM.2008.4483728
25. Zhengbing, H., Dychka, I., Onai, M., Zhykin, Y.: Blind payment protocol for payment channel networks. Int. J. Comput. Netw. Inform. Secur. (IJCNIS) **11**(6), 22–28 (2019). https://doi.org/10.5815/ijcnis.2019.06.03
26. Urgent and express delivery, logistic service in Ukraine and abroad. https://novaposhta.ua/
27. Jun, S., et al.: A cost-efficient software based router and traffic generator for simulation and testing of IP network. Electronics **9**(1), 40 (2019). https://doi.org/10.3390/electronics9010040
28. Xu, H., Przystupa, K., Fang, C., Marciniak, A., Kochan, O., Beshley, M.: A combination strategy of feature selection based on an integrated optimization algorithm and weighted k-nearest neighbor to improve the performance of network intrusion detection. Electronics **9**(8), 1206 (2020)
29. Hrusha, V., Kochan, R., Kurylyak, Y., Osolinskiy, O.: Development of measurement system with remote access based on Internet. In: 4th IEEE Workshop on Intelligent Data Acquisition and Advanced Computing Systems: Technology and Applications, pp. 126–128. Dortmund (2007). https://doi.org/10.1109/IDAACS.2007.4488388
30. Chen, X., et al.: Forecasting short-term electric load using extreme learning machine with improved tree seed algorithm based on Lévy flight. Eksploatacja i Niezawodność **24**(1), 153–162 (2022)
31. Fang, M.T., Chen, Z.J., Przystupa, K., Li, T., Majka, M., Kochan, O.: Examination of abnormal behavior detection based on improved YOLOv3. Electronics **10**(2), 197 (2021)
32. Xiong, G., et al.: Online measurement error detection for the electronic transformer in a smart grid. Energies **14**(12), 3551 (2021)
33. Yatsuk, V., Bubela, T., Pokhodylo, Y., Yatsuk, Y., & Kochan, R.: Improvement of data acquisition systems for the measurement of physical-chemical environmental properties. Paper presented at the Proceedings of the 2017 IEEE 9th International Conference on Intelligent Data Acquisition and Advanced Computing Systems: Technology and Applications, IDAACS 2017, pp. 1 41–46 (2017). https://doi.org/10.1109/IDAACS.2017.8095046
34. Kochan, V., Lee, K., Kochan, R., Sachenko, A.: Approach to improving network capable application processor based on IEEE 1451 standard. Comput. Stan. Interfaces **28**(2), 141–149 (2005). https://doi.org/10.1016/j.csi.2005.01.015
35. Kochan, V., Lee, K., Kochan, R., Sachenko, A.: Approach to improvement the network capable application processor compatible with IEEE 1451 standard. In: Proceedings of Second IEEE International Workshop on Intelligent Data Acquisition and Advanced Computing Systems: Technology and Applications, pp. 437–441. Lviv (2003). https://doi.org/10.1109/IDAACS.2003.1249602

36. Stepanenko, A., Lee, K., Kochan, R., Kochan, V., Sachenko, A.: Development of a minimal IEEE 1451.1 model for microcontroller implementation. In: Proceedings of the 2006 IEEE Sensors Applications Symposium, pp. 88–93. Houston, Texas (2006). https://doi.org/10.1109/SAS.2006.1634243

37. Hrusha, V., et al.: Distributed Web-based measurement system. In: 2005 IEEE Intelligent Data Acquisition and Advanced Computing Systems: Technology and Applications. pp. 355–358. Sofia (2005). https://doi.org/10.1109/IDAACS.2005.283002

38. Kochan, R., Kochan, O., Chyrka, M., Vasylkiv, N.: Precision data acquisition (DAQ) module with remote reprogramming. In: 2005 IEEE Intelligent Data Acquisition and Advanced Computing Systems: Technology and Applications, pp. 279–282 (2005)

39. The RC link and ASK environment. https://www.electroschematics.com/rc-link-ask-enviro nment/

40. Saxena, P.: OSI reference model–a seven layered architecture of OSI model. Int. J. Res. (IJR) 1(10), 1145–1156 (2014)

41. Maykiv, I., Stepanenko, A., Wobschall, D., Kochan, R., Kochan, V., Sachenko, A.: Software–hardware method of serial interface controller implementation. Comput. Stan. Interfaces 34(6), 509–516 (2012). https://doi.org/10.1016/j.csi.2011.10.009

Environmental Pollution from Nuclear Power Plants: Modelling for the Khmelnytskyi Nuclear Power Plant (Ukraine)

Igor Vasylkivskyi[1], Vitalii Ishchenko[1], Orest Kochan[2,3($\boxtimes$)], and Roman Ivakh[3]

[1] Vinnytsia National Technical University, Vinnytsia, Ukraine
[2] Lublin University of Technology, Lublin, Poland
[3] Lviv Polytechnic National University, Lviv, Ukraine
`orest.v.kochan@lpnu.ua`

Abstract. It is known that there is emission of radionuclides during normal operation of nuclear power plants. The paper analyses sources of radioactivity and environmental pollution for the case study of Khmelnytskyi nuclear power plant in Ukraine. Based on the radionuclides concentrations in air emissions, the highest concentrations are found for Nitrogen-16, Krypton-83m, Argon-41, Tritium, and Xenon-131m. Among non-radioactive chemicals, only nitrogen dioxide concentration in the air is found to be below the permissible limit. The highest concentrations of suspended particles in the air near the Khmelnytskyi power plant were measured in the range 3.4–7.7 mg/m^3. Up to 90% of non-radioactive air emissions are generated in the boiler room. The main contribution to the gas-aerosol emission of Khmelnytskyi nuclear power plant is made by inert radioactive gases, which do not directly migrate in ecosystem along food chain. The expected contribution of iodine radioisotopes and activated corrosion products to the total emission is very small. Crop contamination with aerosol radionuclides from nuclear power plant is analysed. Contamination of agricultural products with aerosol radionuclides in case of maximal projected accident at Khmelnytskyi nuclear power plant (Ukraine) is calculated.

Keywords: Modelling · Crop Contamination · Radionuclide · Nuclear Power Plant

1 Instruction

Reliable electricity sources are necessary in modern world [1] to maintain critical infrastructure [2, 3]. Heat power plants generate most of electricity in many countries. Despite many efforts intended to improve their operation [4–7] (because their efficiency has both economic and ecological effect [8, 9]), there are still many environmental issues. Under this circumstance, nuclear power becomes the priority electricity source for many countries. However, nuclear power plants also emit environmental pollution like any other industry [6]. There are gaseous, liquid, and solid emissions from nuclear power plants. Obviously, such emissions include radioactive elements [10]. According to the energy strategies of many countries, construction of more nuclear power plants is planned.

© The Author(s), under exclusive license to Springer Nature Switzerland AG 2023
Z. Hu et al. (Eds.): ICCSEEA 2023, LNDECT 181, pp. 815–826, 2023.
https://doi.org/10.1007/978-3-031-36118-0_71

In Ukraine, more nuclear units at existing nuclear power plants are projected with the total capacity of up to 22 GW [11]. Similar strategies are accepted in USA, China, India, and many other countries [12–16]. National safety, including reliable energy supply, is always of the highest priority. But, environmental risks should not be underestimated. There are studies [17, 18] revealing radionuclides emission from nuclear power plants. Radionuclides can be disseminated and precipitated at large areas leading to soil and crop pollution. Modelling of radionuclide emission for the Fukushima nuclear power plant was done by Leelőssy et al. [19]. Morino et al. [20] have found 22% of cesium-137 and 13% of iodine-131 distributed over 500,000 km^2 area after the Fukushima accident. Other studies [21–24] also reported about long distances of radionuclides dissemination from nuclear power plants. Besides, impact on humans and environmental issues resulting from air emissions of nuclear power plants are analysed in [25]. There is a research [26] dedicated to the nuclear power contribution to climate protection due to decreased gaseous emissions in comparison to energy production from fossil fuels. Eisma et al. [27] have revealed that methane isotope $^{14}CH_4$ is emitted from nuclear power plants in higher amounts than generally assumed (known sources: pressurized and boiling water reactors). Also, there is a risk of transboundary transferring of radionuclides. Kim et al. [28] have calculated effective dose to Korea considering all China's nuclear power plants. The results showed that the effective dose should be expected to increase to 1.1E-08 Sv/year, about a 500-fold increase from the 2000 level and a 2.6-fold increase from the 2010 level. One of the most common radionuclides emitted from nuclear power plants is xenon [29], but other elements are also important to measure.

This research analyzes radionuclides dissemination from nuclear power plant for the case study of Khmelnytskyi nuclear power plant in Ukraine. The objectives of this study are as follows: assessment of radionuclides and non-radioactive chemicals concentration in the air near Khmelnytskyi nuclear power plant; modeling of agricultural products contamination during the maximal projected accident at Khmelnytskyi nuclear power plant.

2 Materials and Methods

Radionuclides emission from Khmelnytskyi nuclear power plant (Netishyn, Khmelnytskyi region, Ukraine) was measured in the laboratory of the Department of Environmental Protection at the power plant aforementioned. Air pollution at the area of the power plant was analysed using high-altitude radio-sensing of the atmosphere. Air emissions from the turbine condenser ejectors was registered at Shepetivka research station, some 25 km from Khmelnytskyi nuclear power plant. The concentration of radionuclides was measured by radiometers RAC-02, РГГ-03, РГБ-101–01, РГБ-101–02 combined in Automatic System of Radiological Control. The results used in this paper are average for 2014–2016 years. Besides, contamination of agricultural products with aerosol radionuclides for a maximal projected accident at Khmelnytskyi nuclear power plant was calculated based on radionuclides emission and sedimentation.

Radionuclides concentrations in agricultural products were calculated in following way [30]:

$$q = q_a + q_r,$$ (1)

where q_a is direct intake of radionuclides to the plant from atmosphere; qr is intake of radionuclides to the plant through roots.

We can define q_a as follows:

$$q_a = (0.25/\rho) \cdot r \cdot c \cdot V \cdot t, \tag{2}$$

where ρ is the amount of aboveground plant biomass, kg/m^2; r is the ratio of radionuclide concentration in the commercial and aboveground part of plant (average 0.1); c is average annual radionuclide concentration in atmosphere (Bq/m^3); V is the rate of radionuclide sedimentation (m/s); t is the time passed since the beginning of growing season (s).

Thus, q_r can be defined as follows:

$$q_r = (1/m) \cdot c_s \cdot k, \tag{3}$$

where m is the mass of soil layer of 1 m^2 square, where radionuclide uptake occurs, (average 300 kg/m^2); c_s is soil contamination with radionuclide, Bq/m^2; k is the factor of radionuclide accumulation in plant, can be obtained from [31].

Radionuclide concentration in atmosphere (c) and soil contamination with radionuclide (c_s) can be assessed according to the methodology presented in [32] and based on the Gaussian air pollutant dispersion and sedimentation equations. The input data include:

- category of stability – F;
- wind velocity – 1 m/s;
- distance from the nuclear power plant – 2–25 km;
- soil roughness – 10–100 cm;
- altitude of layer mixing – 200 m;
- season – summer;
- rate of aerosol sedimentation – 0.001 m/s.

3 Results

3.1 The Sources of Radionuclides Emission from the Khmelnytskyi Nuclear Power Plant

For a typical nuclear power plant, radionuclides can be found in products of nuclear fuel decomposition, activation products, and products of constructions corrosion [22]. Even if a nuclear power plant operates as usual, some fission products can be transferred into the heat carrier if damage of the fuel element coverage happens [10]. One of the main fission products in the heat carrier of the primary circuit is Tritium (around 12 years half-life). It is a radioactive isotope of hydrogen [11]. Tritium is generated in the reactor of a nuclear power plant as a result of following processes: i) radioactive decay of nuclear fuel; ii) combination of the neutron and deuterium nuclei; iii) interactions of the fast neutrons and reactor constructions; iv) boric acid activation in the heat carrier of the primary circuit. Water leakages from the first circuit into the water tanks can lead to tritium emission from the water of the first circuit. The dissolved fission and activation products are extracted from the heat carrier using ion exchange. This results in the ion-exchange resin generation in water treatment equipment [26]. These resins are periodically replaced, which leads to solid and liquid radioactive waste generation. Leakages from the steam generator of the primary circuit lead to radioactive pollution of water in the secondary circuit. There are also gaseous emissions generated in the primary circuit: aerosols, tritium water vapour, noble gases, and many other gases [22]. When the reactor is stopped for a revision, the pressure of the cooling system is decreased, the reactor lid is removed, and some fuel elements are moved to the storage pool. When removing spent fuel elements, air emissions and liquid radioactive waste may be generated in the storage pool, construction revision mine, and revision mine of protecting pipes. Radioactive aerosols are emitted into the air mainly from the following equipment: i) ventilation tubes of the reactor (emission height is over 100 m); ii) turbine ejector [11]. Such aerosols contain short-lived radionuclides (up to 3 h half-life) and long-lived radionuclides (over 3 h half-life) [11]. The half-life influences the consequences for the human and exposure time. The radioactive aerosols may be inhaled with air or consumed along with water and food.

Table 1. Average pollution with radionuclides from the Khmelnytskyi nuclear power plant [33]

Isotope	Half-life time	Emission [Ci/day]	Isotope	Half-life time	Emission [Ci/day]
Antimony-129	4.4 h	1.25×10^{-8}	Niobium-101	7.1 s	3.04×10^{-8}
Argon-41	1.82 h	1.05	Nitrogen-16	7.13 s	2.14
Barium-137 m	2.552 m	1.02×10^{-5}	Nitrogen-17	4.17 s	2.98×10^{-4}
Bromine-83	2.39 h	3.34×10^{-6}	Potassium-42	12.36 h	1.00×10^{-5}
Carbon-14	5730 y	1.12×10^{-7}	Praziodim-144m	7.2 m	$1.62 \times \times 10^{-11}$
Cerium-143	33.0 h	2.36×10^{-8}	Rhodium-103m	56.11 m	1.87×10^{-7}
Cesium-137	30.20 y	2.74×10^{-6}	Rubidium-88	17.8 m	7.96×10^{-2}
Chrome-51	27.7 d	6.72×10^{-8}	Ruthenium-103	39.25 d	2.06×10^{-9}
Cobalt-60	5.27 h	4.68×10^{-9}	Selenium-83	22.4 m	2.76×10^{-8}
Ferrum-55	2.68 y	2.34×10^{-9}	Sodium-24	14.97 h	3.34×10^{-7}
Iodine-131	8.01 d	1.91×10^{-4}	Strontium-89	50.62 d	3.68×10^{-8}
Ittree-90	64.26 h	4.12×10^{-11}	Technetium-101	14.2 m	9.84×10^{-7}

(*continued*)

Table 1. (*continued*)

Isotope	Half-life time	Emission [Ci/day]	Isotope	Half-life time	Emission [Ci/day]
Krypton-83m	1.83 h	2.66	Tellurium-129m	33.6 d	1.55×10^{-10}
Lanthanum-141	3.92 h	2.14×10^{-7}	Tin-130	3.7 m	9.44×10^{-8}
Manganese-54	312.2 d	1.66×10^{-9}	Tritium	12.33 y	3.22×10^{-1}
Molybdenum-99	66.02 h	5.88×10^{-10}	Xenon-131m	11.97 d	8.28
Niobium-95m	3.61 d	8.04×10^{-11}	Zirconium-95	64.02 d	4.76×10^{-7}

The data on radionuclides emission from the ventilation tubes of Khmelnytskyi nuclear power plant (average in 2010–2015 years) can be found in Table 1.

Suspended aerosol particles may also emit into the air during transferring the ash and dust. The concentration depends on the surface and meteorological parameters. The highest concentrations of suspended particles in the air near the Khmelnytskyi power plant were measured in the range 3.4–7.7 mg/m^3 (Netishyn, 4 km to the north from the plant) and 1.65 mg/m^3 (Komarivka, 5 km to the east from the plant) [33].

3.2 Non-radioactive Emissions of the Khmelnytskyi Nuclear Power Plant

Non-radioactive emissions may be released into the atmosphere by the objects and the equipment generating harmful gases during technological processes. At the Khmelnytskyi nuclear power plant, up to 90% of non-radioactive air emissions are generated in the boiler room [33]. Emissions from other sources are low due to the low capacity of gas treatment equipment. Concentrations of some chemicals measured in the air (just over the surface) near the Khmelnytskyi nuclear power plant may be seen in Table 2.

At the Khmelnytsjyi nuclear power plant, the monitoring of meteorological and aerial characteristics of the atmosphere is carried out in order to assess the impact of air emissions. These parameters directly influence the radionuclides transportation. Some meteorological parameters reduce the air self-cleaning mechanisms and contribute to the chemicals accumulation in the air. These parameters are as following: temperature stratification of the atmosphere, the velocity and the direction of wind; regime of precipitation, clouds and fogs formation [23].

3.3 Crop Contamination with Aerosol Radionuclides from Nuclear Power Plant

The level of radioactive contamination of crop products and its influence on elements of food chain (animals and humans) are determined by the following [22]: total amount of

Table 2. The maximum concentrations of some chemicals measured in the air near the Khmelnytskyi nuclear power plant [33].

Chemical	Limit, [mg/m³]	Concentration at the border of the protection area [mg/m³]
abrasive-metal dust	0.4	<0.01
ash	0.15	0.015
butyl acetate	0.1	0.05
carbon monoxide	5	0.03
inorganic dust	0.3	0.05
nitrogen dioxide	0.2	0.21
sawdust	0.1	<0.01
sulphur dioxide	0.5	0.22
toluene	0.6	<0.05

radionuclides released into the environment; the nature of these radionuclides dissemination; physicochemical properties of radionuclides; properties of contaminated soil and plants.

The main contribution to the gas-aerosol emission of Khmelnytskyi nuclear power plant is made by inert radioactive gases, which do not directly migrate in ecosystem along food chain. The expected contribution of iodine radioisotopes and activated corrosion products to the total emission is very small. Many of these radionuclides have several hours half-life. When considering radionuclides migration in ecosystems, a primary attention should be given to radioisotopes of so-called biogenic chemical elements. To assess the consequences of impact, the estimation of possible values of the territory contamination with radionuclides should be one of the main criteria along with data on soil and climatic conditions of the territory, features of agricultural manufacturing, etc.

A lot of radionuclides emitted into the atmosphere create aerosols and fall to the earth surface as rain, fog, and snow under the influence of gravitation. The sedimentation of radionuclides on the surface of plants also occurs on days without precipitation. Sometimes dry sedimentation is quite high (depends on the atmosphere conditions). The contact of radioactive aerosols with plant surface leads to their accumulation in aboveground parts of the plant. The retention of radioactive aerosols and their subsequent transformation depend on following factors [22]: density of phytomass per area, plant species, aerosol particles size, humidity before and after precipitation, etc. Immediately after the sedimentation, the process of removing of radioactive particles from the plant surface begins. The maximal excretion of radionuclides from plants occurs as early as the first day. An average elimination half-time of different meadow-grassland up to 70–90% of losses occur within the first 7–10 days [26]. The average half-life of meadow-grassland plants is longer and ranges from 7 to 17 days [26]. In the case of continuous precipitation, the contribution of this process to the total contamination of plants is determined by the intensity of sedimentation of radioactive substances. The intensity of radionuclides sedimentation at the expected emission level is very low (for example, it is 10–10 Bq/(m^2·s) for ^{137}Cs) and therefore plant contamination is also very low. Under continuous precipitation, the soil surface is constantly contaminated by radioactive substances precipitated from the atmosphere. Some radionuclides are directly accumulated by soil surface and others are delayed by plants. The process of nuclides removal from plants is very dynamic. Within few weeks after the emission, radionuclides almost completely move to the soil surface and are included in the subsequent migration processes in the ecosystem. These processes include vertical/horizontal migration of radionuclides and their intake by plants. The driving forces that cause radionuclide migration in soils are as follows [34]: filtration of atmospheric precipitation deep into the soil, capillary water flow to the surface as a result of evaporation, thermal transfer of moisture due to the temperature gradient, water flows on the soil surface, diffusion of free and adsorbed ions, radionuclides transfer by migrating colloidal particles and plant roots, sorption and desorption by soil substance. The intensity of migration depends very much on the physico-chemical properties of radionuclides and the soil conditions. For example, radionuclides ^{137}Cs and ^{90}Sr behave in completely different way under the same conditions [17]. More than 90% of cesium radionuclides emitted due to Chornobyl disaster are found in the upper five-centimeter layer of undisturbed soils [11]. Strontium radionuclides are more mobile. It is now disseminated at 40 cm depth [11] due to the vertical migration. In special exclusion zone around Chornobyl nuclear power plant, maximal concentrations of strontium radionuclides in some sands are found at depths of more than one meter [11]. Taking into account the fact that groundwater in Chornobyl region is mainly located at a depth of more than three meters, we can conclude that vertical migration slightly contributes to the radionuclides transferring to the groundwater and further dissemination in open water reservoirs. Even if radionuclides migrate to water objects, their subsequent horizontal migration is very slow due to low groundwater velocity.

The soil has a high capacity of radionuclides absorption and limits their spatial dissemination and plants intake by roots [34]. The root system of plants acts as a selective barrier preventing the movement of biologically inert radioactive elements to the above-ground phytomass. Thus, only biologically mobile radionuclides should be taken into account when considering the root path of nuclides. On the one hand, the sorption of radionuclides by soil restricts their flow to plants. On the other hand, this process forces the radionuclides to retain in the soil for a long time being a source for plants. The intensity of radionuclides transfer to the plants depends on many factors. These include agrochemical properties of soil, physico-chemical properties of radionuclides, biological characteristics of plants, etc. [34].

There are several ways of radionuclides intake by the human body with food (Fig. 1) [11].

One of these ways is the chain: arable land – plants – food products. The intensity of radionuclide migration along this chain is primarily determined by soil type, crop type, physico-chemical properties of radionuclides, and contamination density of soil surface. The highest contamination of plants is found for sod-podzolic soils (especially soils with light granulometric composition) [35]. Low contamination is found for grey forest soils, while the lowest – for black soil (chornozem). Another way of radionuclides

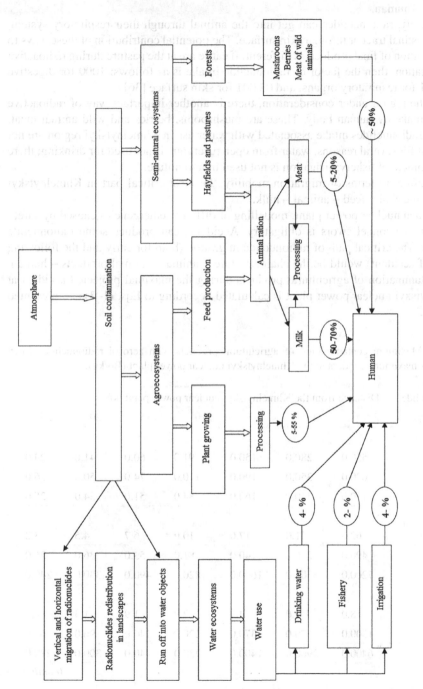

Fig. 1. Main ways of radionuclides intake by human body along with food

intake by human with food is related to the following chain: feed – animal – animal products – human.

Basically, radionuclides can get into the animal through their respiratory system, gastrointestinal tract or through skin surface. The potential contribution of these ways to contamination of final products is different. If cattle are at the pasture during radioactive sedimentation, then the factor of radionuclide intake is as follows: 1000 for digestive system, 1 for respiratory organs, and 0.0001 for skin surface [36].

For the region under consideration, there is another important way of radioactive cesium intake by human body. These are mushrooms, berries and wild animal meat. Ways of radionuclides intake associated with water use in Khmelnytskyi region are not important for several reasons: water from open reservoirs is not used for drinking; there is no commercial fishery; irrigation is not used in agriculture.

Therefore, in terms of migration mobility, the most critical part in Khmelnytskyi region is the chain: feed – animal – milk.

For each nuclear power plant, modelling of different emergencies caused by safety failures or personnel errors is obligatory. Accidents can produce some radionuclide emission. The critical path of radionuclide migration (both for early and the following phases of accident) would be the chain: pasture – animal – animal products – human. The contamination of agricultural products during the maximal projected accident at Khmelnytskyi nuclear power plant is calculated according to Eqs. (1–3) and presented in Table 3.

Table 3. Maximum contamination of agricultural products with aerosol radionuclides when modelling maximum accident at the Khmelnytskyi nuclear power plant (Bq/kg).

Radionuclide	Distance from the Khmelnytskyi nuclear power plant, km						
	2	4	6	10	15	20	25
Bread							
Sr^{90}	550.0	280.0	150.0	91.0	60.0	41.0	29.0
Cs^{137}	690.0	350.0	190.0	110.0	74.0	50.0	36.0
I^{131}	740.0	340.0	160.0	84.0	51.0	34.0	25.0
Milk							
Sr^{90}	62.0	31.0	17.0	10.0	6.7	4.5	3.2
Cs^{137}	490.0	250.0	140.0	81.0	53.0	36.0	26.0
I^{131}	7200.0	3300.0	1600.0	820.0	490.0	330.0	240.0
Meat							
Sr^{90}	18.0	9.0	4.9	2.9	1.9	1.3	0.9
Cs^{137}	1300.0	670.0	370.0	220.0	140.0	96.0	69.0
I^{131}	6400.0	2900.0	1400.0	730.0	440.0	300.0	220.0

(*continued*)

Table 3. (*continued*)

Radionuclide	Distance from the Khmelnytskyi nuclear power plant, km						
	2	4	6	10	15	20	25
Vegetables							
Sr^{90}	770.0	390.0	220.0	130.0	83.0	57.0	40.0
Cs^{137}	960.0	490.0	270.0	160.0	100.0	70.0	50.0
I^{131}	12000.0	5500.0	2600.0	1400.0	820.0	550.0	410.0
Fruits (apples, pears)							
Sr^{90}	19.0	9.6	5.3	3.1	2.0	1.4	1.0
Cs^{137}	24.0	12.0	6.5	3.9	2.5	1.7	1.2
I^{131}	300.0	140.0	66.0	34.0	21.0	14.0	10.0

As can be seen from the Table 3, main attention should be given to iodine nuclide I^{131}. The highest concentrations of selected radionuclides in agricultural products are found in vegetables and meat. The contamination level is significantly lower than the estimates given above if urgent measures are done in time.

4 Conclusion

Nuclear power plants generate air emissions from the equipment. Air emissions may contain radionuclides and non-radioactive aerosols causing environmental risks and harmful effects on the human. Majority of the sources of air emission operate periodically. Therefore, the total emission is relatively small. The chemicals concentration in the air near the Khmelnytskyi nuclear power plant are found to be below the Ukrainian limits. Only nitrogen dioxide concentration is found to slightly exceed the limit. The highest concentrations are measured for Argon-41, Nitrogen-16, Xenon-131m, Tritium, and Krypton-83m radionuclides. In case of a projected accident, the concentrations of radionuclides in different agricultural products are found to be very high even at the distance of 20 km and more from power plant. Therefore, people living at the territories, where pollution from nuclear power plants may occur, have to know the rules of behaviour in emergency situations. Despite the low level of air pollution during usual exploitation, the environmental risks of nuclear power plant operation have to be thoroughly assessed. The results obtained may be used when assessing environmental performance of nuclear power plants. Besides, the data on pollution of agricultural products with radionuclides is crucial for providing environmental safety.

References

1. Przystupa, K.: Selected methods for improving power reliability. Przegliad Elektrotechniczny **94**(12), 270–273 (2018)

2. Hu, Z., Khokhlachova, Y., Sydorenko, V., Opirskyy, I.: Method for optimization of information security systems behavior under conditions of influences. Int. J. Intell. Syst. Appl. 9(12), 46–58 (2017)
3. Hu, Z., Tereikovskyi, I., Chernyshev, D., et al.: Procedure for processing biometric parameters based on wavelet transformations. Int. J. Modern Educ. Comput. Sci. 13(2), 11–22 (2021)
4. Lederer, T.: Metrology for improved power plant efficiency. http://surl.li/betue. Accessed on 20 Jan 2022
5. Jun, S., Kochan, O., Kochan, R.: Thermocouples with built-in self-testing. Int. J. Thermophys. 37(4), 37 (2016)
6. Jun, S., Kochan, O., Chunzhi, W., Kochan, R.: Theoretical and experimental research of error of method of thermocouple with controlled profile of temperature field. Measur. Sci. Rev. 15(6), 304–312 (2015)
7. Mehra, M., Jayalal, M.L., Arul, A.J., Rajeswari, S., Kuriakose, K.K., Murty, S.A.V.S.: Study on different crossover mechanisms of genetic algorithm for test interval optimization for nuclear power plants. Int. J. Intell. Syst. Appl. 6(1), 20–28 (2014)
8. Wang, T., Bediones, D., Swirla, P., Henrikson, H., Janhunen, E., Bachalo, K.: Stabilized metal sheathed type K and E thermocouples improve turbine efficiency. In: Proceedings of ISA, pp. 439–448 (1997)
9. Ishchenko, V., Pohrebennyk, V., Kochanek, A., Przydatek, G.: Comparative environmental analysis of waste processing methods in paper recycling. Proc. Int. Multidisc. GeoConf. SGEM 17(51), 227–234 (2017)
10. Ishchenko, V., Styskal, O., Vasylkivsky, I.: Air pollution with heavy metals compounds in Vinnytsia region. Struct. Env. 1(6), 33–37 (2014)
11. Zvorykin, A.. Pioro, I., Panchal, R.: Study on current status and future developments in nuclear-power industry of Ukraine. In: Proceedings of the 2016 24th International Conference on Nuclear Engineering. American Society of Mechanical Engineers. V005T15A020-V005T15A020 (2016)
12. Guo, X., Guo, X.: Nuclear power development in China after the restart of new nuclear construction and approval: a system dynamics analysis. Renew. Sustain. Energy Rev. 57, 999–1007 (2016)
13. Hart, D.: Nuclear Power in India: A Comparative Analysis. Routledge (2019). https://doi.org/10.4324/9780429278303
14. Gattie, D.K., Darnell, J.L., Massey, J.N.: The role of US nuclear power in the 21st century. Electr. J. 31(10), 1–5 (2018)
15. Kobayashi, M.: Nuclear development status in the World (4): four new emerging countries (China, Russia, India, and South Korea) leading global nuclear development. J. At. Energy Soc. Japan 59(11), 638–643 (2017). https://doi.org/10.3327/jaesjb.59.11_638
16. Ming, Z., Yingxin, L., Shaojie, O., Hui, I.S., Chunxue, L.: Nuclear energy in the post-Fukushima era: research on the developments of the Chinese and worldwide nuclear power industries. Renew. Sustain. Energy Rev. 58, 147–156 (2016)
17. Barros, C.P., Managi, S.: French nuclear electricity plants: productivity and air pollution. Energy Sources. Part B: Econ. Plann. Policy 11(8), 718–724 (2016)
18. Vilardi, I., et al.: Large-scale individual monitoring of internal contamination by gamma emitting radionuclides in nuclear accident scenarios. J. Radiol. Prot. 40(1), 134 (2019)
19. Leelőssy, Á., Mészáros, R., Lagzi, I.: Short and long term dispersion patterns of radionuclides in the atmosphere around the Fukushima Nuclear Power Plant. J. Environ. Radioact. 102(12), 1117–1121 (2011)
20. Mathieu, A., et al.: Atmospheric dispersion and deposition of radionuclides from the Fukushima Daiichi nuclear power plant accident. Elements 8(3), 195–200 (2012)

21. Morino, Y., Ohara, T., Nishizawa, M.: Atmospheric behavior, deposition, and budget of radioactive materials from the Fukushima Daiichi nuclear power plant in March 2011: radioactive materials from Fukushima. Geophys. Res. Lette. **38**(7), L00G11 (2011). https://doi.org/10.1029/2011GL048689

22. Eichholz, G.: Environmental aspects of nuclear power. Advances in Astronomy and Space Physics (1976)

23. Jeong, H.J., Kim, E.H., Suh, K.S.: Determination of the source rate released into the environment from a nuclear power plant. Radiat. Prot. Dosimetry. **113**(3), 308–313 (2005)

24. Mahura, A.G., Jaffe, D.A., Andres, R.J., Merrill, J.T.: Atmospheric transport pathways from the Bilibino nuclear power plant to Alaska. Atmos. Env. **33**(30), 5115–5122 (1999)

25. Kharecha, P., Hansen, J.: Prevented mortality and greenhouse gas emissions from historical and projected nuclear power. Environ. Sci. Technol. **47**(9), 4889–4895 (2013)

26. Rashad, S.M., Hammad, F.H.: Nuclear power and the environment: comparative assessment of environmental and health impacts of electricity-generating systems. Appl. Energy **65**(1–4), 211–229 (2000)

27. Eisma, R., Vermeulen, A.T., Van Der Borg, K.: $^{14}CH_4$ emissions from nuclear power plants in northwestern Europe. Radiocarbon **37**(2), 475–483 (1995)

28. Kim, H.J., et al.: Radiological impact assessment of radioactive effluents emitted from nuclear power plants in Korea and China under normal operation. Prog. Nucl. Energy **153**, 104434 (2022)

29. Kalinowski, M.B., Tatlisu, H.: Global radioxenon emission inventory from nuclear power plants for the calendar year 2014. Pure Appl. Geophys. **178**(7), 2695–2708 (2021)

30. Gaziev, I.Y., Kryshev, A.I.: Radioactive contamination of air. underlying surface. agriculture products and population exposure doses near the Novovoronezh NPP site: model computations. Radiatsiya i risk–Radiation and Risk **19**(1), 48–59 (2010)

31. Staven, L.H., Napier, B.A., Rhoads, K., Strenge, D.L.: A compendium of transfer factors for agricultural and animal products. (No. PNNL-13421). USA: PNNL. Richland. WA (United States) (2003)

32. Georgievskyi, V.B.: Environmental and dosing models under radiation accidents. Kyiv, Ukraine (1994) (in Russian)

33. Przystupa, K., Vasylkivskyi, I., Ishchenko, V., Pohrebennyk, V, Kochan, O., Jun, S.: Assessing air pollution from nuclear power plants. In: Proceedings of the 12th International Conference "Measurements 2019", pp. 232–235. Smolenice, Slovakia (2019)

34. Kirchner, G.: Applicability of compartmental models for simulating the transport of radionuclides in soil. J. Environ. Radioact. **38**(3), 339–352 (1998)

35. Askbrant, S., et al.: Mobility of radionuclides in undisturbed and cultivated soils in Ukraine, Belarus and Russia six years after the Chernobyl fallout. J. Env. Radioact. **31**(3), 287–312 (1996)

36. Howard, B.J., Beresford, N.A., Barnett, C.L., Fesenko, S.: Quantifying the transfer of radionuclides to food products from domestic farm animals. J. Environ. Radioact. **100**(9), 767–773 (2009)

Location Optimization of Vessel Traffic Service Radar Station in Expansion Process

Chuan Huang, Changjian Liu[✉], Jia Tian, Hanbing Sun, and Xiaoqing Zhang

Transport Planning and Research Institute, Ministry of Transport, Beijing 100028, China
{huangchuan,huangchuan}@tpri.org.cn

Abstract. Vessel Traffic Service (VTS), as the main tool of China's maritime authorities to maintain and control the waterway traffic order, has played an important role in ensuring the safety of shipping vessels, improving the efficiency of vessel traffic, and protecting the waterway environment since its introduction. And in order to meet the new emerging waterway supervision demand caused by the large-scale ship size and the growing ship traffic flow, the new round of VTS system construction is particularly important. In this paper, first, we study the optimization of VTS radar station location and configuration problems in the expansion process considering environmental occlusion, electromagnetic wave attenuation, and variable coverage radius. Then, consider the obstacles such as mountains and forests may affect electromagnetic wave propagation, we propose an environmental occlusion judgment method based on spatial geometry in three-dimensional space, and construct a multi-objective VTS radar station location and configuration optimization model for the VTS radar station expansion. Lastly, the feasibility and validity of the environmental occlusion judgment method and the location and configuration optimization model are verified by the example based on ArcGIS software with the VTS radar station project in Zhaoqing City.

Keywords: VTS radar station · Environmental occlusion · Radar attenuation · Variable coverage radius · Multi-objective optimization

1 Introduction

Up to 2021, China has 135,679 civil transport vessels, including 126,805 motor vessels and 8,874 barges. And the cargo throughput of national coastal and inland rivers reached 145,499.91 million tons, including 948,002 million tons of coastal cargo throughput and 506,989 million tons of inland river cargo throughput [1, 2]. At the same time, the vessel traffic flow in China's coastal and inland water area has also emerged rapid growth. However, the rapid growth of vessel traffic flow makes the number of waterway transport accidents increase. The frequent occurrence of waterway traffic accidents not only endangers the safety of ships, people and property, but also damages the environment and has a serious impact on both the development and use of water areas. In 2019, 137 waterway traffic accidents occurred among Chinese ships in domestic waters, resulting in 155 deaths or missing persons, and 46 shipwrecks. And direct economic losses of 17.987

million yuan [3]. In order to ensure the safety of ship navigation, maritime authorities need to rely on vessel traffic service (VTS) systems to regulate and guarantee water transportation safety [4].

The main role of the VTS system is to use communication facilities such as radar stations, automatic ship identification system base stations, VHF calls, and shipboard terminals to realize the monitoring of ships in the supervised water area and provide ships with navigation safety information required for the navigation to ensure the safety of the ship. Since its introduction into China, it has played an important role in maintaining waterway traffic order and ensuring navigation safety.

As an important tool for the maritime authorities to achieve waterway traffic safety management, the construction of the VTS system can effectively maintain the stability of the waterway traffic safety situation, improve the efficiency of vessel navigation, and promote the good and rapid development of the shipping economy. Among them, the VTS radar station, as the core component of the VTS system, plays a vital role in monitoring the ships in navigation and inbound and outbound, and its location deployment and radar configuration affect the function of the whole system, so it is necessary to study the site selection and radar configuration of VTS radar station [4].

However, the number of risky water areas in China has gradually increased, and the existing VTS radar stations cannot meet the new emerging monitoring demand. As a result, it's necessary to analyze the location optimization problem of the VTS radar station in the expansion process.

Abundant domestic and international research has been conducted on VTS radar station location and configuration optimization. And considering that the result of VTS radar station location and optimization is essentially a decision to locate a set of VTS radar stations subject to certain constraints and objective functions, the VTS radar station location and configuration optimization problem studied in this paper is similar to the facility location problem. The objective of facility location problems such as VTS radar station location is to determine the optimal location or combination of locations of facilities in the pursuit of objectives such as minimum cost or maximum service area [5]. The facility location problem has been applied in several research areas, such as the location of logistics warehouses, logistics distribution centers [6] and emergency material storage [7]. Therefore, a large number of location optimization models have been proposed successively, and these models can be classified into p-center model, p-median model, and the coverage model according to the way of location model architecture[8].The field of VTS radar station location optimization research based on the coverage model has also been richly researched by related scholars. Cao et al. [9–11] first constructed an evaluation index system for the initial screening of VTS radar station candidate points, and then constructed and solved a two-objective location optimization model for VTS radar stations based on the ensemble coverage model. On this basis, Cao et al. [12] considered the VTS center and line laying cost from the VTS center to the VTS radar station, and constructed and solved a bi-objective location optimization model for the VTS radar station based on the ensemble coverage model. Ai et al. [13–15] considered the radar type in the model construction and constructed a bi-objective model for VTS radar station location and configuration based on the ensemble coverage model.

It can be seen that in the existing research, only the relatively ideal VTS radar station location optimization model is constructed. And the factors of VTS radar configuration, environmental impact, and VTS radar electromagnetic wave attenuation in the actual environment are not fully considered. Thus, the existing VTS radar station location model is difficult to be applied and promoted in a complicated environment, so in the context of the upcoming new round of VTS system construction, it is urgent to carry out research on the optimization of VTS radar station location and configuration.

This paper fills the gap and constructs the location optimization model under the consideration of environmental obstacles and VTS radar electromagnetic wave attenuation, which aims at enriching the existed VTS radar station location optimization model and proposing a novel location optimization solution method for the construction of VTS radar station in the actual world. In the following, Sect. 2 investigates the problem and relevant judgment and evaluation method, Sect. 3 constructs the mathematical model, Sect. 4 describes the algorithm and proposes the novel method to improve the effectiveness and efficiency of the algorithm, Sect. 5 introduces the experimental results, Sect. 6 concludes the article.

2 Problem Description

This section investigates the VTS radar station location and configuration problem in the expansion process and constructs a two-stage mathematical model to optimize the VTS radar station location.

2.1 Problem Description

There are two major approaches to achieving VTS radar station expansion ideas. One is to select new locations to settle down new VTS centers and radar stations, and the other is to increase the number of VTS radar stations to be built on the basis of the existing VTS system. In the actual decision-making process, maritime authorities prefer to choose to increase the number of VTS radar stations based on the spatial geographic location of the built VTS system, so this section adopts the latter expansion method to research the location optimization problem in the expansion process. To improve the overall supervision effect of the VTS system, taking into account the distance between the expanded VTS radar station and the VTS center, the expanded VTS radar station is built within the effective radius of the built VTS radar station.

Figure 1 depicts a schematic diagram of the VTS radar station location optimization in the expansion process. Figure 1(a) assumes that three VTS radar stations 1, 3, and 5, have been built, taking into account radar electromagnetic wave attenuation and environmental obstacles, and that these three VTS radar stations can meet the monitoring needs of water area A-E. However, as the number of risky water areas increases and new monitoring demands of water areas F and G emerge, as shown in Fig. 1(b), the three VTS radar stations already built cannot provide effective coverage for a total number of seven water areas A-G in the figure. To meet the new demand, a certain number of VTS radar stations must be built, and environmental occlusion caused by environmental obstacles and electromagnetic wave attenuation must be considered in the expansion construction process. Figure 1(b) depicts the final results of the expansion process, VTS radar station 2 is constructed within the radius centered on radar station 1, and VTS radar station 4 is constructed around VTS radar station 3.

However, during the construction of VTS radar stations, obstacles such as mountains and forests can block the propagation of radar electromagnetic waves, and there is also attenuation when radar electromagnetic waves propagate in air, water and other media. As a result, the aim of this study is to find the optimal location and configuration type in the expansion process under the consideration of electromagnetic wave attenuation, environmental occlusion, and variable coverage radius of different radar type.

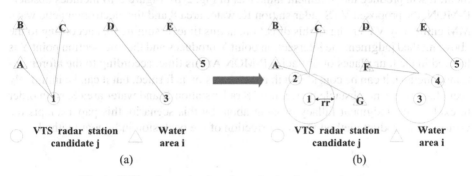

Fig. 1. VTS radar station location selection in expansion process

2.2 Electromagnetic Wave Attenuation Evaluation

The electromagnetic wave will show an attenuation phenomenon when propagating in the actual VTS radar station monitoring environment, so in order to effectively measure the VTS radar electromagnetic wave attenuation, this section introduces the attenuation function for VTS radar electromagnetic wave attenuation measurement study, the commonly used attenuation functions are linear (linear), exponential (exp), Gauss (gauss), and logarithmic (log) functions.

To determine which kind of attenuation function is better suited for measuring the attenuation of VTS radar electromagnetic waves, the path loss function of the Okumura-Hata electromagnetic wave propagation model is used to evaluate the attenuation degree of VTS radar station electromagnetic waves, and finally the logarithmic function is determined as the appropriate type of attenuation function.

2.3 Environmental Occlusion Judgment and Evaluation

This section proposes a judgment method about whether the obstacle forms an occlusion to the VTS radar electromagnetic wave in three-dimensional space. Only when the judgment is accurately applied, a suitable formula can be adopted for water area coverage calculation.

The environment shown in Fig. 2 (a) includes the mountain obstacle T-RSQ, the proposed VTS radar station K, the water area 8, and the electromagnetic wave MM that the VTS radar emits in this direction. The actual elevation of the VTS radar station is calculated by its own built height and geographical elevation, and the actual elevation of water area 8 is its geographical elevation, as well as the obstacle T-RSQ. In the three-dimensional space, the method that judges whether the occlusion forms is determined by whether the straight-line MM and the four faces of the cone T-RSQ TSQ, TRQ, TRS, and RSQ form an intersection point when the electromagnetic wave MM propagates through the obstacle T-RSQ. If the intersection point exists and is located within the four planes, then the occlusion is figured out to be formed.

However, if we completely rely on this method for environmental occlusion judgment, it will produce the judgment fallacy as in Fig. 2(b). Figure 2(b) includes obstacle P-MON, the proposed VTS radar station K, water area 8 and the electromagnetic wave MM emitted by VTS radar in this direction, at this time, assuming that according to the above method judgment the intersection point Y produces and the intersection point Y is located in the four planes of obstacles P-MON At this time, according to the aforementioned method, it can be considered that the occlusion is formed, but it can be intuitively seen that there is no obstacle between VTS radar station K and water area 8, so in order to exclude the judgment fallacy brought about by this scenario, this paper adopts the vector inner product method for the correction of the occlusion judgment method.

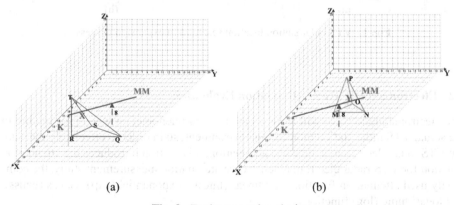

(a)   (b)

Fig. 2. Environmental occlusion

3 Mathematical Modeling

In this section, the mathematical model of the location and configuration optimization problem of VTS radar stations in the expansion process is divided into two phases. The first phase is the optimal location and configuration of VTS radar stations with no expansion. Based on the results of the first phase, the second phase sets the effective expansion radius and the number of expanded VTS radar stations, and carries out the modeling in the second phase. Thus the overall goal of the second phase is to achieve the maximum coverage of the total water area based on the combination of radar stations formed by the built radar stations and the new VTS radar stations in the pursuit of minimizing the total station construction cost of the second phase.

3.1 Assumptions

1) The external environment will remain constant during the optimization decision process.
2) The number, area, and elevation of water area to be monitored are known; the number and elevation of VTS radar station candidate points are known.
3) The number and penetration rate of obstacles and the formula of the attenuation function have been predetermined.
4) Each candidate point can only build and configure at most one VTS radar station and one type of radar.
5) VTS radar stations equipped in the expansion process have the same range of radar type selection as the original built VTS radar station.
6) This paper assumes that the expansion radius of VTS radar station is known.

3.2 Parameters and Variables

3.2.1 Parameters and Variables

I: set of water areas.
J: set of VTS radar station candidate points.
K: set of radar types available for configuration.
M_j: fixed construction cost of VTS radar station candidate point $j, j \in J$.
Ek: the cost of configuring type radar $k, k \in K$.
λ_j^k: the probability of monitoring a VTS radar station candidate point j when configured with a type radar $k, j \in J, k \in K$.
D: distance matrix.
D_{ij}: the Euclidean distance between the desired coverage area i and the VTS radar station candidate point $j, i \in I, j \in J$;
C: the coverage matrix.
f: attenuation function.
C_{ij}: the result of the coverage of the water area i by the VTS radar station candidate point j, calculated by the attenuation function $f(D), i \in I, j \in J$.
Q denotes the total number of VTS radar station candidates to be built.
G: the set of obstacles.
Og: the penetration rate of the obstacle $g, g \in G$;
Q: coverage count matrix.
q_{ij}: the number of times a VTS radar station candidate point j covers the desired water area i, and $i \in I, j \in J$.
μ_i: area coverage per water, $i \in I$;
θ_i: area coverage threshold for each body of water with knowledge of radar surveillance requirements, $i \in I$.
s_i: the area of the water area, $i \in I$;
S: the total area of all water area.
rr: radius of expansion.
N: the set of VTS radar station candidates in the expansion phase, which can be obtained based on the expansion radius.
L: the total number of VTS radar stations to be built during the expansion phase.

M_n: the fixed construction cost of the VTS radar station candidate point n, and $n \in N$.

E_k: the cost of configuring type radar k, $k \in K$.

λ_n^k: the monitoring probability of the VTS radar station candidate point n when configured with type radar k, and $n \in N$, $k \in K$.

D_{in}: the Euclidean distance between the desired coverage area i and the VTS radar station candidate point n, and $i \in I$, $n \in N$.

C_{in}: coverage matrix, representing the results of the coverage of the water area i at the VTS radar station candidate point n during the expansion phase.

$C_{i(j+n)}$: the coverage matrix of the expanded combined VTS radar station over the water area i, calculated from C_{ij} and C_{in}.

3.2.2 Decision Variables

y_j: binary variable, if the VTS radar station candidate point j is selected, then $y_j = 1$, otherwise $y_j = 0$; $j \in J$.

z_j^k: binary variable, if the radar station candidate point j is selected and the model radar k is configured, then $z_j^k = 1$, otherwise $z_j^k = 0$; $j \in J$, $k \in K$.

y_n: binary variable, if the VTS radar station candidate point n is selected, then $y_n = 1$, otherwise $y_n = 0$; $n \in N$.

z_n^k: binary variable, if the VTS radar station candidate point n is selected and the radar type k is configured, then $z_n^k = 1$, otherwise $z_n^k = 0$; $n \in J$, $k \in K$.

3.3 Mathematical Model

3.3.1 First Phase

Objectives: The minimal total construction cost, formed by the fixed construction cost and radar configuration cost:

$$\min H = \sum_{j=1}^{J} y_j M_j + \sum_{j=1}^{J} yj \left(\sum_{k=1}^{K} z_j^k E_k \right) \tag{1}$$

The maximum water area coverage:

$$\max X = \left(\sum_{i=1}^{I} \left(\left(1 - \prod_{j=1}^{J} \left(1 - C_{ij} \lambda_j^k \right) \right) * S_i \right) \right) / S \tag{2}$$

Constraints: Each water area is required to be covered at least 2 times:

$$\sum_{j=1}^{J} y_j Q_{ij} \geq 2 \tag{3}$$

Area coverage constraint of each water area:

$$\mu_i \geq \theta_i \tag{4}$$

Each VTS radar station candidate point, regardless of whether it is built or not, can only be equipped with one type of radar for that corresponding VTS radar station candidate point:

$$\sum_{k=1}^{K} z_j^k = 1 \tag{5}$$

Water area equation:

$$S = \sum_{i=1}^{I} s_i \tag{6}$$

Water area coverage calculation formula:

$$\text{water area coverage} = \begin{cases} 0, & r1 \leq r \text{ or } r_1 \geq R \\ C_{ij}, & \text{no occlusion exists} \\ C_{ij} \cdot O_g, & \text{occlusion exists} \end{cases} \tag{7}$$

Binary variable constraints:

$$y_j \in \{0, 1\} \tag{8}$$

$$z_j^k \in \{0, 1\} \tag{9}$$

Second Phase Objectives: The minimal total construction cost, formed by the fixed construction cost and radar configuration cost in the expansion process:

$$\min \ HH = \sum_{n=1}^{N} y_n M_n + \sum_{n=1}^{N} y_n (\sum_{k=1}^{K} z_n^k E_k) \tag{10}$$

The maximum water area coverage:

$$\max \ XX = \sum_{i=1}^{I} ((1 - \prod_{n=1}^{N} (1 - C_{i(j+n)} z_n^k \lambda_n^k)) S_i)/S \tag{11}$$

Constraints: The number constraints of VTS radar stations in the expansion process:

$$\sum_{n=1}^{N} y_n = L \tag{12}$$

Each VTS radar station candidate point within the expansion area, regardless of whether it is built or not, the corresponding VTS radar station candidate point can only be equipped with one type of radar:

$$\sum_{k=1}^{K} z_n^k = 1 \tag{13}$$

Binary variable constraints:

$$y_n, z_n^k \in \{0, 1\} \tag{14}$$

4 Algorithm Description

Given the improved performance of the adaptive particle swarm algorithm (ASPSO) in solving problem, this section introduces the chaos mechanism and the spiral position update mechanism of the moth flame algorithm to improve the adaptive particle swarm algorithm in order to improve the algorithm's operation speed and the ability to find optimal solutions. The following is an introduction to chaos mechanism and the spiral position update mechanism. Other procedures are the same as ASPSO.

4.1 Chaos Mechanism

In order to increase the algorithm's solving ability to find the best local and global solutions and the speed of algorithm solution, considering the characteristics of chaotic graphs such as randomness and ergodicity, this paper uses chaotic adaptive weights for algorithm improvement, which helps to improve the performance of the algorithm. The chaotic adaptive weights is calculated as follows.

$$x_{tt} = A \cdot x_{tt-1} \cdot (1 - x_{tt-1}), \; x_{tt} \in (0, 1) \tag{15}$$

$$\omega_{tt} = (\omega_{max} - \omega_{min}) \cdot \frac{(T_{max} - tt)}{T_{max}} + \omega_{min} \cdot x_{tt} \tag{16}$$

tt is the number of current iterations, x_{tt} is the random number of iterations tt, ω_{tt} is the weight of iterations tt, and A is a constant, where $A = 4$, T max represents the maximum number of iterations, ω_{max} and ω min are 0.8 and 0.3, respectively.

4.2 Spiral Position Update Mechanism

The position update strategy of the standard particle swarm algorithm makes the particles always move toward the best position of the previous algebra, which reduces the ability to search for neighborhoods around known optimal solutions, so this section introduces the spiral flame catching approach of the moth flame algorithm for particle position update. The formula is as follows.

$$\lambda = \frac{\exp(x_i^d(tt))}{\exp(\frac{1}{N} \sum_{i=1}^{N} x_i^d(tt))} \tag{17}$$

$$x_i^d(tt+1) = \begin{cases} D1 \cdot \exp(b \cdot l) \cdot \cos(2\pi l) + gbest^d(tt), & \lambda < rand \\ x_i^d(tt) + v_i^d(tt+1), & other \end{cases} \tag{18}$$

$x_i^d(tt)$ is the position value of the dimension d of the first i particle of the iterations tt, N is the total number of particles in each generation, $D1 = |gbest^d(tt) - x_i^d(tt)|$ is the absolute value of the difference between the position value of the dimension d of the global optimal particle of the generation tt and the dimension d of the first i particle of the generation tt, b is the custom constant, l is an arbitrary constant between $[-1,1]$, and $v_i^d(tt+1)$ is the velocity of the dimension d of the first i particle of the generation $(tt+1)$.

5 Experimental Results

5.1 Example Introduction

In this section, the water area of Xijiang River from Fengkai County to Yunan County in Zhaoqing City, Guangdong Province, is used as an example to verify the effectiveness and practicality of the proposed method and model. And the environmental digital elevation map of this section of water area is obtained based on LocaSpaceviewer software, as shown in Fig. 3(a). And the overall fluctuation range of elevation is [0,255]. In order to obtain the coordinates of the water and the candidate VTS radar stations, the image is divided into 20 and 10 segments along the horizontal and vertical axes, as shown in Fig. 3(b), and the coordinate system is established around the fishing net grid line. The waters are divided into 12 parts according to the results of fishing net division, as shown by the red circles in Fig. 3(b), and the divided waters are labeled for the convenience of differentiation.

Through the elevation scale in Fig. 3(b), we can figure out the red and white areas have a high elevation, so this paper set the red and white parts as obstacles, so the three obstacles occupy the range and plane coordinates are shown in Fig. 4(a), in order to simplify the calculation, the environment of the water area is simplified to Fig. 4(b) shows the three-dimensional coordinate map. This three-dimensional space range is $XYZ \in \{(0, 0, 0) - (20, 10, 10)\}$, there are obstacles D-ABC, H-EFG and L-JIK, used to simulate the mountain obstacles in the actual environment. According to the established coordinate system, the coordinates of the obstacles are shown in Table 1. And the table is also set around parameters such as obstacle penetration rate, water area, and the number of times the water needs to be covered, where the Z-axis coordinates indicate the elevation of these obstacles, and in addition as stated in Table 2. This model assigns a penetration rate to each obstacle to measure the VTS radar electromagnetic wave attenuation, and in this example the penetration rates for each obstacle in this example are also shown together in Table 1. The water coordinates are shown in Table 2. The specific coordinates, elevation and water area of these 12 waters are also set, where the Z-axis coordinates indicate the geographical elevation of these water area in that environment, and the required water area coverage threshold for each water under the regulatory needs of the maritime authority is also set.

In this three-dimensional coordinate system, the VTS radar station candidate points are all integer coordinates in the coordinate system. In the actual environment, the VTS radar station candidate points often have geographical elevation values, so this example applies three-dimensional environment in the terrain elevation change cross-sectional diagram as shown in Fig. 5(a) to simulate the actual environment of rolling terrain changes. For the convenience of calculation, the elevation value of each point is rounded to the nearest whole as shown in Fig. 5(b), so the actual elevation value of each VTS radar station candidate can be obtained. In addition, the height of VTS radar station is assumed as 0.5. From the perspective of variable coverage radius, this section sets the VTS radar parameters shown in Table 3, there are three types of VTS radar available for configuration, the radius of action is 5, 7 and 9, and the corresponding cost is therefore different; the VTS radar electromagnetic wave attenuation function adopts the logarithmic function for attenuation measurement, the attenuation formula in this example is shown in formula (19).

$$f(x) = \begin{cases} 0, 0 \leq x < 0.5 \\ 1, 0.5 \leq x < 1 \\ \lg(10 - x), 1 \leq x < 9 \\ 0, x \geq 9 \end{cases} \quad (19)$$

In this paper, the mathematical model is constructed based on the maximum coverage model and the aggregate coverage model. Considering that the total number of water area to be supervised is 12, so the total number of VTS radar stations to be built needs to be less than 12, otherwise this location selection result will lack economy. In order to meet the model constraints, after repeated tests, the model has no feasible solution when the number of stations to be built is lower than 4. Thus, the proposed number interval of VTS radar stations for each particle in the particle initialization of the algorithm is [4, 11]. In order to improve the algorithm operation efficiency, the location of the VTS radar station candidate points occupied by obstacles need to be excluded when the particle initialization, and the cost of VTS radar station construction is set to 500,000 yuan in this example.

The algorithm operation parameters are set as follows: the particle swarm size is set to 100, the number of iteration generations is 100, the particle dimension is $21 \times 11 = 231$, the variation probability is random number variation, the upper limit of the number of external archives is 50, and the particles that do not meet the coverage number and area coverage requirements are applied with $[X + P, 10 - Z + P]$, $P = 10^8$.so that they are always in the dominated position and reduce the probability of entering the non-dominated solution set.

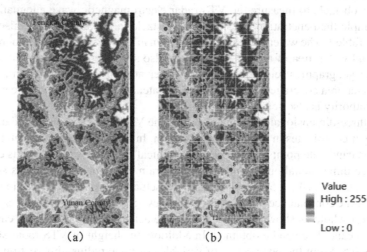

(a)   (b)

Value
High : 255

Low : 0

Fig. 3. Environmental digital elevation model image

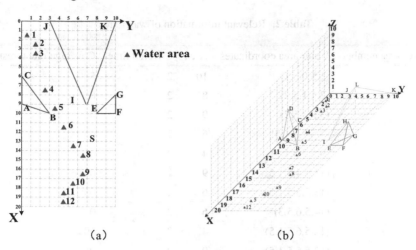

Fig. 4. Environment about West River in three-dimensional space

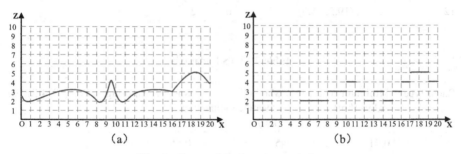

Fig. 5. Geographic elevation variation

Table 1. Relevant information on obstacles

Obstacle number	Obstacle vertex	Vertex coordinates	Obstacle penetration rate
1	A	(9,0,0)	0.35
	B	(10,3,0)	
	C	(6,0,0)	
	D	(8,0,4)	
2	E	(10,8,0)	0.4
	F	(10,10,0)	
	G	(8,10,0)	
	H	(9,10,3)	
3	I	(9,7,0)	0.3
	J	(0,3,0)	
	K	(0,10,0)	
	L	(4,7,5)	

Table 2. Relevant information of water area

Water area number	Water area coordinates	Area	Covered time	Coverage rate threshold
1	(1.5,0.5,1)	10	2	50%
2	(2.5,1.5,2)	8	2	0%
3	(3.5,1.5,2)	9	2	0
4	(7.5,2.5,1.5)	6	2	0
5	(9.5,3.5,1.5)	4	2	0
6	(11.5,4.5,1.5)	9	2	0
7	(13.5,5.5,1.5)	6	2	65%
8	(14.5,6.5,3)	6	2	0
9	(16.5,6.5,1.5)	4	2	60%
10	(17.5,5.5,1.5)	9	2	0
11	(18.5,4.5,1.5)	6	2	0
12	(19.5,4.5,3)	6	2	0

Table 3. VTS Radar type

Radar type	Minimum coverage radius (km)	Maximum coverage radius (km)	Radar monitoring rate	Cost(10^4 RMB)
1	0.01	5	0.85	2
2	0.01	7	0.9	4
3	0.01	9	0.95	6

5.2   Results Analysis

5.2.1   First Phase

Based on the parameters set above, MATLAB 2018b is used to run and simulate the program, and all calculations are performed using Ryzen R7-4800H, RAM 16G memory, and Windows 10 operating system.

The final Pareto front is calculated as shown in Fig. 6, which shows that the final Pareto front is smoother and more uniform. And the information of the objective function of each dominated solution is obtained as shown in Table 4. Then the fuzzy set method is used to get the compromise solution selected from these non-dominated solutions. The sum of each non-dominated solution has been calculated and shown in the table, so the particle number 7 is finally selected as the compromise solution for this example.

Based on this compromise solution, the final location and configuration plan achieves 98.286% coverage rate of the total water area when building 7 VTS radar stations, and the total station construction cost is 3.76 million RMB. After model optimization, the

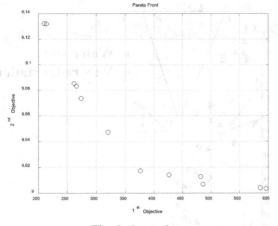

Fig. 6. Pareto front

Table 4. Non-dominated particles information

Solution number	Objective function 1	Objective function 2	u_1	u_2	$u_1 + u_2$
1	210	9.13220441	1	0	1
2	214	9.13201045	0.989637	0.001508	0.99
3	262	9.085224554	0.865285	0.365151	1.230
4	266	9.08325795	0.854922	0.380437	1.235
5	274	9.07362639	0.834197	0.455298	1.29
6	320	9.04718788	0.715026	0.660792	1.38
7	376	9.01713791	0.569948	0.894356	1.46
8	426	9.013999999	0.440415	0.918745	1.36
9	482	9.0126867	0.295337	0.928953	1.22
10	486	9.0067964167	0.284974	0.974735	1.26
11	586	9.0040043	0.0259067	0.996437	1.02
12	596	9.0035458796	0	1	1

spatial distribution of these 7 VTS radar stations is shown in Fig. 7. The coordinates of each VTS radar station and the VTS radar configuration are shown in Table 5.

5.2.2 Second Phase

The Pareto front diagram for the expansion phase is shown in Fig. 8, which shows that the final Pareto front contains three non-dominated solutions, including all location and configuration optimization options in the case of only one VTS radar station expansion.

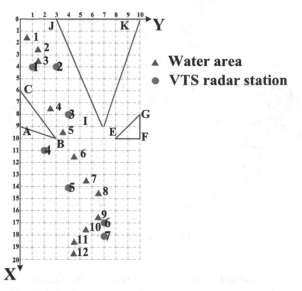

Fig. 7. Distribution map of VTS radar station to be built

Table 5. Location information

Radar station number	Coordinates	Configured radar type
1	(4, 1)	1
2	(4, 3)	1
3	(8, 4)	2
4	(11, 2)	3
5	(14, 4)	2
6	(17, 7)	3
7	(18, 7)	1

The fuzzy set method is used to choose the compromise solution for these three non-dominated solutions, and the second non-dominated solution is finally chosen as the compromise solution. According to which the total cost of station construction in the second stage is 540,000 yuan, so the total cost of station construction in the two stages is $376 + 54 = 4.3$ million yuan, and the total area coverage of 14 waters reaches 95.02% with the construction of 8 VTS radar stations. After model optimization, the expansion phase of VTS radar station construction site is shown in Fig. 9, and finally choose to expand VTS radar station 8 around VTS radar station 2 and configure radar type 2.

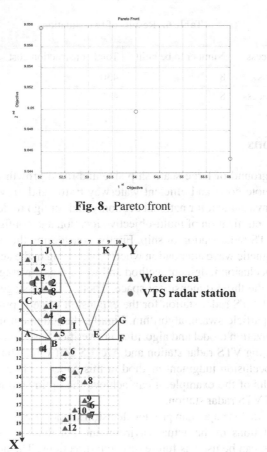

Fig. 8. Pareto front

Fig. 9. VTS radar station

5.3 Further research

In summary, it can be seen that a total number of 8 VTS radar stations are built in the two phases, with a total construction cost of 4.3 million yuan and a coverage rate of 95.02%. While the result is 4.26 million yuan and a coverage rate of 98.60% if 8 VTS radar stations are built directly without considering the expansion phase according to the results of phase 1. These two situations are shown in the table below. It can be seen that the latter achieved a higher coverage rate with less cost. Therefore, in the actual construction process, it can be built ahead of time based on the current demand for water area supervision, so as to achieve the optimization of construction cost and water area coverage (Table 6).

Table 6. Results of two situations

Construction process	Number to be built	Total construction cost	Total coverage rate
Expansion process	8	430	95.02%
No expansion process	8	426	98.60%

6 Conclusions

Under the background of large-scale ship size and rapid growth of vessel traffic flow, in order to promote good and efficient waterway traffic and provide a safe and stable transportation environment for national economy and foreign trade, this paper conducts research on the optimization of multi-objective location and configuration in the expansion stage for VTS radar station of ship. Firstly, considering the obstacle blockage and radar electromagnetic wave attenuation when propagating in the actual environment, the environmental occlusion judgment method and water area coverage calculation method are proposed in the three-dimensional space. Then, the two-stage location and configuration model of VTS radar station for the expansion stage is proposed and solved by multi-objective particle swarm algorithm. Finally, the judgment method, location configuration optimization model and algorithm are verified based on the location selection project of Zhaoqing VTS radar station and ArcGIS software. The results show that the environmental occlusion judgment method in this paper is effective. According to the calculation results of the example, it can achieve the optimization of the location and configuration of VTS radar station.

The research in this paper can provide decision support for the location optimization of VTS radar stations in the actual environment, but there are still shortcomings in this paper, which can be used as future research directions. The article only focuses on the optimization of VTS radar station location and configuration, while many maritime authorities are already trying to deploy oil spill detection radar in the VTS system, so it is of great practical importance to study the optimal location optimization of the combination of VTS radar station and oil spill detection radar station.

References

1. National Bureau of Statistics of China. http://www.stats.gov.cn/.2021
2. Ministry of Transport of the People's Republic of China. https://www.mot.gov.cn/.2021
3. Igor, R., Vlado, F., Marko, V., et al.: Early detection of vessel collision situations in a vessel traffic services area. Transport **35**(2), 121–132 (2020)
4. Gan, L., Yu, F., Zheng, Y., et al.: Research on modeling and simulation in overshadowing influence of coastal building on vessel traffic service radar. Adv. Mech. Eng. **10**(10), 1–14 (2018)
5. Karatas, M.: A multi-objective facility location problem in the presence of variable gradual coverage performance and cooperative cover. Eur. J. Oper. Res. **262**(3), 1040–1051 (2017)
6. Wan, S., Chen, Z., Dong, J.: Bi-objective trapezoidal fuzzy mixed integer linear program-based distribution center location decision for large-scale emergencies. Appl. Soft Comput. **10**, 107757 (2021)

7. Bochen, W., Qiyuan, Q., Jiajing, G., et al.: The optimization of warehouse location and resources distribution for emergency rescue under uncertainty. Adv. Eng. Inform. **48**, 101278 (2021)

8. Du, B., Zhou, H., Leus, R.: A two-stage robust model for a reliable p-center facility location problem. Appl. Math. Modell. **77**(1), 99–114 (2010)

9. Cao, D., Lu, J., Ai, Y., et al.: Optimization model of VTS radar station location problem. J. Beijing Univ. Aeronaut. Astronaut. **40**(06), 727–731 (2014). (in Chinese)

10. Cao, D., Lu, J., Jiang, X..: Coastal VTS classification based on hierarchical clustering analysis. J. Wuhan Univ. Technol. **38**(03), 576–579+584 (2014). (in Chinese)

11. Cao, D., Lu, J., Ai, Y., et al.: Evaluation of VTS performance in China based on cloud model. Port & Waterway Eng. (10), 18–22+33 (2014). (in Chinese)

12. Cao, D., Lu, J., Ai, Y., et al.: Optimization location model of VTS radar stations based on set covering theory. Trans. Beijing Instit. Technol. **34**(07), 752–756 (2014). (in Chinese)

13. Ai, Y., Lu, J., Zhang, L., et al.: Optimization model of VTS radar station allocation and radar system configuration. Navig. China **37**(04), 54–58 (2014). (in Chinese)

14. Ai, Y., Lu, J., Zhang, L.: The optimization location-allocation model of water emergency supplies repertories. J. Dalian Maritime Univ. **41**(02), 62–66 (2015)

15. Ai, Y., Cao, D., Shen, B., et al.: Bi-level optimization model of VTS center layout and radar station location-configuration. J. Dalian Maritime Univ. **43**(03), 107–111 (2017)

Research on the Development Path of Domestic Trade Container Transport in China's Ports Under the New Situation

Changjian Liu[1,2], Qiqi Zhou[1,2], Hongyu Wu[1,2(✉)], Chuan Huang[1,2], and Xunran Yu[1,2]

[1] Planning and Research Institute of the Ministry of Transport, Beijing 100028, China
[2] Digital Laboratory of Integrated Transportation Planning, Beijing 100028, China
wuhy@tpri.org.cn

Abstract. In order to crystallize development opportunities and potentials of domestic trade containers, which are under the new economic development pattern and industrial development background, this thesis analyzes the development of domestic trade container throughput and domestic trade container hub ports. Besides, it clarifies the new driving force for the growth of the port's domestic trade container throughput. The thesis gives an in-depth study on the new driving force, development trend and future development path of domestic trade container development, and it also puts forward suggestions on accelerating the high-quality development of domestic trade containers in China. The results show that China's domestic trade container has great development potential in the future, whose throughput will be about 220 million TEU in 2035. The results also illustrate that building a smooth and efficient transport channel, strengthening the construction of hub ports, improving the transport network system and cultivating first-class comprehensive service providers are the essential way to high-quality development of China's domestic trade container transport. The research results have important reference value for decision-making of governments at all levels, and they are both significant to the relevant ports and shipping enterprises in China.

Keywords: Container · Domestic trade · Port throughput · Forecast · Development path

1 Instruction

In the future, China will build a new development pattern that focuses on domestic circulation and promotes both domestic and international circulation. The new development pattern will take cultivating a strong domestic market as the strategic starting point, and take stimulating consumption, expanding effective investment, and promoting infrastructure construction as important measures, which has brought direct impetus to the development of domestic trade containers. As one of the modern transportation modes, the port domestic trade container transportation has prominent advantages in economy, efficiency, and environmental protection, and plays an important role in unblocking the domestic circulation. China needs to recognize the new features and changes, clarify the new situation and new requirements, and plan the new tasks and new paths for the development of China's port domestic trade container transport under the new development pattern.

Domestic trade promotes the rapid growth of the container transport market, and the relevant research is also very rich [1, 2]. On the research of domestic trade containers in China's ports, the representative documents mainly include Xing Husong and others, who analyzed the advantages and development prospects of domestic trade container transport in China [3, 4]. Ding Yichun et al. predicted the domestic trade container throughput of China's major ports [5]. The time series, grey prediction and other methods are comprehensively used to forecast the demand for domestic trade container transport and put forward development suggestions [6]. Liu Liyao studied the development of domestic container ports under the COVID-19 [7]. Zhu Xinhai studied the transformation of domestic trade containers in China [8]. Lin Weimeng and others studied the domestic trade container shipping market in China [9]. Tong Mengda proposed to establish a domestic trade container transport sharing system [10]. Yan Li et al. analyzed the factors affecting the container throughput of China's major ports, providing research support for the prediction of domestic trade container throughput [11]. In general, the research on the development of domestic trade containers in China is relatively rich, but the existing research fails to fully consider the requirements under the new development concept and new development pattern, and the judgment on the future development trend is not accurate enough.

In this paper, firstly, it is proposed that the domestic trade characteristics are obvious in the growth of container transport in China's ports. At the same time, it is proposed that in the context of the COVID-19, the container throughput of domestic trade in China's ports still maintains a good growth momentum, providing new momentum for the development of port container transport. Then, under the background of analyzing the situation and opportunities faced by China's domestic trade container transport, the domestic trade container throughput of China is predicted by using various forecasting methods. Finally, the paper puts forward the main path and policy recommendations for the development of China's domestic trade container transport in the context of the new development pattern. The research in this paper is of great significance for the government, ports and shipping enterprises to grasp the development trend of domestic trade containers in China in the future, and to support government policy formulation, enterprise development strategy formulation and port and shipping infrastructure construction decision-making.

2 The "Domestic Trade" Characteristics of China's Port Container Transport Growth

2.1 Overview of Domestic Trade Container Development

A strong domestic market is an important driver for the growth of domestic trade container transport. Since the international financial crisis in 2008, China's economy has been changing to take the domestic big cycle as the main body. China's domestic demand has made a prominent contribution to the economy, and the degree of economic extroversion has declined. In this context, China's port container transport also presents similar characteristics. Since 2006, the domestic trade container throughput of the national coastal ports has grown rapidly, with an average annual growth rate of 12.6% during 2006–2019, which is 5.2 and 5.3% points higher than the growth rate of the national coastal port cargo throughput and container throughput respectively, and 4.7% points higher than the national GDP growth rate; The growth rate of domestic trade container throughput in 2020 and 2021 was 3.7% and 6.2% respectively, and the average annual growth rate in 2019–2021 was 5.0%. During the same period, the average annual growth rate of national container throughput was 4.1%. Affected by epidemic and other factors, the growth rate of domestic trade container throughput slowed down in 2022, with a year-on-year growth rate of 3.6%, 1.1% points lower than the average annual growth rate of the national port container throughput. The econometric analysis shows that China's domestic trade container throughput is highly correlated with China's GDP and total retail sales of consumer goods. The strong growth of domestic demand is an important reason for the growth of domestic trade container throughput. Relationship between China's GDP and the throughput of domestic trade containers in ports is shown in Fig. 1.

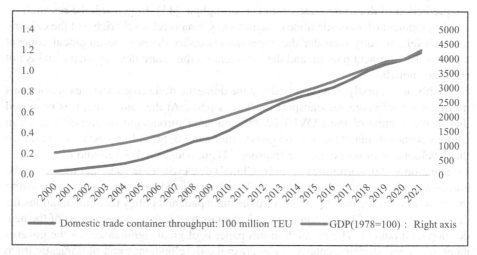

Fig. 1. Relationship between China's GDP and the throughput of domestic trade containers in ports

2.2 Overview of Contribution of Domestic Trade Containers

The rapid growth of domestic trade container throughput has become an important driving force for the growth of port container throughput. The contribution of domestic trade container throughput to container throughput growth was 17% in 2003, 31% in 2006, 56% in 2019, and 81% in 2020–2021; The proportion of domestic trade container throughput in container throughput was 22% in 2006 and increased to 41% in 2019; From 2006 to 2019, domestic trade containers contributed 52% to the growth of container throughput; In 2019, the domestic trade container throughput accounted for 56% of the net increase of container throughput by 41%, becoming an important driving force for the growth of container throughput; In 2020, domestic trade container throughput accounted for 124% of the net increase of total container throughput with 42% (foreign trade container throughput was negative growth); In 2021, domestic trade container throughput accounted for 42% of the total container throughput net increase of 38%; In 2022, it contributed 32% of the total container throughput net increase with 42%; In 2021 and 2022, the growth of foreign trade containers was relatively fast. Influenced by the epidemic and other factors, the contribution of the port's domestic trade container throughput decreased slightly. Change in the proportion of domestic trade container throughput to container throughput in China's ports is shown in Fig. 2.

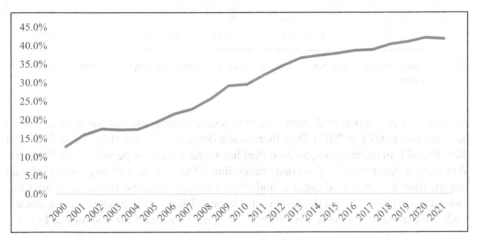

Fig. 2. Change in the proportion of domestic trade container throughput to container throughput in China's ports

2.3 Overview of Domestic Trade Container Hub Port

Driven by strong domestic demand, the domestic trade container throughput of regional important domestic trade container ports has shown a rapid growth momentum, forming a number of domestic trade container hub ports. Tianjin Port in the Beijing-Tianjin-Hebei region completed 8.56 million TEU of domestic trade container throughput in 2019. Compared with 2007, the average annual growth rate was 13.8%, 6.1% points faster than the total container throughput growth rate, and the average annual growth rate in 2019–2021 was 15.4%, 7.1% points faster than the total container throughput growth rate; The proportion of domestic trade in the total container throughput increased from 26% in 2007 to 50% in 2019, 56% in 2021 and 54% in 2022, with domestic trade accounting for more than half. Change in the proportion of domestic trade container throughput to container throughput in Tianjin Port is shown in Fig. 3.

Suzhou Port in the Yangtze River Delta completed 3.42 million TEU of domestic container throughput in 2019, with an average annual growth rate of 12.8% compared with 2007, 2.3% points faster than the total container throughput growth rate; 3.53 million TEU will be completed in 2021, with an average annual growth rate of 1.6% from 2019

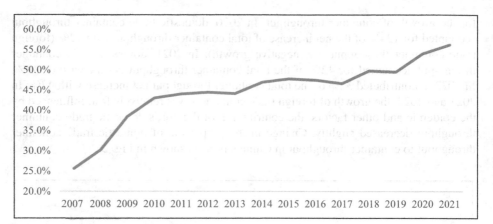
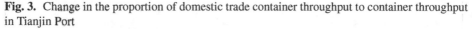

Fig. 3. Change in the proportion of domestic trade container throughput to container throughput in Tianjin Port

to 2021; The proportion of domestic trade in total container throughput increased from 42% in 2009 to 66% in 2013; Then there was a decline, 55% in 2019, 44% in 2021 and 40% in 2022. In recent years, Suzhou Port has increased its cooperation with Shanghai Port and the development of internal branch lines. The container throughput of internal branch lines has increased rapidly, and the growth of domestic trade throughput has slowed down relatively, but the scale is still large. Change in the proportion of domestic trade container throughput to container throughput in Suzhou Port is shown in Fig. 4.

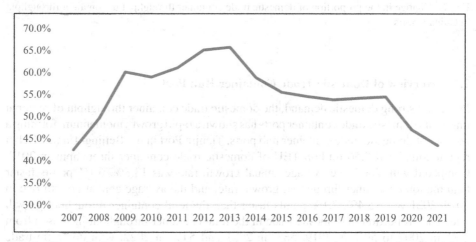
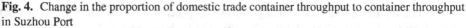

Fig. 4. Change in the proportion of domestic trade container throughput to container throughput in Suzhou Port

Dongguan Port in Guangdong-Hong Kong-Macao Greater Bay Area completed 3.43 million TEU of domestic trade container throughput in 2019. Compared with 2007, the average annual growth rate was 82%, 46% points faster than the total container throughput growth rate; 3.13 million TEU will be completed in 2021, with an average annual growth rate of - 4.4% from 2019 to 2021; The proportion of domestic trade in total container throughput increased from 4.9% in 2009 to 93.2% in 2019, 92.2% in 2021 and 86.7% in 2022; In the past two years, affected by external environment and other factors, the container throughput of Dongguan Port has declined, and the domestic trade throughput has also shown a downward trend, but the domestic trade throughput still accounts for more than 90%; In Guangzhou Port in the Great Bay Area, the proportion of domestic trade container throughput has always maintained at about 60%, 59.6% in 2021 and 58.4% in 2022. Change in the proportion of domestic trade container throughput to container throughput in Dongguan Port is shown in Fig. 5, Change in the proportion of domestic trade container throughput to container throughput in Guangzhou Port is shown in Fig. 6.

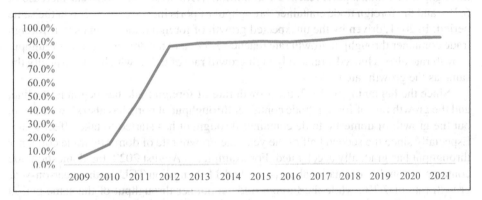

Fig. 5. Change in the proportion of domestic trade container throughput to container throughput in Dongguan Port

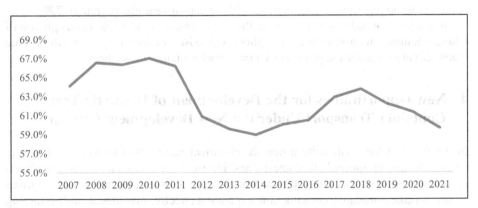

Fig. 6. Change in the proportion of domestic trade container throughput to container throughput in Guangzhou Port

3 The New "Domestic Trade" Driving Force of Port Container Throughput Growth in the Context of the Epidemic

Since 2020, the COVID-19 has seriously affected port production. Due to the effective control of the domestic epidemic, the large-scale outbreak of the international epidemic, and the weak international market demand, the domestic demand improved quarter by quarter, and achieved growth in the third quarter of 2020. In this context, China's domestic trade container transport has become a ballast for the steady growth of port container transport production.

In 2020, the domestic trade container throughput of China's ports reached 110 million TEU, 3.7% growth on a – year-on-year growth, and the foreign trade container throughput of ports increased by - 0.5% in the same period. In 2021, the domestic trade container throughput of China's ports reached 120 million TEU, with a year-on-year increase of 6.2%, and the foreign trade container throughput of ports increased by 7.5% in the same period. In 2021, driven by the unexpected growth of foreign trade, China's port foreign trade container throughput growth rate reached 7.5%, and the domestic trade throughput growth rate also achieved a relatively high growth rate of 6.2%, which was basically the same as the growth rate in 2019.

Since the beginning of 2022, the growth rate of foreign trade has begun to decline, and the growth rate of foreign trade container throughput of ports has also slowed down, but the growth of domestic trade container throughput has started to take effect again. Especially since the second half of the year, the growth rate of domestic trade container throughput has gradually accelerated. For example, in August 2022, the domestic trade container throughput of national ports reached 11.06 million TEU, with a year-on-year growth rate of 7.7%, while the foreign trade container throughput of the same period reached 14.67 million TEU, with a year-on-year growth rate of 1.0%; By December, China's ports had completed 10.85 million TEU of domestic trade container throughput, with a year-on-year growth rate of 12.1%, and 14.45 million TEU of foreign trade container throughput in the same period, with a year-on-year growth rate of 7.8%.

It can be expected that in the future, the driving force for the continuous growth of China's domestic trade container throughput will exist for a long time, and will become a new driving force for the growth of China's port container throughput.

4 New Opportunities for the Development of Domestic Trade Container Transport Under the New Development Pattern

In the future, China will build a new development pattern that focuses on domestic circulation and promotes both domestic and international circulation. The new development pattern will take cultivating a strong domestic market as the strategic starting point, and take stimulating consumption, expanding effective investment, and promoting infrastructure construction as important measures, which has brought direct impetus to the development of domestic trade containers. The new opportunities brought by the construction of new development pattern for the development of waterway container transport are mainly reflected in the following aspects:

4.1 The Need to Build a Strong Domestic Market

A strong domestic market is the key prerequisite for the development of domestic trade containers. In the future, China will comprehensively promote consumption, strengthen the fundamental role of consumption in economic development, improve traditional consumption, cultivate new consumption, and appropriately increase public consumption; Improve the modern circulation system and reduce the circulation cost of enterprises; Cultivate an international consumption center city. The scale of domestic demand directly determines the scale of domestic trade container demand. The continuous growth of the strong domestic market, the continuous growth and upgrading of domestic consumption, and the continuous increase of consumer demand for consumer goods and high-end manufacturing products will directly promote the development of domestic trade containers. The domestic trade container transportation mode has many modern characteristics. It is easy to realize the whole logistics transportation, and it is an important support and foundation for improving the modern circulation system. The cultivation of international consumption center cities and the improvement of consumption levels in urban agglomerations and urban circles will bring internal impetus to the construction of regional domestic trade container hub ports, as well as new demand for the construction of fast and efficient domestic trade container transport corridor between urban agglomerations.

4.2 The Need to Build a Modern Circulation System

The dual cycle of domestic trade and foreign trade needs to build an efficient modern container logistics system. China will accelerate the development of modern service industry, the development of R&D and design, modern logistics and other service industries; and promote the deep integration of modern service industry with advanced manufacturing industry and modern agriculture. Domestic container logistics connects the production end and the consumption end, which is an important part of the smooth domestic economic cycle. Accelerating the development of domestic trade containers is also an important breakthrough in promoting the integrated development of modern service industry and advanced manufacturing industry. Focusing on the layout of manufacturing and agricultural production, focusing on the characteristics of logistics organizations such as manufacturing raw materials, semi-finished products, finished products, grain, etc., accelerating the layout of domestic trade container route network and hub network, and accelerating the construction of modern container logistics system is to further give full play to the advantages of long distance, large volume, low cost, green and environmental protection, optimize and improve the modern comprehensive transportation system, and better help reduce logistics costs The objective need to smooth the circulation of the national economy.

4.3 Optimization of Industrial Spatial Layout Brings New Requirements

The optimization and adjustment of industrial layout under the new development pattern needs the support of waterway container transportation. China will vigorously promote the modernization of the industrial chain supply chain, promote the orderly transfer of industries in China, optimize the regional industrial chain layout, and promote the diversification of the industrial chain supply chain. Nationwide industrial transfer and industrial chain optimization will lead to significant changes in the spatial distribution characteristics of corresponding production activities and corresponding adjustments in the flow direction of logistics activities, many of which are cross-regional and long-distance. At the same time, the diversified layout of the industrial chain supply chain needs to give play to the advantages of their respective modes of transportation and build a logistics supply chain system with multiple modes of transportation and multiple routes. As a long-distance, efficient, economic, green and environmentally friendly modern mode of transport, water transport is bound to usher in a broad space for development.

4.4 Infrastructure Investment Brings Transportation Demand

Major domestic infrastructure construction and other investment activities have brought continuous impetus to the development of domestic trade container transport. China will further optimize the investment structure and maintain a reasonable growth of investment; We will speed up efforts to supplement the weaknesses in infrastructure, municipal engineering and other fields, promote the construction of major projects such as new infrastructure, new urbanization, transportation and water conservancy, implement major projects such as the Sichuan-Tibet Railway and the new land-sea corridor in the west, and promote the construction of a number of major projects with strong foundation, added functions and long-term benefits, such as major scientific research facilities, and coastal transportation along the border and rivers. The above fields have huge potential for construction investment in the future, which is the key direction for the country to optimize the investment structure and achieve precision investment. It will also generate a large number of infrastructure construction activities, as well as corresponding domestic trade container transport demand for machinery and equipment, parts, steel, mining and construction materials, cement, etc., which provides a strong support for the container supply.

4.5 Green Development Brings New Space

Green and low-carbon development has brought huge space for the development of domestic trade containers. China will accelerate the promotion of green and low-carbon development; Support green technology innovation, promote cleaner production, develop environmental protection industry, and promote green transformation of key industries and important fields. With the in-depth implementation of green environmental protection, air pollution prevention and other measures, the adjustment of transportation structure is accelerated, and the transformation of goods from bulk to centralized is accelerated, which has excavated new space for the increment of domestic trade containers in the stock of waterway transportation, especially, some of the goods originally transported in bulk, such as grain, steel, coal, will be replaced by container transportation, which will be an important breakthrough for domestic trade containers to excavate the space of existing goods and achieve rapid growth, It is also the requirement and important direction of the national green development policy.

5 Prediction of the Development Trend of Domestic Trade Container Transport in China

The growth trend of domestic trade container throughput is closely related to economic growth and industrial structure. The method of GDP forecast is relatively mature [12]. This paper assumes that China's economy will continue to grow in the future, with an average annual growth rate of 5% from 2021–2035. Based on the regression prediction of China's economic growth and the growth trend of domestic trade container throughput since 2000, it is predicted that the domestic trade container throughput of China's ports will be about 265 million TEU in 2035, and the average annual growth rate will be 5.9% during 2021–2035.

Considering the future adjustment and transformation of China's industrial structure, the domestic trade container throughput per unit of GDP will show a trend of first growth and then decline. Based on the characteristics of historical changes in strength and the freight intensity prediction method, it is predicted that the domestic trade container throughput of China's ports will be about 213 million TEU by 2035, and the annual growth rate will be 4.3% during 2021–2035. Change in domestic trade container throughput per unit of GDP in China is shown in Fig. 7.

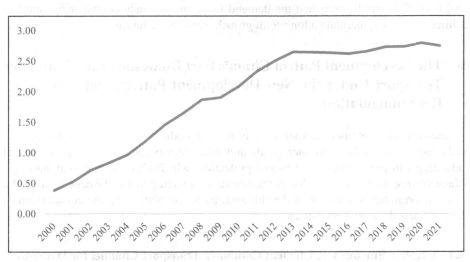

Fig. 7. Change in domestic trade container throughput per unit of GDP in China

Based on the historical change law of the GDP elasticity coefficient of China's domestic trade container throughput, it is predicted that the elasticity coefficient will show a downward trend in the future; Using the elasticity method, it is predicted that the domestic trade container throughput of China's ports will be about 178 million TEU in 2035, with an average annual growth rate of 3.0% from 2021–2035. Change of GDP elasticity coefficient of China's domestic trade container throughput is shown in Fig. 8.

To sum up, using the mathematical average method and comprehensive analysis and prediction [13–15], the domestic trade container throughput of China's ports will be

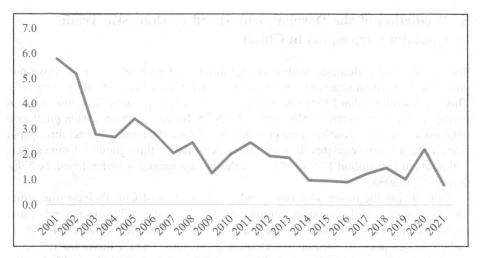

Fig. 8. Change of GDP elasticity coefficient of China's domestic trade container throughput

about 220 million TEU in 2035, and the average annual growth rate will be 4.5% during 2021–2035. It can be seen that the demand for domestic trade container transport in China's ports will maintain a long-term growth trend in the future.

6 The Development Path of China's Port Domestic Trade Container Transport Under the New Development Pattern and Policy Recommendations

Domestic trade container transport is an important mode of transport for trans-regional bulk industrial goods, consumer goods and other transportation, and has prominent advantages in green and environmental protection, which plays an important and irreplaceable role in the construction of the new development pattern. We need to grasp the new opportunities to plan a new development path, and better help the construction of China's new development pattern.

6.1 Create a Smooth and Efficient Container Transport Channel for Domestic Trade

The North-South coastal transport corridor and the east-west transport corridor such as the Yangtze River and Xijiang River play an important role in the domestic trade container transport. For example, about 94% of the domestic trade container throughput of Tianjin Port in 2018 needs to be transported through the north-south coastal transport corridor; In 2019, the domestic trade container throughput of the Yangtze River trunk ports (excluding Shanghai Waigaoqiao) reached 11.54 million TEU, accounting for 58% of the total container throughput. Therefore, it is necessary to face the overall situation of smooth domestic and international double circulation, focus on building a safe, efficient and smooth domestic trade container transport channel system mainly composed of the coastal north-south transport channel and the east-west transport channel of the Yangtze River and the Xijiang River, encrypt the domestic trade container route network, form three main channels of "two horizontal and one vertical", connect the important coastal urban agglomeration and metropolitan area, and connect the Beijing-Tianjin-Hebei, the Yangtze River Economic Belt, the Yangtze River Delta Guangdong, Hong Kong and Macao Greater Bay Area and other national strategic regions ensure the flexibility and resilience of the inter-regional container logistics supply chain, and help build the modern logistics supply chain.

6.2 Build a Domestic Trade Container Hub with Outstanding Distribution and Radiation Capability

At present, the number of domestic trade container port hubs have been initially formed in key coastal and riverside areas. Tianjin Port in the north accounted for 68% of the domestic trade container throughput of Tianjin and Hebei coastal ports in 2019, 71% in 2021 and 70% in 2022; The domestic trade container throughput of Yingkou Port accounts for 48% of the domestic trade container throughput of Liaoning coastal ports; The proportion will be 38% in 2021 and 58% in 2022. The domestic trade container throughput of Guangzhou Port in the Pearl River Delta accounted for 58% of the domestic trade container throughput of Guangdong coastal ports in 2019, and the total domestic trade container throughput of Guangzhou and Dongguan Port accounted for 72% of the domestic trade container throughput of Guangdong coastal ports; The proportion will be 70% in 2021 and 74% in 2022. In order to better connect the functions of the North-South container transport channel and realize the rapid distribution and better radiation of containers at both ends of the channel, it is necessary to further centralize and intensively arrange several domestic trade container hubs in the key areas of the North-South and East-West routes, and form 1–2 domestic trade container hub ports in the Beijing-Tianjin-Hebei, Changsanjiao, Guangdong-Hong Kong-Macao and other regions, respectively, to undertake the transfer, distribution and other tasks of the North-South and East-West trunk domestic trade container transport. Select 1–2 ports in the middle and upper reaches of the Yangtze River, focus on building inland container transport hub, and efficiently connect with the Yangtze River Delta hub ports. In Guangxi along the Xijiang River, select 1–2 ports to build a domestic trade container hub in the upper reaches of the Xijiang River, and realize the connection with ports in the Greater Bay Area of Guangdong, Hong Kong and Macao.

6.3 Improve the Domestic Trade Container Transport System Linking the Trunk and Branch with Regional Network

The efficient regional domestic trade feeder transport and the connection of trunk and branch transit transport network is an important prerequisite for maintaining efficient and smooth trans-regional domestic trade container transport channels. At present, some regions are already forming this kind of transportation system connecting trunk and branch. For example, Tianjin and Hebei ports, domestic and foreign trade branch line transportation has developed rapidly. In 2020, the traffic volume of Tianjin Port's Bohai Rim branch line exceeded 1 million TEUs, with a year-on-year increase of 67%. In each region, highlight the integrated hub function of 1–2 domestic trade container ports, form an efficient and dense regional container feeder route network between other ports and integrated hub ports, and efficiently and closely connect with the cross-regional trunk transportation channel routes, and finally form a transportation organization system of trunk and branch connection, cross-regional large channel connection, and regional domestic trade container network, so as to better protect the regional urban agglomeration The economic development of the metropolitan area needs to better assist the construction of an international consumption center city.

6.4 Cultivate Domestic Trade Container Transport Comprehensive Service Providers

At present, there are many enterprises engaged in domestic trade container transportation, and the group enterprises such as COSCO Shipping and China Merchants Bureau have a relatively high market share. At present, the domestic trade container transport chain is mainly concentrated in the maritime link, and the land-based business is properly expanded through agency, contract entrustment, etc. The real whole-process logistics service system has been built for some goods such as automobiles and heavy goods, while the entire logistics service system for most container transport goods is not perfect; There are few domestic trade container transport logistics operators that really provide one-stop services. In order to provide more efficient logistics supply chain services and better support the construction of modern supply chain and industrial chain, it is necessary to further accelerate the transformation and upgrading of traditional domestic trade container transport enterprises; Increase the connection between the sea channel and the land-based channel, strengthen the resource integration of the land-based logistics channel transportation, station operation and terminal customers, and provide door-to-door logistics services. It is suggested that the enterprise should strengthen the integration ability of the mainland to resources and cooperate with logistics participants at both ends of the production and sales places through mergers and acquisitions, joint ventures and cooperation; The connection between the mainland and the sea transportation should be strengthened, and the integration of port nodes should be strengthened to better realize the efficient connection; Increase the expansion of logistics value-added services, especially in logistics informatization, logistics financial services, logistics big data applications, etc.

6.5 Policy Recommendations

First, it is suggested to study and issue the development plan of domestic trade container transport under the background of the new development pattern. From the aspects of transportation channel pattern, hub port layout, domestic trade container port network improvement, transportation organization optimization, first-class enterprise cultivation, etc., the system layout will improve the adaptability of domestic trade container transport supply side, better support the cultivation of domestic strong market, and better help the construction of new development pattern.

Second, it is suggested to study and issue policies to support the development of container transport in port domestic trade. Support will be provided from the aspects of route and flight development, finance, taxation and finance, public infrastructure project approval and financial support of hub ports, domestic trade informatization and financial service upgrading, so as to better cultivate a number of domestic trade container transport integrated logistics service providers and domestic trade container hub ports.

Third, it is suggested to continue to accelerate the reform and consolidation of some goods, increase the research and development of relevant transportation standards, information technology standards and other aspects, support the construction of a smooth and efficient container logistics system, and better help the construction of a new development pattern.

7   Summary and Conclusion

1) The expansion of the domestic market, the construction of a modern circulation system, the optimization of industrial spatial layout, the demand for infrastructure investment and the demand for green development have brought new opportunities for the development of domestic trade containers in China.

2) The demand for domestic trade container transport in China's ports has great potential and will maintain a long-term growth trend in the future. It is predicted that the domestic trade container throughput of China's ports will be about 220 million TEU in 2035.

3) Building a smooth and efficient transport channel, strengthening the construction of hub ports, improving the transport network system and cultivating first-class integrated service provider are the essential way to highly develop domestic trade container transport.

4) Since policy is crucial to promote the high-quality development of China's domestic trade container transport, a development plan of domestic trade container transport under the background of the new development pattern should be studied and issued. Meanwhile, a series of policies which can better support the development of container transport in port domestic trade should be issued.

References

1. Yang, J.: Development of domestic trade container transportation market in China. In: Liu, B.-L., Wang, L., Lee, S.-j, Liu, J., Qin, F., Jiao, Z.-L. (eds.) Contemporary Logistics in China. CCERS, pp. 159–177. Springer, Heidelberg (2016). https://doi.org/10.1007/978-3-662-477 21-2_9

2. Wang, L., Su, S.I., Ruamsook, K.: Analysis of the round-trip cost of road container transportation in China. Transp. J. **50**(2), 204–217 (2011)

3. Xing, H., Du, L.: Prospects for the development of domestic container shipping in China. China Port **04**, 42–46 (2022). (in Chinese)

4. Yu, M., Zhang, Y.: Multi-agent-based fuzzy dispatching for trucks at container terminal. Int. J. Intell. Syst. Appl. **2**, 41–47 (2010)

5. Yichun, D., Liangduo, S., Ying, L., Hua, S.: Forecast of domestic trade container throughput in China's major ports. Containerization **29**(03), 4–7 (2018). (in Chinese)

6. Comprehensive Research on China's domestic trade container transport network. Dalian Maritime University (2017). (in Chinese)
7. Liyao, L.: Measures to break the situation of domestic container port development under the COVID-19. Containerization **32**(11), 5–6 (2021). (in Chinese)
8. Xinhai, Z.: Special study on the transformation of domestic trade container industry. Transp. Manager World **07**, 159–160 (2021). (in Chinese)
9. Weimeng, L., Minfang, H.: Prospects for the development of domestic trade container shipping in China. Containerization **31**(Z1), 9–12 (2020). (in Chinese)
10. Mengda, T.: Building a domestic trade container transport sharing service system with ports as the hub. China Ports **11**, 1–4 (2020). (in Chinese)
11. Li, Y., Shunquan, H.: Analysis of factors affecting domestic trade container throughput in China's major ports based on static panel data model. Water Transp. Manage. **38**(12), 18–20 (2016). (in Chinese)
12. Hassan, M.M., Mirza, T.: Using time series forecasting for analysis of GDP growth in India. Int. J. Educ. Manage. Eng. **3**, 40–49 (2021)
13. Taylor, S.J., Letham, B.: Business time series forecasting at scale. PeerJ. Prepr. **35**(8), 48–90 (2017)
14. Mgandu, F.A., Mkandawile, M., Rashid, M.: Trend analysis and forecasting of water level in mtera dam using exponential smoothing. Int. J. Math. Sci. Comput. **4**, 26–34 (2020)
15. Egrioglu, E., Aladag, C.H., Yolcu, U., et al.: Fuzzy time series forecasting method based on Gustafson-Kessel fuzzy clustering. Expert Syst. Appl. Int. J. **38**(8), 10355–10357 (2011)

The Significance and Project Planning of Building a Shipping System in Heilongjiang Province

Xunran Yu[1], Lingsan Dong[2(✉)], Changjian Liu[1], and Hongyu Wu[1]

[1] Transport Planning and Research Institute, Ministry of Transport, Beijing 100028, China
[2] School of Economics and Management, Harbin Institute of Technology, Harbin 150006, China
13611058357@163.com

Abstract. Currently, Heilongjiang Province suffers from an imbalanced transport structure, high logistics cost, scattered shipping resources, lack of market players, and insufficient systematization ability of shipping logistics organization. Meanwhile, water resources in the province have potential value to be released, and water transport port capacity needs to be upgraded. Integrating resources to clarify the main body, improving infrastructure network construction, and promoting the integrated development of material and trade ports is the crucial way to solve the above problems. From the perspective of transport, this paper puts forward the conception of building a shipping system in Heilongjiang Province whereby contributes to first, stimulate the effectiveness of water transport in Heilongjiang province, optimizing the transport structure, realizing the goal of water transport "capacity of over ten million tons, service trade volume over one hundred billion yuan." Secondly, promoting low-carbon green development and enhancing logistics development in Heilongjiang Province at par with that of the country. Thirdly, manifest the advantages of resources and location, realize deep interiorization of the strategy to Russia, and give birth to the development of industrial clusters.

Keywords: Waterway transport · Shipping system · Modern logistics · Industrial development

1 Introduction

According to the Ministry of Transport 2021 statistical report, China's inland waterway shipping cargo volume accounted for only 8% of the total, and the average five-year waterway freight volume of Heilongjiang Province 2017–2021 accounted for about 1.28% [1]. We can see that Heilongjiang Province inland waterway freight accounted for a meager ratio, whether from a domestic or international perspective. In addition, the total social logistics cost in Heilongjiang Province has remained between 1.08% and 1.24% of the national total social logistics cost for five consecutive years. The whole social logistics cost has remained between 1.37% and 2.21% of the national social logistics cost [1], indicating that the entire social logistics cost in the province is high. The switch from highway transport to waterway transport is efficient in south China.

Z. Hu et al. (Eds.): ICCSEEA 2023, LNDECT 181, pp. 860–872, 2023.
https://doi.org/10.1007/978-3-031-36118-0_74

However, Heilongjiang Province's shipping logistics efficiency is as low as a northern province and suffers from an imbalanced transport structure and high logistics costs (Table 1).

Meanwhile, the shipping resources of Heilongjiang Province are scattered, and the market's main body is missing. Heilongjiang Province has 15 national first-class water ports (including 9 first-class water ports to Russia) and 154 berths. By the end of 2021, the province registered 3,316 enterprises in border trade with Russia, with only 150 key foreign trade enterprises accounting for less than 5%. The overall performance is "many and scattered, small and weak," homogeneous competition, disorderly operation, weakening the port's ability to attract and gather productivity factors. The main body of shipping resources is scattered and fails to form a synergistic effect, resulting in the provincial location advantages not being effectively transformed into development advantages (Fig. 1).

Third, the potential value of water resources has yet to be released. The Heilongjiang River system is a virtual inland transport channel in China. The main navigable rivers in the territory are the Heilongjiang River, Songhua River, Ussuri River, and Nengjiang River. Currently, the province's river navigation mileage is 5528 km, with 4295 km of channel maintenance, 1924 km of Class III and above channels, and 1213 km of Class IV channels [2]. The Heilongjiang River system has many open water ports along the border between China and Russia, providing convenient transportation for trade between China and Russia.

Fourth, Water transport ports need to be upgraded. From the "Thirteenth Five-Year Plan" onwards, the state canceled the central subsidy for constructing inland ports. Due to the limited financial resources of the Heilongjiang provincial government at all levels, the port infrastructure construction and daily operation and maintenance funds to protect all pressed to the enterprise [3]. The financing ability of the relevant enterprises is inferior, and the port construction funds generally gap is significant. Part of the port infrastructure is not perfect; its function and role are far from fully released and play due to the low utilization rate of resources, the lack of professional management team operation, and the long-term shutdown.

Finally, the capacity of shipping logistics organization systemization is insufficient. Heilongjiang province belongs to the inland river economy less developed area with poor infrastructure and non-significant scale effect. Small and medium-sized enterprises on the inland river lack organizational design. They are not highly motivated to develop water transport-related industries. The multimodal transport system of iron and water, river and sea are not yet sophisticated, making the conversion rate of "highway transport to waterway transport " and " railway transport to waterway transport" too low. The last mile problem still exists, and the coordination among port layout, logistics channels, and integrated transport hubs is scant. The role of the junction is not enough, and its function is relatively single, weakening the port to attract and gather productivity factors of the location advantage. The objectives of this paper are to analyze the essential causes of the above five problems that currently exist in Heilongjiang Province and to find fundamental solutions. In general, there is abundant research on the shipping system in China, but they mainly focus on the Yangtze River Delta and Pearl River Delta, and lack of attention to Northeast China. Therefore, it is necessary to carry out research on the

Table 1. Freight volumes by mode of transport 2017–2021 (ten thousand tons)

Year	Rail		Road		Waterway		Aviation		Pipelines	
	Volume	Proportion	Volume	Proportion	Volume	Proportion	Volume	Proportion	Volume	Proportion
2017	11161	18.1	44127	71.9	1110	1.8	13	0.02	4996	8.1
2018	11357	18.2	42943	68.7	890	1.4	13	0.02	7329	11.7
2019	12073	20.8	37624	64.9	780	1.3	14	0.02	7552	13.0
2020	12603	22.5	35521	63.4	538	1.0	12	0.02	7357	13.1
2021	12512	20.1	420086	67.6	519	0.8	11	0.02	7189	11.5
Average	11941	19.9	40460	67.4	767	1.3	12	0.02	6885	11.5

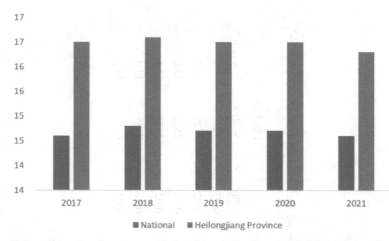

Fig. 1. Heilongjiang Province versus National Logistics Data 2017–2021 (Ratio of total social logistics costs to GDP)

construction of a shipping system based on the shipping characteristics of Heilongjiang Province.

2 Overview of Solution Ideas

2.1 Integrate Resources, Clarify the Market Subject and Lead the Development of Market Scale and Organization

Recommended to support the promotion of the province's port resources for the integration of work, the implementation of the "one province, one port" policy, a clear subject, the province's port construction. Focus on supporting the construction of Heihe, Fuyuan, and Tongjiang ports to form the advantages of port scale, improve port production capacity and optimize port resources. Let some influential Chinese enterprises with specific investment capacities enter foreign markets first. Through the integration of resources, industrial upgrading, and capacity adjustment, these enterprises can complete the construction of overseas material trade organization capacity and logistics infrastructure control capacity and the structure of overseas material trade organization and guarantee system.

2.2 Improve the Construction of the Infrastructure Network, Enhance the Capacity of the Shipping Market, and Serve the Integration of the Supply Chain, Industry Chain, and Value Chain

Improve the construction of the shipping infrastructure network in Heilongjiang Province, forming a waterway network among the Songhua River, Nengjiang River, Black River, and Ussuri River. Then build a "complete, safe and reliable, intensive and efficient, green and intelligent, orderly regulation water system network [4]. Relying on the location of Heilongjiang Province and the rich resources of the Russian Far East attract Russian ore, coal, timber, grain, oil, chemical raw materials, and other "original" products in Heilongjiang Province production and processing. Promote the development of one, two, and three industries to achieve the supply chain industrial chain value chain three chain integration.

2.3 Promote the Integrated Development of Animal Port and Trade, Cultivate Industrial Corridors, and Serve the Opening up to the North

2.3.1 Simultaneous Allocation of Russian and Chinese Resources for the Integration of Goods, Ports, and Trade

Implement particular actions for developing and opening up along the border in the new era around the boundary river. Relying on petrochemical, grain and oil processing, ship, aviation facilities, equipment manufacturing, modern logistics, shipping services, and other port industries, ports in Heilongjiang Province can exert their intensive synergistic amplification to build an integrated industry cluster of material, port, and trade[5].

2.3.2 Differentiated Development of Industries and Cultivation of Industrial Corridors

Attract freshwater biology, marine fishery, non-GMO soybean processing, and other particular industries to cluster in the border area, forming an industrial pattern with differentiated development from overseas enterprises. At the same time, use differentiated sectors to complement each other, enlarge the market scale and drive more market players to expand foreign trade business. In addition, the modern logistics system led by science and technology is the most critical link in the whole industrial chain, which is the fundamental support for China's economic development to achieve global buying and selling. To sum up, taking the transfer of industry and technology transformation with high quality as the grasp, construct the modern industrial system of shipping economic belt and build the shipping "industrial corridor".

2.3.3 Connecting the Market and Supply Side Through Logistics Corridors to Serve the Opening up to the North

Heilongjiang Province has unique logistics channels and infrastructure network resources, which can easily be integrated into the extensive logistics system that can reflect the national will. We should seize the opportunity to complete the layout of the local and overseas channel infrastructure in advance, lengthen the logistics industry chain, innovate the cooperation mode of the river and sea intermodal transport channel, and form the water-rail intermodal transport domestic and overseas extensive circulation system. Guide foreign trade enterprises, cross-border e-commerce, and logistics enterprises to strengthen business synergy and resource integration, rely on border ports, and accelerate the layout of overseas warehouses, warehouses, distribution centers, and other logistics infrastructure networks. Improve the efficiency of logistics operation and asset utilization, reduce the cost of foreign trade commodity circulation, promote the formation of an efficient and accessible, large-scale development of domestic and international markets, and serve the opening up to the north.

3 Methodology of Building a Shipping System in Heilongjiang Province

3.1 Build a Cross-Border Waterway Transport System

Improve the iterative upgrading of transport organization modes and equipment at border crossings and cross-border channels in Heilongjiang Province and actively participate in constructing transport infrastructure at the Russian counterpart. Realize direct access to cities in the Russian Far East, such as Blagoveshchensk, Nizhniy Leninskoye, Khabarovsk, and Komsomolsk, through river-sea intermodal transport. The railroad into the Russian territory takes the Eurasian Continental Bridge and can be connected with the countries of Eastern Europe. From the lower reaches of the Heilongjiang River to the sea, through the Tatar Strait, can reach Japan, South Korea, North Korea, and Southeast Asian countries and China's southeastern coastal port cities, Hong Kong region, forming a river and sea international trade channel [6]. Thus, Northeast China, the Russian Far East, Japan, Korea, and North Korea are linked into a circular water transport network, running through entire northeast Asia, connected to the Arctic Ocean route, and then reaching Europe and the United States.

3.2 Access to a Major Waterway for Grain Transportation Between the Songhua and Nenjiang Rivers

Songhua River and Nengjiang River connect most grain-producing cities and counties in Songnun Plain and Sanjiang Plain. The grain water transport corridor can pass ships of over 1,000 tons. Outbound grain is loaded into standard containers and boarded through the port collection and distribution system. Thus, now the grain waterway transportation, the formation of grain logistics waterway corridor, to improve quality, reduce costs and increase efficiency, innovative grain storage and transportation pattern.

3.3 Promote the Construction of "Four Rivers" Linking the Shipping System

Heilongjiang Province is rich in water resources. The principal rivers are Heilongjiang River, Songhua River, Ussuri River, and Nengjiang River. To actively promote the Heilongjiang River and other critical international navigable rivers channel comprehensive development and utilization and maintenance management, and orderly promote the Ussuri River and other global navigable rivers channel construction, maintenance, and management [7]. Besides, promote the Heilongjiang international navigable rivers and waterways essential obstruction beach remediation [8]. In practice, connecting the Heilongjiang River, the Songhua River, the Ussuri River, and the Nengjiang River is recommended to create a network of four rivers connected to the extensive shipping system (Table 2).

Table 2. Basic information on the main rivers in Heilongjiang Province [9]

	Heilong river	Songhua River	Ussuri River	Nenjiang River
place of origin	Erguna River	North source of the Nengjiang River and the south source of the Changbai Mountains	Southwest slope of Sikhot Mountains	South Urchin River
Convergence site	Amur River	Heilong river	Heilong river	Heilong river
Flowing places	Mongolia, China, Russia	China	China, Russia	China
Total length (km)	4440	1927	909	1370
Length of Boundary River (km)	3000	-	492	-
Total watershed area (km2)	186	55	19	30
Watershed area in the territory (10 thousand km2)	8.91	-	6.15	-
Frozen period (month)	6	5	5	5
Average runoff (billion m3)	346.5	33.5	62.4	22.7
Hydraulic Resources (10 thousand kw)	304	292	-	227

With the Heilongjiang River, the Songhua River, the Ussuri River, and the Nenjiang River along the port as the axis, the highway collection and distribution for the width to build a "port axis public spokes" of the shipping system. Based on the comprehensive development and utilization of the Heilongjiang River, the Songhua River, the Ussuri River, and the Nenjiang River channel, northward through Temple Street into the Tartar Strait, southward to operate the mouth into the Bohai Bay, to build a "direct water to water" shipping system. Taking the Heilongjiang River, Songhua River, Ussuri River, and Nengjiang River as the base and Jilin Daan Port and Vladivostok as the medium, the shipping system of "water-rail intermodal transportation" is built. Improve the infrastructure network of the Longjiang River shipping system, effectively play the effectiveness of the shipping system in Heilongjiang Province, and support the construction of a new highland of China's northern opening.

3.4 Plan the Northeast Asia International Waterway

3.4.1 Connect to Northeast Asia and the Arctic Northeast Passage

The revitalization of Northeast China needs a grand economic carrier. Opening up the Songliao Canal and creating a tremendous financial page, the Northeast Asia Waterway is the fundamental solution to this problem.

The Northeast Asia International Waterway will cover 1.45 million square kilometers of land in Northeast China, nourish 110 million people, and build a river network of nearly 7,000 km and an ecological, economic corridor with huge potential [10]. Further, it will promote the docking and interaction between Northeast China and Beijing-Tianjin-Hebei, Yangtze River Economic Belt, Guangdong-Hong Kong-Macao Greater Bay Area and other city clusters, and connect the Yangtze River industrial clusters and radiate the southeast coast. The canal will become a hot spot for investment, forming a developed economic belt similar to the Yangtze River [11]. A historical span of international water transport corridor in Northeast Asia, thus enhancing the connected radiation function of the Northeast region as the geographical center of Northeast Asia.

3.4.2 Serve the Regional Economy Along the Route, Pulling the Industry to Reduce Costs and Increase Efficiency

Northeast Asia's international water transport channel through the "southward river and sea transport" and "northward river and sea transport" directly realize the "Chinese and foreign" and "foreign and foreign "The two markets and two kinds of resources of the great cycle. The Northeast waterway network will become an important trade network and economical carrier to promote the Northeast to become the import and export trading center of the bulk commodities of China and Russia and an important logistics hub for Northeast Asia.

Relying on convenient and cheap water transportation and abundant arable land resources, the Northeast can develop into a granary and agricultural processing base like the Mississippi River Basin in the United States [8]. Extending the radiation capacity of city clusters and linking node cities along the corridor will enhance the stickiness of the development of the northeast region with interactive resources, interoperable industries, mutual trust in humanities, and interconnected transportation. It further drives the delta development in the Liaohe Delta and the Three Rivers Plain area, forming a new pattern of comprehensive development in Northeast China with lower costs, more robust mobility, and greater industrial stickiness (Fig. 2).

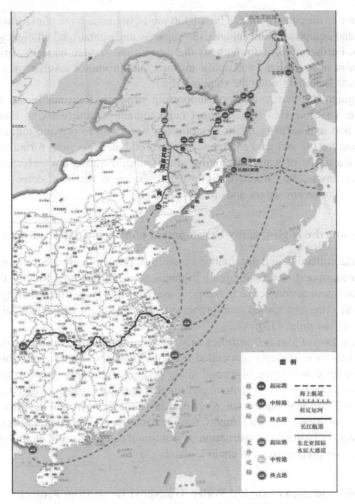

Fig. 2. Schematic diagram of the Northeast Asia Waterway

4 Practical Project Planning

4.1 Create a Smooth and Efficient Waterway

Accelerate the construction of high-grade waterways. Following the general requirements of the "four rivers" link shipping, open up the Songhua River Nengjiang Heilongjiang Ussuri River blockage. The implementation of upstream and downstream waterway improvement improves the waterway's navigability, accelerates the construction of the gradient hub, and sets aside space for long-term expansion and upgrading.

Optimize the functional layout of the port. Plan port layout and development and construction timing along the waterway with the product and utilization of the towns and port industry layout along the canal [12]. In turn, it will accelerate the growth of port areas, promote the construction of broad port areas, implement the transformation of port facilities and equipment, and enhance the capacity and operational efficiency of existing terminals. Furthermore, it strengthens the port's total service capacity, prioritizes the development of specialized and public airports, and promotes several town and village convenience terminals along the shipping route.

Strengthen the construction of supporting projects. Reasonably set up mooring anchorage, water service area, channel management station, working boat dock and other support and security facilities along the waterway to provide temporary mooring, waiting for the lock and emergency shelter, and other services for navigable ships. To meet the needs of waterway maintenance, administrative law enforcement, water emergency rescue, etc., improve the level of service to the boat, crew, and shipping management to ensure safety and smooth passage.

4.2 Promote the High-Quality Development of Multimodal Transport

Improve the collection and distribution network of ports. Accelerate the construction of the port area, operation area dredging highway, and railroad line. Realize the highway and railroad connection between the port area and highway, railroad freight channel, factory and mine enterprises, comprehensive logistics hub, and effectively solve the problem of "last mile" [13].

Construction of logistics facilities network. Comprehensively promote the construction of multi-level, strongly associated logistics facilities system along the waterway. Coordinate the construction of several integrated logistics parks, promote the construction of professional logistics centers, and support the development of logistics organization needs of Heilongjiang Province's particular industries. Around the primary agricultural products origin, sale, and import, the construction of cold chain logistics facilities, and actively declaring the construction of a national backbone cold chain logistics base to create a comprehensive cold chain logistics cluster.

4.3 Implementation of Digital Water Transport Innovation Project

Construction of digital waterways. Accelerate the construction of an electronic channel chart system and unified scheduling information systems for navigable buildings. Port and thus functionally realize water transfer and quality monitoring, channel maintenance, water traffic safety supervision, water traffic law enforcement, waterway transport management, and other port and navigation business refinement and digitalization.

Create a digital port. Actively support port enterprises to improve the automation of essential equipment, such as terminal front loading and unloading equipment, transport vehicles, yard loading, and unloading machinery, to further enhance the efficiency of port loading and unloading operations [14]. Promote the application of intelligent technologies such as yard systems, wireless dispatching communication systems, and intelligent cargo handling systems for container bridges.

Improve the digital platform. Actively support the shipping center wisdom logistics comprehensive service platform to achieve the goal of "extensive information interconnection, optimal allocation of resources, business cooperation and linkage, port and industry development". Improve the water traffic information system such as water traffic safety and environmental protection communication, electronic cruise, convenient ship crossing, safe bridge crossing, etc., to provide timely and intelligent information service for ships.

5 Policy Recommendations

5.1 Stimulate the Effectiveness of Water Transport, Optimize the Transport Structure and Achieve the Goal of Doubling the Thousand

In response to the requirements of the "Work Plan for Optimizing the Adjustment of Transport Structure by Promoting the Development of Multimodal Transport (2021–2025)" issued by the General Office of the State Council, we will accelerate the construction of water transport infrastructure such as port logistics hubs, vigorously develop and improve waterways, and open up the 1,000-ton waterways. Accelerate the development of water transport, increase the proportion of water transport, grain, containers, coal, and other bulk materials, "highway transport to waterway transport" "railway transport to waterway transport", "river-sea transport", and "land-sea transport". Promote the deep integration of various modes of transport, further optimize and adjust the transport structure, improve the efficiency of integrated transport, and strive to achieve the water transport "capacity of more than 10 million tons, service trade volume of more than 100 billion yuan" double thousand goals.

5.2 Activate the Potential of Water Transport, Green Low-Carbon Development, to Achieve Parity with the National Average Level of Logistics

Improve the quality and efficiency of water transport supply, accelerate the construction of Heilongjiang Province's large shipping system, significantly reduce shipping costs, and release the potential of water transport. The amount of investment required for the construction of waterway transport compared to the investment in railroad transport saves about half, and the cost of water transport is 1/2 of the cost of railroad transport, is about 1/3 of the cost of road transport [15]. Compared with other modes of transport, water transport has the advantages of large capacity, low cost and small emissions, and is the greenest mode of transport. Create superior water transport conditions, attract a large number of industrial capital to build factories along the river, the formation of industrial clusters, water transport to provide more and more sources of goods, the increase in freight volume to promote the development of water transport and the cost of continuous decline, thereby reducing the total cost of social logistics. Increase the proportion of water transport in the freight volume of Heilongjiang Province, narrowing the gap of 0.8% between it and the national average, to achieve parity with the national average logistics level.

5.3 Highlight the Advantages of Resources and Location, Realizing the Strategic Melanization with Russia, and Giving Birth to the Development of Industrial Clusters

China-Russia economic and trade relations are stable, China's industrial chain is more complete, affected by the international situation, the export of Russian energy and the import and export of consumer goods, equipment, etc. are more dependent on China. As the geographical frontier of cooperation with Russia, Heilongjiang Province, by building a new system of big shipping, providing efficient and convenient professional services and playing a function beyond the channel itself, is able to fix the trade increment brought by this part of order transfer in the province, and then pull the natural resources of Russian Far East and surrounding regions into the province and continue southward. Attract the industrial capital and manufacturing enterprises overflowing from the developed regions in the south of China to settle in the north. Play the efficiency, stability and cost advantage formed by the new system of big shipping, gradually form the comparative advantage of Heilongjiang Province, attract the import and export processing industry to gather along the port of the province, the formation of shipping "industry corridor".

6 Discussion and Conclusion

Integrating resources to clarify the main body, improving the infrastructure network construction, and promoting the integrated development of the physical trade port is a direct way to solve the imbalance of the transport structure of Heilongjiang Province, high transport costs, and the bottleneck of the development of the shipping industry. However, from a practical point of view, the lack of growth momentum of water transport in Heilongjiang Province, therefore, the construction of a shipping system in Heilongjiang Province is the fundamental way to solve the above problems.

This paper starts with five aspects of the current problems of the shipping industry in Heilongjiang Province, firstly targeting direct solutions and further researching to arrive at the fundamental solution of building a shipping system in Heilongjiang Province. Finally, constructive suggestions are made in the concrete implementation aspects of the solution. The research in this paper focuses on the development of shipping in Northeast China and helps to address the problem of unbalanced and inadequate development in the various regions of China. At the same time, the findings of this paper are important for grasping the development trend of the water transport industry, national policy formulation, enterprise development strategy formulation, and shipping infrastructure construction decisions.

Meanwhile, the research in this paper has some limitations, as the perspective of this paper mainly focuses on the transportation field, and the subsequent project implementation also needs to consider various aspects such as environmental and financial issues.

References

1. Ministry of Transport of the People's Republic of China: Statistical bulletin on the development of the transport industry in 2021. (in Chinese)
2. Guo, L.: Annual Report on Northeast China (2020). Social Sciences Academic Press (China), Beijing (2021). (in Chinese)
3. Alop, A.: The main challenges and barriers to the successful "smart shipping." TransNav Int. J. Mar. Navig. Saf. Sea Transp. **13**(3), 521–528 (2019)
4. Shi, W., Xiao, Y., Chen, Z., et al.: Evolution of green shipping research: themes and methods. Marit. Policy Manag. **45**(7), 863–876 (2018)
5. Wang, L., Zhu, Y., Ducruet, C., Bunel, M., et al.: From hierarchy to networking: the evolution of the "twenty-first-century Maritime Silk Road" container shipping system. Transp. Rev. **38**(4), 416–435 (2018)
6. You, Y.: The "New Songliao Canal" and "the northeast Asia international waterway channels." Liaoning Econ. **01**, 26–31 (2016). (in Chinese)
7. Ministry of Transport of the People's Republic of China: Outline of Inland Waterway Navigation Development (2020). (in Chinese)
8. Ministry of Transport of the People's Republic of China: Water transport "fourteen five" development plan interpretation (2022). (in Chinese)
9. Heilongjiang Statistical Bureau: Heilongjiang Statistical Yearbook. China Statistic Press, Beijing (2018). (in Chinese)
10. Liang, S., Cui, Q.: A study on north-south water diversion and inland waterway network construction in northeast China. Water Plan. Des. **01**, 1–8 (2022). (in Chinese)
11. Xv, X., Zhang, H., Xiao, J., et al.: Study on the development system of clean energy application in the old industrial base of northeast China. China's Strateg. Emerg. Ind. **44**, 1–4 (2017). (in Chinese)
12. Willems, J.J., Busscher, T., Woltjer, J., et al.: Co-creating value through renewing waterway networks: a transaction-cost perspective. J. Transp. Geogr. **69**, 26–35 (2018)
13. Asaul, A., Malygin, I., Komashinskiy, V.: The project of intellectual multimodal transport system. Transp. Res. Procedia **20**, 25–30 (2017)
14. Monks, I., Stewart, R.A., Sahin, O., et al.: Revealing unreported benefits of digital water metering. Lit. Rev. Expert Opin. Water **11**(4), 838 (2019)
15. Jing, M., Zheng, W.: Economic development advantages of low-carbon economy under waterway transportation. IOP Conf. Ser. Mater. Sci. Eng. **780**(6), 062029 (2020)

Advances in Technological and Educational Approaches

Mindfulness and Academic Achievements of University Students: A Chain Mediation Model

Jing Tao and ...

School of Business, Sichuan University, Chengdu 610065, China

Abstract. ... Academic Achievements ... Mindfulness and Academic Achievements ... Academic Achievements—Academic Achievement.

Keywords: Mindfulness · ... · Academic Achievements · Chain Mediation effect

1 Introduction

© The Author(s), under exclusive license to Springer Nature Switzerland AG 2023
X. Li et al. (Eds.): ICCSM 2023, LNBCIP 191, pp. 371–384, 2023.
https://doi.org/10.1007/978-3-031-36117-5_28

Mindfulness and Academic Achievements of University Students: A Chain Mediation Model

Ying Gao and Yi Chen[✉]

School of Business, Sichuan University, Chengdu 610065, China
273204008@qq.com

Abstract. As one of the main references for Chinese employers and graduate schools to select talents, great value is set on Academic Achievements by universities and students. This study mainly uses the questionnaire survey to explore the impact of students' personal characteristics on their Academic Achievements and the relationship between the variables. The research sample involves undergraduates in different provinces across China, majoring in literature, management, science, engineering and medicine, etc. It is hoped that this study can provide reference for students' management practice, help university students' self-management, and provide ideas for relevant research in the future. The conclusion includes the following three parts: Firstly, except Mindfulness and Academic Self-efficacy, university students' Mindfulness Level, Emotional Intelligence, Academic Self-efficacy and Academic Achievement are related to each other. However, there is no significant correlation between Mindfulness and Academic Self-efficacy. Secondly, Emotional Intelligence plays a mediating role in the influence of Mindfulness on Academic Achievement. Furthermore, the influence of Mindfulness on Academic Achievement is partly realized through the chain intermediary of Mindfulness—Emotional Intelligence—Academic Self-efficacy—Academic Achievement.

Keywords: Mindfulness · Emotional Intelligence · Academic Self-efficacy · Academic Achievement · Chain Mediation effect

1 Introduction

At university, academic achievements can be considered as "the performance of students". Both employers and graduate schools pay close attention to the academic achievements of candidates to predict their future work or research performance. Influenced by this, universities and their students attach great importance to academic achievements, too. In fact, academic achievements are not a single dimensional concept, but an evaluation of students from multiple perspectives. For example, Wang et al. believed that university students' academic achievements should include two parts: behavioral performance and objective performance. Among them, behavioral performance includes three dimensions: learning performance, interpersonal facilitation and learning dedication. Objective performance includes intellectual education score, cultural education score, physical education score, moral education score and comprehensive score.

In recent years, management research has found that mindfulness can positively predict employees' performance. Is this conclusion also true on university students' academic achievements? Through literature review, it is found that though there is no direct research, some scholars have proposed that mindfulness affects students' creativity, executive power and other abilities which can be considered as sub-dimensions of academic achievement. Therefore, mindfulness is likely to have an impact on university students' academic achievements. In addition, some pedagogical research has shown that emotional intelligence and self-efficiency could affect university students' academic achievements. At the same time, management scientists have found a positive correlation between employees' mindfulness and their emotional intelligence and self-efficacy. This implies that mindfulness may indirectly affect academic achievements through emotional intelligence and self-efficacy.

Based on previous research and empirical research, this study aims to build a model of the relationship between university students' mindfulness, emotional intelligence, self-efficacy and academic achievements. It is hoped that the conclusions of this study could help the management and development of university students and improve their academic achievements.

2   Literature Review and Hypothetical Inference

2.1   Mindfulness and Academic Achievement

Mindfulness was first introduced into the field of psychology by Jon Kabat-Zinn, defined as "the awareness that emerges through paying attention in a particular way: on purpose, in the present, and nonjudgmentally" [1]. With the development of this concept in academia, some scholars regard mindfulness as a state-like trait, meaning that there are intraindividual variations in mindfulness which could be trained. Through theory carding and empirical research, K. W. Brown & Ryan described mindfulness as a trait of individuals' strengthened attention and cognition to the current experience or real life and compiled the Mindful Attention Awareness Scale (MAAS) to measure mindfulness [2]. In recent years, this professional concept has been extended and generalized. Some scholars have introduced the concept of mindfulness into the field of pedagogy. Rusadi et al. pointed out that mindfulness-based cognitive therapy can improve university students' academic grit [3]. José Gallego et al.'s study on Spanish university students confirmed that mindfulness program can alleviate anxiety and stress [4].

The boundaries of academic achievement vary according to the stage of study. University students' academic achievements are multi-dimensional, including not only GPA, but also improvements of their ability, which is a comprehensive evaluation. In the context of Chinese education, Chinese scholars have given more comprehensive definitions. Wang et al. viewed academic achievement as the sum of students' learning achievements, learning behaviors and learning attitudes over a period of time [5]. Zhong believed that university students' academic achievements referred to the comprehensive development of individuals in the stage of higher education, including the admission opportunities themselves and the employment results after receiving education [6]. On this basis, Yang considered academic achievements to be the holistic development of university students in the whole university stage, which is in essence a comprehensive quality and ability [7], and compiled the Academic Achievement Scale for University Students with Li [8].

Based on conservation of resources theory (COR), individuals must constantly protect existing resources from loss through resource investment, recover from resource loss faster, and obtain new resources. Mindfulness, as a kind of trait with individual differences which can be trained, conforms to the nature of individual characteristic resources: individuals with high mindfulness traits have higher ability to obtain resources. From the concept of mindfulness, if an individual has a high level of mindfulness, he or she will have a stronger objective perception of the current experience, so that he or she could be able to alter goals or plans to adjust learning state and methods in a timely manner, that is, better use his or her own resources to exchange for higher output, which is higher academic achievements for students. Chapman-Clarke's book describes how mindfulness be used in management to improve job performance [9]. This may also work in education.

H1: University students' mindfulness is positively correlated with their academic achievements.

2.2 The Relationship Among Mindfulness, EI and Academic Achievement

Emotional intelligence (EI) was first proposed by Salovey & Mayer, meaning "the subset of social intelligence that involves the ability to monitor one's own and others' feelings and emotions, to discriminate among them and to use this information to guide one's thinking and actions" [10]. They conceptualized EI as a combination of four distinct dimensions: 1. Appraisal and expression of emotion in the self (self-emotional appraisal, SEA); 2. Appraisal and recognition of emotion in others (others' emotional appraisal, OEA); 3. Regulation of emotion in the self (regulation of emotion, ROE); 4. Use of emotion to facilitate performance (use of emotion, UOE) [11]. Based on Salovey & Mayer's research, Wong & Law conducted many empirical studies to show the effects of EI on job performance and compiled the Wong and Law Emotional Intelligence Scale (WLEIS) for Chinese respondents which is a better predictor of job performance on Chinese [12, 13].

There is evidence that mindfulness can affect EI or act as a antecedent variable of EI [11, 14]. As mentioned, Wong & Law's studies showed how EI affected job performance. Some scholars have also studied whether EI affects academic achievement, but the results were different [7, 15]. At present, university students' learning forms and contents are becoming more diverse while interpersonal cooperation and communication play an increasingly important role in learning. According to the emotional contagion theory, an individual's emotional state and behavior attitude will have an impact on another individual or the group. The transmission of interpersonal emotional experience is an unconscious process with individual differences. Based on this theory, a strong ability to perceive and regulate emotions should be able to improve the effectiveness of interpersonal cooperation and communication, thereby improving output, that is, individuals with high EI have high academic achievements.

Based on the above discussion, the following hypothesis are proposed:

H2: University students' EI is positively correlated with their academic achievements.

H3: University students' mindfulness is positively correlated with their EI.

H4: University students' EI plays an intermediary role between mindfulness and academic achievements.

2.3   The Relationship Among Mindfulness, Self-efficiency and Academic Achievement

Self-efficacy theory (SET) was first developed by A. Bandura and already been recognized by academia. Self-efficacy refers to the individual's belief in his or her own ability, which enables him or her to achieve certain achievement through organization and implementation. Bandura believed that self-efficacy was based on activities and behaviors in specific areas. If the work areas or situations differ, the self-efficacy of individuals would be different [16]. Therefore, this study selects academic self-efficacy in the learning field as a variable. Obviously, according to SET, academic self-efficacy affects university students' academic achievements, which has been confirmed [17].

Some scholars have studied the relationship between mindfulness and self-efficacy. Hanley et al.'s research explained that university students with higher dispositional mindfulness have higher academic self-efficacy after perceiving academic failure [18]. The research of Lu in China also showed that mindfulness training could effectively improve university students' academic self-efficacy [19].

Based on the above discussion, the following hypothesis are proposed:

H5: University students' self-efficiency is positively correlated with their academic achievements.

H6: University students' mindfulness is positively correlated with their self-efficiency.

H7: University students' self-efficiency plays an intermediary role between mindfulness and academic achievements.

2.4   The Chain Mediation Effect of EI and Self-efficiency

Many scholars believe that EI is an antecedent variable of self-efficiency, which means that the higher the EI, the higher the self-efficiency. For example, Shagini Udayar found that self-efficacy completely mediates the relationship between EI and university students' subjective and objective performance in a stressful task [20]. Saeed Wizra & Ahmad Riaz more clearly pointed out that EI has a significant role in promoting university students' academic self-efficacy [21].

Furthermore, individuals with high level of mindfulness can tolerate and uncritically accept their fear of difficulties when facing academic tasks. After objectively recognizing their emotions, they are more likely to adjust and use their emotions, thus generating reasonable estimate about their learning ability and the learning outcomes, so as to achieve high academic achievements. It is not difficult to find that there may be a chain mediation in the relationship between mindfulness and academic achievements. Cheng and his colleagues' research on kindergarten teachers has confirmed that there is a chain mediation effect of EI and self-efficacy between teaching mindfulness and job burnout, which provides a reference for the model [22].

H8: University students' EI is positively correlated with their self-efficiency.

H9: University students' EI and self-efficiency play a chain mediation role in the relationship between mindfulness and academic achievements (Fig. 1).

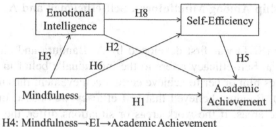

H4: Mindfulness→EI→Academic Achievement
H7: Mindfulness→Self-Efficiency→Academic Achievement
H9: Mindfulness→EI →Self-Efficiency →Academic Achievement

Fig. 1. Research Hypotheses

3 Data Collection

3.1 Questionnaire Design and Chosen Scale

We use questionnaires to collect data, inviting respondents to fill in the self-report scale voluntarily. There are 78 questions in the questionnaire, divided into the following parts:

1) Explanation. Explain the purpose of this questionnaire, and ask the respondents to sign the informed consent form.
2) Mindful Attention Awareness Scale (MAAS). This study adopts the MAAS compiled by K. W. Brown & Ryan, which has shown good reliability and validity in many countries and regions including China. The scale used in this study was translated and revised by Chen Siyi et al. [23].
3) Wong and Law's Emotional Intelligence Scale in Chinese (WLEIS-C). This study adopts the WLEIS-C compiled by Wong & Law in the context of Chinese culture, which has been widely used in Chinese research.
4) Academic Self-efficacy Scale. Different from the general sense of self-efficacy, this study focuses on the sense of self-efficacy in the field of learning. So, we adopt the Chinese Academic Self-Efficacy Scale compiled by Liang Yusong [24], which performs well in many empirical studies.
5) Academic Achievement Scale for University Students. Considering the comprehensiveness of university students' academic achievements, this study adopts the four-dimensional academic achievement scale compiled by Yang Na et al. This scale evaluates from four aspects: learning cognitive ability, communication ability, self-management ability and interpersonal promotion.
6) Demographic information. Collect demographic information such as gender, major, grade, and respondents' universities.

3.2 Respondents

A total of 245 questionnaires were distributed, 200 valid questionnaires (*81.63%*) were recovered. Among the participants, 82 were male (*41%*), and 118 were female (*59%*); 36 were freshmen (*18%*), 49 were sophomores (*24.5%*), 50 were junior students (*25%*), and 65 were senior students (*32.5%*); 148 were from regular universities (*74%*), and 52 were from top universities (*26%*). There were more females among the respondents, which may reduce the explanatory power of the data. About 3/4 of the respondents were from ordinary universities, which accorded with reality.

4 Data Analysis

4.1 Common Method Bias Test and Collinearity Test

Because self-report scales were used, check common method deviation by Harman's one-factor test before analyzing the data. *27.48%* of the variance was explained by the largest factor, which was less than the critical value of *40%*, indicating that there was no significant common method bias in this study.

   To ensure the accuracy of the results, variance inflation factor (VIF) method was used for collinearity test. Generally, if VIF > 10, it means that there is a serious collinearity problem between variables. The test showed that the VIFs of mindfulness (*1.145*), EI (*1.771*), self-efficacy (*1.593*) and other virtualized categorical variables like gender were all less than 5, meaning the non-existence of collinearity problem and indicating that they were suitable for further mediation effect tests.

4.2 Correlation Analysis for Mindfulness, Emotional Intelligence, Self-Efficiency, and Academic Achievements

After controlling the 4 variables of gender, grade, major category and university, Pearson's product moment correlation analysis was used to analyze mindfulness, emotional intelligence, self-efficiency and academic achievements (Table 1).

   According to the above correlation analysis results, 1) mindfulness is significantly positively correlated with emotional intelligence and academic achievement ($r = 0.30$, $p < 0.01$; $r = 0.34$, $p < 0.01$). But the correlation with self-efficiency is not significant ($r = 0.18$, $p > 0.01$). H1 and H3 are supported while H6 is not supported; 2) emotional intelligence is significantly positively correlated with self-efficiency and academic achievement ($r = 0.57$, $p < 0.01$; $r = 0.72$, $p < 0.01$) H2 and H8 are supported; 3) self-efficiency is significantly positively correlated with academic achievement ($r = 0.78$, $p < 0.01$). H5 is supported.

4.3 Relationship Between Mindfulness and Academic Achievements: A Chain Mediation Effect Test

Use the bias correction percentile Bootstrap method compiled by Hayes to test the significance of the mediation, and implement it through Hayes's SPSS plug-in Process.

Table 1. Descriptive statistics and correlation matrix for each variable

Variable	Mean	Standard Deviation	1	2	3	4
Mindfulness	4.00	0.74	1			
Emotional Intelligence	4.74	1.04	0.30**	1		
Self-Efficiency	3.22	0.71	0.17	0.57**	1	
Academic Achievements	3.45	0.53	0.34**	0.73**	0.78**	1

** means $p < 0.01$.

After controlling the 4 variables of gender, grade, major category and university, the mediation between mindfulness and academic achievement was tested (Fig. 2).

The results show that mindfulness could significantly positively predict emotional intelligence ($\beta = 0.36$, $p < 0.001$, $R^2 = 0.23$). But the predictive effect of mindfulness on self-efficacy was not significant ($\beta = 0.02$, $p > 0.05$). Emotional intelligence could also significantly positively predict self-efficacy ($\beta = 0.52$, $p < 0.001$, $R^2 = 0.38$). Both mindfulness, emotional intelligence and self-efficacy could significantly positively predict academic achievement ($\beta = 0.17$, $p < 0.001$; $\beta = 0.36$, $p < 0.001$; $\beta = 0.51$, $p < 0.001$; $R^2 = 0.76$).

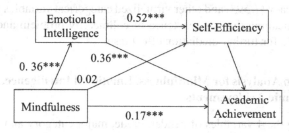

Fig. 2. The Chain Mediation Model (***means $p < 0.001$)

Further test of the mediation effect (Table 2) shows that the total effect of mindfulness on university students' academic achievement is 0.38, of which the direct effect accounts for 42.08%. The total indirect effect of emotional intelligence and self-efficacy in the impact of mindfulness on university students' academic achievements is 0.22, accounting for 57.92% of the total effect. And its bootstrap 95% CI is [0.12, 0.33], which excludes 0, indicating that emotional intelligence and self-efficacy are the mediating variables in the impact of mindfulness on university students' academic achievements.

This mediation effect is mainly composed of the following two paths: 1) mindfulness → EI → academic achievement (95% CI = [0.08,0.17], SE = 0.02). The mediation effect is 0.12, accounting for 31.82% of the total effect. H4 is supported; 3) mindfulness → EI → self-efficacy → academic achievement (95% CI = [0.05,0.13], SE = 0.02). The mediation effect is 0.09, accounting for 23.61% of the total effect. H9 is supported.

Note that the 95% CI of the mediation effect path 2) mindfulness → self-efficacy → academic achievement is [-0.06,0.08] (SE = 0.04), including 0, which indicates that the null hypothesis cannot be rejected. This mediation effect path is not significant. H7 is not supported.

Table 2. Bootstrap analysis of the mediation effect test

Dependent variable	Paths	Effect	Boot SE	Boot CI	Relative effect
Direct effect	$1 \rightarrow 4$	0.16	0.04	[0.08,0.24]	42.08%
Indirect effect1	$1 \rightarrow 2 \rightarrow 4$	0.12	0.02	[0.08,0.17]	31.82%
Indirect effect2	$1 \rightarrow 3 \rightarrow 4$	0.01	0.04	[−0.06,0.08]	/
Indirect effect3	$1 \rightarrow 2 \rightarrow 3 \rightarrow 4$	0.09	0.02	[0.05,0.13]	23.61%
Total indirect effect		0.22	0.05	[0.12,0.33]	57.92%
Total effect		0.38	0.06	[0.26,0.50]	100.00%

5   Conclusion and Future Directions

By linking mindfulness of university students with their academic achievements, we confirmed that mindfulness, as a trait, relates to students' academic achievements, which has certain innovation. Different from other personal traits, mindfulness can be cultivated and strengthened through mindfulness training. This means that university students can actively carry out systematic mindfulness training to improve their mindfulness and then improve their academic achievements.

This study also confirmed the correlation between mindfulness, EI and academic achievement. We proved the intermediary role of EI in the impact of mindfulness on academic achievement. This conclusion reminds universities to pay attention to the cultivation of students' non-intelligent abilities. Some students may be unable to get high academic achievements, not because of intellectual defects, but because of bad habits or emotional problems. Universities can introduce mindfulness training to help the worst performers adapt to learning and improve their self-confidence and self-management ability.

One conclusion of this study is that there is no significant correlation between university students' mindfulness and self-efficacy, which is different from the research results of other scholars. The reasons for the difference may be some variables limited by other studies, such as the situation of academic failure or mindfulness training, which is not limited in this study. Therefore, the interpretation of the result is that we can't prove the relevance between the mindfulness level and academic self-efficacy of usual students who do not receive any mindfulness training. Furthermore, this study tests the mediation effect of self-efficacy, which is also not significant.

Anyway, this study verifies the significance of the chain mediation effect of path mindfulness → emotional intelligence → self-efficacy → academic achievement, hoping this conclusion can enlighten future research. This result also implies that though mindfulness has no direct effect on self-efficacy, it may have an indirect effect through emotional intelligence, which needs to be confirmed by further studies.

Although this study has made some achievement, there are still many deficiencies.

Firstly, in terms of research methods, this study uses the self-report scale to measure university students' academic achievements, which gets more from university students' self-perception. If possible, we would like to use more comprehensive evaluation methods in the future, such as 360°Feedback.

Secondly, in terms of research type, the data collected in this study is static data at a certain time. But individual's mindfulness is changeable and can be improved through mindfulness training. It is hoped that the future research can be dynamic and explore the effect of mindfulness training by collecting time series data.

Finally, in terms of data collection methods, the respondents of this study participated voluntarily, which may lead to the problem that the sample cannot effectively represent the whole. It will be better if future research could reduce the error in data collection by stratified sampling.

References

1. Kabat-Zinn, J.: Wherever You Go, There You are: Mindfulness Meditation in Everyday Life. Hyperion, New York (1994)
2. Kirk-Warren, B., Richard-M, R.: The benefits of being present: mindfulness and its role in psychological well-being. J. Pers. Soc. Psychol. 4(84), 822–848 (2003)
3. Rusadi, R.M., Sugara, G.S., Isti'adah, F.N.: Effect of mindfulness-based cognitive therapy on academic grit among university student. Curr. Psychol. 42, 1–10 (2021)
4. Gallego, J., Aguilar-Parra, J.M., Cangas, A.J., Langer, Á.I., Mañas, I.: Effect of a mindfulness program on stress, anxiety and depression in university students. Span. J. Psychol. e109(17), 1–6 (2014)
5. Wang, Y., Li, Y., Huang, Y.: A study on the relationship between university students' psychological capital, achievement goal orientation and academic achievement. Explor. High. Educ. 06, 128–136 (2011). (in Chinese)
6. Zhong, Y.: An empirical analysis of the influence of class background on university students' academic achievements. Dev. Eval. High. Educ. 02(28), 108–115 (2012). (in Chinese)
7. Na, Y.: A Study on the Relationship Between Emotional Intelligence, Self-Efficacy and Academic Achievements of University Students. Qufu Normal University, Rizhao (2016). (in Chinese)
8. Li, X., Yang, N.: A study on the structure of the academic achievement scale for university students: its development, reliability and validity. Da Xue (Res. Ed.) 03, 41–53 (2016). (in Chinese)
9. Chapman-Clarke, M.: Mindfulness in the Workplace. Kogan Page, London (2016)
10. Salovey, P., Mayer, J.D.: Emotional intelligence. Imagin. Cogn. Pers. 3(9), 185–211 (1990)
11. Schutte, N.S., Malouff, J.M.: Emotional intelligence mediates the relationship between mindfulness and subjective well-being. Personality Individ. Differ. 7(50), 1116–1119 (2011)
12. Wong, C.-S., Law, K.S.: The effects of leader and follower emotional intelligence on performance and attitude: an exploratory study. Leadersh. Quart. 13(3), 243–274 (2002)
13. Law, K.S., Wong, C.-S., Huang, G.-H., Li, X.: The effects of emotional intelligence on job performance and life satisfaction for the research and development scientists in China. Asia Pac. J. Manage. 25(1), 51–69 (2008)
14. Heidari, M., Morovati, Z.: The causal relationship between mindfulness and perceived stress with mediating role of self-efficacy, emotional intelligence and personality traits among university students. Electron. J. Biol. 4(12), 357–362 (2016)

15. Jenaabadi, H.: Studying the relation between emotional intelligence and self esteem with academic achievement. Procedia. Soc. Behav. Sci. **114**, 203–206 (2014)
16. Bandura, A.: Self-efficacy: The Exercise of Control. Worth Publishers, Incorporated, New York (1997)
17. Chen, X., Ryung, K.J.: The effects of academic self-efficacy and academic emotion regulation on academic achievement among Chinese students in Korea: mediating effect of achievement goal orientation. Korean Assoc. Learner-Centered Curriculum Instr. **18**(15), 361–382 (2018)
18. Hanley, A.W., Palejwala, M.H., Hanley, R.T., Canto, A.I., Garland, E.L.: A failure in mind: dispositional mindfulness and positive reappraisal as predictors of academic self-efficacy following failure. Personality Individ. Differ. **86**, 332–337 (2015)
19. Lu, C., Pan, F., Fang, F.: The relationship between active and passive procrastination, mindfulness and self-efficacy of university students. J. Shandong Univ. (Med. Ed.) **10**(59), 108–113 (2021). (in Chinese)
20. Udayar, S., Fiori, M., Bausseron, E.: Emotional intelligence and performance in a stressful task: the mediating role of self-efficacy. Personality and Individ. Differ. **156**(C), 109790 (2020)
21. Saeed, W., Ahmad, R.: Association of demographic characteristics, emotional intelligence and academic self-efficacy among undergraduate students. J. Pak. Med. Assoc. **3**, 70 (2020)
22. Cheng, X.-L., Zhang, H., Ma, Y.: The relationship between kindergarten teachers' teaching mindfulness and job burnout: the chain mediation effect of emotional intelligence and self-efficacy. Res. Preschool Educ. **03**, 65–78 (2022). (in Chinese)
23. Chen, S., Cui, H., Zhou, R., Jia, Y.: Revision of mindful attention awareness scale (MAAS). Chin. J. Clin. Psychol. **20**, 148–151 (2012). (in Chinese)
24. Liang, Y.: Study on Achievement Goals, Attribution Styles and Academic Self-efficacy of College Students. Central China Normal University, Wuhan (2000). (in Chinese)

Evaluation System of College Students' Key Abilities Development in the Context of "Integration Education of Specialization, Creation and Morality"

Liwei Li[✉], Dalong Liu, and Yanfang Pan

Transportation College of Nanning University, Guangxi 530200, China
819395370@qq.com

Abstract. In recent years, with the rise of reform in China's education industry, it has become a major trend to create "golden courses" and eliminate "defective courses". All kinds of curriculum reform models and high-level academic evaluation methods came into being. In this context, this paper deeply analyzes the current situation of academic evaluation research under the teaching innovation reform of "integration of professional education and innovation", "ideological and political curriculum", and combines the development trend of high-level academic evaluation to build a developmental evaluation principle and evaluation system based on the three integration of "professional education, innovation and entrepreneurship education, and ideological and political curriculum". Finally, using the analytic hierarchy process to quantify the weight of each index in the evaluation system, a high-level academic evaluation system suitable for the current curriculum innovation model is established, which provides a certain reference for the training and assessment of comprehensive talents in colleges and universities.

Keywords: Specialized and creative integrated education · Ideological and political education · Development evaluation system · Academic assessment

1 Introduction

Improving the quality of higher education is a lofty task and historical responsibility entrusted to educators by the times [1]. Since the Ministry of Education of China specially issued the Notice on Implementing the Spirit of the National Undergraduate Education Conference in the New Era in August 2018, "high-level, innovative and challenging" has become the theme vocabulary of undergraduate education in Chinese universities [2]. Under the east wind of this reform, curriculum innovation models such as "integration of expertise and innovation" and "ideological and political curriculum" have become research hotspots, and effective academic evaluation has become a new pain point of curriculum reform in such colleges and universities [3–5].

Academic evaluation is a feedback on students' development and an important part of teaching quality feedback [6]. The key to academic evaluation is to make a good

Z. Hu et al. (Eds.): ICCSEEA 2023, LNDECT 181, pp. 885–900, 2023.
https://doi.org/10.1007/978-3-031-36118-0_76

diagnosis of the quality of education and teaching, guide students to develop in their comprehensive quality and personality, and provide evidence for the development of in-depth education and teaching. The developmental evaluation of college students' key abilities, as the core link of high-level academic evaluation, will become the mainstream trend of research.

Thereupon, this paper aims to study the principles and indicator system of the developmental evaluation of college students' key abilities in the context of the integration of "professional education, innovation and entrepreneurship education, and ideological education", simply called three integration. It uses the analytic hierarchy process to design the indicator weights to form an effective evaluation system, which provides some ideas for diagnosing students' growth and ensuring teaching quality.

2 The Evaluation of College Learning and Its Development in the Context of Teaching Innovation

2.1 Current Situation of Education Reform under the Three Integration

In recent years, seeking for the coordinated development of innovative education and professional education, implementing the implementation details and evaluation mechanism of the integration of professional and creative education, has gradually become the focus and difficulty of higher education.

From the current research literature, there is little difference between its assessment model and professional education assessment model. Among them, the evaluation model with strong innovation is as follows: Yang Qiuling, Huang Gaoyu and others proposed to establish a credit flexibility system. For example, students who start businesses abroad or study entrepreneurship and innovation courses abroad can convert their entrepreneurial achievements into innovation and entrepreneurship credits [7]; Ni Yu put forward an innovation incentive evaluation mechanism, and carried out incentive evaluation on innovative products, patent applications, awards in competitions, participation in exhibitions, etc. [8]. Gao Yixia and others proposed to combine short-term evaluation with long-term evaluation. The short-term evaluation is based on the spirit of innovation, entrepreneurial awareness and innovation and entrepreneurship ability. The long-term evaluation needs to evaluate the students' working status, ability and performance in their posts [9].

2.2 The Status Quo of Academic Evaluation under the Background of "Moral Education"

"Curriculum Ideological" refers to taking curriculum as a carrier and moral cultivation as a fundamental educational activity and teaching subject. At present, Xiong Xiaoyi and others have built a quantifiable whole process assessment and evaluation system based on curriculum ideological and political education, and inspected students' thinking ability through stage tests, research results, case studies, group discussion contribution rate, cooperation and cooperation, and acceptance of new knowledge [10]. Wang Hui and others discussed the significance and design principles of formative assessment applied to the teaching effect evaluation of "curriculum ideological and political education" [11]. Lu Daokun and others solved the problems in the organization of evaluation activities, the construction of evaluation standards and the construction of evaluation method system in the ideological and political curriculum, and used the form of "text evaluation + t eaching observation + customer evaluation" to conduct ideological and moral evaluation through anecdotal records, investigation reports, classroom records, symposiums, logs and other methods related to ideological and political elements [12].

2.3 Development of High-Level Academic Assessment

In the context of continuous innovation in higher education, the curriculum reform has gradually set up high-level goals, while the traditional learning evaluation mechanism will not be able to effectively detect high-level goals. This makes it a new trend of research to focus on the direction of students' academic development, help them formulate development goals, and achieve self-recognition, self-education and self-development of high-level academic assessment [13–17].

To sum up, there are few studies on the evaluation system of the existing innovative courses of the "integration of professional innovation" and "ideological and political curriculum" types, mainly focusing on teaching design and teaching methods. More innovation was made on the basis of the traditional professional curriculum academic evaluation method, which failed to break through the critical point of the subversive reform of academic evaluation. Looking forward to the future research trend, the curriculum innovation combining "specialty, creativity and thinking" will therefore become an important form of curriculum innovation in China [18, 19], and its high-level academic evaluation system will also become an effective way to test the quality of such courses.

3 Evaluation Principles of Key Abilities of College Students in the Context of "Three Integration"

The traditional evaluation method mainly based on "transcripts" leads to "high scores but low abilities" and "incomplete quality" [20]. In particular, in the face of the background of teaching reform such as the integration of specialty and creativity and the ideological and political education of curriculum, the traditional evaluation principles are no longer applicable, and the student evaluation system must be reformed to highlight the developmental function of evaluation and the cultivation of professional ability. Therefore, the new evaluation system should comply with the following basic principles.

3.1 Principle of Diversified Evaluation Contents

What deserves full attention is the students' professional ability, quality and innovation and entrepreneurship ability, while the general memory content can be diluted. In terms of evaluation content, it is necessary to try to incorporate professional knowledge and skills, methodological ability and social ability, emotional attitude and values, innovation awareness and entrepreneurial ability into the evaluation system [20]. The evaluation of students should abandon the worship of knowledge, not blindly pursue academic achievements, but also focus on the cultivation and evaluation of students' comprehensive quality, pay attention to students' learning interest, emotional experience, psychological quality, values, innovation spirit and practical ability, and return to the essence of seeking college education [21].

3.2 Principle of Diversity of Evaluation Methods

In the teaching of higher education, appropriate testing methods must be adopted according to the characteristics of specialties, disciplines and courses. In addition to the traditional written examination, for the assessment of theoretical knowledge, classroom performance, quick answer and discussion, reports, small papers, structured five question reflection reports, symposiums and other methods can be used. For practical courses, skills operation assessment, virtual simulation, evaluation scale, awards and other methods can be used. Textual assessment methods such as self-evaluation log and personal deeds can be used for moral quality that is difficult to quantify.

3.3 Principle of Open Evaluation Subject

It is a desirable method to combine self-assessment with his assessment. The academic assessment covers a wide range of areas and has a long time span. The evaluation of students requires the joint participation of teachers, individuals, classmates, parents, schools and enterprises [22].

3.4 Principle of Dynamic Evaluation Process

The result of teaching is static, while the teaching process and students' growth are dynamic. Only dynamic process evaluation can effectively evaluate and regulate students' learning behavior. The evaluation process must combine process evaluation, formative evaluation and summative evaluation, run evaluation through daily teaching, objectively and fairly reflect students' ability performance in various periods, and give students the opportunity to evaluate many times.

3.5 Principle of Effectiveness of Evaluation Feedback

In order to understand the real ability level of students through the original score of the representation, the adjustment model based on latent variables can be used to form a phased feedback mechanism with the ability level as a reference, adjust the scores obtained by students based on the ability factors, and analyze their real ability level through the score representation [23, 24]. Through the understanding of students' real ability level, immediate feedback and academic intervention are carried out to form an effective and high-quality staged feedback mechanism.

4 Construction of the Developmental Evaluation System of College Students' Key Abilities

4.1 Evaluation System

The developmental education evaluation system was proposed by scholars from the School of Education of the British Open University in the 1980s [25]. They believe that the fundamental purpose of evaluation is to promote the development of students.

Table 1. Developmental evaluation system of key abilities of college students in the context of three integration

Level I indicators	Level II indicators	Assessment subject				Evaluation carrier
		Indi-vidule	classmate	teacher	enterprise	
Working ability	Professional knowledge	√	√	√	√	score/comments
	Related knowledge	√		√		comment
	Humanities and social knowledge	√				comment
	Professional skill	√	√	√	√	score / Video evaluation
	Information collection and processing capacity	√		√		score/ comment
	Problem solving ability	√	√	√		score / comment
	Learning ability		√	√		comment
	Communication ability		√	√	√	Video evaluation
	Interpersonal skills		√	√	√	Video evaluation
	Team cooperation skills		√	√	√	Video evaluation
Ideolo-gical literacy	Responsibility	√	√	√		comment
	Morality Cultivation	√		√		comment
	Humanistic accomplishment	√		√		comment
	Professional ethics	√		√	√	score / comment
	Professional norms	√		√	√	score / comment

(*continued*)

Table 1. (*continued*)

Level I indicators	Level II indicators	Assessment subject				Evaluation carrier
		Indi-vidule	classmate	teacher	enterprise	
	Realistic and pragmatic	√		√		score / comment
	Rigorous and meticulous	√	√	√		score / comment
	Self-control	√	√			Video evaluation
	Compressive capacity	√			√	Video evaluation
Innovation and Entrepre-neurship	Critical thinking	√		√		comment
	Observation and judgment ability	√	√	√		comment
	Divergent thinking	√		√		Video evaluation
	Entrepreneurship awareness	√		√		score / comment
	Innovation and entrepreneurship achievements	√		√		score / comment

Based on the goal mentioned above, a developmental evaluation system for students' key abilities in the context of three integration will be developed, as shown in Table 1.

4.2 Index System and Weight Design Based on Analytic Hierarchy Process

After the establishment of the developmental evaluation system for students' key abilities, it is necessary to further scientifically plan the weight of evaluation indicators [26] in order to clarify the evaluation priorities and highlight the emphasis of the evaluation system on talent cultivation under the new curriculum reform model. Based on the characteristics that AHP can deal with complex decision-making problems with multiple objectives, multiple criteria or no structural characteristics, this method is selected to calculate the index weight and verify its effectiveness [27]. The specific steps are described below.

The first step is to establish a ladder structure according to the characteristics of talents' abilities under the curriculum reform mode of "specialty, creativity and thinking", as shown in Fig. 1.

The second step is to establish the judgment matrix A of each step through expert scoring method, as shown in Formula (1).

$$A \overset{\text{def}}{=} (a_{ij}) \tag{1}$$

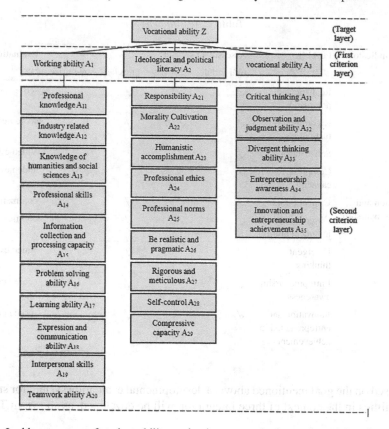

Fig. 1. Ladder structure of student ability evaluation system in the context of three integration

In Formula (1) a_{ij} refers to the importance scale of element i compared with element j. The 1, 3, 5, 7, 9, 2, 4, 6, 8 and their reciprocal in the scale are used to judge the importance of adjacent elements respectively. The larger the value is, the higher the importance of the former is than that of the latter.

Table 2. Basic Level Judgment Matrix

Z	A_1	A_2	A_3
A_1	1	3	3
A_2	1/3	1	1
A_3	1/3	1	1

The scoring experts were scored by 30 students, 8 teachers and 4 enterprise technicians, totaling 42 people. Finally, the following four judgment matrices were obtained, as shown in Tables 2, 3, 4 and 5.

Table 3. A1 Judgment Matrix

A_1	A_{11}	A_{12}	A_{13}	A_{14}	A_{15}	A_{16}	A_{17}	A_{18}	A_{19}	A_{20}
A_{11}	1	3	3	1/5	1/3	1/5	1/5	4	2	2
A_{12}	1/3	1	1	1/5	1	1/7	1/5	2	2	1
A_{13}	1/3	1	1	1/5	1	1/7	1/5	1	1	1
A_{14}	5	5	5	1	3	1/5	1	3	3	3
A_{15}	3	1	1	1/3	1	1/5	1/5	1	1	1
A_{16}	5	7	7	5	5	1	2	3	3	3
A_{17}	5	5	5	1	5	1/2	1	4	4	4
A_{18}	1/4	1/2	1	1/3	1	1/3	1/4	1	1	1
A_{19}	1/2	1/2	1	1/3	1	1/3	1/4	1	1	1
A_{20}	1/2	1	1	1/3	1	1/3	1/4	1	1	1

Table 4. A2 Judgment Matrix

A_2	A_{21}	A_{22}	A_{23}	A_{24}	A_{25}	A_{26}	A_{27}	A_{28}	A_{29}
A_{21}	1	1/3	1	1/5	1/3	1	1	1	1
A_{22}	3	1	2	1	1	2	2	3	2
A_{23}	1	1/2	1	1/5	1/3	1/3	1/3	1	1
A_{24}	5	1	5	1	1	1	2	4	2
A_{25}	3	1	3	1	1	2	1	1	1
A_{26}	1	1/2	3	1	1/2	1	1/3	1	1
A_{27}	1	1/2	3	1/2	1	3	1	3	2
A_{28}	1	1/3	1	1/4	1	1	1/3	1	1
A_{29}	1	1/2	1	1/2	1	1	1/2	1	1

Table 5. A3 Judgment Matrix

A_3	A_{31}	A_{32}	A_{33}	A_{34}	A_{35}
A_{31}	1	2	1	1/3	1/3
A_{32}	1/2	1	1/3	1/3	1/5
A_{33}	1	3	1	2	1/5
A_{34}	3	3	1/2	1	1/3
A_{35}	3	5	5	3	1

The third step is to use SPSSPRO software to complete the calculation of the subsequent AHP, and calculate the eigenvector, weight value, maximum eigenvalue, CI value, CR value, etc. The calculation results are shown in Tables 6, 7, 8 and 9.

Table 6. AHP hierarchy analysis and consistency test results of the first criterion layer

Items	Eigenvector	Weight (%)	Maximum characteristic root	CI	RI	CR	Consistency inspection
Working ability A_1	2.08	60	3	0	0.525	0	Pass
Ideological literacy A_2	0.693	20					
Innovation and Entrepreneurship A_3	0.693	20					

Note: The calculation result of the analytic hierarchy process shows that the maximum characteristic root is 3.0. According to the RI table, the corresponding *RI* value is 0.525, so CR = CI/RI = 0.0 < 0.1, passing the one-time test

Table 7. AHP hierarchy analysis and consistency test results of the second criterion level (working ability)

Items	Eigenvector	Weight (%)	Maximum characteristic root	CI	RI	CR	Consistency inspection
Professional knowledge A_{11}	0.909	6.986	10.987	0.11	1.486	0.074	Pass
Related knowledge A_{12}	0.614	4.721					
Humanities and social knowledge A_{13}	0.535	4.11					
Professional skill A_{14}	2.141	16.461					
Information collection and processing capacity A_{15}	0.725	5.572					
Problem solving ability A_{16}	3.564	27.4					

(continued)

Table 7. (*continued*)

Items	Eigenvector	Weight (%)	Maximum characteristic root	CI	RI	CR	Consistency inspection
Learning ability A_{17}	2.692	20.697					
Communication ability A_{18}	0.568	4.364					
Interpersonal skills A_{19}	0.608	4.677					
Team cooperation skills A_{20}	0.652	5.013					

Note: The calculation result of the analytic hierarchy process shows that the maximum characteristic root is 10.987. According to the RI table, the corresponding RI value is 1.486, so CR = CI/RI = 0.074 < 0.1, passing the one-time test

Table 8. AHP level analysis and consistency test results of the second criterion level (ideological literacy)

Items	Eigenvector	Weight (%)	Maximum characteristic root	CI	RI	CR	Consistency inspection
Responsibility A_{21}	0.655	6.61	9.572	0.072	1.451	0.049	Pass
Morality Cultivation A_{22}	1.737	17.526					
Humanistic accomplishment A_{23}	0.537	5.416					
Professional ethics A_{24}	1.946	19.633					
Professional norms A_{25}	1.379	13.91					
Realistic and pragmatic A_{26}	0.857	8.649					

(*continued*)

Table 8. (*continued*)

Items	Eigenvector	Weight (%)	Maximum characteristic root	CI	RI	CR	Consistency inspection
Rigorous and meticulous A_{27}	1.335	13.473					
Self-control A_{28}	0.672	6.775					
Compressive capacity A_{29}	0.794	8.008					

Note: The calculation result of the analytic hierarchy process shows that the maximum characteristic root is 9.572. According to the RI table, the corresponding RI value is 1.451, so CR = CI/RI = 0.049 < 0.1, passing the one-time test

Table 9. AHP level analysis and consistency test results of the second criterion level (Innovation and entrepreneurship capability)

Items	Eigenvector	Weight (%)	Maximum characteristic root	CI	RI	CR	Consistency inspection
Critical thinking A_{31}	0.74	11.896	5.37	0.093	1.11	0.083	Pass
Observation and judgment ability A_{32}	0.407	6.534					
Divergent thinking A_{33}	1.037	16.667					
Entrepreneurship awareness A_{34}	1.084	17.428					
Innovation and entrepreneurship achievements A_{35}	2.954	47.475					

Note: The calculation result of the analytic hierarchy process shows that the maximum characteristic root is 5.37. According to the RI table, the corresponding RI value is 1.11, so CR = CI/RI = 0.083 < 0.1, passing the one-time test

The fourth step is to calculate the total weight. Fill the weight value calculated by AHP into Table 10, and calculate the final specific weight value of each secondary index to get the calculation result in Table 10.

Table 10. The weight of developmental evaluation indicators of college students' key abilities in the context of three integration

Target layer	Criterion layer I	Criterion layer II	Weight	Weight ranking
vocational ability Z	working ability $A_1(0.6)$	Professional knowledge A_{11} (0.06986)	0.042	5
		Related knowledge A_{12} (0.04721)	0.028	10
		Humanities and social knowledge A_{13} (0.0411)	0.025	13
		Professional skill A_{14} (0.16461)	0.099	3
		Information collection and processing capacity A_{15} (0.05572)	0.033	8
		Problem solving ability A_{16} (0.274)	0.164	1
		Learning ability A_{17} (0.20697)	0.124	2
		Communication ability A_{18} (0.04364)	0.026	12
		Interpersonal skills A_{19} (0.04677)	0.028	10
		Team cooperation skills A_{20} (0.05013)	0.030	9
	Ideological literacy $A_2(0.2)$	Responsibility A_{21} (0.0661)	0.013	18
		Morality Cultivation A_{22} (0.17526)	0.035	7
		Humanistic accomplishment A_{23} (0.05416)	0.011	19
		Professional ethics A_{24} (0.19633)	0.039	6
		Professional norms A_{25} (0.1391)	0.028	10

(*continued*)

Table 10. (*continued*)

Target layer	Criterion layer I	Criterion layer II	Weight	Weight ranking
		Realistic and pragmatic A_{26} (0.08649)	0.017	15
		Rigorous and meticulous A_{27} (0.13473)	0.027	11
		Self-control A_{28} (0.06775)	0.014	17
		Compressive capacity A_{29} (0.08008)	0.016	16
	Innovation and entrepreneurship A_3(0.2)	Critical thinking A_{31} (0.11896)	0.024	14
		Observation and judgment ability A_{32} (0.06534)	0.013	18
		Divergent thinking A_{33} (0.16667)	0.033	8
		Entrepreneurship awareness A_{34} (0.17428)	0.035	7
		Innovation and entrepreneurship achievements A_{35} (0.47475)	0.095	4

4.3 Analysis on the Results of Developmental Evaluation of Students' Key Abilities

It can be seen from the weight value of the first level indicators that, for the evaluation of college students' professional ability, "work ability A1", as the latest basic ability in the workplace, has been widely recognized and is still the main parameter and indicator to measure students' ability development. However, the requirements for "ideological and political literacy A2" and "innovation and entrepreneurship ability A3" are relatively low. On the one hand, there are few requirements for this from the job functions of enterprises. On the other hand, colleges and universities have only begun to pay attention to how to integrate moral education and mass entrepreneurship education into professional education in recent years. The implementation path is still in the process of exploration, and the training mechanism for these two types of abilities is not yet mature. With the development of connotation construction of higher education and the change of enterprises' demand for employees' ability, the proportion of ideological and political literacy and mass entrepreneurship and innovation ability should be gradually increased.

From the perspective of secondary indicators, for the 10 indicators under "working ability A1", the most concerned ones are "problem-solving ability A16", "learning ability A17" and "professional skills A14", which indicates that in terms of working ability, higher education is mainly recognized to cultivate students' ability to deal with problems independently, followed by professional skills. In addition to the above three most important indicators, "Innovation and Entrepreneurship Achievement A35", "Professional Knowledge A11", "Professional Ethics A24" and "Morality A22" are the most concerned. This also poses a new challenge to the development of higher education. The essence of education urges us to gradually abandon the worship of knowledge, and upgrade education evaluation from the once narrow knowledge evaluation to "meaning generation". The diversified and comprehensive value evaluation standard has become the mainstream trend.

5 Conclusion

This paper mainly studies the developmental evaluation system of students' key abilities under several new curriculum reform models, and introduces the analytic hierarchy process to design the index weight. The main research work includes the following aspects.

(1) This paper analyzes the current situation of academic evaluation in colleges and universities in the context of teaching innovation, such as "special integration" and "ideological and political curriculum".
(2) According to the characteristics of the curriculum reform of "professional education, innovation and entrepreneurship education, and ideological and political curriculum", the evaluation principles of key abilities of college students are constructed.
(3) This paper constructs the evaluation system of college students' key abilities, and uses the analytic hierarchy process to scientifically define the index weights. In order to promote the development of students' comprehensive professional ability, a targeted evaluation system has been developed.

There are still some deficiencies in this study. The background is relatively limited, mainly focusing on the curriculum innovation model of "specialization, creativity and thinking". And the indicator system constructed only focuses on key capabilities.

Acknowledgment. This paper is supported by (1) Guangxi Higher Education Undergraduate Teaching Reform Project "Research and Practice of High-Level Academic Evaluation based on the integration of "specialty, creativity and thinking"" (2021JGB431) and (2) Nanning University's College Level Specialty Innovation Integration Teaching Reform Project "the Teaching Reform and Practice of Modern Logistics Equipment Course Combining the Work Process Integration and Specialty Innovation Integration Under the Mixed Teaching Mode" (2020XJZC05).

References

1. Wang, F.: The course design for algorithm analysis and research on improving the students' comprehensive ability. Int. J. Educ. Manage. Eng. (IJEME) **2**(2), 42–46 (2012)
2. Hu, T., Li, X.: The value orientation, blocking factors and promotion strategies of the "integration of expertise and innovation" of application-oriented undergraduate universities. Heilongjiang High. Educ. Res. **40**(12), 127–131 (2022). (in Chinese)
3. Dai, Y.: Research on innovative basic case design path of "thinking, expertise and innovation" integration. Dalian Cadre J. **38**(11), 47–52 (2022). (in Chinese)
4. Boysen, M.S.W., Jansen, L.H., Knage, M.: To share or not to share: a study of educational dilemmas regarding the promotion of creativity and innovation in entrepreneurship education. Scand. J. Educ. Res. **64**(2), 211–226 (2020)
5. Hameed, I., Irfan, Z.: Entrepreneurship education: a review of challenges, characteristics and opportunities. Entrepreneurship Educ. **2**(3), 135–148 (2019)
6. Zhang, Z.: Research on the construction of "accounting foundation" golden curriculum from the perspective of integration of thinking, professional training and innovation. Sci. Technol. **31**, 97–99 (2022). (in Chinese)
7. Liu, X., He, S., Liu, G.: The construction and practice of the talent training mode of higher vocational education with the triple integration of "thinking and innovation, professional innovation and scientific innovation" in the context of smart cities - taking computer application technology as an example. J. Guangdong Commun. Vocat. Tech. Coll. **21**(03), 53–57+61 (2022). (in Chinese)
8. Da, L., Chen, T., Wang, L.: The construction and practice of "thinking, expertise and innovation" integrated innovation and entrepreneurship education system based on OBE concept. Comput. Educ. **08**, 128–132 (2022). (in Chinese)
9. Ji, D., Liu, H.: Research on the design of the teaching work page of metal materials, which integrates professional, creative and ideological education. Ind. Technol. Forum **21**(15), 165–166 (2022). (in Chinese)
10. Wang, J.: The development and practice of the loose leaf teaching work page of the integration of specialty, creativity and education – taking the course "Welding Structure Production" as an example. Sci. Technol. **21**, 41–43 (2022). (in Chinese)
11. Ji, D., Zhang, H.: Research on the micro organization teaching design of metal materials based on the integration of professional, creative and ideological education. Ind. Technol. Forum **21**(14), 232–233 (2022). (in Chinese)
12. Zhu, L., Duan, Z.: From knowledge worship to meaning generation: a new interpretation of students' academic evaluation. Contemp. Educ. Sci. **05**, 27–33 (2022). (in Chinese)
13. Yan, B., Zhang, X.: Research on characteristics of applied undergraduate education quality evaluation system – based on the theoretical background of developmental education evaluation. Teachers' Expo **36**, 4–7 (2021). (in Chinese)
14. Jiang, S.: Green evaluation: transformation and exploration of academic evaluation. Jiangsu Educ. **87**, 61–63 (2021). (in Chinese)
15. Kuang, S., Lan, Y., He, M., Lu, Y., He, Z.: Assessment of academic intelligence: current situation and trend. Educ. Inf. Technol. **Z2**, 8–14 (2021). (in Chinese)
16. Wang, J.: Core literacy oriented academic assessment framework: Australia national science literacy assessment project. China Exam **04**, 60–67 (2021). (in Chinese)
17. Tan, L., Zhang, Y.: Preliminary exploration and practice of scientific modeling ability examination in large-scale academic assessment. Shanghai Educ. Res. **03**, 27–32 (2021). (in Chinese)

18. Dong X.: Research on the integration of curriculum ideological and political education, innovation and entrepreneurship education into professional curriculum system in higher vocational colleges under the new engineering background. In: MATEC Web of Conferences. EDP Sciences, p. 355 (2022)

19. Zhao, X., Zhang, J.: The analysis of integration of ideological political education with innovation entrepreneurship education for college students. Front. Psychol. **12**, 610409 (2021)

20. Liu, H., Zhang, P., Pan, J.: How to make the results of academic evaluation more effective - Research on the adjustment model based on latent variables. J. East China Normal Univ. (Educ. Sci. Ed.) **36**(03), 87–98+169 (2018). (in Chinese)

21. Huang, J.: The new trend of international basic education science assessment – a comparative study based on PISA and TIMSS. Training Prim. Secondary Sch. Teachers **09**, 74–78 (2017). (in Chinese)

22. Feistauer, D., Richter, T.: How reliable are students' evaluations of teaching quality? A variance components approach. Assess. Eval. High. Educ. **42**(8), 1263–1279 (2017)

23. Sun, G.: Research on academic assessment of Australian Higher Education – a case study of the Royal Melbourne University of Technology. Explor. High. Vocat. Educ. **16**(04), 11–17 (2017). (in Chinese)

24. Annabestani, M., Rowhanimanesh, A., Mizani, A., et al.: Fuzzy descriptive evaluation system: real, complete and fair evaluation of students. Soft Comput. **24**(6), 3025–3035 (2020)

25. Zeng, Y., Qu, Y., Chen, T., Liu, R.: The academic assessment system of logistics management professional ability evaluation and key ability training. China Market **41**, 163–166 (2010). (in Chinese)

26. Shen, L., Yang, J., Jin, X., et al.: Based on Delphi method and analytic hierarchy process to construct the evaluation index system of nursing simulation teaching quality. Nurse Educ. Today **79**, 67–73 (2019)

27. Shi, J., Dong, C., Hou, J.: Research and design of teaching evaluation system based on fuzzy model. Int. J. Educ. Manage. Eng. (IJEME) **2**(10), 45–51 (2012)

Exploration and Practice of the Evaluation System of College Teaching Quality Based on MSF Mechanism in OBE Perspective

Yanzhi Pang and Jianqiu Chen[✉]

School of Transportation, Nanning University, Nanning 530200, Guangxi, China
153862839@qq.com

Abstract. In response to the requirements of "all-staff, whole-process and all-round" education, colleges and universities have increasingly higher requirements for the quality of classroom teaching. From the former "teaching" on the standpoint of teachers to the present student-centered, improve students' autonomous learning ability, and pay increasing attention to the quality of curriculum teaching. The data of supervision and evaluation of teaching, peer assessment and student evaluation of teaching are also receiving increasingly more attention. Based on the OBE concept and focusing on the development of students, this paper constructs the evaluation system of college curriculum teaching quality, and optimizes the evaluation indicators. It should not only show solicitude for the achievement of knowledge and ability objectives, but also pay attention to quality objectives, emphasize the combination of teaching with the development of industry and enterprises, optimize the curriculum design of talent training programs, and reflect and improve students' learning interest and enthusiasm in the classroom.

Keywords: OBE · MSF mechanism · Teaching quality · Evaluation system

1 Introduction

Quality is the guarantee of education success [1]. It is not enough to call teaching quality as the lifeline of schools, which has been agreed by all countries in the world [2–4]. Teaching quality plays an important role in ensuring the sustainable development of the school, shaping the successful life of students, and strengthening the comprehensive national strength of the country [5, 6]. Focusing on the growth of students, the OBE concept has been widely accepted and promoted in different disciplines [7–9].

The so-called "golden class" construction requirements of "high-level, innovative and challenging - abbreviated as HIC" need to be completed by teachers and students. It is urgent to establish a multi-dimensional evaluation system of college curriculum teaching quality [10–12]. At present, the teaching quality evaluation system pays more attention to teachers' teaching, neglects students' learning, emphasizes knowledge teaching and ignores ideological guidance, lacks logic rigor, teachers' teaching characteristics are not obvious, and the evaluation system is not sustainable.

Z. Hu et al. (Eds.): ICCSEEA 2023, LNDECT 181, pp. 901–912, 2023.
https://doi.org/10.1007/978-3-031-36118-0_77

In order to deal with the above problems, this paper intends to optimize the under-graduate course teaching quality evaluation table by establishing the framework of the course teaching quality evaluation system based on the OBE concept, introduce the MSF mechanism, and continuously improve the teaching quality evaluation indicators from both teachers and students. Through the improvement of teaching mode, teaching method and teaching design, improve the quality of teachers' "teaching" and the degree of students' "learning".

According to the strategic goal of "students are satisfied with the learning effect, and employers are satisfied with the quality of training talents", the students' practical ability, innovation ability, and the ability to serve enterprises in the industry and the society as the main criteria for evaluating the teaching quality, so as to set the evaluation indicators and control the classroom teaching quality in an all-round way [13–15]. According to the specific requirements for the specific measures of teaching quality evaluation in colleges and universities in the Overall Plan for Deepening the Reform of Education Evaluation in the New Era issued in 2020, as the basis for revising the teaching evaluation indicators, improving the efficiency of education and teaching evaluation, and reforming the current management of teaching quality in higher education [16–18].

2   Main Problems and Cause Analysis of Current Classroom Quality Evaluation

With the specialty and curriculum construction as the core, and the integration of pro-duction and education as the center, we should reform the training mode of application-oriented talents and cultivate high-quality application-oriented talents. Therefore, the main indicator of the evaluation of classroom teaching quality in colleges and uni-versities is the quality of talent training. For curriculum construction, the Ministry of Education proposed that colleges and universities should eliminate "water courses" and build "gold courses". And high order, innovation and challenge. To ensure the teaching quality of the "Golden Class" project, it is necessary to propose quantifiable evaluation criteria, optimize the evaluation indicators, and reflect the complementary process of teacher and student growth [19]. At the same time, the evaluation system of teaching quality is related to the quality of students, teachers, orientation of running a school and characteristics of talent cultivation in colleges and universities. With the gradual deep-ening of teaching reform, the existing evaluation system also highlights some common problems.

2.1   Main Problems in Current Classroom Quality Evaluation

The main problems in the evaluation of classroom quality currently are summarized as follows.

(1) Focus on "teaching" rather than "learning"

The traditional teaching process and methods have been gradually changed. Despite the teaching reform in recent years, cramming teaching has been reduced, but in the selection of classroom teaching quality evaluation indicators, too much attention is still paid to teachers. The selection of indicators also mainly depends on the preparation of teaching materials, the implementation of teaching plans, the relationship between teaching content and the forefront of the industry, and other aspects, which does not fully reflect the "student development centered" teaching philosophy, The real sense of teaching students according to their aptitude has not been realized. The selection of classroom quality evaluation indicators is one-sided, ignoring the impact of teaching objectives, teaching content, teaching methods, etc. on students' learning interest and enthusiasm, and lacking the measurement of students' effectiveness.

(2) Value knowledge over thought

In the process of teaching, teachers usually define knowledge objectives, but with the introduction of "HIC", the requirements for students' quality objectives and emotional objectives are also on the agenda. In terms of knowledge objectives, the teaching content is required to have certain depth and extensibility; The ability goal requires a breakthrough in career selection and ability expansion; Emotional objectives For engineering students, it is required to add moral elements to the teaching content, including the integration of patriotism, craftsmen of big countries, etc. At present, there is little consideration of the ideological and political content of the curriculum in the evaluation indicators of classroom teaching quality, and the effectiveness is basically lack of examination points.

(3) Insufficient layering and logic

The establishment of classroom teaching quality evaluation system indicators needs certain scientific basis. Most of the previous evaluations were based on pre-vious experience, and there was no clear basis for setting the weights of indicators. At the same time, the correlation and level of indicators were not obvious.

(4) The teaching characteristics of teachers are not prominent

Some teachers rely too much on PPT when teaching, and the teaching content is not updated in time. They can not actively find, find and fully reflect new ideas, new ideas, new methods and new technologies in the classroom. The content of teaching materials is relatively lagging behind, resulting in some teaching content being out of touch with the reality of social development in varying degrees. There are also some teachers who are not active enough in using modern information technology, resources and means and have weak operational ability. The research on new classroom teaching modes such as "Internet plus", "micro class" and "flipped classroom" is not deep enough, and some high-quality online teaching resources introduced by the school do not give full play to their effectiveness. At the same time, teachers have different teaching styles because of their different professional backgrounds. The improvement of teaching quality needs to guide and encourage teachers to form their own teaching characteristics. Among the current evaluation indicators of classroom teaching quality, there is less evaluation on teachers' teaching method innovation.

(5) Lack of long-term, standardized and scientific evaluation and improvement system

At the end of the semester, most colleges and universities will require students to evaluate the teaching of all courses anonymously, which has a certain promotion effect on the improvement of teaching quality, but because the evaluation indicators are not accurate enough, the process is relatively lagging behind, and the students' subjectivity is too large, the results can not objectively reflect the teaching effect, and the follow-up improvement measures are not continuous and closed loop, and there is a lack of feedback on the improvement results.

2.2 Cause Analysis

Although the teaching quality evaluation system of colleges and universities has been continuously improved, the degree of attention needs to be strengthened. The evaluation indicators of classroom teaching quality in colleges and universities are closely related to the school's top objectives, professional training objectives, and curriculum achievement objectives, but the degree of fit and goal achievement is not high enough, so it can not fully support the achievement of learning objectives. The evaluation object is still mainly teachers, and the results can only be fed back to relevant teachers after finishing the evaluation at the end of the semester. The course of the next semester is different from that of the current semester, which leads to the open-loop teaching evaluation. Problems found in the teaching process cannot be fed back and improved in time, and the second evaluation result of a certain course by the same group of students cannot be obtained. Such evaluation of teaching is not sustainable. Without the interaction between teachers and students, it is difficult to really play its role.

3   Construction of Teaching Quality Evaluation System Based on OBE Concept

3.1   Significance of Construction of Teaching Quality Evaluation System

To build the evaluation system of classroom teaching quality, we must follow the concept of OBE, adhere to the student-centered principle, timely communicate the information of teaching and learning, give full play to the role of students in the evaluation of classroom teaching effect, establish a scientific view of quality, and form a more scientific and effective evaluation system [20]. At the same time, it can urge teachers to find the deficiencies in the teaching process in time, pay more attention to teaching, research teaching, constantly optimize the teaching process, and improve the teaching quality [21].

By optimizing the evaluation indicators and meeting the requirements of higher education teaching quality management policy guidelines in the new era, the implementation of the system can be improved, so that the key factors and key links affecting the teaching quality are always under control in the whole process of talent training, so as to survive with quality and seek development with quality. Build and gradually improve the teaching quality evaluation system, and form a long-term mechanism to continuously improve the quality of talent training [22–24].

3.2 Formation Process of Classroom Teaching Quality under OBE Concept

The OBE concept emphasizes output-oriented, and based on the orientation of the school, requires the development of a reasonable talent training program. Each course should serve the professional training objectives. The teaching objectives of each class and the learning objectives of students need to support the achievement of professional talent cultivation. The formulation and implementation of teaching plans must also be consistent with them. The selection of classroom teaching evaluation indicators should be consistent with the teaching implementation, and the established requirements should be finally completed, as shown in Fig. 1.

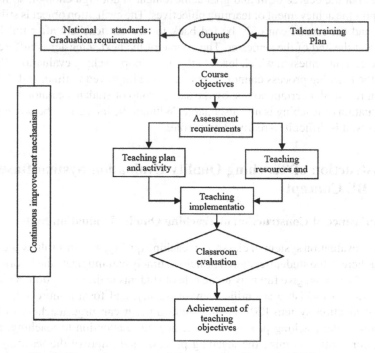

Fig. 1. Formation process of classroom quality under OBE concept

3.3 Evaluation Indicators of Undergraduate Classroom Teaching Quality

Optimize the existing undergraduate course teaching quality evaluation form from the aspects of teaching concept, teaching content, teaching mode, teaching effect, teaching characteristics, etc. The design principles of evaluation indicators are shown in Table 1.

Table 1. Evaluation form of undergraduate class teaching quality

Teaching quality evaluation	Teaching philosophy	Meet the requirements of disciplines and courses
		Pay attention to the ideological education in curriculum, and reflect the cultivation of morality and talents
		Keep "student-centered, output-oriented and continuous improvement"
	Resources material	Carry teaching materials; Master teaching hour to consistent with the schedule
	Teaching attitude	Take classes on time; Lecture with full spirit; Be strict with students and keep the classroom in good order
	Content of courses	With HIC characteristics, in line with the requirements of the syllabus
		The teaching content has depth and breadth, reflects the frontier of the discipline, and penetrates professional ideas
		Highlight key and difficult points, reasonable logic, clear structure and complete teaching links
		Clear learning objectives, reflecting Bloom's educational ideas, and multi-level curriculum design
		The goal of moral education is clear, guiding students to establish ideals and beliefs as well as correct outlook on world, life and values
	Teaching implementation	Effectively integrate ideological and moral content into the curriculum
		Clear teaching, combined with industry development trends
		The teaching methods are rich, and various media and means are used to stimulate students' interest in learning, with good results

(*continued*)

Table 1. (*continued*)

		Observe students' reactions and carry out interactions, discussions, exercises, activities, etc.; Cultivate students' autonomous learning ability
		Reasonable and effective use of modern information technology to support teaching innovation
	Teaching effective-ness	The teaching is attractive, the classroom atmosphere is harmonious, the students are active in thinking and deeply participate in the classroom
		Students have a high attendance rate, listen carefully, take notes, and actively participate in learning activities such as interaction and practice
		Realize the preset learning objectives and improve students' knowledge, ability and quality; The assessment method is reasonable
	Teaching features	Effectively reflect the OBE concept
		Has obvious innovative characteristics
		Make full use of modern information technology to carry out course teaching activities and learning evaluation

4   Test and Evaluation of Teaching Quality Under MSF Mechanism

4.1   Meaning and Role of MSF

MSF is the abbreviation of Midterm Student Feedback. It is a service provided by the teaching management department for teachers in the whole school. It aims to help teachers find problems and find solutions from the perspective of students, so as to improve teachers' teaching level and classroom teaching effect.

MSF is a tool and method for teaching evaluation and diagnosis, which is generally carried out in the early or middle stage of the course. The main body of participation is teachers. Through the teaching consultation activities, the relevant professional teaching consultants and teachers communicate one-on-one, and participate in the teaching activities of teachers synchronously. They collect the information of teachers and students in a multi-dimensional way, collect the real data of students, collect the concerns of students, help teachers reflect on teaching, improve teaching strategies and methods, and optimize the quality of classroom teaching. The flow of MSF is shown in Fig. 2.

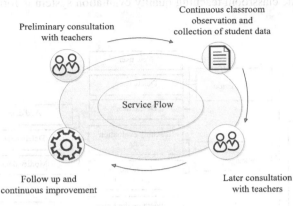

Fig. 2. MSF process

4.2 MSF Process and Steps

The implementation process of MSF mainly includes four steps: early talks, course observation, late talks and late tracking.

4.2.1 Preliminary Talks (30 m)

The counselor understands the basic information of the course and class, informs the teacher of the specific practice of MSF, lets the teacher understand the whole process, and establishes a relationship of mutual trust. This process takes about 30 min.

4.2.2 Class Observation (55–90 m)

The counselor will go to the classroom for 1–2 lessons (perhaps 50–90 min) at the appointed time. The teacher will set aside about 15 min. The counselor will organize students to talk and collect students' feedback on the course and class (anonymous) when the teacher leaves.

4.2.3 Later Talks (30 m)

During the 30-min conversation, the counselor will submit an analysis report to the teacher on the classroom observation and students' feedback, and discuss with the teacher how to improve the teaching and classroom effect.

4.2.4 Post Tracking

At the end of the semester, the counselor will further understand the classroom improvement from teachers and students, evaluate the satisfaction of MSF service, and put forward suggestions for improvement.

4.3 Construction of Classroom Teaching Quality Evaluation System Model

Based on the existing data and combined with the required basic functions, the overall framework of the classroom teaching quality evaluation system is formed as shown in Fig. 3.

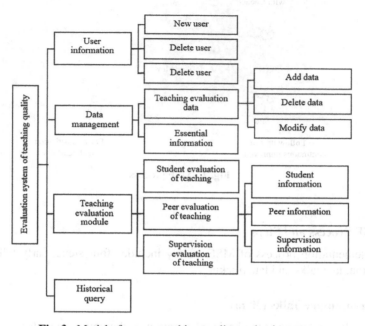

Fig. 3. Model of course teaching quality evaluation system

The curriculum teaching quality evaluation system model mainly includes four parts: user information, data management, teaching evaluation module and history query. Query and modify the data of students' evaluation of teaching, peer evaluation of teaching and supervision evaluation of teaching in order to help teachers improve the quality of curriculum teaching.

Through research and material collection, analyze the problems raised in Sect. 2.1, organize experts to discuss and deduce, and propose the implementation plan of education reform based on OBE concept. Introduce the MSF mechanism, implement it in the classroom and promote it in an all-round way to improve the degree of achievement of teachers' "teaching" and students' "learning". The specific implementation process is shown in Fig. 4.

4.4 Case Analysis

Optimize the curriculum system from the following four aspects: (1) Combine the current informatization characteristics, add courses related to transportation informatization, learn the frontier knowledge of disciplines, and reflect the interdisciplinary characteristics, such as intelligent transportation system. (2) Combined with the development trend

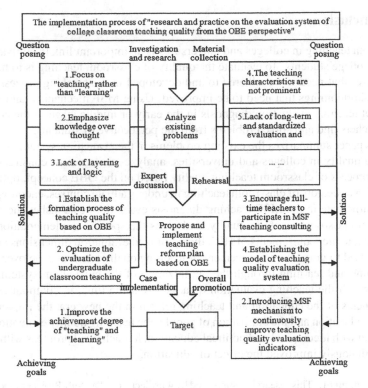

Fig. 4. Specific implementation process

of the industry, add comprehensive transportation courses, such as comprehensive transportation. (3) Integrating geographical features, Thai courses are offered to prepare for serving ASEAN and the the Belt and Road, such as applying Thai and ASEAN culture. (4) A template for achieving carbon peak and carbon neutrality, and green transportation courses, such as traffic safety, are added.

Combined with curriculum optimization, the introduction of MSF mechanism has greatly improved teachers' teaching level, mobilized students' learning interest and initiative, and improved students' evaluation scores for project quality, as shown in Table 2.

Table 2. Data of students' evaluation of teaching in recent three years

Academic year	Total number of courses	Number of "excellent" courses	Number of "good" courses	Number of "medium" courses	Number of "poor t" courses	Excellence rate
2018–2019	62	51	11	0	0	82.26%
2019–2020	59	50	9	0	0	84.75%
2020–2021	219	195	20	4	0	89.04%

5 Conclusion

Classroom teaching in colleges and universities is an important link in cultivating high-quality college students. To evaluate the quality of classroom teaching is to find out the advantages that need to be adhered to and developed in the teaching process, and find out the shortcomings that need to be improved. Using MSF mechanism and method to carry out teaching evaluation diagnosis in the early or middle stage of the course is to help teachers find and solve problems from the perspective of students.

This paper summarizes the existing problems in the evaluation system of classroom teaching quality in colleges and universities, analyzes the reasons, constructs the formation process of classroom teaching quality based on the OBE concept, optimizes the evaluation indicators of classroom teaching in undergraduate courses, encourages teachers to participate in the MSF teaching diagnosis evaluation, establishes the evaluation system of classroom teaching quality, and gives the specific implementation process. According to the requirements of HIC, the optimized evaluation dimensions and indicators are used to guide the improvement of teaching quality. Improve the overall quality of teaching and learning through the optimization of the curriculum system, and verify it through the teaching evaluation data. At the same time, in compliance with the requirements of the education and teaching reform in the new era, the implementation of moral education and the cultivation of people, the teaching quality evaluation system, combined with ideological and political education, has achieved progress with the times and continuously improved the effect of education.

Acknowledgment. This research is supported by 4 projects: (1) The Sub-Project of the Construction of the China-ASEAN International Joint Laboratory for Integrated Transportation (Phase I): "Research on the Technology of Fully Automatic Operation Training" (GK-AA21077011-7); (2) Guangxi Higher Education Undergraduate Teaching Reform Project: "Research and Practice on the Evaluation System of Curriculum Teaching Quality in Applied Universities from the OBE Perspective" (2022JGA390); (3) Nanning University Curriculum Ideological and Political Demonstration Specialty Construction -Transportation Specialty- Project" (2022SZSFZY03); and (4) Core Undergraduate Course of Nanning University: "Train Operation Control System" (2020BKHXK16).

References

1. Hamada, H.: Action research to enhance quality teaching. Arab World Engl. J. **1**, 4–12 (2019). Conference Proceedings
2. Zhiqin, L., Jianguo, F., Fang, W., Xin, D.: Study on higher education service quality based on student perception. Int. J. Educ. Manag. Eng. (IJEME) **2**(4), 22–27 (2012)
3. Kong, L.A., Wang, X.M., Yang, L.: The research of teaching quality appraisal model based on AHP. Int. J. Educ. Manag. Eng. (IJEME) **9**(29), 29–34 (2012)
4. Weiss, K.A., McDermott, M.A., Hand, B.: Characterising immersive argument-based inquiry learning environments in school-based education: a systematic literature review. Stud. Sci. Educ. **58**(1), 15–47 (2022)
5. Nawaz, R., Sun, Q., Shardlow, M., et al.: Leveraging AI and machine learning for national student survey: actionable insights from textual feedback to enhance quality of teaching and learning in UK's Higher Education. Appl. Sci. **12**(1), 514 (2022)

6. Lindgreen, A., Di Benedetto, C.A., Brodie, R.J., Zenker, S.: Teaching: how to ensure quality teaching, and how to recognize teaching qualifications. Ind. Mark. Manag. **100**, 1–5 (2021)
7. Shi, G., Niu, D., Zhou, X.: Construction of curriculum teaching quality evaluation system based on AHP in OBE perspective. J. Changchun Inst. Eng. (Soc. Sci. Edn.) **22**(04), 81–84 (2021). (in Chinese)
8. Zhu, H., Cao, X., Cao, N., et al.: Exploration and practice of teaching quality evaluation system based on OBE concept. Financ. High. Educ. Res. **6**(02), 117–128 (2021). (in Chinese)
9. Meng, F., Wen, X., Gao, L., Yang, Y.: Research on teaching quality evaluation model based on OBE concept. Sci. Technol. Wind **16**, 37–38 (2021). (in Chinese)
10. Wu, Z., Cao, J., Fan, D.: Construction of teaching quality evaluation model in universities based on Kano model. Heilongjiang Educ. (High. Educ. Res. Eval.) **12**, 21–24 (2021). (in Chinese)
11. Wang, Y.: Research on the evaluation index system of teachers' connotative teaching in colleges and universities. J. Shenyang Agric. Univ. (Soc. Sci. Edn.) **22**(06), 720–725 (2020). (in Chinese)
12. Zhuang, W., Xing, F., Fan, J., Gao, C., Zhang, Y.: An integrated model for on-site teaching quality evaluation based on deep learning. Wirel. Commun. Mob. Comput. **2022**, 1–13 (2022)
13. Skedsmo, G., Huber, S.G.: Measuring teaching quality, designing tests, and transforming feedback targeting various education actors. Educ. Assess. Eval. Account. **32**(3), 271–273 (2020)
14. Chen, B., Liu, Y., Zheng, J.: Using data mining approach for student satisfaction with teaching quality in high vocation education. Front. Psychol. **12**, 1–8 (2022)
15. Wang, C.: Research and analysis of teaching quality evaluation system of Ningbo Dahongying College. Yunnan University, Kunming (2015). (in Chinese)
16. Chen, S., Chen, M.: Research on the optimization and upgrading of the operating mechanism of the university teacher development center - based on the perspective of the "Internet plus" era. J. Jimei Univ. (Educ. Sci. Edn.) **20**(05), 1–8 (2019). (in Chinese)
17. Yang, J.: Cultivating the ability of scientific inquiry and evidence reasoning in experimental teaching. Minnan Normal University, Zhangzhou (2021). (in Chinese)
18. Yin, S., Ren, Y.: Research on the "3-all" teaching quality monitoring model of local colleges and universities. Inner Mongolia Educ. **02**, 103–104 (2020). (in Chinese)
19. Lee, H., Lee, G.: Reconstructing education and 'knowledge' in the twentieth and twenty-first century: scientific management, educational efficiency, outcomes based education, and the culture of performativity. Educ. Soc. **32**(2), 63–96 (2014)
20. Priya Vaijayanthi, R., Raja Murugadoss, J.: Effectiveness of curriculum design in the context of outcome based education (OBE). Int. J. Eng. Adv. Technol. **8**(6), 648–651 (2019)
21. Lin, M.: Curriculum reform of employment-oriented "design of machinery" based on OBE education concept. Front. Educ. Res. **5**(9), 55–59 (2022)
22. Wei, Y., Xu, Y.: Research on the function and operation mechanism of the teacher development center in higher vocational colleges. Career **28**, 98–99 (2016). (in Chinese)
23. Hao, S., Liu, R., Xie, X.: Optimization and exploration of training programs for transportation majors based on OBE - taking application-oriented undergraduate universities as an example. Educ. Teach. Forum **51**, 137–140 (2022). (in Chinese)
24. Kaviyarasi, R., Balasubramanian, T.: Exploring the high potential factors that affects students' academic performance. Int. J. Educ. Manag. Eng. (IJEME) **8**(6), 15–23 (2018)

Research on Teaching Application of Course Resources Based on Bloom's Taxonomy of Educational Objectives

Hao Cai[✉], Jing Kang, Qiang Zhang, and Yue Wang

Wuhan Railway Vocational College of Technology, Wuhan 430205, Hubei, China
caihao@wru.edu.cn

Abstract. With the continuous construction of digital campus in Chinese universities, campus information technology and classroom teaching are increasingly integrated, and there are fundamental changes in teachers' teaching methods and students' learning methods and various teaching methods and theories have also been popularized and applied on a larger scale. With the goal of improving teaching quality and relying on the construction and application of campus information platform, Wuhan Railway Vocational and Technical College has applied the concept of student-centered teaching and practiced student-centered teaching applications and research. This paper puts forward a multi-factor regression analysis method of teaching objectives to analyze and evaluate teaching effects by using students' learning cognitive process data. This method firstly uses container technology to quickly deploy course resources suitable for teaching needs, and collects student learning sample data of classroom teaching. Then, Bloom's taxonomy of educational objectives is used for reference, which sums up the relationship of students' learning cognitive process. Finally, this paper fits the sample data and proves the effectiveness and feasibility of achieving the "optimal" teaching objects classification relationship between different courses and classes based on the multiple regression mathematical model.

Keywords: Taxonomy of educational objectives · K8S · Digital campus · Course resources · Teaching effects

1 Introduction

The continuous development of network information technology has expanded more space for the application of education and teaching, and provided more convenience for the sharing of educational resources. At present, in the classroom teaching of Chinese universities, the total amount of information transmitted by teachers to students is often greater than the learning development ability students can match. As a result, classroom teaching focuses on knowledge transfer, neglects students' subjective initiative, and doesn't pay attention to the growth of students' learning ability, and doesn't match with the achievement of teaching objectives.

With the in-depth development of education information reform and the increasingly integration of network information technology and classroom teaching, the teaching methods of teachers and learning methods of students in higher vocational colleges in Hubei Province have also changed fundamentally [1]. Taking the opportunity for building the "Double High-levels Plan" colleges and universities of the Department for Education, Wuhan Railway Vocational College of Technology has launched Information construction of digital campus in an all-round way. Relevant application systems involve all aspects of daily teaching, scientific research, management, and life services of the college.

In the construction and development of campus informatization, our college has fully considered the needs of teachers and students for educational informatization integration. In the classroom teaching reform, with the goal of improving the teaching quality, based on the student-centered teaching concept, the promotion of the application of digital campus system platform has made a beneficial attempt to promote students to effectively aggregate the fragmented time of campus life.

Since 2019, our college has promoted and used the Vocational Education Cloud Platform of Higher Education Press in teaching resource management and classroom teaching services. In the application of course resources teaching based on information technology, our college guides teachers to carry out teaching method reform activities, evaluates teaching effects and summarizes teaching experience based on Bloom's taxonomy of educational objectives.

However, due to various factors such as course resources, class members, teachers' personalized differences in teaching etc., it is difficult for those methods and theories to be used to analyze and evaluate students' learning effects in a short time, and to help teachers make teaching adjustments suitable for such courses, as well as to form "rules" of effect evaluation that are easy to adjust and observe. It is also difficult to "precipitate" teaching methods and experiences that can be shared and used for reference. At the same time, there are few studies on using multi-factor regression analysis techniques to find a method of Bloom's taxonomy of educational objectives. Therefore, through the key activities of students' learning cognitive process, we recorded quantifiable phased evaluation data item by item, and proposed a multi-factor regression analysis method of educational objectives to explore the data correlation of cognitive activities and evaluate teaching effects and students' learning ability.

This method firstly uses container technology to quickly deploy course resources suitable for teaching needs, and collects student learning sample data in classroom teaching. Then, it rationally draws on Bloom's taxonomy of educational objectives, and concludes the relationship of students' learning cognitive process. Finally, this paper fits the sample data of learning cognitive process, and verifies the feasibility of achieving the "optimal" taxonomy relationship of educational objectives for different courses and classes based on the multiple regression mathematical model.

2 Theory and Technology

With the help of K8S container technology, our college has widely implemented and applied the digital campus platform among all teachers and students, continuously paying attention to the experience of teachers and students' course application, actively responding to the teaching needs of teachers and students for course resources, timely adjusting the arrangement and configuration of various resources in classroom teaching, and rapidly realizing the deployment of course resources. In the teaching of course resources, the basic theory of teaching methods based on Bloom's taxonomy of educational objectives is practiced. With the help of information technology, students' learning sample data has been effectively collected in all teaching links, providing data support for subsequent data analysis and evaluation using multiple regression models.

2.1 Bloom's Taxonomy of Educational Objectives

Bloom, a famous American psychologist, and his team published the book Taxonomy of educational objectives: The Cognitive Domain in 1956. They put forward the theory of classified teaching, and divided educational objectives into three parts, namely, cognitive domain, emotional domain and motor skill domain [2, 3]. In 2001, Anderson and Krathowohl revised Bloom's Taxonomy theory and divided the cognitive field into two dimensions: knowledge dimension and cognitive process dimension. Namely, verbs are used to describe the cognitive process; Use nouns to describe the knowledge that learners are required to master. According to the development law of personal cognitive ability from simple to complex, from low to high, cognitive fields are divided into six basic categories: remembering, understanding, applying, analyzing, evaluating and creating [4, 5] (Table 1).

Table 1. Bloom's taxonomy table

The Knowledge Dimension	The Cognitive Dimension					
	Remembering	Understanding	Applying	Analyzing	Evaluating	Creating
Factual knowledge	Listing	Summarizing	Classifying	Ordering	Ranking	Combining
Conceptual knowledge	Describing	Interpreting	Experimenting	Explaining	Assessing	Planning
Procedural knowledge	Tabulating	Predicting	Calculating	Differentiating	Concluding	Composing
Meta-cognitive knowledge	Using Appropriately	Executing	Constructing	Achieving	Taking action	Actualizing

In 2015, Nancy E. Adams also believed that teachers can encourage students to learn by themselves through setting teaching goals with verbs, and helped them to carry out high level cognitive thinking activities. It highlights the characteristics of the integration of learning, teaching and evaluation, and is conducive to improving the classroom teaching effect [6].

2.2 K8S Container Management Platform

Kubernetes is a portable, extensible, open source platform. The name Kubernetes originates from Greek, meaning helmsman or pilot. K8S as an abbreviation results from counting the eight letters between the "K" and the "S". Google open-sourced the Kubernetes project in 2014 [7] (Fig. 1).

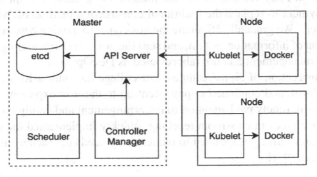

Fig. 1. K8S Global Architecture

In early time, organizations ran applications on physical servers. There was no way to define resource boundaries for applications on a physical server, and this caused resource allocation issues. Because each application runs on a different physical server, it doesn't scale as resources were underutilized, and it is expensive for organizations to maintain many physical servers. Virtual Machine (VM for short) technology can make good use of physical server resources, and because it can easily add or update applications, it can have higher scalability and lower hardware costs. Containers are similar to VMs, but they have relaxed isolation properties to share the Operating System (OS) among the applications. Therefore, containers are considered lightweight. Similar to a VM, a container has its own file system, share of CPU, memory, process space, and so on. As they are decoupled from the underlying infrastructure, they are portable across clouds and OS distributions (Fig. 2).

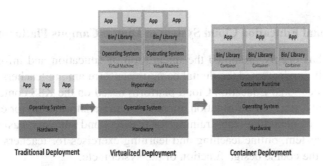

Fig. 2. Evolution process of application deployment mode

In the previous promotion and use experience of classroom teaching, our college usually constructs a business system based on virtual machine resources to carry out teaching activities. However, this mode is often faced with problems such as the delayed response to the actual application of teachers and students, poor scalability, and waste of hardware resources.

Based on the realistic situation of digital campus, our college selects Vocational Education Cloud Platform and adopts the method of "container + micro service" for system deployment to replace the traditional construction method of "virtual machine + single server". We gradually eliminate the non-synchronization between construction and application of information system, and start the practice of integrating school system construction, management and use (abbreviated as DevOps) [8].

In the construction of digital campus, our college has built a K8S container management platform, and continued to pay attention to the user experience of the digital campus platform, timely paid attention to the arrangement and configuration of various information resources in classroom teaching, quickly implemented the deployment of applications, and actively responded to the teaching needs of teachers and students for course resources [9] (Fig. 3).

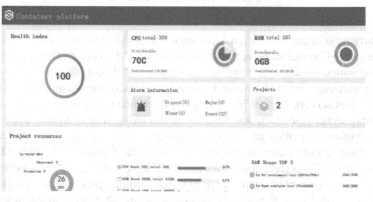

Fig. 3. K8S container platform

2.3 Vocational Education Cloud System of Digital Campus Platform

Our college has deeply promoted the integration of education and information technology, widely implemented the digital campus platform among teachers and students, applied the Vocational Education Cloud platform based on the K8S container systems, and carried out a series of information upgrading and transformation for existing online course resources including team training for teachers and students, basic operation of the learning system, online teaching and learning exercises for teachers and students, familiarizing the course design function of the system, etc.

2.4 Multi-factor Regression Analysis Method of Education Objective

2.4.1 Educational Objective Factors Analysis Model

In the teaching work for students, we gradually realize that there is a correlation and restriction between various "verbs" in the dimension of students' learning cognitive process. However, there will be some differences in the teaching of each course. It often takes dozens or even hundreds of times of experimental teaching to get the teaching experience that is more in line with expectations. In order to achieve the goal of "quickly" aggregating the optimal taxonomy relationship of teaching objectives in relevant courses and minimizing the workload of experiments, we adopt the method of Multi-factor regression analysis, select a multiple system that can handle a large number of random data, and establish a reasonable regression model using function transformation.

With the popularization of computer application, the Multi-factor regression analysis, which has a lot of calculation work, becomes easy to realize. Multi-factor regression analysis method can determine whether there is a correlation between specific variables. For many independent variables that jointly affect a dependent variable, this method can find out which are important factors, which are secondary factors, and what is the relationship between these factors. Moreover, it can predict or control the value of another variable according to the value of one or several variables, and can predict or control the possible accuracy.

Multivariate linear regression considers the linear relationship:

$$Y = \beta_0 + \beta_1 X_1 + \beta_2 X_2 + \cdots + \beta_m X_m + \varepsilon \tag{1}$$

Between dependent variables Y and multiple independent variables $X_1, X_2, \cdots, X_m$. Among them $\beta_0, \beta_1, \cdots, \beta_m$ are unknown parameters, $X_1, X_2, \cdots, X_m$ are m known variables that can be accurately measured and controlled, and ε are random errors.

The following data relationships can generally be assumed:

$$E(\varepsilon) = 0, \quad Var(\varepsilon) = \sigma^2 \tag{2}$$

In many cases, such as significance test or Bayes analysis, stronger assumptions can be made:

$$\varepsilon \sim N(0, \sigma^2) \tag{3}$$

In order to estimate the regression coefficients $\beta_0, \beta_1, \cdots, \beta_m$, Observe variable n for times, Get n groups of observation data items $(Y_i, X_{i1}, X_{i2}, \cdots, X_{im})$, $i = 1, \cdots, n$. In general, $n > m$ is required. It is expressed in matrix form:

$$Y = \begin{bmatrix} Y_1 \\ Y_2 \\ \vdots \\ Y_n \end{bmatrix}, \quad X = \begin{bmatrix} 1 & X_{11} & X_{12} & \cdots & X_{1m} \\ 1 & X_{21} & X_{22} & \cdots & X_{2m} \\ \cdots & \cdots & \cdots & \\ 1 & X_{n1} & X_{n2} & \cdots & X_{nm} \end{bmatrix}, \quad \beta = \begin{bmatrix} \beta_0 \\ \beta_1 \\ \vdots \\ \beta_0 \end{bmatrix}, \quad \varepsilon = \begin{bmatrix} \varepsilon_1 \\ \varepsilon_2 \\ \vdots \\ \varepsilon_n \end{bmatrix} \tag{4}$$

Then the multiple linear regression model:

$$Y = X\beta + \varepsilon \tag{5}$$

Among them $n \times (m + 1)$ matrix is called regression design matrix X. Generally, X is assumed to a full column rank, namely $rk(X) = m + 1$.

Assumptions about errors keep up with (2):

$$E(\varepsilon) = 0, \quad \text{Var}(\varepsilon) = \sigma^2 I_n \tag{6}$$

Among them I_n is the unit matrix.

It corresponds to the above data association (3):

$$\varepsilon \sim N(0, \ \sigma^2 I_n) \tag{7}$$

All data relationships (5), (6) and (7) together are called multivariate linear models.

2.4.2 Solving Model

Main steps of the method are listed:

1) Arrange and adjust the course resources through the K8S container systems.
2) Collect the phased data of students' cognitive process generated by the course.
3) Refer to Multi-factor regression analysis method.
4) Through the fitting and calculation, obtain the factor data of various cognitive activities of educational objectives on the dependent variables of evaluation results.
5) Repeat steps 1), 2), 3) and 4) until the factor data converges.

3   Research on Teaching Application of Course Resources

3.1   Process Implementation and Method Summary

This research takes the teaching of Career Planning as an example, and the implementation methods and processes are as follows:

3.1.1   Determine Teaching Objectives

The teaching objectives of this course are defined according to Bloom's taxonomy of education objectives. For example, conceptual knowledge can guide students to enumerate and explain the policies and situations for our college students' employment; Procedural knowledge can guide students to complete an interview with professionals by using the occupational data collection method.

3.1.2   Compile Course Outline

Design the overall teaching objectives of the course, sort out the teaching objectives of each learning unit, and then complete the writing of the course. Through the Vocational Education Cloud Platform, the syllabus is presented to teachers and students in the form of handouts, including: course introduction, introduction of teachers, course learning objectives, requirements for students' knowledge and skills, course teaching content, reference materials, and assessment methods, etc.

Before the first course starts, teachers will issue the course outline. Guide students to learn about the course objectives, learning methods, learning activities, assessment methods and other course requirements by previewing, so as to reach a consensus on teaching and learning.

3.1.3 Improve Course Design

In order to carry out teaching activities, teachers design courses according to the teaching requirements and teaching contents of each unit's knowledge points and skills points in the course outline. Through the three stages, namely before class, during class and after class, the online learning activities of students are arranged by adopting Bloom's taxonomy of education objectives.

3.1.4 Carry Out Teaching Activities

Before class, students can carry out independent learning, complete the task of topic preview and put forward learning problems through videos, rich texts and other relevant learning materials provided by teachers on the Vocational Education Cloud system. In class, teachers can carry out online and offline teaching procession through the network, based on the problem analysis of students' preview before class; Students can follow teachers to participate in classroom teaching activities and complete learning tasks through laptops and intelligent terminals. Each classroom learning task takes 15 min as a cycle to carry out a presentation activity. Teachers arrange group discussion, brainstorming and other teaching interactions according to the classroom teaching objectives. Students submit the project learning materials, including learning summary, grouping conclusion and test results to the system.

3.1.5 Teaching Feedback and Evaluation

Through the system, teachers evaluate the classroom learning effects, including homework, tests, group discussions, topic messages, etc., and guide students to continue to pay attention to their personal learning status to achieve the course teaching goals (Figs. 4 and 5).

Teachers will feedback the learning effect to students, and students can intuitively obtain the current course learning situation.

3.1.6 Learning Effect and Data Analysis

Students can view the course scores and corresponding grades through the system, and teachers can guide students to gradually achieve the course teaching goals in the process of continuous correction through it.

At the end of the semester, teachers will evaluate students' learning effect of this course. The evaluation of students' course learning consists of 20% of their usual performance (class attendance, participation in discussion, question messages, etc.), 40% of their practical performance (learning tasks completed in class and after class), and 40% of their final course design.

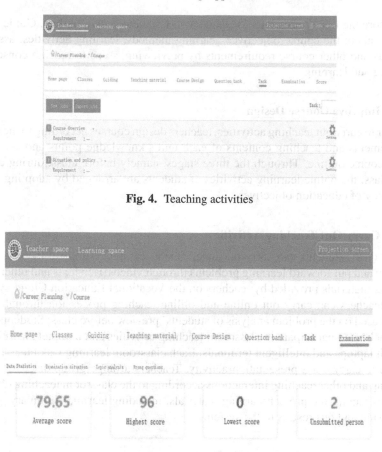

Fig. 4. Teaching activities

Fig. 5. Classroom test

4 Simulation Case

According to The Latest Theory of taxonomy of educational objectives, the important goals of classroom teaching include "retention" and "transfer" [10, 11]. 1) Many methods of learning during a course can be remembered and reused by students in the later learning process; 2) Students can use what they have learned to master new knowledge, answer new questions and solve new situations [12].

Generally, there are 19 cognitive processes that students use to enhance learning effects: 1) "Remember" includes recognizing and recalling; 2) "Understand" includes interpreting, exemplifying, classifying, summarizing, inferring, comparing and explaining; 3) "Apply" includes executing and implementing; 4) "Analyze" includes differentiating, organizing and attributing; 5) "Evaluate" includes checking and critiquing; 6) "Create" includes generating, planning and producing [13–15].

4.1 Teaching Statistics

In 2021, 216 students from four parallel classes taught in two semesters of this course were selected for tracking and analyzing the teaching effect. For two classes, 110 students in total, we carried out classroom teaching activities dominated by this research method, and learned about learning effects and follow-up feedback from students. Through the platform data analysis, the class participation of the students who adopted the research method was close to 88%, and was 13% higher than the other ones of the students who did not carry out the research method (Table 2).

Table 2. Tracking table of students' learning cognitive process

No.	Cognitive process Dimension	Cognitive Levels	Traditional Data	Application Data	Standard Data
1	Creating	6	70%	96%	100%
2	Evaluating	5			
3	Analyzing	4	2 * 30%	2 * 46%	>80
4	Applying	3	75%	84%	100
5	Understanding	2	75%	88%	100
6	Remembering	1	72%	122%	100%

84% of the surveyed students said that they had achieved the learning objectives described in this course; In the traditional method, the former statistical proportion was about 75%. 96% said they used the learning methods mastered by this course in the follow-up study of other courses, and maintained a high interest and state of learning; In the traditional method, the former statistics accounted for about 70%.

According to the statistics of the platform, for the classes that use research methods to carry out teaching, the number of students' after class learning hours every week is 122% of the required after class learning hours in this course. Students who did not use this research method invested about 72% of the required after-school learning hours in the course every week (Fig. 6).

According to previous teaching experience, the outstanding rate of students' performance after the course teaching is the number of people with a course evaluation score greater than 80 divided by the total number of students. Under the same performance evaluation method, the outstanding rate of student classes without research method is about 30%; the outstanding rate of final results of the student class using the research method is 16% higher than that of other classes. When carrying out the research on the taxonomy of educational objectives, We have considered the score data of students' comprehensive analysis of questions when they usually participate in classroom tests, group discussions and assessments, and made a quantitative treatment of the cognitive processes of "analysis", "evaluation" and "creation".

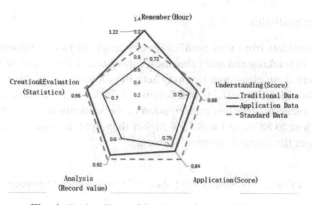

Fig. 6. Radar Chart of Students' Cognitive Process

4.2 Establishment of Regression Model

We randomly selected 50 samples from the tracking data of students' teaching effects for analysis. At the same time, according to the theoretical results of the classification of educational goals, the important goals of classroom teaching are "retention" and "transfer". The important cognitive processes used to strengthen learning achievements are "creation" and "evaluation". To facilitate analysis and reduce regression error, we set "creation & evaluation" as the dependent variable Y, and other cognitive process activities as the independent variable: X1 (Learning time), X2 (Understanding activities score), X3 (Application effect), X4 (Analysis effect), and add quadratic items.

The regression model is:

$$Y = \beta_0 + \beta_1 X_{i1} + \beta_2 X_{i2} + \beta_3 X_{i3} + \varepsilon_i, i = 1, \cdots, 50 \quad \varepsilon_i \sim N(0, \sigma^2)$$

And the variables $\varepsilon_1, \varepsilon_2, \cdots, \varepsilon_{50}$ are independent of each other [16].

4.3 Experimental Data Acquisition

According to the taxonomy of educational objectives, we selected the teaching effect data of 50 students randomly from the student classes, and analyzed them using the research method (Table 3).

4.4 Result Analysis in MATLAB

Through Matlab and Excel fitting, we obtain the regression linear equation of the teaching objective classification method of this course [17].

As follow:

y = −0.5978 + 0.18155x1 + 0.02101x2 + 0.04177x3 + 0.41188x4. It shows in the blue line in the figure below.

Table 3. 50 groups of measured data

No.	X1	X2	X3	X4	Y	No.	X1	X2	X3	X4	Y
1	8.9	89	94	TRUE	TRUE	26	9.2	87	89	TRUE	TRUE
2	4.8	90	95	TRUE	TRUE	27	7.2	85	67	FALSE	FALSE
3	3.2	62	74	FALSE	FALSE	28	6.5	79	99	TRUE	TRUE
4	6.5	90	98	TRUE	TRUE	29	6.4	95	98	TRUE	FALSE
5	5.8	95	69	FALSE	TRUE	30	6.8	94	99	TRUE	FALSE
6	6.9	88	71	FALSE	TRUE	31	7.9	93	73	FALSE	TRUE
7	7.5	84	100	TRUE	TRUE	32	9	91	97	TRUE	FALSE
8	6.3	85	75	FALSE	TRUE	33	6.1	87	77	FALSE	TRUE
9	7.9	80	77	FALSE	TRUE	34	6.2	89	78	FALSE	TRUE
10	7.7	78	79	FALSE	TRUE	35	6.3	90	76	FALSE	TRUE
11	5.8	89	86	TRUE	TRUE	36	9.7	86	69	FALSE	TRUE
12	8.8	83	74	FALSE	TRUE	37	7.2	94	74	FALSE	TRUE
13	9.5	69	61	FALSE	TRUE	38	8.5	78	98	TRUE	TRUE
14	6.2	93	100	TRUE	TRUE	39	9.4	92	79	FALSE	TRUE
15	7.7	84	76	FALSE	TRUE	40	11.5	96	76	FALSE	TRUE
16	3.5	75	88	TRUE	TRUE	41	8.8	97	79	FALSE	TRUE
17	5.8	85	95	TRUE	TRUE	42	7.3	86	76	FALSE	FALSE
18	7.1	88	78	FALSE	TRUE	43	8.9	82	89	TRUE	TRUE
19	4.7	90	64	FALSE	TRUE	44	7.9	95	98	TRUE	FALSE
20	6.3	92	65	FALSE	TRUE	45	7.4	95	78	FALSE	TRUE
21	3.9	89	100	TRUE	TRUE	46	9.5	87	87	TRUE	TRUE
22	5.5	87	95	TRUE	TRUE	47	7.5	84	96	TRUE	TRUE
23	7.7	91	78	FALSE	TRUE	48	7.2	98	77	FALSE	TRUE
24	8.4	95	97	TRUE	TRUE	49	6.3	92	97	TRUE	FALSE
25	10.6	98	62	FALSE	TRUE	50	9.4	86	100	TRUE	TRUE

Those variables $\hat{\beta}_1, \hat{\beta}_2, \hat{\beta}_3, \hat{\beta}_4$, in this equation have clear meanings. $\hat{\beta}_1 = 0.18155$ means that for each additional unit of student learning time, the contribution to "creation & evaluation" effect is equal to 0.18155; $\hat{\beta}_2 = 0.02101$ means that the contribution to "creation & evaluation" effect is 0.02101. Similarly, other situations as $\hat{\beta}_3, \hat{\beta}_4$ can be understood (Fig. 7).

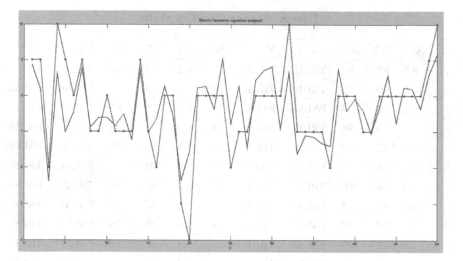

Fig. 7. Measured value, estimated value and residual analysis

5 Summary and Outlook

Based on the classification method of teaching objectives, this paper proposes a multi-factor regression analysis scheme of educational objectives to analyze and evaluate the teaching effect by using students' learning cognitive process data. It provides an operable tool for analyzing course resources, teaching activities, and students' cognition to evaluate the teaching effect.

The scheme is as follows:

1) Evaluate the teaching effect from the perspective of students' learning cognition. The multi-dimensional teaching of learning cognitive activities based on Bloom's taxonomy of educational objectives which is carried out and the corresponding data is collected.
2) The multi-factor regression analysis formula is used to link teachers' teaching, students' learning and evaluation, highlighting the consistency of teaching effect evaluation.
3) According to the fitting and calculation of the phased impact factors, teachers can choose teaching strategies and evaluation methods more targeted. Teachers and students in teaching activities can know more clearly about how to invest in learning to achieve better learning results.

With the help of information technology, this research based on the theory of Bloom's taxonomy of educational objectives shifts the focus of classroom teaching from teachers' teaching to students' learning and mastering. It has made a beneficial attempt in promoting students' learning interest, improving students' learning engagement and participation, and improving teaching effects. The application of this research in the early stage of classroom teaching will increase the workload of teachers. While, after five semesters of running in and application, teachers generally feedback the application scheme, providing convenience for teachers to carry out student-centered teaching, plan classroom teaching content activities (including correcting course assignments, evaluating students' learning process, etc.), and reduce teachers' subsequent workload. This application research scheme can be used to promote teaching in different disciplines and different courses in various colleges and universities that have well solidified digital campus achievements and have developed online course teaching.

Acknowledgment. This project is supported by Projects of China University IUR innovation fund (2020ITA01008).

References

1. Zhang, X., Huang, X., Li, D.: The investigation and study on learning style and behavior habits of college students. Educ. Res. Mon. **2021**(1), 92 (2021)
2. Bloom, B.S.: Taxonomy of Educational Objectives: Handbook I: The Cognitive Domain. David Mckay Co., Inc., New York (1956)
3. Bloom, B.S.: Taxonomy of Educational Objectives Book 1: Cognitive Domain, 2nd edn. Addison Wesley Publishing Company, New York, London (1984)
4. Anderson, L.W., Krathowohl, D.R.: A Taxonomy for Learning, Teaching, and Assessing: A Revision of Bloom's Taxonomy of Educational Objectives Complete. Allyn & Bacon, Boston, MA (2001)
5. Chang, J., Lan, W.: New development of bloom's taxonomy of educational objectives. J. Nanyang Normal Univ. **7**(05), 84–85 (2008)
6. Adams, N.E.: Bloom's taxonomy of cognitive learning objectives. Med. Lib. Assoc. **103**(3), 152 (2015)
7. The Kubernetes Authors [EB/OL]. https://kubernetes.io/docs/concepts/overview
8. Sankaranarayanan, H.B., Rathod, V.: Airport merchandising using micro services architecture. Int. J. Inf. Technol. Comput. Sci. **6**, 52–59 (2016)
9. Mamo, R.: Service-oriented computing for effective management of academic records: in case of Debre Markos University Burie Campus. Int. J. Mod. Educ. Comput. Sci. **4**, 48–53 (2020)
10. Krathwohl, D.R., Bloom, B.S., Masia, B.B.: Taxonomy of Educational Objectives, the Classification of Educational Goals. Handbook II: Affective Domain. David McKay Company, New York (1973)
11. Simpson, E.J.: The Classification of Educational Objectives in the Psychomotor Domain. Gryphon House, Washington, DC (1972)
12. Pohl, M.: Learning to Think, Thinking to Learn: Models and Strategies to Develop a Classroom Culture of Thinking. Hawker Brownlow, Cheltenham, Vic. (2000)
13. Gunaratana, B.H.: Mindfulness in Plain English: Revised and Expanded Edition. Wisdom Publications, Boston (1996)
14. Karunananda, A.S., Goldin, P.R., Talagala, P.D.: Examining mindfulness in education. Int. J. Mod. Educ. Comput. Sci. **12**, 23–30 (2016)
15. Kabat-Zinn, J.: Mindfulness-based interventions in context: past, present, and future. Clin. Psychol. Sci. Pract. **10**(2), 144–156 (2003)
16. Larsen, R.J., Marx, M.L.: An Introduction to Mathematical Statistics and Its Applications. China Machine Press, Beijing
17. Yasir, M., Shah, Z.S., Memon, S.A., Ali, Z.: Machine learning based analysis of cellular spectrum. Int. J. Wirel. Microwave Technol. **2**, 24–31 (2021)

Reform and Practice of Virtual Simulation Practice Course for Logistics Engineering Specialty Based on OBE Concept and School-Enterprise Linkage

Yanfang Pan, Dalong Liu(✉), and Liwei Li

School of Transportation, Nanning University, Nanning 530200, China
281340537@qq.com

Abstract. The talent training of logistics engineering majors focuses on cultivating the ability of students to solve complex engineering problems. The students should adapt to the rapid development of industry and technology under the economic transformation, be able to collect and process the basic data of logistics economic operation, find scientific methods and ways to solve problems through data processing and analysis, and carry out systematic optimization, demonstration and evaluation. This puts forward higher requirements for the logistics virtual simulation practice course. Based on the concept of "results-oriented" + "university-enterprise linkage", relying on the multi-field, cross-industry, cross-regional off-campus production and teaching integration base, and combining the "cross-curriculum" + "multi-stage" + "in and off-campus and class" curriculum system, the article builds an integrated platform of "virtual reality integration" + "innovation integration" logistics virtual simulation teaching system, and uses AHP analysis method to analyze the evaluation indicators, Strive to find the most critical factor for the construction of virtual simulation practice course in the cooperation mechanism.

Keywords: Virtual simulation · Logistics engineering specialty · OBE · School-enterprise linkage

1 Introduction

Outcome-based education (OBE) is an advanced educational concept, and also a construction philosophy of curriculum system that focuses on the requirements of society for talents and adopts the way of reverse thinking [1]. In order to adapt to the new normal development of higher education and meet the needs of the development of "Industry 4.0" and "Internet plus", China's logistics industry has ushered in the development of intelligent upgrading and transformation, and has also driven the undergraduate education of logistics engineering to "smart", "creative" and other fields [2]. As an enabling and value-added entry for the development of industries in the new era, logistics enterprises require logistics professionals to be able to adapt to the form of the times and provide enterprises with the ability and quality of the ecological industry chain of automated, intelligent and intelligent whole-process integrated solutions [3].

© The Author(s), under exclusive license to Springer Nature Switzerland AG 2023
Z. Hu et al. (Eds.): ICCSEEA 2023, LNDECT 181, pp. 927–938, 2023.
https://doi.org/10.1007/978-3-031-36118-0_79

With the rapid changes in the world today, virtual simulation teaching came into being [4]. It is the product of the deep integration of modern education and virtual simulation technology. With the help of "AR/VR", intelligent big data, Internet of Things and other cutting-edge information technologies, it builds a highly simulated training environment and experimental conditions [4, 5]. Students can carry out training and experiments in the virtual simulation environment, and can operate from multiple angles and repeatedly. It is urgent to solve the problems of "high investment, high consumption, high risk, difficult to implement, difficult to observe, and difficult to reproduce" in traditional training and experimental teaching [6, 7].

In the teaching of logistics courses, virtual modeling and simulation is also an important content [8]. In the virtual laboratory environment, it is undoubtedly of obvious value to use simulation experiments as the basis of production capacity acquisition [9].

2 The Crux of Logistics Virtual Simulation Course

In the construction of logistics specialty, the construction of virtual simulation course group and its experimental teaching platform can realize the sharing of experimental teaching resources inside and outside the school, in the region and in a wider range, and establish a sustainable virtual simulation experimental teaching service support system [10–12]. The study found that the crux affecting the construction of the curriculum system and the quality of talent training is mainly reflected in the following three aspects.

2.1 The Virtual Simulation Course is Difficult to Learn

Due to the influence of textbooks and the limitation of course hours, teaching is inevitably a mere formality. In the training program for logistics engineering professionals, virtual simulation courses such as Logistics System Modeling and Simulation, Logistics System Planning and Design, and Supply Chain Optimization Program Design need to focus on training students' ability to solve complex engineering problems. Students should not only master solid professional theoretical knowledge, have logistics technical ability, but also understand certain basic knowledge such as computer language, engineering drawing, data statistics and analysis, It also needs to have certain data processing ability and text expression ability [13]. Virtual simulation course is a comprehensive, difficult to operate, and highly applicable course category, which is very difficult for students to learn. The existing textbooks lack innovative guidance and application value of drawing inferences from one example.

2.2 The Training Results are Not Ideal and Lack of Practical Significance

In the teaching process of the virtual simulation course, students complete the establishment of the whole model according to the experimental steps. Although the model can be built according to the fixed procedures, it cannot cultivate students' ability to find, analyze and solve problems, which is not conducive to cultivating students' creative thinking and pragmatic spirit.

In essence, the virtual simulation experiment teaching course of logistics engineering undergraduate specialty should focus on comprehensive design, with the goal of cultivating students' ability to independently design and solve practical problems through practical training, and pay attention to the degree to which students can use modern analysis tools to carry out independent design and independent analysis, so as to meet the requirements of talent training of logistics engineering specialty [14]. Considering the support and importance of professional core courses for professional talent training, we should take the virtual simulation experiment teaching course as the carrier to cultivate students' realistic, scientific and rigorous learning and working style [15]. However, due to the influence of the teacher-oriented concept, the teaching content of the virtual simulation course needs to be further adjusted and improved.

2.3 The Assessment is Not Objective and Cannot Reflect the Level of Students

The traditional virtual simulation course teaching adopts the course assessment mode of "model" + "experiment report". The students' understanding and mastery of knowledge reflected in the experiment report is fragmented, and cannot form a systematic cognition. The causes and consequences of simulation optimization, overlap and conflict with reality, and inadequate consideration of model simplification and hypothesis have greatly reduced the goal and significance of model establishment [16].

3 Reform Ideas of Virtual Simulation Practice Course for Logistics Engineering Specialty

Based on the OBE concept, the development and reform of virtual simulation courses for logistics engineering specialty are comprehensively carried out by school-enterprise linkage, which is conducive to the overall integration of the curriculum system for logistics engineering specialty. The development of virtual simulation course projects, from topic selection, design, development to optimization, verification and application, pays equal attention to both virtual and real, and helps students further consolidate their professional skills, understand the professional frontier, and improve their interdisciplinary ability [17].

3.1 Basic Ideas of Virtual Simulation Practice Course Reform

The students' comprehensive and systematic knowledge integration from language basis, mathematical logic, professional skills, off-campus practice and in-school experimental courses will help to achieve the objectives and requirements of training logistics engineering professionals in the context of new engineering [18]. Thus, the basic framework of virtual simulation practice course system for logistics engineering specialty based on OBE concept and school-enterprise linkage is constructed, as shown in Fig. 1.

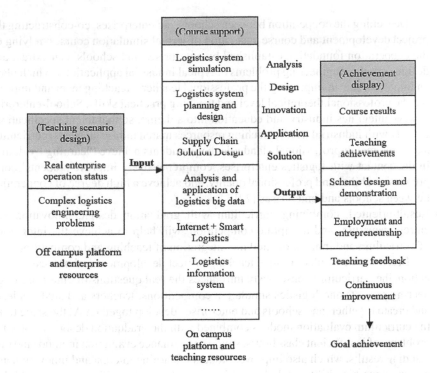

Fig. 1. Virtual Simulation Practice Course System for Logistics Engineering

3.2 Specific Measures for the Reform of Virtual Simulation Practice Course

The virtual simulation practice courses involved in the training program for logistics engineering professionals are mainly concentrated in professional core courses and centralized practice courses, such as Logistics System Simulation, Logistics System Planning and Design and other courses. These courses are organized by knowledge modules in traditional teaching and textbook design. The cases are independent, not coherent, and not easy to understand. Adopt the results-oriented approach, reconstruct the teaching content, excavate the teaching situation from the real enterprise operation status, extract the complex logistics engineering problems as the teaching items of the course, take students as the center, teachers as the guidance and assistance, and lead students to complete the analysis, design, innovation, application and problem solving in the form of project development and promotion, and put the learning into practice [19–21].

The course results are in the form of project reports and team results, giving full play to students' subjective initiative and innovative spirit; The teaching results take solving the actual problems of enterprises as a reference, and carry out the scheme design and demonstration, which can achieve the goal of integration of specialty and innovation and integration of curriculum and innovation to a certain extent, and is conducive to the realization of the goal of virtual simulation course. The specific methods are as follows.

(1) The integration of schools and enterprises, and the close connection and high cooperation between logistics engineering and logistics enterprises.

Deepening the cooperation between schools and enterprises, co-constructing the project development and course evaluation of virtual simulation courses, relying on the cooperation foundation established by enterprises and schools, can extract and design complex engineering problems of typical industrial applications, which plays an important role in improving the professional teachers' teaching team and improving the professional theoretical level and teaching practical skills. School-enterprise linkage integrates industry and education into a theme, so that talent specifications comply with industrial needs, teaching resources match industrial needs, curriculum standards match professional standards, help students achieve "learning by doing" in the contact with logistics enterprises, comprehensively improve students' comprehensive ability and professional quality, and achieve a high degree of cooperation between schools and enterprises [22].

(2) Result-oriented, combining curriculum with graduation design, innovation and entrepreneurship, and discipline competition, will help to achieve the integration of curriculum and innovation, and the integration of teaching and competition.

Based on the results-oriented teaching project development and process organization, the curriculum consciously introduces the real questions of enterprise engineering projects and logistics simulation competitions, teachers and students learn and create together, and schools and enterprises develop together. At the same time, the curriculum evaluation mode is combined with the graduation design to solve the problems of insufficient class hours, single performance evaluation method, and outstanding results, which also improves students' learning interest and innovation and entrepreneurship ability, and also cultivates students' scientific expression Ability required by engineering certification, such as teamwork and self-study.

(3) The joint construction of schools and enterprises and the penetration of curriculum ideology and politics will help students establish good professional ethics and professional quality

According to the development requirements of the logistics industry, sort out the knowledge structure and application level of the virtual simulation course, effectively excavate and screen the ideological and political elements through the participation of enterprises and the combination of points, lines and areas, cultivate students to establish a correct professional view of logistics development, and do a good job in the responsibility of logistics personnel; Establish a correct concept of logistics development and stimulate students' awareness of efficiency, service, cost, safety and low-carbon environmental protection. Through enterprise research and data collection, through modeling and optimization simulation, students are trained to develop a scientific spirit of scientific rigor, truth-seeking and pragmatic, and continuous exploration.

(4) Multiple evaluation, establish the curriculum results of "model" + "scheme" preliminary evaluation and "course" + "final design" inspection

Follow the principle of real combat, adhering to the scientific, rigorous, normative and effective ideological and political concept of curriculum, establish a dual-dimensional and two-level curriculum achievement system of "model" + "program" preliminary evaluation and "curriculum" + "final design" test, and comprehensively evaluate students' problem-solving ability, logical thinking ability, innovation ability and independent learning ability [23].

4 Construction of Evaluation Index System of Logistics Virtual Simulation Course Based on AHP Method

The purpose of evaluating the virtual simulation course of logistics engineering specialty with the method of AHP is to find the key points of course reform [24, 25]. The evaluation needs to focus on the coherence and complementarity between "truth" and "virtual". What called "truth" means that the teaching design of the virtual simulation course requires true data and true scene, and relies on the real enterprise background and enterprise environment, close to the reality of the enterprise; "Virtual" means to reproduce the scene in the school classroom through scientific teaching methods and means with the help of advanced computer networks and platforms. Students do not need to go to the enterprise site.

Table 1. Evaluation Index and Description of Virtual Simulation Course of Logistics Engineering

Items	Evaluating indicator	
Teaching achievements	The assistance of the course to the project enterprises	solve the key problems of the enterprise
	Originality of student work output	Different results from student-centered
	Classroom teaching feedback	Supervision and student evaluation
Course organization	ideological and moral education of the curriculum	Design embodiment in moral education
	Construction of course teaching resource library	Established network teaching resource
	Conforms to the actual production of the enterprise	Curriculum designed from real enterprises
	Use of third-party teaching platform in teaching process	Use third-party platform for teaching
	The difficulty degree of course scenario design	The teaching situation is progressive
	Compliance with the requirements of course objectives	Meet the goal of certification education
Teaching resources	Textbook selection	Choose planning textbook
	Teaching team	Gradient of teaching team is reasonable
	Practical class hours	Practice class hours meet teaching needs

According to the above analysis, fourteen evaluation indicators are summarized from three aspects of course scene mining, course organization and achievement evaluation output to assess the virtual simulation course. See Table 1 for the assessment indicators of course evaluation.

The evaluation indicators of logistics engineering virtual simulation course in Table 1 include both quantitative evaluation indicators and qualitative indicators, and the impact of these indicators on the logistics engineering virtual simulation course is different. In order to better measure the importance of each indicator for the training support of logistics professionals, the analytic hierarchy process (AHP) is used to set each indicator.

4.1 Establishment of Hierarchy Model

Based on the evaluation index of virtual simulation course of logistics engineering specialty in Table 1, the hierarchy model of evaluation index of virtual simulation course of logistics engineering specialty is established, as shown in Fig. 2.

4.2 Evaluate and Calculate the AHP Model

In view of the hierarchical structure model of the evaluation index of the virtual simulation course of logistics engineering specialty, the judgment matrix is established by comparing two different elements at the same level. Through the form of questionnaire and expert interview, the bank and enterprise expert consultation is carried out, the importance of each evaluation index is expressed in numbers, the qualitative index is converted into quantitative numbers, the relative weight is obtained, the consistency of the judgment matrix is checked, and then the total ranking weight of each level is calculated, and the ranking is carried out in order. Get the relative weight of each index. Table 2 shows the judgment matrix and relative weight of criterion layer A. Tables 3, 4 and 5 shows the judgment matrix and relative weights of teaching achievement B_1, curriculum organization B_2 and teaching resources B_3.

4.3 Consistency Inspection

Check the consistency of the judgment matrix of the criterion layer and the target layer.
Maximum characteristic value:

$$\lambda = \frac{1}{n} \sum_{i=1}^{1} \frac{(A\omega)_i}{\omega_i} \quad (n : Order) \tag{1}$$

Consistency indicators:

$$CI = \frac{\lambda - 1}{n - 1} \tag{2}$$

Consistency ratio:

$$CR = \frac{CI}{RI} \tag{3}$$

The consistency test and results obtained by calculation are shown in Table 6.

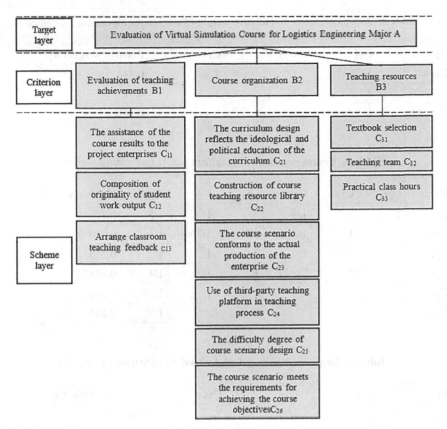

Fig. 2. Hierarchical structure model of evaluation indicators for virtual simulation courses of logistics engineering

Table 2. Criterion level judgment matrix and relative weight

A	B_1	B_2	B_3	Relative weight
B_1	1	1/4	3	0.231
B_2	4	1	5	0.665
B_3	1/3	1/5	1	0.104

4.4 Summary and Correction

From Tables 2, 3, 4 and 5, the relative weights of the indicators at the criterion level and the scheme level are obtained. From Table 6, the results of the consistency test are obtained. Then the relative weights of the scheme can be obtained by the weighted calculation method. The synthetic weight and ranking results of virtual simulation course evaluation are shown in Table 7.

Table 3. Judgment matrix and relative weight of teaching achievement evaluation

B_1	C_{11}	C_{12}	C_{13}	Relative weight (B_1)
C_{11}	1	1/2	3	0.334
C_{12}	4	1	3	0.525
C_{13}	1/3	1/3	1	0.142

Table 4. Course organization judgment matrix and relative weight

B_2	C_{21}	C_{22}	C_{23}	C_{24}	C_{25}	C_{26}	Relative weight (B_2)
C_{21}	1	3	1/4	2	1/2	1/3	0.116
C_{22}		1	1/4	1/3	1/2	1/4	0.060
C_{23}			1	3	1	1	0.265
C_{24}				1	1/2	1/4	0.095
C_{25}					1	1	0.201
C_{26}						1	0.263

Table 5. Judgment matrix and relative weight of teaching resources

B_3	C_{31}	C_{32}	C_{33}	Relative weight (B_3)
C_{31}	1	1/2	1	0.263
C_{32}	1	1	1	0.413
C_{33}	2	1	1	0.324

Table 6. Results and conclusions of consistency inspection

Consistency inspection	Result	Conclusion
Criterion layer	λ max = 3.087; CR = 0.0836 < 0.1	Pass
Course scenario mining	λ max = 3.054; CR = 0.0517 < 0.1	
Course organization	λ max = 6.311; CR = 0.0494 < 0.1	
Evaluation and judgment	λ max = 3.054; CR = 0.0516 < 0.1	

Due to the use of expert scoring, the analytic hierarchy process is greatly affected by subjective factors and has many qualitative components. Therefore, it can also be properly modified according to the actual situation of professional talent training, so that the curriculum evaluation is closer to the professional curriculum objectives and professional talent training objectives.

In the light of the results shown in the last column of Table 7, among the total 12 evaluation items, the first one is the situation that "the curriculum scenario conforms to the actual production of the enterprise". This is in line with the concept of OBE. We should design the talent training system based on the actual situation of the enterprise. The second is that "the curriculum scenario meets the requirements of the degree of achievement of the curriculum objectives", which further explains that the final requirements of talent cultivation play a decisive role in the design of the curriculum system. "The difficulty of curriculum scenario design" ranks the third, which tells us that we should carry out all-round balance and coordination in curriculum design and organization.

Table 7. Evaluation weight and ranking of virtual simulation courses for logistics engineering

Items	Evaluating indicator	Composite weight	Sorting
Teaching achievements	The assistance of the course to the project enterprises	0.077	6
	Originality of student work output	0.121	4
	Classroom teaching feedback	0.033	11
Course organization	ideological and moral education of the curriculum	0.077	5
	Construction of course teaching resource library	0.040	9
	Conforms to the actual production of the enterprise	0.176	1
	Use of third-party teaching platform in teaching process	0.063	7
	The difficulty degree of course scenario design	0.134	3
	Compliance with the requirements of course objectives	0.175	2
Teaching resources	Textbook selection	0.027	12
	Teaching team	0.043	8
	Practical class hours	0.034	10

The construction of virtual simulation course for logistics engineering specialty should focus on the course organization, especially in the case of platform co-construction and school-enterprise cooperation, pay more attention to the relevant standards that conform to the actual production of enterprises and engineering education certification, and pay attention to the progressive teaching organization from easy to difficult.

5   Conclusion

Applying the concept of OBE to the teaching and talent training of logistics engineering is to meet the requirements of social and economic development for talent team. The teaching of virtual simulation practice course needs to combine theoretical teaching with production practice from the actual situation of enterprises. The way of school-enterprise interaction is undoubtedly an effective way to promote the teaching design of schools to really focus on the reality of enterprises and society.

School-enterprise interaction can effectively develop the teaching scene of logistics virtual simulation course, return "virtual" to "real", and the course teaching is more practical and valuable. Achievement orientation can effectively carry out the teaching organization of logistics virtual simulation course, cultivate students' practical and pragmatic scientific attitude, develop students' rational thinking and innovative thinking, and improve students' ability to use modern tools to solve practical problems. Only by integrating the two courses can we better achieve the goal of the virtual simulation course of logistics engineering specialty in the new engineering background, and the course construction and professional development can be more close to the development needs of industry enterprises.

Acknowledgment. The research of this paper is supported by two projects: (1) The teaching reform project of Nanning University "Reform and Practice of Virtual Simulation Practice Course for Logistics Engineering Specialty Based on OBE Concept and School-Enterprise Linkage" (2022XJJG28); (2) The teaching reform project of the integration of specialty and innovation of Nanning University "The teaching reform and practice of the course" Modern Logistics Equipment "combining the systematization of work process under the mixed teaching mode and the integration of specialty and innovation" (2020XJZC05).

References

1. Zamir, M.Z., Abid, M.I., Fazal, M.R., et al.: Switching to outcome-based education (OBE) system, a paradigm shift in engineering education. IEEE Trans. Educ. **65**, 1–8 (2022)
2. Zhiqin, L., Jianguo, F., Fang, W., Xin, D.: Study on higher education service quality based on student perception. Int. J. Educ. Manag. Eng. (IJEME) **2**(4), 22–27 (2012)
3. Al-Hagery, M.A., Alzaid, M.A., Alharbi, T.S., Alhanaya, M.A.: Data mining methods for detecting the most significant factors affecting students' performance. Int. J. Inf. Technol. Comput. Sci. (IJITCS) **12**(5), 1–13 (2020)
4. Laghari, A.A., Jumani, A.K., Kumar, K., Chhajro, M.A.: Systematic analysis of virtual reality & augmented reality. Int. J. Inf. Eng. Electron. Bus. **13**(1), 36–43 (2021)
5. Peng, Y., He, X., Huang, Y.: A Virtual simulation experiment system for requirement analysis. J. Phys.: Conf. Ser. **1757**(1), P012194 (2021)
6. Lin, Z.: The construction of the virtual simulation experiment platform under the background of education informatization. In: Jansen, B.J., Liang, H., Ye, J. (eds.) International Conference on Cognitive based Information Processing and Applications (CIPA 2021). LNDECT, vol. 85, pp. 718–724. Springer, Singapore (2022). https://doi.org/10.1007/978-981-16-5854-9_92
7. Dubovi, I.: Learning with virtual reality simulations: direct versus vicarious instructional experience. Interact. Learn. Environ. **2022**, 1–13 (2022)

8. Xie, X., Guo, X.: Influencing factors of virtual simulation experiment teaching effect based on SEM. Int. J. Emerg. Technol. Learn. **17**(18), 89–102 (2022)
9. Bakhtadze, N., Zaikin, O., Żylawski, A.: Simulation experiment in a virtual laboratory environment as a ground for production competencies acquiring. In: Dolgui, A., Bernard, A., Lemoine, D., von Cieminski, G., Romero, D. (eds.) APMS 2021. IAICT, vol. 630, pp. 535–545. Springer, Cham (2021). https://doi.org/10.1007/978-3-030-85874-2_57
10. Wang, X., Tan, Y.: Exploration of logistics virtual simulation experiment teaching in the training of innovative practical talents. Logist. Technol. **44**(11), 172–174+178 (2021). (in Chinese)
11. Peng, C.: Construction of logistics management virtual simulation experiment teaching system based on post capacity. J. Taiyuan City Vocat. Tech. Coll. **06**, 119–121 (2020). (in Chinese)
12. Hong, M.: Build virtual simulation logistics laboratory and cultivate innovative logistics talents. Electron. Technol. **39**(01), 65–66 (2012)
13. Özgen, K.: The effect of project-based cooperative studio studies on the basic electronics skills of students' cooperative learning and their attitudes. Int. J. Mod. Educ. Comput. Sci. (IJMECS) **2018**(5), 1–8 (2018)
14. Wang, X.D., Wu, Y.: Exploration and practice for evaluation of teaching methods. Int. J. Educ. Manag. Eng. (IJEME) **3**(29), 39–45 (2012)
15. Li, J., Zhou, Y.: The exploration and practice in innovative personnel training of computer science and technology. Int. J. Educ. Manag. Eng. (IJEME) **2**(6), 47–51 (2012)
16. Khan, M.H.: A unified framework for systematic evaluation of ABET student outcomes and program educational objectives. Int. J. Mod. Educ. Comput. Sci. (IJMECS) **11**(11), 1–16 (2019)
17. Wei, X.: Discovery and practice of EDA experimental teaching reform. Int. J. Educ. Manag. Eng. **1**(4), 41–45 (2011)
18. Liu, G., Li, Q., Chen, H., et al.: Development and practice of cold chain logistics virtual simulation course. Packag. Eng. **42**(S1), 191–195 (2021). (in Chinese)
19. Zhang, B., Yuan, Y., Huang, J., et al.: Virtual simulation of production logistics based on Plant Simulation. Sci. Technol. Innov. Appl. **11**(21), 6–9+12 (2021). (in Chinese)
20. Huang, Y., Ji, L.: Exploration on the construction of virtual simulation experimental teaching center for logistics management based on VR technology. Exp. Technol. Manag. **37**(08), 238–242 (2020). (in Chinese)
21. Ren, J.: Construction and practice of industrial fresh e-commerce and cold chain logistics virtual simulation experimental teaching center. Logist. Technol. **41**(03), 125–129 (2022). (in Chinese)
22. Liu, Z., Li, Y., Chen, P., Zhao, L.: Research on the construction of virtual simulation logistics laboratory. Univ. Educ. **07**, 54–56 (2020). (in Chinese)
23. Liu, H.: The application of virtual simulation technology in the practical teaching of logistics management specialty – taking the course of "warehouse and distribution management" as an example. J. Taiyuan City Vocat. Tech. Coll. **10**, 160–161 (2017). (in Chinese)
24. Fu, J., Wu, Y., Yan, H.: The Application of database analytic hierarchy process (AHP) in teacher-to-teacher assessment. Int. J. Educ. Manag. Eng. (IJEME) **3**(1), 27–33 (2013)
25. Kong, L.A., Wang, X.M., Yang, L.: The research of teaching quality appraisal model based on AHP. Int. J. Educ. Manag. Eng. **9**(29), 29–34 (2012)

Application of Google Workspace
in Mathematical Training of Future Specialists
in the Field of Information Technology

Olena Karupu[1], Tetiana Oleshko[1], Valeria Pakhnenko[1], and Anatolii Pashko[2]([✉])

[1] National Aviation University, Kyiv, Ukraine
[2] Taras Shevchenko National University of Kyiv, Kyiv, Ukraine
aap2011@ukr.net

Abstract. Education in the field of information technology in Ukraine should introduce a special innovative approach in the training of specialists in computer science and information technology. Problems of personality-oriented and competence-based models of education in the process of training future specialists in information specialties become topical. All professionals in the field of information technology must have specific personality traits and professional competencies formed in the studying different disciplines. Educational process has to result the formation of both hard and soft skills of students. Future IT specialists require a deep knowledge of the basic mathematical theoretical foundations and possession of the skills of applying mathematics. The study of the educational process in multinational academic groups seems to us quite interesting, since it has some peculiarities. The aim of this article is the consideration of new learning technologies and the features of their implementation in the context of distance education. We study methodological problems of collaborative approach in online teaching of mathematical disciplines applying Google Workspace for Education tools.

Keywords: Teaching mathematics to information technology specialists · Teaching mathematics in multinational academic groups · Collaborative approach in education · Online and blended learning

1 On Mathematical Component in Professional Competence of IT Specialists

In the professional development of future specialists of all technical specialties it is necessary to pay enough attention to forming of the basic mathematical knowledge and the skills of applying mathematical theory. One of the authors who most consistently apply this approach is N. O. Virchenko [1].

Education of future IT specialists should be problem-oriented. Broad and deep training of students should provide preparation of professionals able to improve their skills continually. This can be achieved by the fundamentalization of the educational process and the use of STEM technologies. Problematizing teaching and learning mathematics

Z. Hu et al. (Eds.): ICCSEEA 2023, LNDECT 181, pp. 939–949, 2023.
https://doi.org/10.1007/978-3-031-36118-0_80

in STEM education were considered in [2–4]. Some problems of application of IT in teaching mathematical disciplines to future specialists in information technologies were considered in [5].

Research-oriented students of IT specialties need excellent mathematical knowledge and skills. Industry-oriented students of IT specialties need good mathematical knowledge and skills. On that account, the mathematical training of the first-year students of FCCSE of NAU is well above average level. The mathematical training of students of FCSC of KNU has very high level. It should be noted that in National Aviation University students of most specialties require fairly deep mathematical background. This is especially true for professionals in the field of information technology. The curricula of these specialties contain various mathematical disciplines.

Problems encountered by students in the study of mathematical disciplines are often associated with their school background in mathematics and differ for students from different countries, although there are general characteristics of considered process.

In addition, we consider it necessary to note the existence of certain difficulties in teaching students to solve applied problems having a technical orientation. In our opinion, this is due to the fact that in secondary schools in many countries more attention is paid to the consideration of applied problems having an economic nature. Insufficient level of previous school background in geometry, manifested when students study multivariable functions and multiple integrals, can be partially compensated by the use of CAS. Most students become aware of the existence of CAS and receive initial skills in using CAS while still studying in secondary school. We consider it desirable to provide guidance to students on the use of the various CAS, paying attention to the restrictions on their use.

2 Intensification of Integration Processes and Teaching to Mathematical Disciplines in English in Ukraine

In recent years, we have seen an increase in the number of citizens of other countries who receive higher education in the universities of our country. Medical, engineering and IT professions are among the most popular for foreigners.

Some foreigners are native English speakers or speak it at a fairly high level. A significant number of foreign students plan to work outside of Ukraine in the future. That is why these students choose education in English. In addition, the intensification of integration processes in the global educational and scientific space and the development of academic mobility significantly increases the interest of Ukrainian students in learning in foreign languages.

Problems of implementation of English language in the educational process within the framework of internationalization of Ukrainian universities were analyzed in [6]. Trends and issues of provision of English-medium instruction were investigated in [7]. As a result, the current trend of modern national education in Ukraine is the continuous growth of the contingent receiving higher education in English. Some Ukrainian universities teach some students to some disciplines in English. Students of English-speaking groups may be divided into two categories: foreigners who receive higher education in Ukraine, and Ukrainian students who desire to study in English. And although students of both these categories have their own specific learning needs, the common denominator is that the vast majority of participants in the considered educational process are not native English-speakers. Leading national universities of Ukraine are actively involved in the development of the education system, the introduction of new technologies, methods and models of organization of the educational process, educational and scientific work of students. Within the framework of the Higher Education Program of NAU in a foreign language, a team of experienced teachers of the Department of Higher and Computational Mathematics since 2006 has began conducting research on teaching mathematics in English to foreign and Ukrainian students.

In particular, past few years, teachers of two Kyiv universities, NAU and KNU, have been implementing a project approach to the organization of educational and scientific student's work. Also certain problems of teaching of mathematical disciplines to foreign students in English were investigated in [8]. Teaching Higher Mathematics in English to foreign students of construction specialties was considered in [9]. Some general issues of teaching to higher mathematics in English were considered in [10]. Methods of classifying foreign language communicative competences were investigated in [11]. Some features of teaching higher mathematics in English to students of computer specialties were investigated in [12].

When teaching mathematics to foreign and Ukrainian students teachers have to mind that English is not a native language for these students and that students had received secondary school education in their mother tongues. Separately, it should be noted that most foreign students come from countries using the traditional British system of notations for trigonometric functions, while Ukrainian students (and some foreign students) used a system of notations traditional for Ukraine during secondary school. Preparation of master's theses in English at Taras Shevchenko National University of Kyiv and at National Aviation University allows students to demonstrate not only the level of proficiency in the specialty, but also their ability to join the international scientific community. Our experience of teaching in English of IT specialists in our universities confirms the need for the implementation of such projects. It should be noted that the engineering orientation and the scientific orientation of education at different universities are based on the same curricula, differing only in content.

We also believe that it is interesting to assess the practical skills of professional English proficiency for graduates of both universities. We plan to consider these issues in our further research.

3 Application of Collaborative Approach in Multinational Academic Groups on Practical Training in Mathematical Disciplines

In addition to scientific and practical competencies (hard skills), the professional competence of IT specialists also includes social, including communicative, competences (soft skills). Application of collaborative approach will help students develop both hard and soft skills. The general concept of collaborative approach and technologies in mathematics education was investigated in [13–15]. Preconditions for implementing of technique of CLIL (Content and Language Integrated Learning) in teaching mathematics were analyzed in [16, 17]. Approaches to modeling methods considered in [18–20] are used in the guidance of term papers and scientific work of students. Features of the application of collaborative approach for mathematical education in international academic groups were investigated in [8, 21].

The investigation of the efficiency of certain methods of teaching educational material of mathematical disciplines and the organizing the educational process during lectures, practical training and individual work of students was carried out by traditional methods. That is, the current and the semester grades of different groups were compared; the subjective grades of students obtained through anonymous questionnaires and open discussions were analyzed.

It should be noted that the availability of supporting materials is very important for most students. It should also be noted that supporting materials in the form of logical diagrams or flowcharts of the corresponding algorithms are better perceived by students studying in all specialties related to computer science and informatics. In addition, it should also be noted that to conduction lectures in multimedia classrooms is very useful for all students, since the use of various technical means provides excellent opportunities for visualizing educational material.

Implementing the project approach in study of mathematical disciplines, we apply collective forms of work organization. For this purpose we practice dividing the staffs of academic groups into few teams for joint work with complex tasks, cross-checking the assimilation of the education material, creating presentations, etc., followed by discussion and comparison of results. In our opinion, especially interesting results were shown by the formation of teams consisting of students from different countries. Due to this organization of students' work in multinational academic groups the level of academic success increases, English and interpersonal relationships improve. In the process of teaching to mathematical disciplines also it is necessary to give sufficient attention to the clarification of the peculiarities of the use of terminology in the solving of applied research problems. Also, teacher should give methods for the use of CAS (Maple, Mathematica, MATLAB, Scilab, MathCAD etc.).

4 Application of Google Workspace Tools in Mathematical Training of Future IT Specialists in Online and in Blended Learning

Different problems of online and blended learning were investigated in [22–25]. The impact of COVID-19 on the academic performance of students was investigated by in [26]. Some problems of teaching in Computer Science were investigated in [27–29]. Some problems of online and blended learning for mathematical education in multinational academic groups were investigated in [30, 31].

It should be noted that implementation of collective forms of training in practical classes in the context of the coronavirus epidemic has become very difficult. In recent years within the English-language project of NAU joint research to study the features of the project approach to the organization of educational and scientific work of students while distance learning has been conducted. In the 2019/2020 academic year, the conditions of study at many universities not only in Ukraine but also in many other countries have changed significantly: the first semester and beginning of the of the second semester were held as usually under standard conditions, and then the second semester was continued in remote form. In the next 2020/2021 academic year, teachers and students of Kyiv universities faced new difficulties related to the introduction of a blended form of education at the beginning of the semester and the introduction of full distance learning (online) from October 19. During the second semester both lectures and practical classes were held entirely in remote form (online in Google Meet). This situation was extreme for everyone and had particularly catastrophic consequences for the first-year students. Unfortunately, in 2021/2022 academic year in Ukraine conditions for offline, online and blended learning worsened.

Online learning in many universities is organized by means of cloud-based software platforms Zoom, Moodle and Meet. In NAU teachers realize distance learning in Google Workspace for Education. Tools of Google Workspace (formerly Google Apps and later G Suite) give teachers possibility to design and implement the educational process. We conduct classes online on Google Meet using the Jamboard. This requires special work of the teacher to identify the subjective experience of each student. Organization of the educational process in this way allows to more fully demonstrating skills and abilities of each student, to develop the creative potential of the student and his professional abilities, to gain experience in teamwork, which is important for information technology professionals.

In particular, the use of the Google Drive cloud storage, which allows users to store, jointly edit their data on servers in the cloud and share it with other users on the Internet, is very useful for teachers. Application of Google Drive facilitates the effective interaction of participants in the educational process and teaches a digital style of work. Teacher can effectively plan time and optimize management processes, thus forming information and digital competence as a component of designing a new educational environment. Application of considered cloud technologies make possible to organize better joint work of teachers and students that will increase students' achievement. The introduction of cloud technologies helps to effectively and quickly organize the methodical work of the teacher, improves the quality and efficiency of the educational process, as well as prepares students for life in the modern information society.

From the very beginning of the 2021/2022 academic year at the university, teachers have created Google classes for all groups and streams, in which we place learning materials: theoretical statements and examples of problem solving. In addition to our developments, we offer students various online resources for additional use, such as Math24.net, Math is Fun and others.

Distance learning (online and blended) in some universities is realized in the Google Workspace using Google Meet (Fig. 1). Effective organization of practicals, from our point of view, is especially difficult.

Since in National Aviation University distance learning is realized in Google Workspace, then practical classes are conducted using Google Classroom and Google Meet. We also use Google Jamboard. Solving problems in online practical classes in mathematical disciplines, implemented with the application of Google Jamboard, in general was quite effective. However, it should be noted that it places quite high demands on both the computers on which students and teachers work, and the quality of the Internet.

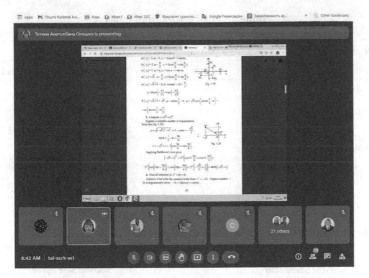

Fig. 1. Practical Classes on Mathematical Analysis

In our opinion, the use of Google Jamboard to organize the work of student teams on practicals, workshops and consultations was especially interesting. During last two academic years this method to organize collective work on online practical classes in mathematical disciplines, in particular higher mathematics, was implemented for English-speaking groups of certain specialties of the Faculty of Cybersecurity, Computer and Software Engineering.

5   Experimental Results and Discussion

We have analyzed the use of collaborative approach applying technologies of online learning based on Google Workspace Tools, classic online and blended learning by examining the results of studies of National aviation university students in the subjects "Higher Mathematics" and "Mathematical Analysis" in the first and second semesters of the 2021–2022 academic year.

In the first semester we used collaborative approach applying technologies of online learning based on Google Workspace Tools, and in the second semester we used classic online and blended learning.

The results of studies of students of English-speaking groups of "Software engineering" and "Computer Systems and Networks" specialties were investigated.

Table 1 contains the analysis of the academic performance of English-speaking students majoring in "Computer Systems and Networks" in section "Mathematical Analysis" of subject "Higher Mathematics" and English-speaking students majoring in "Software engineering" in subject "Mathematical Analysis".

Table 1. Final grades from subjects

Semester	Final Grades (as a percentage)			
	excellent	good	satisfactory	bad
"Higher Mathematics"				
First	21	21	58	0
Second	21	21	37	21
"Mathematical Analysis"				
First	25	54	21	0
Second	21	43	11	25

Figure 2 reflects the results presented in Table 1.

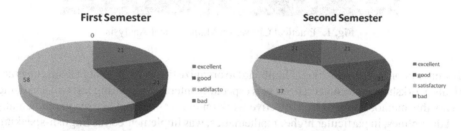

Fig. 2. Final Semester Grades (in percentages)

The samples under consideration are dependent. The same students study the same subject in different semesters and using different methods. This makes it possible to evaluate the effectiveness of different approaches to learning.

The scheme of the algorithm for comparing two technologies is presented in Fig. 3.

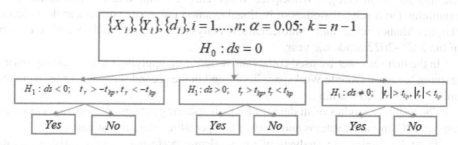

Fig. 3. Algorithm for Comparing Two Technologies, *Yes* - we accept the null hypothesis, *No* - we reject the null hypothesis.

The results of the calculations.

We consider ordered pairs (X_i, Y_i), where X_i are study results in the first semester, Y_i are study results in the second semester. Let's calculate $d_i = Y_i - X_i$, $ds = \frac{\sum_{i=1}^{n} d_i}{n}$.

We will test the hypothesis $H_0 : ds = 0$, about the equality of the mean zero. Alternative hypothesis with one-sided criterion $H_1 : ds > 0$ or $H_1 : ds < 0$, with a two-sided criterion $H_0 : ds \neq 0$.

Let's calculate the statistics according to the formula

$$t = \frac{ds\sqrt{n}}{Sd}, \, Sd = \sqrt{\frac{n\sum_{i=1}^{n} d_i^2 - \left(\sum_{i=1}^{n} d_i\right)^2}{n(n-1)}}$$

The statistic has Student's distribution with $k = n - 1$ degrees of freedom.

Subject "Higher Mathematics" for students of specialty "Computer Systems and Networks". Sample size $n = 29$. Estimated value of the criterion $t_r = 2.5359$ for $\alpha = 0.05$, critical value for the two-sided criterion $t_{kp} = 2.049$. Since $t_r > t_{kp}$, the null hypothesis is rejected.

Subject "Mathematical Analysis" for students of specialty "Software Engineering". Sample size $n = 28$. Estimated value of the criterion $t_r = 2.8007$ for $\alpha = 0.05$, critical value for the two-sided criterion $t_{kp} = 2.052$. Since $t_r > t_{kp}$, the null hypothesis is rejected. The test result shows that the approach proposed in the first semester is more effective.

6 Summary and Conclusion

The need to combine distance and mixed forms of learning and to ensure the formation of the creative personality of the student and the development of student's professional abilities in these conditions is the peculiarity of today educational process. Availability of many platforms for online classes requires finding new approaches to organize the educational process, both in its form and content. The research presented in the paper was based on a collaborative approach to organizing collective work. Conducting offline and online practical classes in mathematical disciplines was implemented in English-speaking multinational academic groups of certain specialties. Online classes and blended learning were conducted using Google Workspace tools. In particular, the implementation of a collaborative approach in online practicals with usage of Google Jamboard turned out to be very effective. This tool allows few teams to work simultaneously on different pages with further discussion and comparison of results.

In general, results of this implementation in online practical classes were more successful for FC CSE than for other faculties. These results may be related to the specifics of students majoring in IT. Therefore, it is recommended to use this approach primarily in the training of students of IT specialties. The collaborative approach to organize the educational process for IT specialties allowed more fully identifying abilities and skills of each student, developing students' creative and professional abilities. Future IT specialists also need skills of team work. Especially important is experience of team work in international teams. In our opinion, the obtained results allow to make some generalizations, confirm the successful choice of the direction of organizing the educational process and encourage further study of this approach.

We have analyzed and compared the effectiveness of various approaches in the educational process when studying mathematical analysis. The testing result shows that the collaborative approach with using Google Jamboard is more effective compared to other forms of organizing online learning based on Google Workspace Tools. In the future work, we plan to investigate the application of this approach in teaching other mathematical disciplines and different approaches to teams forming.

Acknowledgment. This project is supported by Key Projects of China Southern Power Gird (GZ2014-2-0049).

References

1. Virchenko, N.O.: Selected Questions of the Methods of Higher Mathematics. Zadruga, Kyiv (2003). (in Ukrainian)
2. Li, Y., Schoenfeld, A.H.: Problematizing teaching and learning mathematics as "given" in STEM education. Int. J. STEM Educ. **6**, Article no. 44 (2019). https://doi.org/10.1186/s40 594-019-0197-9
3. Abramovich, S., Grinshpan, A.Z., Milligan, D.L.: Teaching mathematics through concept motivation and action learning. Educ. Res. Int. **2019**, Article ID 3745406 (2019). https://doi.org/10.1155/2019/3745406
4. Biziuk, A., Biziuk, V., Shakurova, T.: Analysis of teaching elements on technical and mathematical disciplines in modern distance education. Technol. Audit Prod. Reserves **4**(2(60)), 28–32 (2021). https://doi.org/10.15587/2706-5448.2021.237455
5. Karupu, O.W., Oleshko, T.A., Pakhnenko, V.V., Pashko, A.O.: Applying information technologies to mathematical education of IT specialists in English-speaking academic groups. Bull. Taras Shevchenko Natl. Univ. Kyiv Ser. Phys. Math. **4**, 70–75 (2019). https://doi.org/10.17721/1812-5409.2419/4.9
6. Bolitho, R., West, R.: The Internationalisation of Ukrainian Universities: The English Language Dimension. Stal, Kyiv (2017)
7. Kvasova, O., Westbrook, C., Westbrook, K.: Provision of English-medium instruction: trends and issues. Ars Linguodidacticae **5**, 11–21 (2020). https://doi.org/10.17721/2663-0303.2020.5.02
8. Karupu, O.W., Oleshko, T.A., Pakhnenko, V.V.: About teaching of mathematical disciplines in English to foreign students. East.-Eur. J. Enterp. Technol. **2**(2(56)), 11–14 (2012). (in Ukrainian). https://doi.org/10.15587/1729-4061.2012.3657
9. Fedak, S.I., Romaniuk, L.A., Fedak, S.A.: Teaching the subject "Higher Mathematics" in English to foreign students of construction specialties. Acad. Commentaries. Ser.: Pedagogical Sci. **156**, 106–111 (2017). (in Ukrainian)
10. Snizhko, N.V.: On the problems of teaching the course of higher mathematics in English. In: Science Week: Abstract Annual Science Practical Conference on Zaporizhzhia: ZNTU, pp. 292–293 (2018). (in Ukrainian)
11. Pashko, A., Pinchuk, I.: Methods of classifying foreign language communicative competence using the example of intending primary school teachers. In: Babichev, S., Lytvynenko, V., Wójcik, W., Vyshemyrskaya, S. (eds.) ISDMCI 2020. AISC, vol. 1246, pp. 98–113. Springer, Cham (2021). https://doi.org/10.1007/978-3-030-54215-3_7
12. Rybalko, A.P., Stiepanova, K.V.: Features of teaching higher mathematics in English to students of computer specialties. Profession. Teach.: Theoret. Methodol. Aspects **12**, 33–44 (2020). (in Ukrainian)
13. Clarke, P.A.J., Kinuthia, W.A.: Collaborative teaching approach: views of a cohort of preservice teachers in mathematics and technology courses. Int. J. Teach. Learn. High. Educ. **21**(1), 1–12 (2009)

14. Seidouvy, A., Schindler, M.: An inferentialist account of students' collaboration in mathematics education. Math. Educ. Res. J. **32**, 411–431 (2020). https://doi.org/10.1007/s13394-019-00267-0

15. Vlasenko, K., Chumak, O., Sitak, I., Chashechnikova, O., Lovianova, I.: Developing informatics competencies of computer sciences students while teaching differential equations. Revista Espacios **40**(31), 11 (2019)

16. Tarasenkova, N., Akulenko, I., Kulish, I., Nekoz, I.: Preconditions and preparatory steps of implementing CLIL for future mathematics teachers. Univers. J. Educ. Res. **8**(3), 971–982 (2020). https://doi.org/10.13189/ujer.2020.080332

17. Kyrpychenko, O., Pushchyna, I., Kichuk, Ya., Shevchenko, N., Luchaninova, O., Koval, V.: Communicative competence development in teaching professional discourse in educational establishments. Int. J. Mod. Educ. Comput. Sci. **13**(4), 16–27 (2021). https://doi.org/10.5815/ijmecs.2021.04.02

18. Pashko, A., Oleshko, T., Syniavska, O.: Simulation of fractional Brownian motion and estimation of Hurst parameter. In: Proceedings of the IEEE 15th International Conference on Advanced Trends in Radioelectronics, Telecommunications and Computer Engineering, pp. 632–637. IEEE (2020). https://doi.org/10.1109/TCSET49122.2020.235509

19. Pashko, A.O., Lukovych, O.V., Rozora, I.V., Oleshko, T.A., Vasylyk, O.I.: Analysis of simulation methods for fractional Brownian motion in the problems of intelligent systems design. In: Proceedings of the 2019 IEEE International Conference on Advanced Trends in Information Theory, pp. 373–378. IEEE (2019). https://doi.org/10.1109/ATIT49449.2019.9030478

20. Pakhnenko, V.V.: About some features of teaching ordinary differential equations to future specialists in the aviation industry. In: Proceedings of the XIII International Conference Modern Education – Accessibility, Quality, Recognition, pp. 136–128. DSMA, Kramatorsk (2021). (in Ukrainian)

21. Karupu, O.W., Oleshko, T.A., Pakhnenko, V.V.: Modeling future aviation and IT specialists' professional skills development on mathematical practical training with application of information technologies. In: Proceedings of the 2021 IEEE 3rd International Conference on Advanced Trends in Information Theory (ATIT), pp. 215–220. IEEE (2021). https://doi.org/10.1109/ATIT54053.2021.9678904

22. Lim, D.H., Morris, M.L., Kupritz, V.W.: Online vs. blended learning: differences in instructional outcomes and learner satisfaction. J. Asynchronous Learn. Netw. **11**(2), 27–42 (2007)

23. Borba, M.C., Askar, P., Engelbrecht, J., Gadanidis, G., Llinares, S., Aguilar, M.S.: Blended learning, e-learning and mobile learning in mathematics education. ZDM Math. Educ. **48**(5), 589–610 (2016). https://doi.org/10.1007/s11858-016-0798-4

24. Means, B., Toyama, Y., Murphy, R., Baki, M.: The effectiveness of online and blended learning: a meta-analysis of the empirical literature. Teach. Coll. Rec. **115**(3), 1–47 (2013)

25. Asarta, C.J., Schmidt, J.R.: The effects of online and blended experience on outcomes in a blended learning environment. Internet High. Educ. **44**, Article no. 100708 (2020). https://ejournal.undiksha.ac.id/index.php/JERE

26. Sakpere, A.B., Oluwadebi, A.G., Ajilore, O.H., Malaka, L.E.: The impact of COVID-19 on the academic performance of students: a psychosocial study using association and regression model. Int. J. Educ. Manag. Eng. **11**(5), 32–45 (2021). https://doi.org/10.5815/ijeme.2021.05.04

27. Kuzminykh, Ie., Yevdokymenko, M., Yeremenko, O., Lemeshko, O.: Increasing teacher competence in cybersecurity using the EU security frameworks. Int. J. Mod. Educ. Comput. Sci. **13**(6), 60–68 (2021). https://doi.org/10.5815/ijmecs.2021.06.06

28. Nuga, O.A.: An application of the two-factor mixed model design in educational research. Int. J. Math. Sci. Comput. **5**(4), 24–32 (2019). https://doi.org/10.5815/ijmsc.2019.04.03

29. Adebayo, E.O., Ayorinde, I.T.: Efficacy of assistive technology for improved teaching and learning in computer science. Int. J. Educ. Manag. Eng. **12**(5), 9–17 (2022). https://doi.org/10.5815/ijeme.2022.05.02

30. Karupu, O.W., Oleshko, T.A., Pakhnenko, V.V.: On peculiarities of teaching linear algebra to future IT specialists within the program "Education in English" of the National Aviation University. Phys. Math. Educ. **4**(26), 21–26 (2020). https://doi.org/10.31110/2413-1571-2020-026-4-003

31. Karupu, O.W., Oleshko, T.A., Pakhnenko, V.V.: About teaching certain sections of higher mathematics to students of technical specialties in multinational academic groups at the National Aviation University. Sci. Educ. New Dimension. Pedagogy Psychol. **IX(97)**(246), 17–20 (2021). (in Ukrainian). https://doi.org/10.31174/SEND-PP2021-246IX97-04

The Current Situation and Development Trend of the Evaluation of Moral and Ideological Curriculum in China-Visual Analysis Based on CNKI Publications

Yunyue Wu[✉], Yanzhi Pang, and Xiaoli Teng

School of Transportation, Nanning University, Nanning 530200, China
W17736621947@163.com

Abstract. The implementation of moral education in teaching is to adhere to the combination of knowledge impartation and value orientation, and cultivate students into talents with both morality and talent and all-round development. This is the focus of teaching reform in colleges and universities at this stage. This paper takes curriculum moral education as the research object. This paper uses Citespace software to visually analyze the knowledge map of 474 curriculum ideological and moral evaluation documents in CNKI database from 2017 to 2021 through the number of documents, co-occurrence of authors, co-occurrence of institutions, co-occurrence of keywords and clustering. The results show that the current research focuses on ideological and political courses, teaching reform, teaching design, moral and ideological education and professional courses. The future research should focus on the construction of curriculum moral and ideological indicators and evaluation system, integrate the spirit of ideals and beliefs into knowledge learning, and effectively improve the comprehensive quality and ability of college students.

Keywords: Moral and ideological curriculum · Teaching evaluation · Citespace · Reform in education · Knowledge graph

1 Introduction

At the National Ideological and Political Conference of Colleges and Universities in 2016, China's leadership proposed that professional teachers should carry out Curriculum Ideological and Political Education, which means to promote education in ideology, morality and ethics in curriculum teaching. In December 2017, the Outline for the Implementation of the Quality Improvement Project of Ideological and Political Work in Colleges and Universities was issued, which clearly pointed out that the "ideological and political curriculum", namely, moral and ideological education, should be included in teacher performance assessment. In June 2020, the Guiding Outline for the Construction of Moral and Ideological Education in Colleges and Universities issued by the Ministry of Education once again emphasized that its construction should be comprehensively promoted in all universities and disciplines.

Z. Hu et al. (Eds.): ICCSEEA 2023, LNDECT 181, pp. 950–961, 2023.
https://doi.org/10.1007/978-3-031-36118-0_81

The evaluation of teaching effect is not only a tool to check the effect of teaching reform, but also a reference for continuous improvement [1–4]. As a "baton" to test the implementation of "curriculum ethical and ideological education", the evaluation of "curriculum ideological and moral education" is a problem that schools and teachers are generally confused and urgently need to solve in the process of comprehensive implementation of ideological and moral education.

As an important tool for visualization research, Citespace is widely used in the analysis of research status and research trends. For example, Rawat K S and other scholars used the software to analyze the application of computers in the field of education [5]. For another example, scholars such as Zhang Sheng Liu sorted out the literature on curriculum ideological and political research [6]. Li Xiangxiang and others also used Citespace to draw a map of research hotspots and development trends of Moral and ethical education in university courses [7]. However, most of the existing studies have been carried out from the theme of curriculum Moral and ethical evaluation, and the research focus and trend of curriculum ideological and moral evaluation need to be further deepened. In this context, this paper uses CNKI database literature as the data source to visually analyze the current situation of curriculum Moral and ethical evaluation, and forecasts the research trend, hoping to provide reference for curriculum ideological and political evaluation.

2 Data Sources and Research Methods

2.1 Data Source

The sample data of this paper are all from the CNKI database, with the theme of "Moral and ethical evaluation of courses", the time range from 2017 to 2021, and the source category is the core of Peking University, CSSCI. A total of 474 journal papers were searched. After manual verification by the author and two other scholars, it was finally determined that 474 sample literature data were obtained, which are all valid and valuable references.

2.2 Research Methods

In the work of scientific research, we need to find useful information in a large number of documents and read as comprehensively as possible. In the face of numerous vast literature databases, it is a hard job to select targeted and high-quality articles to school. It is necessary to excavate the frontiers of the discipline, find out the key documents worthy of intensive reading and research hotspots, which can not only broaden the horizon of authors, but also inspire them to generate new ideas and creativity. This is the first problem to be solved before conducting research. CiteSpace is such an excellent bibliometrics software. It can participate in advanced methods to visually display the relationship between documents in the form of scientific knowledge atlas in front of users, so that users can have a general understanding of future research prospects by combing the past research tracks.

This paper uses Citespace software to visually analyze the sample literature. The Citespace software was developed and designed by Professor Chen Chaomei. Domestic scholars have widely used it in information science, information and library science, computer science, business economics and other fields. Through the generation of maps, the structure, laws and distribution of scientific knowledge are presented intuitively and clearly.

3 Data Analysis

For the data analysis of published literature, it mainly includes several aspects, namely: research space-time distribution, source journal distribution, author statistics and cooperation network analysis, as well as the analysis of the author's affiliations and cooperation network.

3.1 Research on Space-Time Distribution

The number of papers published can show the development trend of research topics. This paper makes statistics on the relevant literature published in China from 2017 to 2021. After searching and selecting, the final sample number of papers is 474, as shown in Fig. 1. It can be seen from the figure that in the past five years, the research on ideological and political evaluation of curriculum in China has shown a significant upward trend. In 2017, the number of papers issued was only 7, while in 2021, it has risen to 279, which also shows that the research on ideological and moral evaluation of curriculum has been paid more and more attention by the academic community.

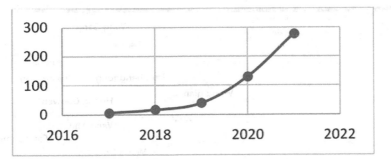

Fig. 1. Distribution of documents published in 2017–2021 on moral and ideological courses evaluation

3.2 Distribution of Source Journals

The source of journals reflects the quality of papers to a certain extent, and also reflects the spatial distribution characteristics of the research field. By analyzing the number of papers issued by journals, the focus and research groups in the research field can be clarified [8]. After analyzing the sample documents of ideological and political evaluation of courses (see Table 1), it is found that Chinese Foreign Languages, Sports Journal, China Audio Visual Education, Journal of Xinjiang Normal University (Philosophy and Social Sciences Edition), Journal of Beijing Sport University and other journals have the most papers on ideological and moral evaluation of courses.

Table 1. List of top 6 journals of ideological and moral evaluation of courses in 2017–2021

number	Journal	number of publications
1	Foreign Languages in China	7
2	Journal of Physical Education	7
3	Journal of Southeast University(Philosophy and Social Science)	4
4	China Educational Technology	3
5	Journal of Xinjiang Normal University(Edition of Philosophy and Social Sciences)	3
6	Journal of Beijing Sport University	3

3.3 Author Statistics and Cooperation Network Analysis

The analysis of the number of papers and cooperation of the authors can clearly understand the key figures in this field and determine the research authority in this field [9]. In this paper, the author used Citespace software to measure the author and the author's cooperation, and obtained the co-occurrence chart of the authors of the course ideology, morality and ethics evaluation study from 2017 to 2021 (see Fig. 2).

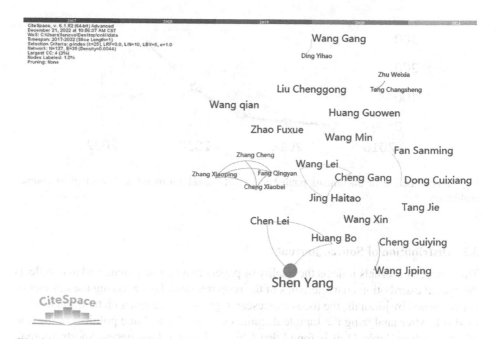

Fig. 2. Co-occurrence of author cooperation in 2017–2021 on the evaluation of moral and ideological curriculum

It can be seen from the figure that the cooperation of scholars is not obvious. Most scholars only have one article on moral and ethical evaluation of curriculum, and Shen Yang has the largest number (3) of articles.

3.4 Analysis of the Cooperation Network of the Sending Agency

The co presentation chart of the publishing houses can show the academic status of the research institution in a certain field. This paper conducts measurement and statistics on the number of documents issued by the issuing institutions, and finally forms a co-occurrence chart of the research institutions (Fig. 3). It can be seen from the figure that the sending institutions are basically from universities, and the institutions with the largest number of documents are Northeast Normal University, Beijing Foreign Studies University, Hehai University, Wuhan University, Southeast University, etc. The above schools have made great contributions to the study of ideological and moral evaluation of courses.

Fig. 3. Co-occurrence of author organizations of 2017–2021 curriculum ideological and moral evaluation

4 Keyword Analysis

4.1 Keyword Co-occurrence Analysis

"Co-occurrence", as the name implies, refers to the phenomenon of the common occurrence of information described by the characteristics of literature. "Co-occurrence analysis" is a quantitative study of co-occurrence phenomenon, which aims to reveal the content relevance of information and the knowledge implied in the characteristic items. The feature items extracted in this paper mainly focus on keywords, including co-occurrence analysis, clustering analysis and emergence analysis of keywords.

Using the keywords in the bibliography, CiteSpace software is used to conduct keyword co-occurrence analysis on the sample literature of curriculum ideological and political evaluation. The author input the sample literature into CiteSpace 6.1 software and got the co-occurrence diagram of research keywords (Fig. 4). It can be seen from the figure that scholars mainly focus on moral and ethical courses, teaching reform, teaching design, ideological and moral education, professional courses, etc. How to carry out moral and ethical evaluation of courses is the focus of scholars' research.

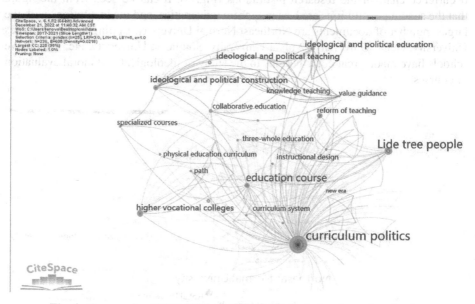

Fig. 4. Key words of 2017–2021 curriculum ideological and moral evaluation

4.2 Keyword Clustering Analysis

The author conducted a cluster analysis of keywords and got Fig. 5. The smaller the number in the figure, the more keywords it contains, and the higher its compactness. According to the summary of the cluster diagram, the author divides it into three aspects.

(1) The teaching reform of ideology, morality and ethics evaluation of professional courses. Curriculum is the foothold of curriculum moral and ethical education. Professional courses should carry out ideological and moral education together with moral and ethical courses, so ideological and political education in professional courses will become the focus of professional curriculum teaching reform. It is not only necessary to integrate moral and ethical elements into professional courses, but also to carry out curriculum ideological and moral evaluation reform. Some scholars have explored the morality, ideology and ethics evaluation of English curriculum, and

believe that the ideological and political evaluation of curriculum should include not only the evaluation of students' acquisition, but also the evaluation of teaching process, as well as the evaluation of teachers [10]; Physical education teachers integrate the spirit of women's volleyball into the volleyball general course, and design the evaluation requirements for ideology, morality and ethics content according to this, including five dimensions of curriculum moral and ideological education, including teacher level and development, curriculum design, teaching content, teaching implementation, student cognition and growth [11]; Teachers of traditional Chinese medicine have designed the evaluation index of "the same standard" for professional teaching evaluation and ideology, morality and ethics education evaluation, with teaching objectives, teaching design, teaching methods and teaching effects as the main evaluation elements [12].

(2) The evaluation principles of ideological and political evaluation of curriculum. The ultimate goal of implementing the curriculum ideology, morality and ethics reform in colleges and universities is to achieve the cultivation goal of "establishing morality and cultivating people". When carrying out the curriculum ideological and moral evaluation, the fundamental requirement of evaluation should be "establishing morality and cultivating people" [13]. The ideology, morality and ethics evaluation of curriculum should follow the basic requirements of effectiveness orientation and student development, and form an evaluation system that avoids rigidity, advocates comprehensiveness, emphasizes perception and views effectiveness [14, 15].

(3) The practical path of ideological and moral evaluation of curriculum. An important system for the implementation of curriculum ideology, morality and ethics evaluation is to build a scientific evaluation system. At present, domestic research on the construction of curriculum ideological and ethical evaluation system mainly focuses on the design of evaluation indicators. The evaluation indicators are mostly based on the experience of experts and scholars, literature and national policy documents. On this basis, scholars have designed different evaluation indicator systems. For example, Tan Hongyan (2020), after a year of research and learning from the concepts and methods of professional evaluation and curriculum evaluation, believes that indicators at the level of schools and secondary units should include six dimensions: top-level design, teacher team building, teaching system, textbook construction, quality assurance and output of results, while indicators at the level of teachers cover five parts: syllabus, teaching content, teaching methods, learning effectiveness and teaching reflection [16]; Sun Yuedong (2021) and other scholars used literature research, questionnaires and expert consultation methods to design an evaluation index system that combines peer evaluation, student evaluation and teacher self-evaluation [17]; The evaluation of students is mostly carried out by teachers according to the characteristics of the curriculum, focusing on the process evaluation and value-added evaluation of students [18, 19].

4.3 Keyword Emergence Analysis

Emergent words are keywords with a sudden increase in citation frequency in a certain period of time, which can be used to reflect the research trend in a certain period of time;

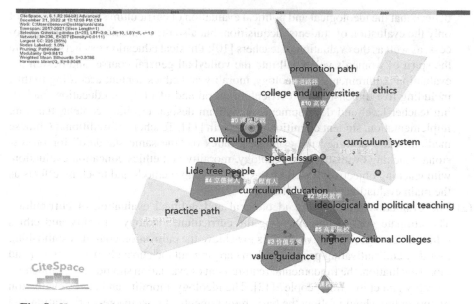

Fig. 5. Keyword cluster diagram of 2017–2021 curriculum ideological and moral evaluation

Judge and identify frontier hotspots, and master the evolution of research topics in the field [20]. Generate the keyword emergence diagram of curriculum ideology, morality and ethics evaluation through citespace (as shown in Fig. 6). It can be seen from the figure that from the timeline, keywords can be divided into two periods. The first period is from 2017 to 2018, including key words such as ideological and moral construction, value guidance, educational function, and curriculum education. In this period, most scholars focus on the role of curriculum ethical and ideological, as well as the theoretical basis of curriculum ideological and moral evaluation [21]; After 2020, the emerging keywords include professional ideology, morality and ethics education, teaching methods, curriculum reform, teaching quality, etc. Scholars pay more attention to the reform and practice of curriculum ideological and political evaluation in the curriculum [22], and the construction of curriculum ideology, morality and ethics evaluation system [23, 24].

5   Results and Enlightenment of CiteSpace Analysis

Ideological and moral curriculum is an educational and teaching concept, which plays a role in cultivating college students' world outlook, outlook on life and values. It is also a way of thinking. The discipline thinking of moral education is used to refine the cultural genes and value paradigms contained in professional courses, transform them into concrete and vivid effective teaching carriers, and integrate the spiritual guidance of ideal and belief into students' knowledge learning in its unique and silent way.

In the ways of combination of infusion and infiltration, theory with practice, history and reality, explicit education and implicit education, commonness and individuality, as

Top 25 Keywords with the Strongest Citation Bursts

Keywords	Year	Strength	Begin	End	2017 - 2021
ideological and political construction	2017	1.55	2017	2017	
educating the person as well as imparting book knowledge	2017	1.28	2017	2017	
knowledge teaching	2017	1.28	2017	2017	
value guidance	2017	1.08	2017	2017	
educational function	2017	1.04	2017	2019	
curriculum education	2017	1.72	2018	2018	
fusion	2017	1.09	2018	2019	
traditional chinese medicine	2017	1.09	2018	2019	
new era	2017	1.05	2018	2019	
education course	2017	2.08	2019	2019	
professional thought and administration	2017	2.02	2019	2021	
collaborative education	2017	1.8	2019	2019	
postgraduate	2017	1.34	2019	2021	
top-level design	2017	1.2	2019	2019	
realization path	2017	0.92	2019	2019	
teaching method	2017	1.29	2020	2021	
value	2017	1.03	2020	2021	
connotation	2017	1.03	2020	2021	
pe teacher	2017	0.86	2020	2021	
general course	2017	0.86	2020	2021	
research status	2017	0.86	2020	2021	
subject moral education	2017	0.86	2020	2021	
curriculum reform	2017	0.86	2020	2021	
predicament	2017	0.86	2020	2021	
higher vocational english	2017	0.86	2020	2021	

Fig. 6. Key words of 2017–2021 curriculum ideological and moral evaluation

well as the combination of positive education and discipline, moral and ethical education in curriculum teaching is capable to be improved, so as to guide students' moral development to the correct and healthy direction.

In this paper, CiteSpace software is used to conduct a quantitative and qualitative analysis of the research literature on curriculum ideological and moral evaluation. It is found that although the research on curriculum ethical and ideological evaluation has achieved fruitful results, it still needs to be further deepened from the following aspects:

(1) In terms of research methods. At present, most of the scholars' research is based on personal experience summary, with more qualitative research and more practical research. However, theory is the basis for guiding practice. For how to evaluate the ideological and moral curriculum, we need to adopt a variety of research methods, and the combination of quantitative and qualitative needs to be further deepened;

(2) As for the research content, the research focuses on evaluation subject, evaluation object, evaluation principle and evaluation index. Although some achievements have been made, a universally recognized evaluation system has not been formed;

(3) From the perspective of system construction, although various schools around the country have carried out a lot of practice, some regions and universities have not yet formed a system and mechanism for how to carry out ideological and moral evaluation of courses, lacking a complete supervision and evaluation mechanism.

6 Conclusion and Outlook

6.1 Conclusion of the Research

In this paper, Citespace software is used to analyze the current literature on ideological and moral evaluation of courses. Through word frequency co-occurrence, word frequency clustering and word frequency emergence, the following conclusions are drawn:

(1) From the above analysis, it can be seen that the evaluation of curriculum ethical and ideological education is a hotspot in the current academic research. The exploration of the theoretical basis and practical logic of curriculum ideology, morality and ethics education will continue for a very long time, and the number of papers in the future will continue to rise.

(2) Through keyword co-occurrence and cluster analysis, we can see that the domestic focus on the evaluation of moral and ideological teaching quality of the curriculum is mainly; The teaching reform of ideological and ethical evaluation of specialized courses, the evaluation principles of ideological and moral evaluation of courses and the practical path of ideological and political evaluation of courses. Through the keyword emergence analysis, we can see that the construction of the evaluation system of the ideological and moral teaching quality of the curriculum will be the focus of the study.

6.2 Outlook for Future Research

Based on the above analysis, we can see that a lot of research has been done on the evaluation of the teaching quality of ideological and moral courses in China, with many achievements. However, the current research on the evaluation of ideological and moral education in curriculum teaching in colleges and universities is not deep enough, and there is still a certain gap with the actual needs of the reform of ideological and ethical courses. Future research should focus on the following two aspects.

(1) The evaluation index of curriculum ideology and politics is the key to the evaluation of curriculum ideology and politics

It is an important part of the construction of the curriculum ideological and moral evaluation system to establish an evaluation index for the teaching quality of ideological and moral education, which has both reliability and validity and is easy to operate. At present, no authoritative evaluation indicators have been formed. The evaluation indicators designed by scholars are only based on a single school, or a major, or a single course. To form high-quality evaluation indicators, a lot of research and data support are needed.

(2) The evaluation system of ideological and moral education in curriculum teaching is the guarantee of the implementation of ideological and political evaluation of curriculum.

The ideological and moral education of the curriculum cannot be achieved alone, but requires the cooperation of "five aspects of education". Then we should do a good job in top-level design and formulate corresponding systems, such as making clear the status of curriculum ideological and political education, making good curriculum ideology, morality and ethics planning, investing funds to ensure the implementation of curriculum ideological and moral education, improving teachers' teaching ability of curriculum ideological and ethical education, and including curriculum ideology, morality and ethics performance into the annual assessment, etc.

Acknowledgment. This paper is supported by three project funds: (1) The 2023 of the 14th Five-Year Plan for Educational Science in Guangxi "Research on the Quality Evaluation System of Ideological and Political Education in Colleges and Universities (2023C691)"; (2) The 2022 Guangxi Vocational Education Teaching Reform Research Project "Research and Practice on the Effective Connection of the Transportation Logistics Major Group Undergraduate Courses and the Construction of the Curriculum System" (GXGZJG2022B181); (3) The 2022 Guangxi Higher Education Undergraduate Teaching Reform Project "Research and Practice on the Evaluation System of the Teaching Quality of Applied Colleges and Universities in the OBE Perspective" (2022JGA390).

References

1. He, Y., Yang, M.: Study on quality evaluation system for university students. Int. J. Wirel. Microwave Technol. (IJWMT) **1**(3), 29–34 (2011)
2. Nicolas Gravel, C.A., Levavasseur, E., Moyes, P.: Eval. Educ. Syst. **53**(45), 5177–5207 (2021)
3. Zhong, L., Qi, C., Gao, Y.: Deep learning-assisted performance evaluation system for teaching SCM in the higher education system: performance evaluation of teaching management. Inf. Resour. Manag. J. **35**(3), 1–22 (2022)
4. Amjad, M., Linda, N.J.: A web based automated tool for course teacher evaluation system (TTE). Int. J. Educ. Manag. Eng. (IJEME) **10**(2), 11–19 (2020)
5. Rawat, K.S., Sood, S.K.: Knowledge mapping of computer applications in education using CiteSpace. Comput. Appl. Eng. Educ. **29**(5), 1324–1339 (2021)
6. Liu, Z., Zhu, S., Liu, G.: Visualization analysis of curriculum moral and ideological research in China from the perspective of bibliometrics. Creat. Educ. **10**(10), 2201–2218 (2019)
7. Li, X., Zhu, F., Sun, Z.: Research hotspot and development trend of ideological and political curriculum in colleges and universities - visual analysis based on Citespace knowledge map. J. Dali Univ. **5**(01), 42–48 (2020). (in Chinese)
8. Du, L., Hao, Z.: The current situation, context and trend of family finance research - a visual analysis based on WoS and CNKI journal papers. J. Southwest Minzu Univ. (Human. Soc. Sci. Edn.) **40**(02), 114–124 (2019). (in Chinese)
9. Zhang, H., Hou, Y.: Summary of research on innovation behavior of domestic employees based on knowledge map. Sci. Technol. Progr. Countermeas. **34**(11), 153–160 (2017). (in Chinese)
10. Liu, B., Feng, L.: Construction of ideological and political system for English major courses: realistic difficulties and breakthrough paths. Foreign Lang. Audio Vis. Teach. **2022**(04), 23–28+112 (2022). (in Chinese)
11. Zhang, M., Yuan, F., Liang, Z.: Research on the ideological and political theory and practice of college volleyball curriculum in the context of the integration of sports and education – the design of the volleyball general course with the integration of women's volleyball spirit. J. Beijing Sport Univ. **44**(09), 156–165 (2021). (in Chinese)
12. Fang, W., Bao, Y.: Exploration of ideological and political education in traditional Chinese medicine. China Vocat. Tech. Educ. **35**, 55–60 (2020). (in Chinese)

13. Ao, Z., Wang, Y.: Research on the value core of "ideological and political curriculum" in colleges and universities and the choice of its practice path. Heilongjiang High. Educ. Res. **37**(03), 128–132 (2019). (in Chinese)
14. Lu, D.: Design and implementation of ideological and political evaluation of courses. Ideologic. Theoret. Educ. **03**, 25–31 (2021). (in Chinese)
15. Du, Z., Zhang, M., Qiao, F.: Principles, standards and operating strategies for teaching evaluation of ideological and political education in science and engineering courses. Ideologic. Theoret. Educ. **2020**(07), 70–74 (2020). (in Chinese)
16. Tan, H., Guo, Y., Wang, J.: Construction and improvement of ideological and political evaluation index system of university curriculum. Res. Teach. Educ. **32**(05), 11–15 (2020). (in Chinese)
17. Sun, Y., Cao, H., Yuan, X.: Research on the construction of the evaluation index system of ideological and political teaching of science and engineering courses. J. Jiangsu Univ. (Soc. Sci. Edn.) **23**(06), 77–88+112 (2021). (in Chinese)
18. Bi, J.: Building the "trinity" of "curriculum, thought and politics" – taking the economics course as an example. J. Shanxi Univ. Finan. Econ. **42**(S2), 57–60+71 (2020). (in Chinese)
19. Xie, X., Wang, L.: Mining of ideological and political elements and teaching design of agricultural extension course based on MOOC. J. Southwest Normal Univ. (Nat. Sci. Edn.) **46**(08), 132–139 (2021). (in Chinese)
20. Li, J., Chen, C.: SiteSpace: Scientific Text Mining and Visualization, 2nd edn. Capital University of Economics and Business Press, Beijing (2017). (in Chinese)
21. Qiu, W.: The value implication and generation path of ideological and political curriculum. Ideologic. Theoret. Educ. **07**, 10–14 (2017). (in Chinese)
22. Dong, C., Fan, S., Gao, Y.: Theoretical basis and structural system construction for the establishment of ideological and political elements in physical education major courses. J. Phys. Educ. **28**(01), 7–13 (2021). (in Chinese)
23. Jia, Z., Han, X.: Construction of evaluation system on multimedia educational software. Int. J. Educ. Manag. Eng. (IJEME) **3**(1), 34–38 (2013)
24. Ning, H.: Analysis and design of university teaching evaluation system based on JSP platform. Int. J. Educ. Manag. Eng. (IJEME) **7**(3), 43–50 (2017)

Target Performance of University Logistics Teachers Based on the Combination of AHP and Support Vector Machine in the Context of "Double First-Class" Project Construction

Lei Zhang[✉]

School of Management, Guangzhou City University of Technology, Guangzhou 510800, China
495997665@qq.com

Abstract. In the future for a long period of time, "double first-class" has become the direction and subject of higher education reform and development. Based on the historical background of "double first-class" construction, the construction of teacher performance evaluation system in colleges and universities will change accordingly. The construction of teacher performance evaluation system is particularly important. As a means and tool of evaluation, it can be applied to the management of teacher performance evaluation in colleges and universities to verify the performance of teachers' responsibilities, improve their initiative, enhance their sense of competition, and improve their teaching level. Thus, a good competition and incentive mechanism can be formed, which has gradually become an important part of university management. Under the background of "double first-class" construction, this paper selects the main indicators that affect the target performance of logistics teachers in universities and colleges, and constructs an evaluation system based on them. At the same time, the principle of analytic hierarchy process is used to build the hierarchy index system, and the support vector machine is used for comprehensive evaluation, in order to further improve the teaching level of logistics teachers in colleges and universities, scientific research ability, team assistance and other aspects of exploration, in order to do their utmost.

Keywords: Logistics teacher · Target performance appraisal · AHP-SVM

1 Introduction

The General Plan for Promoting the Construction of World-class Universities and Disciplines was issued by The State Council in October 2015, with the purpose of enhancing the international competitiveness and comprehensive strength of China's higher education, accelerating the construction of world-class disciplines and a number of world-class universities. At the same time, with the advent of industry 4.0 era and the implementation of the strategy of "Made in China 2025", higher requirements are put forward for higher vocational colleges to create first-class majors, strengthen connotation construction, improve the quality of talent training, strengthen the integration of production and education, and build first-class universities. Among them, teachers are the basic resources for personnel training and school teaching quality [1]. Only when the professional level and comprehensive quality of college teachers are improved, the connotation construction, first-class specialty and quality of personnel training of higher vocational colleges can be improved. At present, higher vocational colleges have entered a stage of rapid development, and performance appraisal is an effective tool for modern enterprises to improve their competitiveness. If it is scientifically applied to teachers' management in colleges and universities, the performance of teachers' responsibilities can be investigated and a good incentive mechanism can be formed [2]. However, there are many problems in the evaluation of teachers in higher vocational colleges. The evaluation system, methods and ways have not been formed and perfected, which is far from the requirements of the new situation. It's very important. This paper puts forward the evaluation index and construction system of teachers' goal performance in higher vocational colleges based on analytic hierarchy process under the new situation, and uses support vector machine to evaluate it, which is a beneficial exploration to give full play to teachers' subjective initiative, enthusiasm and creativity and explore the establishment of an incentive mechanism [3].

Foreign scholars have made research on earlier performance evaluation, Chinese scholars in recent years have also made research in this field and made some achievements, in recent years, the performance evaluation of colleges and universities has been studied [4]. The research methods are mainly literature analysis, induction, investigation and comparative analysis. In terms of research content, there are many researches on the index system and evaluation methods of teacher performance evaluation [5]. In recent years, domestic scholars' researches on how to carry out the reform of teacher performance evaluation under the background of "double first-class" construction are also increasing [6]. However, in general, the current research still has the following problems: First, the breadth and depth of the research is insufficient, there are more researches on general and universal problems, but less special researches [7]. Second, the research field of vision is insufficient. At present, most of the research in China is carried out for ordinary universities, and there are few researches on world-class universities [8]. Third, empirical research is insufficient. There are relatively many literatures based on the methods and index systems of teacher performance evaluation in colleges and universities, but most of them lack the test of empirical cases and lack of case support. Fourth, there are more qualitative studies, but not enough quantitative ones [9].

In the past, only one method was used to study the performance evaluation of teachers in colleges and universities. However, the performance appraisal of college teachers is characterized by complex assessment objectives, difficult to quantify assessment indicators and large subjectivity, so the method combining qualitative analysis and quantitative analysis is more suitable for high vocational school teacher performance assessment. Therefore, this paper puts forward an analytic hierarchy process-support vector machine (AHP-SVM) based performance evaluation method for college teachers.

2 Principles Description

2.1 Principles of the Analytic Hierarchy Process

T.L. Schatty, a well-known operations research scholar, proposed the Analytic Hierarchy Process (AHP) in the 1970s. The analytical hierarchy process (AHP) is a multi-objective decision-making method that combines qualitative and quantitative analysis. It can quantify decision-makers' qualitative judgment. It can produce satisfactory results when the target structure is complex and data is scarce, particularly for complex problems that are difficult to fully quantify [10].

AHP divides complex problems into different elements and merges these elements into different levels by analyzing the factors involved and their interrelationships. This results in a multi-level structure. A judgment matrix can be established at each level by comparing the elements of the layer one by one according to a specified criterion. The weight of the layer elements to the problem can then be calculated by calculating the maximum eigenvalue of the judgment matrix and the orthodontic eigenvector. The combined weights of the elements of each level for the overall objective are calculated on this basis, and the weights of different schemes are obtained, which provides the basis for selecting the optimal scheme [11].

2.2 Principles of Analytic Hierarchy

Statistical learning theory was a machine learning method with a solid theoretical foundation that gradually developed on the basis of traditional statistics, gradually improved, and formed a relatively complete theoretical system in the 1990s. A new pattern recognition method, support vector machine, is proposed on this basis [12].

Based on Statistical Learning Theory (SLT), V. Vapnik proposed the Support Vector Machines (SVM) method [13]. It is a new machine learning method that has demonstrated many unique and exceptional advantages in solving nonlinear, small sample, and high dimensional model problems. It has produced good results in probability density estimation, pattern recognition, function approximation, and many other areas, and it can effectively realize the accurate fitting of high-dimensional nonlinear systems using small samples [14]. Its regression function is as follows:

$$y_i[(w \cdot x_i + b)] - 1 \geq 0 \ i = 1, \ldots, n \tag{1}$$

The optimal classification surface problem of the above formula is transformed into a dual problem by Lagrange optimization method, and the maximum value of the following function is obtained.

$$Q(a) = \sum_{i=1}^{n} a_i - \frac{1}{2} \sum_{i,j-1}^{n} a_i a_j \ y_i y_j k\left(x_i, x_j\right) \tag{2}$$

The Lagrange multiplier corresponds to each sample, and the corresponding sample is the support vector, and the optimal classification function can be obtained [14].

$$f(x) = sgn\left\{\left(W^* \cdot x\right) + b^*\right\} = sgn\left\{\sum_{i=1}^{n} a_i^* y_i k\left(x_i, x_j\right) + b^*\right\} \tag{3}$$

Gaussian radial basis kernel function algorithm has a wide convergence range and only one parameter, which is easy to optimize, so it is the most widely used in support vector machines. In this paper, radial basis kernel function is also used for calculation [15–18, 19].

2.3 Combination of Analytic Hierarchy Process and Support Vector Machine

The AHP (Analytic Hierarchy Process) method decomposes the relevant elements of a problem into objective, standard, scheme and other levels, quantifies people's subjective judgment objectively with a certain scale, and carries out qualitative and quantitative analysis on this basis. This method is suitable for human qualitative judgment, and the result of decision is difficult to be measured directly and accurately (Table 1).

Table 1. Target performance appraisal system index set of vocational college teachers

Target layer	Criteria layer	Indicator layer	Description
Target performance appraisal of teachers in vocational colleges (A)	Basic teaching work(B_1)	Number of class hours(C_1)	Complete the basic teaching workload
		Graduation guidance (C_2)	Guide students in professional comprehensive practice
		Graduation thesis (C_3)	The guidance content is detailed and the paper format is standard
		Final examination paper (C_4)	The proposition is in line with the syllabus
		Teaching related materials (C_5)	File format specification
		Teaching plan (C_6)	The format of teaching plan is standard and the content is designed
		Experimental teaching materials (C_7)	File format specification
		Number of course changes (C_8)	No semester transfer or temporary transfer
		Invigilation times (C_9)	10 times per semester
		Teaching effect (C_{10})	The result of students' evaluation of teaching was > 95

(*continued*)

Table 1. (*continued*)

Target layer	Criteria layer	Indicator layer	Description
	Discipline and specialty construction (B₂)	Discipline construction (C₁₁)	Participate in the discipline special construction plan
		Teaching research and reform work (C₁₂)	Establishment of teaching research and reform projects
		Professional curriculum construction (C₁₃)	Participate in the construction of professional curriculum archives
		Practice work (C₁₄)	Maintain and develop practice base
		Teaching and research office activities (C₁₅)	Carry out more than 5 teaching and research activities per semester
		Instruct students to compete (C₁₆)	Three cup competitions
	Scientific research work (B₃)	Scientific research achievements (papers, research, patents) (C₁₇)	School research performance rating standards
		Academic exchange (C₁₈)	Attend academic conference
		Research institutions and platforms(C₁₉)	Provincial, departmental and university levels
		Scientific research application (C₂₀)	Provincial, departmental and university levels
		Master's program construction (C₂₁)	Teacher's report
		Group activity (C₂₂)	Attendance and participation in group activities

SVM (Support Vector Machines) is a novel machine learning method based on statistical learning theory. It has demonstrated numerous distinct advantages in solving small sample, nonlinear, and high-dimensional pattern recognition problems, as well as good results in pattern recognition, function approximation, and probability density estimation.

The evaluation index system is divided into hierarchical structure by combining analytic Hierarchy Process (AHP) and support vector machine (SVM), and the weight of each index is determined by AHP, and the comprehensive evaluation results are obtained by SVM training.

3 Application and Test of SVM

The objective performance indicators used to evaluate teachers in higher vocational colleges are all non-linear and interconnected. As a result, evaluation methods such as the analytic hierarchy process, the fuzzy comprehensive evaluation method, and the factor analysis method are all based on the independence and linear relationship of the indicators, which cannot fully and efficiently solve the evaluation of teachers' target performance in higher vocational colleges. SVM must be used to train and self-learn the training samples in order to better describe the relationship between each evaluation index and the overall objective evaluation. Only through the statistical learning of the training samples can we constantly discover and explore the internal rules between input and output values. Support Vector Machine (SVM) is a data-driven "black box" modeling method with advantages such as self-learning habit, memorability, good "robustness", simple operation and no more prior knowledge. Compared with the neural network method, it can effectively deal with the small sample problems that cannot be solved by it. Moreover, support vector machine has advantages in the evaluation of nonlinear problems. Using it for data processing, it can better overcome the nonlinear correlation between various indicators and make the evaluation results more accurate and objective.

In the application process of SVM, the determination of kernel function is the most critical step, which needs to transform the input space into a high-dimensional space through the nonlinear transformation defined by kernel function. SVM is actually a linear classifier determined by support vector.

3.1 Selection of SVM Kernel Function

Kernel functions most commonly used in support vector machines mainly fall into the following four types:

(1) Gaussian radial basis kernel function

$$K(x_i, x_j) = exp\left(-\gamma \|x_i - x_j\|^2\right) \tag{4}$$

(2) Polynomial kernel function

$$K(x_i, x_j) = \left((x_i \cdot x_j) + 1\right)^d \tag{5}$$

(3) Linear kernel function

$$K(x_i, x_j) = (x_i \cdot x_j) \tag{6}$$

(4) One-dimensional Fourier kernel function

$$K(x_i, x_j) = \left(1 - q^2\right) \Big/ \left[2\left(1 - 2q\,cos(x_i - x_j) + q^2\right)\right] \qquad (7)$$

Radial basis kernel function is the most widely used among the above four kinds of kernel functions. Because it has a wide convergence domain, contains only one parameter and is easy to optimize, it is a relatively ideal classification basis function. Therefore, this paper adopts radial basis kernel function for calculation.

3.2 Training of SVM Model

The sample data of previous college teachers' performance evaluation were collected and divided into two groups, each of which included those with good performance evaluation and those with poor performance evaluation. The data in this group were taken as training samples, Libsvm2.9 toolbox was adopted and gnuplot tool was used for parameter optimization to determine the final SVM. The training samples are input into SVM for training, and then the test samples are input into the model for testing. If the expected accuracy is not reached, the SVM is trained again, and satisfactory output results are obtained through repeated training.

The 183 sets of performance evaluation data of teachers of logistics major in a vocational college in recent three years were used as sample data set, and the sample data were trained for the SVM model. After the training, 183 groups of samples were judged by the back estimation method. The discriminant results were consistent with the actual situation of gas risk occurrence, and the misjudgment rate of the back judgment was 0, indicating that the training results of the prediction model were highly correct and could be applied in practical practice.

4 The Empirical Evaluation of Teachers' Performance in Vocational Colleges

The complexity and multi-objectives of the qualitative evaluation of university teachers' performance evaluation are difficult problems. This paper proposes a method combining analytic Hierarchy process (AHP) and support vector machine (SVM).

On the basis of literature analysis and expert team, AHP is firstly used to hierarchize assessment indicators, decompose assessment objectives into different factors and form a multi-level assessment structure to build an indicator system. 22 evaluation indicators are made into an assessment scale, and the evaluator scores the performance of the evaluated on each item to form a sample data set. Secondly, SVM method is used to carry out hierarchical evaluation of evaluation indicators, and finally get the comprehensive evaluation of college teachers' performance evaluation. This method combines qualitative and quantitative methods, is practical and easy to operate, and provides objective, accurate, comprehensive and feasible basis for teachers' performance evaluation in vocational colleges.

In this paper, 183 sets of performance evaluation data of teachers of logistics major in a vocational university in recent three years are selected as sample data.

(1) By using ROSETTA software, the continuous indicators among the 22 indicators of evaluation are discretized. Then, the 183 test samples are used to determine C_1, C_3, C_{10}, C_{11}, C_{13}, C_{16}, C_{17}, C_{20} as the main indicators.
(2) Select the appropriate kernel function of support vector machine. At present, there are three types of inner product kernel functions that are widely used. Generally, radial basis function is considered first when selecting kernel function, mainly because of its relatively few parameters and small numerical difference.

Gaussian function in support vector machines, has a wide convergence domain and only contains one parameter, which is easy to optimize. Therefore, radial basis kernel function is adopted in this study.

In the MATLAB6.5 environment, LS-SVMlab1.5 calculation kit was used for calculation analysis. The radial basis function was used as the kernel function. The first 140 groups of 183 sample data were used as the training sample set, and the last 43 groups of samples were used as the test sample set to test the generalization ability of the model.

(3) Sample training is an important task. 140 training samples were input into the SVM-TRAIN tool for training, and the remaining 43 samples were input into the SVM-PREDICT tool for testing.
(4) Sample test is the last job. After the completion of sample training, through the accuracy of the actual output, it was found that the correct rate of discriminating 43 test samples reached 97.6%. Therefore, the model established in this paper has high accuracy.

For the university teacher performance evaluation model constructed in this paper, it has the advantages of simple use, high efficiency, high precision, and not easy to be affected by subjective factors. Based on the accumulation of certain sample data, the model can be continuously trained, so as to further improve the accuracy of the model and provide scientific guarantee for teachers' performance evaluation.

5 Conclusion

The construction of the performance appraisal system for vocational and vocational teachers is a systematic project. The establishment of scientific performance indicators can maximize the work enthusiasm and initiative of the staff, and change the teacher management from form to connotation. It is necessary to bring the establishment of performance appraisal standards, appraisal procedures, and appraisal openness and transparency into the scope of system construction, and strengthen supervision. In order to build the first-class teaching team under the background of "double first-class" in higher vocational colleges, we need to pay more attention to the professional development of teachers and improve their teaching, scientific research and practical abilities through multiple channels. It is necessary for schools to formulate school development strategies from a long-term perspective according to different stages of development, so as to achieve a multi-dimensional and multi-level performance appraisal system for schools, so as to rapidly improve the comprehensive ability of teachers.

To realize the strategic goal of "double first-class" construction, it is necessary to strengthen the construction of first-class teaching staff, which is also the urgent work of all colleges and universities at present. "Double first-class" development needs to be performance-centered and driven by innovation. It requires bold innovation, active exploration and quick implementation. At present, the research on teacher performance evaluation in higher vocational colleges is still in the initial stage of exploration, and there are still a lot of work to be done.

Based on literature analysis and expert group discussion, this paper proposes an evaluation model based on AHP-SVM, and evaluates the established target performance evaluation index system of higher vocational teachers. The accuracy and rationality of this evaluation method are demonstrated by the comprehensive evaluation of the teachers' goal performance in a vocational college. Through this method, it is helpful to explore the vocational teachers' target performance evaluation and provide more extensive ideas.

References

1. Zhong, B.: Firmly promote the construction of world-class universities and disciplines. Educ. Res. **39**(10), 12–19 (2018). (in Chinese)
2. Wang, H., Wu, F.: Research hotspots and development of teacher performance evaluation in Chinese universities. J. Natl. Inst. Educ. Adm. **11**, 45–52 (2016)
3. Zhao, L., Huang, P.: Teacher performance evaluation in colleges and universities: pursuing the balance between quantity and quality. J. Xidian Univ. (Social Science Edition) **26**(3), 91–95 (2016). (in Chinese)
4. Wang, B., He, M.: Salary system of university teachers under the background of "double first-class" construction. China High. Educ. **5**, 14–17 (2017). (in Chinese)
5. Yao, X.: Discussion on university teacher performance management system to promote the development of "double first-class" strategy. Natl. Inst. Educ. Adm. **2**, 57–62 (2017). (in Chinese)
6. Wang, B.: Salary reform in colleges and universities: improving quality and efficiency and stimulating teachers' potential-salary system under the background of comprehensive reform in colleges and universities research. China High. Educ. **7**, 9–13 (2016). (in Chinese)
7. Lu, D.: Research on teacher professional development of Stanford university and its enlightenment. China High. Educ. Res. **3**, 48–54 (2014). (in Chinese)
8. Li, C., Zhang, l, Su, Y.: An empirical study on the relationship between salary structure, job satisfaction and job performance of university teachers. Fudan Educ. Forum **5**, 89–95 (2016). (in Chinese)
9. Lu, W., Song, X.: A study on the performance evaluation of university administrators based on analytic hierarchy process – a case study of a university in Guangdong province. High. Educ. Explor. **10**, 40–46 (2017). (in Chinese)
10. Wang, L., Xu, S.: Introduction to analytic hierarchy process, pp. 37–155. China Renmin University Press, Beijing (1990). (in Chinese)
11. Vapnik, V.: The nature of statistical learning theory, pp. 56–112. Tsinghua University Press, Beijing (2000). (in Chinese)
12. Zhang, X.: On Statistical learning theory and support vector machines. Acta Automatica Sinica **26**(1), 32–34 (2000)
13. Zhang, L.: Evaluation of China's petroleum security system: based on rough set and support vector machine. China Soft Science **11**, 13–19 (2022). (in Chinese)

14. Zhang, Z.: Research on intrusion detection model of wireless sensor networks based on RS-SVM. Intell. Comput. Appl. **3**, 319–320 (2019)
15. Júnior, P.R.M., Boult, T.E., Wainer, J., Rocha, A.: Open-set support vector machines. IEEE Trans. Syst. Man Cybern. Syst. **52**(6), 3785–3798 (2022)
16. Piccialli, V., Sciandrone, M.: Nonlinear optimization and support vector machines. Ann. Oper. Res. **314**(1), 15–47 (2022)
17. Yadav, O.P., Pahuja, G.L.: Bearing fault detection using logarithmic wavelet packet transform and support vector machine. Int. J. Image, Graph. Signal Process. (IJIGSP) **11**(5), 21–33 (2019)
18. Zhou, L., Amoh, D.M., Boateng, L.K., Okine, A.A.: Combined appetency and upselling prediction scheme in telecommunication sector using support vector machines. Int. J. Mod. Educ. Comput. Sci. (IJMECS) **11**(6), 1–7 (2019)

Research and Practice of Outcome-Oriented Innovative Talent Training System of Material Specialty

Hu Yang and Qian Cao(✉)

School of Materials Science and Engineering, Wuhan University of Technology, Wuhan 430070, China

caoqian@whut.edu.cn

Abstract. In the context of emerging engineering, the new education model based on outcome is the direction of teaching reform. Traditional training systems of material professionals can no longer meet the needs of the current society and industry. Thus, it is necessary to reform training systems to promote the balanced development of students' professional competency, engineering literacy and development ability. Aiming at material specialty, this paper analyzes the problems existing in traditional talent training system, and studies the latest talent needs in society by tracking and investigating graduates and employers. Based on the above analysis and research, a new talent training course system, teaching system and evaluation system are constructed. After the implementation of this talent training system, students' participation in innovation and entrepreneurship training, competitions and other extracurricular activities has increased significantly. Meanwhile, students' non-technical ability has also been significantly improved. The acceptance rate of graduate students has reached more than 50%, and 73.97% of graduates are still engaged in material-related work.

Keywords: Material specialty · Outcome-based education · Emerging engineering · Talent training system · Non-technical ability evaluation

1 Introduction

Different themes and connotations of different industrial development era have put forward different needs for engineering and technical talents, and therefore promoted the continuous change and development of engineering education [1, 2]. Cultivating new-type engineering and technological talents to face the industry, the world, the future, and to support economic development of the innovation era, has become the fundamental purpose of Emerging Engineering Education [3–5]. The traditional talent training system of materials majors can no longer meet the development of Emerging Engineering Education. Thus, it is necessary to explore a new talent training system under the guidance of advanced teaching concepts or models.

Outcome-based Education(OBE) is a novel outcome-oriented education model proposed by William G S. in 1994 [6, 7]. It takes learning outcomes that students achieve through the education process as the goal of instructional design and implementation. In recent years, many scholars have carried out teaching reforms around this new education model, and conducted several research and practice [8, 9]. Van M. E. et al. [10]. Conducted a study on the evaluation method of competency-based medical education, and proposed six criteria to evaluate the results of teaching reform. Sasipraba T. et al. [11]. Studied and applied the outcome-oriented education model assessment methods and evaluated student performance in capstone projects. The results showed that this assessment method contributes to the continuous improvement of teaching effectiveness. Shreeranga B. et al. [12]. Used a lean thinking approach to study the effectiveness of outcome-oriented teaching models, and showed that this method could not only enhance teaching and learning efficiency, but also improve student collaboration ability. However, at present, there are few studies on the outcome-based talent training system of material specialty.

Materials science is a discipline with strong experimental and practical nature, and it is always developed under the traction of social needs [13, 14]. With the rapid development of social economy and related disciplines, advanced materials, technologies, and methods are continuously emerging [15–17]. Therefore, technical talents in material specialty should be equipped with strong engineering innovation capabilities to adapt to future development. For this reason, it is very important to study and build an experimental and practical training system that can comprehensively improve engineering innovation capabilities of talents.

This article analyzes the problems of the traditional talent training system of material specialty. In response to the problems, a new training system of innovative talents is constructed based on the OBE concept. The evaluation and implementation methods of non-technical ability teaching is researched and established, and finally the practice has been carried out.

2 Problems Existing in the Traditional Training System of Material Specialty

In recent years, with the rapid development of technology, material majors urgently need a group of outstanding talents that can support the transformation and upgrading of the engineering industry [18, 19]. However, the traditional training mode cannot meet the higher requirements. The existing problems could be summarized as follows.

(1) Professional education and social needs are not highly fit.

Under the background of social transformation, industrial transformation and technological innovation, the demand for talents in new fields and interdisciplinary disciplines is strong. The social needs are no longer a single professional and skilled talent, but an innovative person with a wealth of non-professional and technical capabilities. Traditional engineering students cannot meet the new social needs, and their career competitiveness is insufficient.

(2) Inadequate support for the cultivation of engineering innovation ability of curriculum system.

In the past, the curriculum system for material professionals was complicated. It rarely open innovative and entrepreneurial courses or set up internship and practical training credits. Students do not require to participate in competitions or innovation and entrepreneurial training programs before graduation. The training of engineering knowledge and engineering innovation ability is fragmented. The unreasonable curriculum system leads to insufficient training of college students' ability to systematically design and effectively solve complex engineering problems, especially their engineering literacy and development ability are weak.

(3) Insufficient conversion of engineering practice and competency training.

Experimental and practical training are formalistic, the teaching mode and training content are single, effective resources are scarce, and practical training lacks innovation and challenges. Because students cannot contact actual production, they cannot understand the specific problems that arise in actual production, and cannot propose solutions to problems according to their professional knowledge, let alone make breakthroughs and innovations on the existing basis. In addition, due to the rapid development of technology, the latest technology and scientific research achievements cannot be converted into teaching content in time, thus traditional theoretical knowledge may deviate from actual production practice, which can easily cause students to be armchair strategists. In the long run, students' practical ability gradually deteriorates and are unable to meet the needs of industry innovation.

(4) Insufficient research of outcome evaluation.

The teaching effect assessment is a very important link in the process of talent training [20–22]. Through the effectiveness evaluation of the outcome, continuous improvement of the teaching process can be achieved. However, in the actual process, evaluation is often the most overlooked link. At present, the evaluation system established by current training system of material specialty is not perfect, especially the implementation of non-technical ability training. The assessment and evaluation are vague, lacking effective assessment methods and evaluation standards.

3 Outcome-Oriented Innovative Talent Training System

In view of the above problems in the training of material professionals, the researchers of this paper track and investigate graduates and employers, and detail the connotation of innovative talents in material engineering under the background of emerging engineering according to the adjustment of industrial structure and changes in social demand. Talents should not only possess professional ability and self-development ability of lifelong learning to carry out comprehensive research and development, technology creation and engineering transformation during the whole process of materials from design, preparation to service, but also have correct moral and value orientation, as well as excellent engineering literacy. Through the analysis of the updated requirements for the training of material talents, a new curriculum system, teaching system and evaluation system are formulated.

3.1 Course System

According to the needs of different levels and different ability aims, a curriculum system for the cultivation of innovative talents in materials majors are designed. The curriculum system consists of systematic theoretical teaching, modular experimental teaching, and advanced practical training (see Fig. 1).

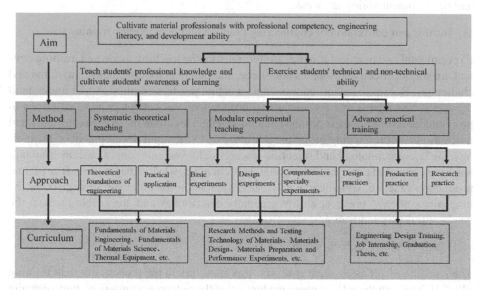

Fig. 1. Outcome-based Innovative talent training system

In the constructed course system, theoretical teaching courses include the basic courses of engineering theory such as Fundamentals of Materials Engineering and Fundamentals of Materials Science. Meanwhile, a practical application courses of Thermal Equipment is also set up. The experimental teaching courses are mainly composed of basic experiments Research Methods and Testing Technology of materials, design experiments Materials Design, and professional comprehensive experiments Materials Preparation and Performance Experiments. According to the students' professional base, the practical training courses apply design practice, production training and research practice, and offer courses such as Engineering Design Training, Job Internship, and Graduation Thesis, respectively. Through the construction of the above curriculum system, students will be instructed professional knowledge, technical ability, and at the same time their learning skills and non-technical ability could be improved. This curriculum system aims at the cultivation of innovative talents to adapts to the development of emerging engineering, to achieve effective support for different levels of ability goals and further promote the balanced development of students' professional competence, engineering literacy and development ability.

3.2 Teaching System

In terms of the new requirements of ability training, in this paper, traditional teaching methods are reformed and a multi-collaborative teaching chain of teaching-learning-evaluation-research is established (see Fig. 2).

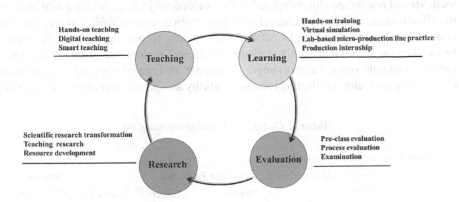

Fig. 2. The multi-collaborative teaching chain of teaching-learning-evaluation-research

Teaching: with the development of science and technology, more and more teaching methods appear [23–25]. This study integrates a variety of teaching methods such as hands-on teaching, digital teaching (digital design training, virtual factory roaming, etc.), and smart teaching (distance online teaching, virtual simulation, etc.), to promote students' learning effect.

Learning: through methods of offline hands-on training, virtual simulation human-computer interaction experiment, lab-based micro-production line practice, production internship, etc., the teaching effect of experimental and practice courses could be improved.

Evaluation: in the context of emerging engineering, the evaluation standards need to assess students' professional ability, cooperation ability, innovation ability, communication ability and other comprehensive ability, so it is necessary to comprehensively evaluate the students' pre-class performance, in-process behavior, and final examination.

Research: transform the latest scientific research into teaching content and make it advanced. Improve teaching effectiveness by researching teaching methods and developing teaching resources, to promote the continuous progress of the teaching system.

3.3 Evaluation System

This study mainly focuses on the reform of the experimental and practical teaching evaluation system, which not only examines students' acquisition of technical skills, but also non-technical abilities such as and social adaptability, professional norms, team cooperation, project management, and lifelong learning. Incorporate experimental data recording, laboratory management system implementation, safe operation, waste liquid disposal, environmental hygiene, and other behaviors into the assessment system for evaluating professional norms, which may enable the cultivation of professional normative capabilities. In the whole process of experimental and practical teaching, the evaluation methods and results are closely related to the outcome objectives, which could improve the effectiveness of the non-technical ability evaluation results. Table 1 shows the evaluation strategy for achieving graduation requirements. To sum up, this new program is formulated for material specialty students of different majors and grades. Throughout the four academic years, it aims to improve students' ability to design and solve complex engineering problems, cultivating innovative ability and project management capability.

Table 1. Competition assessment standard

Evaluation items	Evaluation method					
	Course scores	Questionnaires	Behavioral Evaluation			Extracurricular assessment
			Student self-evaluation	Student peer evaluation	counselor evaluation	
Engineering knowledge	√	√	√			
Analytical skills	√	√	√			
Design & development competency	√	√	√			
Research skills	√	√	√			
modern tools usage	√	√	√			
Social adaptability	√	√	√	√	√	
Professional norms	√	√	√	√	√	
Teamwork	√	√	√	√	√	
Communication skills	√	√	√	√	√	
Project management	√	√	√	√	√	
Lifelong Learning	√	√	√	√	√	√

4   The Practice and Effectiveness of the Training System

The outcome-oriented material professional training system is practiced for students majoring in inorganic non-metallic materials at the School of Materials Science and Engineering of Wuhan University of Technology.

4.1   Analysis of Students' Ability Attainment

In order to ensure that students' ability meet the training requirements, the courses involved in the newly established training system conduct an analysis of students' competency attainment. The following uses an experimental course of material preparation and performance experiment as an example to analyze the degree of students' competency attainment. (See Table 2).

Table 2. Analysis of students' competency attainment

Course objectives	Evaluation basis	Evaluation value					
		Full mark E_1	Conversion ratio F_1	Conversion points $A_1 = (E_1 \times F_1)$	Sample average score G_1	Sample conversion score $B_1 = (G_1 \times F_1)$	Attainment value C $\sum B_1 / \sum A_1$
1. Master the testing principles and methods of material physical and chemical properties, and design feasible experimental schemes	Prelab report	15	1	15	13.12	13.12	$C_1 = 0.8747$
2. Master the material performance test method, and utilize experimental equipment to obtain reasonable test data	Experimental process	25	1	25	22.20	22.20	$C_2 = 0.8880$

(continued)

Table 2. (*continued*)

Course objectives	Evaluation basis	Evaluation value					
		Full mark E_1	Conversion ratio F_1	Conversion points $A_1 = (E_1 \times F_1)$	Sample average score G_1	Sample conversion score $B_1 = (G_1 \times F_1)$	Attainment value C $\sum B_1/\sum A_1$
3. Master experimental data processing and analysis methods, and be able to analyze and interpret experimental results	Experimental report of performance experiments (Data processing section)	45	0.5	22.5	36.91	18.46	$C_3 = 0.8202$
4. Understand and apply the correct method to analyze the relationship between material composition, structure, preparation and performance according to experimental data, and obtain reasonable and valid conclusions	Experimental reports of design experiments (Data processing section)	45	0.5	22.5	39.47	19.74	$C_4 = 0.8771$

(*continued*)

Table 2. (*continued*)

Course objectives	Evaluation basis	Evaluation value					
		Full mark E_1	Conversion ratio F_1	Conversion points $A_1 =$ $(E_1 \times F_1)$	Sample average score G_1	Sample conversion score $B_1 =$ $(G_1 \times F_1)$	Attainment value C $\sum B_1 / \sum A_1$
5. Consciously abide by engineering professional ethics and norms, and ensure the authenticity of experimental data;	Experimental process (data real)	5	1	5	4.95	4.95	C5 = 0.9900
6. Abide by laboratory rules and regulations, and be able to operate experimental instruments and equipment safely and standardly;	Experimental process (6s)	5	1	5	4.92	4.92	C6 = 0.9840
7. Complete experiments in groups to build team spirit. The group members have a clear division of work and can complete their respective tasks independently	Experimental process (teamwork)	5	1	5	4.91	4.91	C7 = 0.9820
Sample information (Number of students)	Excellent(above 90 points): 10; good(80-89points): 21; average(70-79points): 5; pass(60–69 points): 0; fail(below 60 points): 0						
Attainment	$D = \frac{C_1 + C_2 + C_3 + \cdots + C_n}{n} \times 100 = 91.66$						

It can be seen from Table 2, after the reform of the curriculum, students' professional abilities (engineering knowledge, problem analysis, research, etc.), engineering literacy (professional norms, teamwork, etc.) and development skills (communication, lifelong learning, etc.) have been cultivated, and the training aims have been achieved.

This talent training system can not only promote students' professional ability, but also effectively cultivate students' engineering literacy and development ability. After the practice of this training system, students' participation in innovation and entrepreneurship training, competitions and other extracurricular activities has increased significantly. The acceptance rate of graduate students reaches more than 50%, and 73.97% of graduates are still engaged in material-related jobs.

4.2 Questionnaire

In the practice of the talent training system, the researchers in this article conducted a teaching effect questionnaire survey on relevant teachers and students. The results of the questionnaire survey are shown in Table 3.

Table 3. Teaching Effect Questionnaire

Items	Significantly improved	improved	General	Not improved
Engineering knowledge	48.87	10.69	20.36	20.08
Analytical skills	50.23	20.38	10.69	18.70
Design& development competency	62.35	10.36	15.38	11.91
Research skills	70.46	15.25	8.40	5.89
modern tools usage	60.45	20.19	9.65	9.71
Social adaptability	62.68	25.92	8.54	2.86
Professional norms	90.22	5.68	2.76	1.34
Teamwork	86.36	10.23	2.35	1.06
Communication skills	84.64	9.87	4.23	1.26
Project management	93.46	3.45	2.20	0.89
Lifelong Learning	92.53	4.58	1.89	1.00

It can be seen from Table 3 that after the construction of the new talent training system, more than 90% of teachers and students believe that students' professional norms and lifelong learning ability have been significantly improved, more than 80% of teachers and students believe that team cooperation, and communication skills have been significantly improved, and about 70% of teachers and students believe that students' research ability has been significantly improved. Evidently, the establishment of this talent training system has remarkably improved students' non-technical ability.

5 Conclusion

In the stage of transforming material specialty from traditional engineering to emerging engineering, society has put forward higher requirements for the training system of material professionals: balanced development of professional competency, engineering literacy and development ability. In order to cultivate material talents that meet the demands of society and the industry, curriculum, teaching methods, evaluation methods need to be reformed to improve the training system for material professionals. Based on above facts, this paper formulates a new curriculum system, teaching system and evaluation system, respectively. The curriculum system consists of the systematic theoretical teaching, and advanced practical training. A diversified and collaborative teaching chain of teaching-learning-evaluation-research is established, and a methodology for evaluating non-technical competencies was implemented. After the practice of the training system, students' participation in innovation and entrepreneurship training, competitions and other extracurricular activities increased significantly. The acceptance rate of graduate students reached more than 50%, and 73.97% of graduates were still engaged in material-related work. The students' non-technical ability has improved significantly. This training system provides a reference for material majors in colleges and universities.

Acknowledgments. This work has been supported by the teaching reform and research project of Wuhan University of Technology (Grant No. W2022061 and W2022064).

References

1. Jamilurahman, F., Mohammad, S.U.: A conceptual framework for software engineering education: project based learning approach integrated with industrial collaboration. Int. J. Educ. Manage. Eng. (IJEME) **5**, 46–53 (2021)
2. Ozgen, K.: SASMEDU: security assessment method of software in engineering education. Int. J. Educ. Manage. Eng. (IJEME) **7**, 1–12 (2018)
3. Gu, Y., Chen, J., Liu, Z.: Course resources and teaching methods of logistics management major under emerging engineering education. Lecture Notes on Data Eng. Commun. Technol. **83**, 434–443 (2021)
4. Shen, J., Li, T., Wu, M.: The new engineering education in china. Procedia Comput. Sci. **172**, 886–895 (2020)
5. Song, J., Shi, H., Liu, Y.: Exploration of core competence and teaching mode of new engineering. J. High. Educ. **8**(23), 66–69 (2022). (in Chinese)
6. Liu, J., Feng, L., Wang, Z.: Assessment and implementation of college instruction based on OBE. J. North China Univ. Sci. Technol. **22**(6), 79–84 (2022). (in Chinese)
7. William, G.S.: Outcome-Based Education Critical Issues and Answers. American Association of School Administrators, New York (1994)
8. Ines, G., Daniel, D., Belen, B.: Learning outcomes based assessment in distance higher education. A case study. Open Learn. J. Open Distance and e-Learning **37**(2), 193–208 (2022)
9. Liu, L., Feng, W., Li, J., et al.: Application of BOPPPS model in the teaching design of the principle and method of micro-joining course. Adv. Comput. Sci. Eng. Educ. **134**, 509–517 (2022)
10. Van, M.E., Hall, A.K., Schumacher, D.J., et al.: Capturing outcomes of competency-based medical education: the call and the challenge. Med. Teach. **43**(7), 794–800 (2021)

11. Sasipraba, T., Kaja, R., Nandhitha, N.M., et al.: Assessment tools and rubrics for evaluating the capstone projects in outcome based education. Procedia Comput. Sci. **172**, 296–301 (2020)
12. Shreeranga, B., Sathyendra, B., Ragesh, R., et al.: Collaborative learning for outcome based engineering education: a lean thinking approach. Procedia Comput. Sci. **172**, 927–936 (2020)
13. Tian, J., Li, L., Zhou, C., et al.: Research on the construction of talent training Mechanism of materials subject in colleges and universities. Educ. Modernization **9**(19), 60–63 (2022). (in Chinese)
14. Yuan, Y., Liang, S.: Discussion on the connection of undergraduate, postgraduate and doctoral courses in materials discipline. Education Modernization **6**(51), 187–188,197 (2019). (in Chinese)
15. Qian, J., Nam, P.N., Yutaka, O., et al.: Introducing self-organized maps (SOM) as a visualization tool for materials research and education. Results in Materials **4**, 100020 (2019)
16. Elistratkin, M.Y., Lesovik, V.S., Zagorodnjuk, L.H., et al.: New point of view on materials development. IOP Conf. Ser. Mater. Sci. Eng. **327**(3), 032020 (2018)
17. Wibawa, B., Syakdiyah, H., Kardipah, S.: Technology development in thermochemistry materials. J. Phys: Conf. Ser. **1869**(1), 012023 (2021)
18. Fan, Y., Dong, X., Guo, C., et al.: Teaching mode reform based on cultivation of innovative talents in materials science. China Mod. Educ. Equipment **1**, 121–123 (2022). (in Chinese)
19. Wang, Z.: Computer Programming reform and practice for top talents in the material category oriented to engineering certification. The Science Education Article Collects **13**, 85–87 (2022). (in Chinese)
20. Zhao, J., Xu, Y., Pan, H., et al.: Effectiveness of clinical classroom teaching evaluation system. Basic and Clinical Medicine **42**(3), 529–532 (2022). (in Chinese)
21. Wang, S.: Evidence-based evaluation for instruction quality: taking the instruction quality assessment of mathematics in the USA as an example. Journal of China Examinations **2**, 73–80 (2022). (in Chinese)
22. Soumi, M., Soumalya, C., Sayan, C.: Interactive web-interface for competency-based classroom assessment. Int. J. Educ. Manage. Eng. (IJEME) **1**, 18–28 (2023)
23. Tang, H., Qian, Y., Wang, S.: Designing MOOCs with little. Cogent Education **9**(1), 2064411 (2022)
24. Lin, Y., Feng, S., Lin, F., et al.: Adaptive course recommendation in MOOCs. Knowl.-Based Syst. **224**, 107085 (2021)
25. Sun, J., Sun, C., Zhang, Y., et al.: Virtual simulation experiment of impulse voltage generator. Lab. Sci. **20**(2), 64–67 (2022). (in Chinese)

Development of the Online Library of the University Department to Support Educational Activities: A Conceptual Model

Nataliia Vovk[✉] and Daryna Prychun

Lviv Polytechnic National University, Stepana Bandery Street 12, Lviv 79013, Ukraine

{nataliia.s.vovk,daryna.prychun.mdkib.2021}@lpnu.ua

Abstract. The research carried out an analytical review of the sources, which showed that today the issue of information support for the activities of university libraries has been studied at a sufficient scientific and educational-methodological level. Researchers consider this process both from the point of view of creating electronic libraries (electronic resources) in the Internet environment, and from the point of view of using social networks to establish communication and spread information about the activities of book collections. The basis of the research was carried out at the Department of Social Communications and Information Activities of the Lviv Polytechnic National University. Although library service to users at the university level is organized at a high level, library service at the department level is not sufficiently regulated. The proposed solution will allow not only to automate library service at the department level, but also to establish communications with readers - students of the department. The developed website of the Department of Social Communications and Information Activities library allows users to familiarize themselves with the document fund of the department, and registered users to access full-text documents Such e-library allows not only systematizing the works of university teachers (teaching manuals, textbooks, scientific articles, theses, etc.), but also to provide students with more opportunities for online learning, including the following: constant full-text access to such materials; independent preparation for classes; abstract analysis of such documents, etc.. In addition, the study presents general proposals for the development of information activities of the department's library, the problems of implementing the online library and the possibility of avoiding and getting out of conflict situations.

Keywords: university department · library · students · information support · website

1 Introduction

The creation and provision of new library services is the result of transformations caused by technological progress, new document formats, innovative ways of circulating information, and user needs, as well as such social phenomena as the fragmentation of society, the aging of the population, and a very low level of reading information. The impetus in the process of providing library services, among which the book publisher is still the main one, was the wide use of information technologies.

The majority of public, scientific and university libraries today already widely use all available methods and means of online reader service: electronic catalogs, access to Internet resources, online queue, etc. However, in the structure of library service for users in institutions of higher education, emphasis is placed only on the activities of the main university book collections. Thus, leaving out of consideration small educational department's libraries, the fund of which, as a rule, includes the work of department employees: teaching aids, textbooks, lecture notes, other methodical publications, monographs, scientific articles, etc. Therefore, the relevance of the chosen topic is determined by the lack of work on the use of information technologies by the educational department's libraries of higher education institutions.

The purpose of the study is to develop a conceptual model of the online library of the university department. It is envisaged that such a library will allow not only systematizing the works of university teachers (teaching manuals, textbooks, scientific articles, theses, etc.), but also to provide students with more opportunities for online learning, including the following: constant full-text access to such materials; independent preparation for classes; abstract analysis of such documents; etc. According to the goal, there are the following research tasks: analysis of existing works on the research topic; development of a conceptual model of the electronic library of the university department; analysis of available software and its selection; development of the alpha version of the site, its analysis and further improvement.

Today, the issue of the development of modern libraries has been sufficiently studied at the theoretical level, clear ways of their development have been determined and methods of promotion of modern libraries have been chosen. At the stage of implementation of these ideas, there is a kind of imbalance: on the one hand, quite high successes have been achieved (transformation of libraries into information centers and media libraries), the use of information technologies in the process of library service, on the other hand, insufficient attention has been paid to certain types of libraries, in particular to the educational department's libraries of higher education institutions education.

2 Related Work

The issue of information provision of library support for studies at universities is the subject of analysis of works by many researchers. This process is considered from different perspectives: both from the point of view of developing information services and systems, remote databases for library service to readers, and from the point of view of popularizing the activities of university libraries.

Samah N. and others state that currently university students and teachers generally obtain reference materials from online databases and electronic resources accessed by university libraries. In this study, the authors identify the number and names of online databases most frequently searched, viewed, and downloaded by university students and staff for educational purposes. The data were obtained from the annual reports published by university libraries during 2014–2017. The most popular such databases, according to research, are the following: IEEE Xplore Digital Library, ScienceDirect, EdITLib, Datastream and Scopus [16].

The most popular system for e-learning, in which it is possible to mark up electronic manuals and other electronic publications, is the system on the Moodle platfor. Bhupesh Rawat and Sanjay K. Dwivedi point out that Moodle (modular object oriented developmental learning environment) is one of the most widely used open source learning management systems. They can record any student activities including reading, writing, taking tests, performing various tasks, and even communicating with peers through collaborative forums such as chat, forum, glossary, wiki and workshop [3].

Ani O. and Ahiauzu B. investigated the levels of electronic information resources development in Nigerian university libraries. During the national seminar on electronic information for the library network (eIFL.net) in 2007, a survey was conducted to obtain information about libraries' access to electronic information resources. The results of the study are presented in Fig. 1.

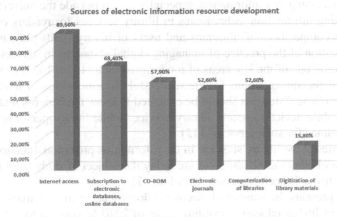

Fig. 1. Sources of electronic information resource development

The study findings show that the Internet is the main source of electronic information resources development in Nigerian university libraries. According to the survey results, there is a high level of electronic information resources development in Nigerian university libraries through direct subscription to electronic information (online databases, CD-ROM, etc.) than converting information into electronic form in the library through computerization and digitization [2]. Other studies show that the use of scientific databases is quite low. Such low indicators are caused by insufficient information about the existence of these library resources [6]. This shows that the participants of the educational process prefer those educational materials that are freely available on the Internet.

Gulara A. Mammadova and other researchers define the concept "personalized learning". Personalized learning is the adaptation of the pedagogy, curriculum and learning environments to meet the needs and styles of the individual students learning. The key elements that are customized in personalized e-learning are: the pace of learning, the learning approach, learning activities, educational content [7]. Such content can be e-journals, textbooks, e-magazines, etc.

Borrego A. and other researchers present the results of a survey on the use of e-journals by university teachers belonging to the Consortium of Academic Libraries of Catalonia. The results show that a significant part of teachers and researchers prefer electronic journals in contrast to printed formats [4].

McMartin F. in the study reports the results of a national survey of 4,678 respondents representing 119 higher education institutions in the United States regarding their use of digital resources for academic purposes. In particular, the following survey results are presented: demographic indicators commonly used in higher education to classify population groups cannot reliably predict the use of online digital resources; the assessment of online digital resources corresponds only to higher levels of use of certain types of digital resources; lack of time was a significant barrier to using the materials [11].

Librarians report an increase in journal reading among scholars and students due to the increase in the number of e-journals available and the improvement of tools for finding and accessing this information, especially access outside the university [14].

When using information technologies in library services for readers of university libraries, the competences of librarians and users of library funds play an important role. The question of the practice of managing digital resources in university libraries has also become one of the key areas of modern research [1, 10, 13, 15, 18]. Ocran T. provides clear recommendations for the successful implementation of library services in the mobile phone: "students should be educated to know the benefits that come with the use of mobile device to access library services while library personnel should be adequately trained for such services" [12].

Peculiarities of the library sites use in the information provision direction of library services [9] and other online resources are today the subject of study from the point of view of their use in branch university libraries [17].

In Fig. 2 presents the number of electronic documents in the electronic catalogs of the libraries of higher education institutions as of 2020 (according to the research of N. Levchenko).

As of 2020, most libraries were at the same level according to the number of electronic documents in the library's electronic catalogs. The scientific library of the Zaporizhia National University differs sharply from most.

The current state of information technology development contributes to the rapid development of various online libraries. It is worth noting that most full-text databases (in the form of an online library) offer fiction. While electronic libraries of higher education institutions are mostly designed as electronic catalogs, where at best, the user can place an online order from a book publisher.

To ensure the successful transition of Ukrainian libraries to the use of new national standards of Ukraine, harmonized with international normative documents, the National Library of Ukraine named after V. Vernadskyi initiated the creation of an interdepartmental working group, which included experts from leading libraries of various systems and departments. From September 1, 2016, new national standards of Ukraine entered into force, in particular ISO 16439:2016 (ISO 16439:2014) «Information and documentation – Methods and procedures for assessing the impact of libraries». This international

The number of electronic documents in the electronic catalog

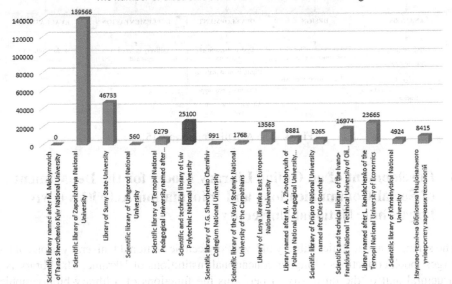

Fig. 2. Number of electronic documents in electronic catalogs of Ukrainian university libraries

standard was developed for the purpose of strategic planning and quality management of libraries; to facilitate comparison of library impact over time between libraries of a similar type and purpose; to increase the value of libraries for learning and scientific research, education and culture, social and economic life; to support policy decisions regarding the strategic goals of libraries.

3 Methodology

It is advisable to use online libraries to develop a ADDIE instructional design model proposed by Peterson [5]. In order to create such an e-library, all design models needs the following stages: analysis, design, development, implementation, and evaluation; these stages are abbreviated as ADDIE. The efficacy of ADDIE is found in its next-phase-informing nature, making it logical and viable for the development any kind of mobile education tool.

Acclaimed for its resounding popularity in the development of educational instruction, it also presents an easy process, which even the amateur analyst, can follow. It also has the advantage of rapid prototyping and possesses the capacity for continual and formative feedback during instructional materials creation. Most importantly, the ADDIE model conserves the precious time of the developer due to the early fixes it allows at the initial phases. The description of the stages of the original model is described with the aid of Fig. 3. – the diagram depicting the ADDIE framework.

In addition, the modeling method was used for the work, which made it possible to depict informational and functional models.

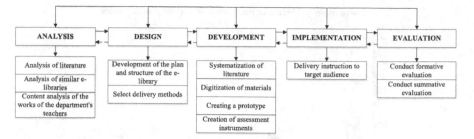

Fig. 3. The ADDIE Framework [5].

4 Development of an Online Library Model for the Department of Social Communications and Information Activities (Lviv Polytechnic National University)

The scientific and technical library of Lviv Polytechnic National University is one of the largest book collections of higher educational institutions of Ukraine. The library is a structural unit of the university. It performs the functions of a library-bibliographic, scientific-informational, educational-supportive, cultural-educational institution, provides the educational process, scientific-pedagogical activity, scientific-research, educational, and cultural-educational work, with books and other information carriers that make up the fund libraries. The book and magazine fund of the library includes about 1 million 800 thousand copies, more than half of which is the fund of scientific literature.

In addition to the library itself, individual copies of literature are located in the departments of the university. Such a department is the graduate department of social communications and information activities of the Institute of Humanities and Social Sciences.

The Department of Social Communications and Information Activities (SCIA) was founded on August 23, 2011 with the aim of training bachelors and masters in the specialty "Information, library and archival affairs" in the field of knowledge "Culture".

In 2016–2022, the teachers of the SCIA department issued 35 study guides and 11 lecture notes (Fig. 4).

In addition to the publication of educational and methodological literature, scientific and pedagogical workers are actively engaged in publishing activities in the direction of scientific publications: articles, monographs, abstracts of conferences, etc. So, for 2016–2022, 31 monographs and sections of monographs were published (Fig. 5).

All scientific research is carried out by the teaching staff of the department on the registered topic "Management of social communication processes in the global information space".

As for the organization of the work of the library of the SCIA department, it is divided into two directions:

- the placement of educational and methodological materials in electronic educational and methodological complexes in the virtual educational environment of the Lviv Polytechnic;

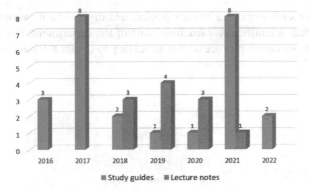

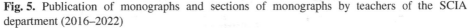

Fig. 4. Publication of educational literature by teachers of the SCIA department (2016–2022)

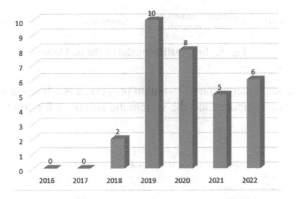

Fig. 5. Publication of monographs and sections of monographs by teachers of the SCIA department (2016–2022)

- the book exhibition of teachers' manuals at the department.

However, taking into account the fact that during the last three years the education has been conducted in a distance format, students' access to such an exhibition, at least for familiarization is practically non-existent.

In addition, users can familiarize themselves with the manuals of the department's teachers, published by the Publishing House of the Lviv Polytechnic, and on the website of the publishing house itself. The manuals of the teachers of the SCIA department are posted on the website of the Publishing House in the section "IT, computers".

But there are two drawbacks: only those editions published by the Publishing House of the Lviv Polytechnic are posted on the website; on the site, users can only purchase a copy, but not view it online.

In order to improve the work of the library of the department of social communications and information activities, it was decided to create a website, which will host an electronic catalog of available manuals, textbooks and other materials, links to Internet resources where such works are already posted, and digitized educational and method-ological materials scientifically - teaching staff of the department. For a better under-standing of the site creation process, it is necessary to depict the information model of this task (Fig. 6).

Fig. 6. Information model of the problem

To reflect the implementation of the main process, it is advisable to develop a goal tree that allows you to understand the systematic solutions for solving the main goal (Fig. 7).

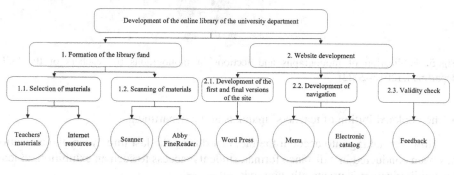

Fig. 7. Goal tree

The main goal "Development of the online library of the department of the higher education institution" is divided into 2 sub-goals: "Formation of the library fund" and "Development of the website".

Regarding the first sub-goal, part of the library fund of the SKID department is already available in electronic format (available from teachers or in the form of Internet resources), the other part needs to be digitized. The second sub-goal is divided into three:

- "Development of the first and final versions of the site" - all versions developed using CMS WordPress must be approved by the head of the department;
- "Navigation development" - it is planned to create a site menu and an electronic catalog on the site;

- "Feedback" - receiving feedback from users by the electronic library.

Using the DFD-representation, in Fig. 8 presents the process of organizing activities for the development of the online library of the Department of Higher Education.

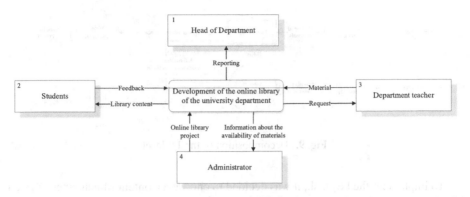

Fig. 8. DFD information support

The main process is the development of the online library of the department of higher education, there are 4 external entities: the head of the department, students, teachers of the department and the administrator of this library. Data flows include the following flows:

- input (feedback, materials, online library project)
- output (information about availability of materials, request, library content, reporting).

The key purpose of creating such an online library is to place on it educational aids and textbooks issued by the teachers of the department. Then most students will be able to use them. For this, it is necessary, first, to receive such materials from the authors, to systematize them, if necessary, to digitize them and then place them on the website (library content). The head of the department must approve all versions of the site. In the case of receiving information about the absence of a certain book on the site, the administrator must, at the time determined by the head of the department, perform actions to download the manual / textbook to the department's online library.

The process of developing the online library of the Department of Higher Education can be divided into seven sub-processes (Fig. 9):

1. Materials structuring
2. Application processing
3. Content systematization
4. Information Systematization
5. Request formation
6. Website development
7. Reporting Formation

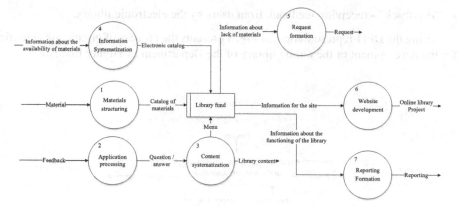

Fig. 9. Decomposition of the 1st level

To implement the key task, it was decided to choose a Content Management System (CMS), which today significantly facilitates the work of web developers.

Today, the most popular freely distributed CMS are the following: WordPress, Joomla, Drupal, MODX. Each of them has its own characteristics, advantages and disadvantages (Table 1).

Table 1. Analysis of CMS

Measurement time difference/s	WordPress	Joomla	Drupal	MODX
Keeping a record of user actions	–	–	+	+
Protection against automatic filling of forms	?	+	+	+
Caching pages	+	+	+	+
Distribution of user rights	-	+	+	+
SSL protocol support	+	+	+	+
Several interface languages	+	+	+	+
Support for multilingual sites	+	+	+	+
Web statistics	+	+	+	+
Chat	–	+	+	?
Forum	–	+	+	–
Photo gallery	+	+	+	+
Catalogue	–	+	+	+
Payment systems	+	+	+	+

To create an online library of the Department of Social Communications and Information Activities of the Lviv Polytechnic, choose the WordPress content management system.

One of the important issues in the development of the e-library of the university library is the determination of the reliability of the selected software and the subsequent protection of the created information resource. To ensure the reliability of the e-library, both at the first stage of its development and in its subsequent functioning, it is advisable to use various protection methods, in particular the following: an SSL certificate, which ensures a secure and confidential connection via the HTTPS protocol; ISO 27001 and 27018 certificates; additional layers of site protection, such as TLS 1.2; protection against DDoS attacks; monitoring the site for any suspicious activity and constantly checking the security of the software; application of two-stage verification of e-library visitors; application of GDPR and CCPA standards, etc.

The developed website of the SCIA department library contains information about the library fund of the department. The main page contains information about the department's new arrivals, which have been digitized and are available online (Fig. 10).

The website menu contains quick links to the relevant sections of the library fund. If the appropriate item is selected, the user will go to the page with materials belonging to this category. The website menu contains quick links to the relevant sections of the library fund. If the appropriate item is selected, the user will go to the page with materials belonging to this category (Fig. 11).

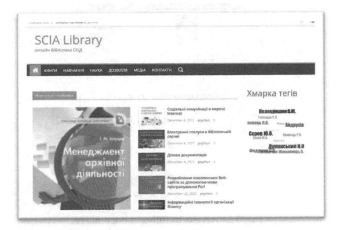

Fig. 10. Main page of the site

Viewing library materials has differences for registered users. Anyone has access to the site and can be acquainted with the library fund of the department, but only reference and descriptive information is available for viewing. Registered users also have access to full-text documents (Fig. 12).

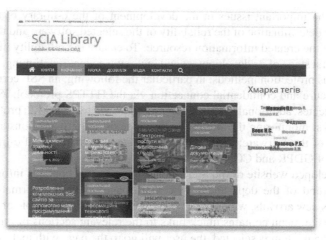

Fig. 11. Section "Education"

Fig. 12. Access to full-text documents

5 Proposals Regarding the Improvement of Document and Information Activities of the Institution

New forms of organizing the functioning of universities stimulate the need to review the structure of libraries, their functional interaction with other units of the university. The strengthening of market trends in the field of education, the need to train competitive specialists make it necessary to provide a high level of information support for the educational process.

With a noticeable increase in the total volume of the document and information fund of the libraries of higher education institutions, the index of renewal of library funds has a tendency to decrease. In the face of financial constraints, libraries need to look for non-traditional methods of stocking them: book exchange, sponsorship, holding various promotions, etc. Considering the problem of insufficient amount of literature, especially educational, libraries should form full-text collections of educational and auxiliary materials.

The use of modern information technologies has changed the standards of library work in general, reference, and bibliographic service to users in particular. The strategy of searching for information has changed, and access to information sources has expanded significantly. Higher education libraries today must develop - the use of digital technologies leads to new forms of working with users, especially in terms of the accumulation, formation and use of information resources.

Libraries of higher education institutions should intensify their activities in the creation of electronic catalogs and thematic databases; provide users with access to numerous global information arrays via the Internet. Today, library users can search for information using both traditional reference and bibliographic equipment, as well as electronic catalogs, databases, etc.

Libraries of higher education institutions implement a comprehensive approach to the formation of information culture of future specialists, using means and methods of correction and management of this process; provide an opportunity to combine cognitive and practical information activities of students. The management of the leading libraries of higher education institutions develops and implements the course "Fundamentals of information culture", and uses other means of individual and group assistance to users. Improving the libraries activities in the formation of users the information culture requires, first of all, a state approach: a transition from episodic, non-systematic work to activities within the framework of a special state program, which involves training personnel, solving problems of coordinating the work of libraries and educational institutions, developing methodological support, creating necessary material base.

The modern university library is not only an informational and educational center, but also a cultural and educational center. The libraries priority activity today is the expansion of cultural, recreational, and leisure-educational activities.

One of the promising areas of activity of pedagogical and engineering-pedagogical university libraries is research and scientific-methodical work. Libraries study the information portfolios of users, conduct scientific work with old prints and rare editions, collections of domestic scientists-pedagogues, create a full-fledged system of secondary information documents, develop instructional and methodological materials, participate in various scientific projects, conduct scientific and practical conferences, etc.

An integral component of modern library service in the libraries of higher education institutions is the expansion of the computer park, the use of multiple technologies, the presence of a local network and an automated library information system. This will allow automating all the main library processes and provides an opportunity for communicative information exchange.

Today, university libraries have gained positive experience in working with library staff. However, the current system of professional development in the conditions of rapid technological changes in libraries cannot fully ensure the updating of theoretical and practical knowledge, mastery of modern information and library technologies, organizational and legal aspects of library management, etc. The rather high cost declared by the organizers for the course participants does not allow this form of advanced training to be practiced widely. Libraries, especially their managers and administrative staff, need advanced training courses under special programs taking into account the specifics of their work.

6 Summary and Conclusion

An analytical review of the sources showed that today the issue of information support for the activities of university libraries has been studied at a sufficient scientific and educational-methodical level. Researchers consider this process both from the point of view of creating electronic libraries (electronic resources) in the Internet environment, and from the point of view of using social networks to establish communication and spread information about the activities of book collections. Analysis of available Internet resources allows us to conclude that today there is a significant number of universally accessible online libraries. However, most of them offer users only fiction books. At that time, the university libraries of Ukraine mostly placed only electronic catalogs on their websites. In addition, it is worth noting that domestic libraries today are guided by various state standards, which are issued based on international ISO standards, which regulate processes of quality compliance in library services.

The analysis of the websites of the leading universities of Ukraine showed that on all such information resources there is only a page (sometimes a separate site) of the library, which contains the following services: electronic catalog, resources for compiling indexes of the unified decimal classification, thematic selection of literature (in the form of a list of literature), access to Scopus and Web of Science databases. Instead, such an analysis showed the lack of full-text access to specialized literature for both students and other visitors to electronic libraries. It is advisable to start the development of full-text online library resources from the e-library of individual departments. They will become the alpha version of the e-library of the entire university.

Library service at the Lviv Polytechnic National University is organized at a high level. The information provision of this activity is regulated by the relevant regulatory internal documents of the University. However, it is worth noting that library service at the level of departments is not sufficiently regulated. This will allow not only to automate library service at the level of the department, but also to establish communications with readers - students of the department. With the help of the development of an informational and functional model, the general concepts of the department's library development were formed.

Implementation of the online library requires the use of additional resources. One of the main ones is the availability of a domain and hosting. Lviv Polytechnic National University has its own servers for hosting the library website and its own domain. Such problems may arise in smaller universities or those with limited resources (personal servers, wide Internet channel).

As of January 2023, students of the Department of Social Communications and Information Activities are testing the developed electronic library. Part of the students (registered users) use full-text access to the electronic library fund for preparing essays, term papers, graduation papers. The testing period will last until June 2023. Based on user recommendations and errors detected during the website's operation, the e-library will be modified and made available to the public. Results will be presented in further research.

References

1. Alphonce, S., Mwantimwa, K.: Students' use of digital learning resources: diversity, motivations and challenges. Inf. Learn. Sci. **120**(11–12), 758–772 (2019). https://doi.org/10.1108/ILS-06-2019-0048
2. Ani, O.E., Ahiauzu, B.: Towards effective development of electronic information resources in Nigerian university libraries. Libr. Manag. **29**(6–7), 504–514 (2008). https://doi.org/10.1108/01435120810894527
3. Bhupesh, R., Sanjay, K.: An architecture for recommendation of courses in e-learning system. Int. J. Inf. Technol. Comput. Sci. **17**, 39–47 (2017). https://doi.org/10.5815/ijitcs.2017.04.06
4. Borrego, À., Anglada, L., Barrios, M., Comellas, N.: Use and users of electronic journals at Catalan universities: the results of a survey. J. Acad. Librarianship **33**(1), 67–75 (2007). https://doi.org/10.1016/j.acalib.2006.08.012
5. ChukwuNonso, N., Ikechukwu, U., Njideka, M.: GeoNaija: enhancing the teaching and learning of geography through mobile applications. Int. J. Educ. Manage. Eng. **6**, 11–24 (2019). https://doi.org/10.5815/ijeme.2019.06.02
6. Dadzie, P.S.: Electronic resources: access and usage at Ashesi university college. Campus-Wide Inf. Syst. **22**(5), 290–297 (2005). https://doi.org/10.1108/10650740510632208
7. Gulara, A., Firudin, T., Lala, A.: Use of social networks for personalization of electronic education. Int. J. Educ. Manage. Eng. **5**, 25–33 (2019). https://doi.org/10.5815/ijeme.2019.02.03
8. Hale, M.: Automated library acquisitions and the internet: a new model for business. New automation technology for acquisitions and collection development, pp. 65–82 (2019). https://doi.org/10.4324/9780429329364-5. www.scopus.com
9. Martin, C.R., Rodriguez-Parada, C., Pacios, A.R., Osti, M.V., Bravo, B.R.: Transparency in the management of Spanish public university libraries through the internet. Libri **70**(3), 239–250 (2020). https://doi.org/10.1515/libri-2019-0051
10. McCoy, M.: Why didn't they teach us that? The credibility gap in library education. Conflicts in reference services, pp. 171–178 (2019). https://doi.org/10.4324/9780429354373-18. www.scopus.com
11. McMartin, F., Iverson, E., Wolf, A., Morrill, J., Morgan, G., Manduca, C.: The use of online digital resources and educational digital libraries in higher education. Int. J. Digit. Libr. **9**(1), 65–79 (2008). https://doi.org/10.1007/s00799-008-0036-y
12. Ocran, T.K., Underwood, E.P.G., Arthur, P.A.: Strategies for successful implementation of mobile phone library services. J. Acad. Librarianship **46**(5), 102174 (2020). https://doi.org/10.1016/j.acalib.2020.102174
13. Okeji, C.C., Mayowa-Adebara, O.: An evaluation of digital library education in library and information science curriculum in Nigerian universities. Digit. Libr. Perspect. **37**(1), 91–107 (2020). https://doi.org/10.1108/DLP-04-2020-0017
14. Ollé, C., Borrego, A.: Librarians' perceptions on the use of electronic resources at Catalan academic libraries: results of a focus group. New Libr. World **111**(1–2), 46–54 (2010). https://doi.org/10.1108/03074801011015685
15. Rahman, M.H.: Review of digital record management needs for academic libraries. Library Hi Tech News **37**(3), 21–22 (2020). https://doi.org/10.1108/LHTN-11-2019-0083

16. Samah, N.A., Karno, M.R., Sharuddin, N., Tahir, L.M., Samah, N.A.: Examining the usage of and access to online databases for academic purposes: a study at an engineering- and technology-based university in Malaysia. Paper presented at the TALE 2019 - 2019 IEEE International Conference on Engineering, Technology and Education (2019). https://doi.org/10.1109/TALE48000.2019.9225991. www.scopus.com
17. Smith, D.: Re-visioning library support for undergraduate educational programmes in an academic health sciences library: a scoping review. J. Inf. Literacy 13(2), 136–162 (2019). https://doi.org/10.11645/13.2.2520
18. Zhao, X., Li, S., Xiao, Y., Huang, H.: Digital humanities scholarly commons at Beijing normal university library. Libr. Trends 69(1), 250–268 (2020). https://doi.org/10.1353/lib.2020.0031

Research on the Influencing Factors of Students' General Competency Improvement in Application-Oriented Universities

Jing Zuo[1] and Tongyao Huang[2(✉)]

[1] Nanning University, Nanning 530022, Guangxi, China
[2] Guangxi University of Foreign Languages, Nanning 530000, Guangxi, China
117991414@qq.com

Abstract. Students' competency cultivation is a very important measure of the effectiveness of talent development. In order to explore the influencing factor of student's general competency in application-oriented Universities, this study take 9652 students from different majors in one application-oriented University as sample, regression analysis has been applied with a dependent variable (general competency improvement) and six independent variables (student's learning behavior: in-class, after-class, interact with classmates; teacher's teaching behavior; evaluation of teaching content; professional identity) to sort out the influencing factors. The results show that teacher's teaching behavior, evaluation of teaching content, student's learning behavior (in-class, after-class, and interact with student), and professional identity is significantly positively related to the improvement of students' general competency; especially, the teacher's teaching behavior, and evaluation of teaching content, these are the key factors of student's general competency improvement.

Keywords: General competency · The influencing factors · Regression analysis

1 Introduction

Application-oriented university, is an important part in the development of higher education, which has the mission of cultivating high-quality and highly skilled talents in various fields, such as production, construction, service, and management. In the process of application-oriented talents development, students' competency cultivation is a very important measure of the effectiveness of talent development, and also the main task of talents development in application-oriented universities [1]. In general, student competency includes general competency and professional competency, and cultivating students' general competency can help students adapt to the changing social requirements and industry needs [2]. In order to help application-oriented universities improve the quality of talent development, this study applies the quantitative research methods to explore the factors that affect the cultivation of students' general competency in application-oriented college.

Z. Hu et al. (Eds.): ICCSEA 2023, LNDECT 181, pp. 1000–1010, 2023.
https://doi.org/10.1007/978-3-031-36118-0_85

2   Literature Review

2.1   Competency and General Competency

In 1973, professor McClelland proposed the concept of competency in Testing for Competency rather than intelligence. Nadler believes that competency refers to a resource that can help improve competitive advantage [3]. Based on Françoise and Jonathan's definition using multidimensional framework, competency includes meta-competence, functional competence, social competence, and cognitive competence [4]; while some researchers propose that competence can be divided into Core Competency (key competences) and Specific Competency [5]. Most studies in the field of competency use general competences as synonyms of Core Competency and key competences, which refer to the competence applying to different work and life [6] without limiting specific majors or fields [7].

Many researches, at home and abroad, clarifies the definition of general competency (core competency). In Core competence of Chinese student, Core Competency contains six parts: cultural heritage, scientific spirit, learning to learn, healthy life, responsibility, practice and innovation [8]. A proposal on core competency for lifelong learning was adopted by the Council of the European Union in 2018, meanwhile core competency was been divided into eight parts: literacy competence, multilingual competence, cultural awareness, expression competence, and so on [9]. Secret's Commission on Achieving Necessary Skills, established in the United States in 1990, they had determined the basic competency required to cope with the work through a lot of research [10], such as resources, interpersonal communication, information, system, technology [11]. Therefore, the definition of core competency in different countries has its own characteristics, which provides a useful reference for the development of the scale in this study.

2.2   The Influencing Factors of General Competency

A large body of literature explores what factors could amplify or mitigate general competency, however, they mainly discuss the influencing factors of a specific part of competency, like cross-culture competency [12], scientific and technological innovation competency [13], critical thinking competency [14], autonomous learning competency [15]. In addition, some researchers focus on the factors that influence the mismatch of engineering students' general competency [16], and the influence of institution and education environment on the formation of general competency [17]. The influencing factors of general competency are different as most of the previous researches were mainly about a specific part of competency in different scenarios, but it can mainly be divided into these three types: learner, influencer (teacher, tutor, classmate), and environment (school, major). Based on the previous researches, this study will also be conducted including these three types of influencing factors (learner' learning behavior, teacher's teaching behavior, teaching content and professional recognition).

3   Methodology

3.1   Questionnaire Design

The questionnaire design of this study includes two parts: personal information (grade, gender, college, major) and five scales with fifty-three questions (general competency, student's learning behavior, teacher's teaching behavior, evaluation of teaching content, professional identity).

The scale of general competency based on the SCANS standard of the United States [10], includes teamwork, information searching and processing, self-awareness, environmental adaptation, communication, design thinking, reading ability, critical thinking, organizational leadership, problem solving, autonomous learning, planning ability. The scales of student's learning behavior, teacher's teaching behavior, evaluation of teaching content, and professional identity are designed according to relevant literature. The scale (Table 1) has been modified according to the pilot test and the results of analysis on Reliability and Validity.

3.2   Data Collection and Sampling

The researchers distributed 9652 questionnaires to freshmen, sophomore and junior in an application-oriented university, 8931 valid questionnaires (recovery rate is 92.5%) were collected (Table 2).

Table 1.   Scales and Samples

Name of scale		Questions	Sample	Remark
General competency		12	Improvement of "teamwork" ability in this year	Likert scale (1–5: not at all improved, neutral, somewhat improved, very improved)
Teacher's teaching behavior		11	Teachers add interaction in the classroom and pay attention to students' participation	Likert scale (1–5: never, rarely, sometimes, often, always)
Student's learning behavior	Before-class	2	Finish reading and homework before class in this year	

(continued)

Table 1. (*continued*)

Name of scale		Questions	Sample	Remark
	In-class	3	Concentrate in class in this year	
	After-class	6	Review notes after class in this year	
	Interact with classmates	5	Work with classmate for course assignment or homework in this year	
Evaluation of teaching content		7	Teaching content give emphasis to the combination of practice and theory	Likert scale (1–5: strongly disagree, disagree, neutral, agree, strongly agree)
Professional identity		7	Pay attention to the hotspots or cutting-edge information of discipline development	

Table 2. Sample Descriptive Statistics Results

Grade	Male		Female		Total	
	Students	Percent	Students	Percent	Students	Percent
Freshman	2346	26.3%	2378	26.6%	4724	52.9%
Sophomore	1061	11.9%	1006	11.3%	2067	23.1%
Junior	1004	11.2%	1136	12.7%	2140	24.0%
Total	4411	49.4%	4520	50.6%	8931	100.0%

4   Results

4.1   Analysis on Reliability

Cronbach's alpha coefficients of the scales can be seen in Table 3. The Cronbach's alpha coefficients of general competency scale, teacher's teaching behavior scale, student's learning behavior scale, and teaching content scale are similar ($\alpha > 0.9$, excellent reliability), while the one in professional identity scale is slightly lower than 0.9 (good reliability).

Table 3. Reliability, Average and Standard Deviation

Scale	Number of students	Mean	Std. Deviation	Cronbach's alpha
General competency	8931	3.89	0.73	0.965
Teacher's teaching behavior	8931	4.01	0.71	0.966
Student's learning behavior	8931	3.40	0.84	0.941
Evaluation of teaching content	8931	3.66	0.77	0.942
Professional identity	8931	3.58	0.75	0.886

4.2 One-Way Analysis of Variance

The researcher conducted a one-way analysis of variance on the general competency of three groups of students from different grades (P < 0.05), differences between groups can be found. The f test statistic cannot check whether the variances of the samples are equal to the same value, so Tamhane has been utilized to check. The results show that there is no significant difference between sophomores and juniors, while a significant difference can be witnessed between freshmen and sophomores, juniors.

4.3 Regression Analysis

In order to find out the factors that influence university students' general competency and whether the factors vary by different grades, first, regression analysis of three different grades have been done, with another regression for all the samples.

4.3.1 Regression Analysis for Three Grades

Linear regression analysis conducted between different grades with dependent variable (general competency improvement) and independent variables (student's learning behavior (in-class) (abbr. in-class), student's learning behavior (after-class) (abbr. After-class), student's learning behavior (interact with classmates) (abbr. Interact); teacher's teaching behavior (abbr. T's behavior); evaluation of teaching content (abbr. Evaluation); professional identity (abbr. Identity)). According to the number of independent variables (Table 4), each regression analysis has seven models. Model 7 with a greater proportion of dependent variable changes ($R^2 = 0.46, 0.47, 0.43$) show that this is true for 40% of student's general competency improvement.

As shown in Table 5, for freshmen, student's learning behavior (before-class) has no significant impact on the general competency improvement (P = 0.08 > 0.05); the coefficient of evaluation of teaching content and teachers' teaching behavior is relatively the largest, with the largest impact on improving student's general competency. For sophomore, student's learning behavior (before-class) and student's learning behavior (in-class) (P > 0.05) indicate no significant impact in general competency improvement,

Table 4. Three Regression Analysis Model

Type	R	R^2	Change R^2	Change Statistics				
				R^2 Change	F Change	df1	df2	Sig. F Change
Freshman	0.68	0.46	0.46	0.00	42.652	1	4716	0.000
Sophomore	0.69	0.48	0.47	0.00	15.903	1	2059	0.000
Junior	0.66	0.44	0.43	0.00	10.844	1	2132	0.001

while the greatest impact is still the evaluation of teaching content and teachers' teaching behavior. For junior, no significant impact can be seen in general competency improvement for the factors of student's learning behavior (before-class), student's learning behavior (in-class), and student's learning behavior (interact with classmate) (P > 0.05), with the largest impact in evaluation of teaching content and teachers' teaching behavior.

By contrast, the influence of students' learning behavior (in class), student's learning behavior (interact with classmates), and professional identity on the improvement of students' general competency in different grades gradually decreases, while the influence of students' learning behavior (after-class) and teachers' teaching behavior gradually increases.

Table 5. Regression Analysis and the Influencing Factors (grades)

Model 7	Freshman		Sophomore		Junior	
	Regression Coefficient	Std. Error	Regression Coefficient	Std. Error	Regression Coefficient	Std. Error
(constant)	0.51***	9.16	0.74***	9.52	0.84***	10.41
Before-class	0.02	1.77	0.00	0.25	0.02	1.29
In-class	0.05***	3.77	0.03	1.81	−0.00	−0.08
After-class	0.05***	3.12	0.09***	3.89	0.11***	5.06
Interact	0.08***	5.93	0.07***	3.34	0.03	1.35
T's behavior	0.28***	20.53	0.30***	14.61	0.36***	15.29
Evaluation	0.32***	18.91	0.26***	11.26	0.31***	9.40
Identity	0.09***	6.53	0.08***	3.99	0.06***	3.29

P.S.: *** shows that the correlation coefficient is significant at 0.01 level, ** shows that the correlation coefficient is significant at 0.05 level.

4.3.2 Regression Analysis for All Sample

Because students' before-class learning behavior is not significant in the regression analysis, this factor is excluded in the all-sample analysis. Linear regression analysis conducted in this study with a dependent variable (general competency improvement) and six independent variables (student's learning behavior: in-class, after-class, interact with classmates; teacher's teaching behavior; evaluation of teaching content; professional identity). Model 6 with a greater proportion of dependent variable changes (R2 = 0.461) suggest that these factors is true for 46.1% of student's general competency improvement.

In Table 6, all factors have a significant impact on general competency, the impact is sorted from large to small as evaluation of teaching content, teacher's teaching behavior, professional identity, student's learning behavior (interact with classmates), student's learning behavior (in-class), student's learning behavior (after-class).

Table 6. Coefficients

Model		Unstandardized Coefficients		Standardized Coefficients	t	Sig.
		β	Std. Error			
6	(constant)	0.51	0.06		9.191	***
	Evaluation	0.32	0.02	0.27	19.025	***
	T's behavior	0.28	0.01	0.28	20.474	***
	After-class	0.05	0.01	0.07	3.736	***
	Identity	0.09	0.01	0.09	6.662	***
	In-class	0.06	0.01	0.07	4.570	***
	Interact	0.08	0.01	0.10	6.027	***

5 Discussion

5.1 The Influence of Evaluation of Teaching Content on the Improvement of Students' General Competency

According to Table 6, evaluation of teaching content is significantly positively related to the improvement of students' general competency ($\beta = 0.32$, $P < 0.01$). In this study, students will evaluate the teaching content from the following seven aspects: "informative major course" and "integrate knowledge, ability and accomplishment" are the basic requirement of course; "give emphasis to the combination of practice and theory", "pay attention to the intersection and integration of disciplines", and "improve problem solving ability" are the features of application-oriented talents development; "expand student's knowledge" and "catch the hotspots or cutting-edge information of discipline development" focus on depth and extension of teaching content. The findings show that evaluation of teaching content reflects the actual situation of curriculum construction, curriculum is a small step of talent development [18], classroom teaching is the main position of talent training, the construction of effective curriculum has a positive impact on students.

5.2 The Influence of Teacher's Teaching Behavior on the Improvement of Students' General Competency

According to Table 6, a significantly relationship can be seen between teacher's teaching behavior and the improvement of students' general competency ($\beta = 0.28$, P < 0.01). Teacher's teaching behavior refers to the actions taken by teachers to achieve certain educational and teaching purposes (group cooperative learning, utilize multimedia, instructional design). As teachers play a leading role in the classroom, the teaching methods adopted by teachers will influence students' learning, and thus affect students' learning effect and competency improvement [19].

5.3 The Influence of Student's Learning Behavior on the Improvement of Students' General Competency

In Table 6, the students' learning behavior in-class, after-class and interaction with classmates is significantly related to the improvement of students' general ability ($\beta = 0.06$, 0.05, 0.08, P < 0.01). Learning behavior is a series of behaviors adopted by students to achieve learning goals, including listen carefully in class, make study plan after class, join in competition with classmates. The learning behavior in-class, after-class and interaction with classmates are taken by students to improve academic achievement. Although academic achievement focuses more on achievement, competency improvement is also a kind of academic outcome [20].

In Table 6, there is no significant relationship between students' before-class learning behavior and their general ability improvement (P > 0.05). Before-class learning behavior includes finishing homework or reading, and preview or organize knowledge points, which are also the behavior input taken by students to achieve their learning goals, but they are not significantly related to the improvement of general competency. The learning behavior focusing on rote memorization, short-term memory, and uncritical acceptance of knowledge [21], are more likely to be surface learning and it has nothing to do with competency improvement.

5.4 The Influence of Professional Identity on the Improvement of Students' General Competency

According to Table 6, there is a significant correlation between professional identity and student's general competency ($\beta = 0.09$, P < 0.01). Professional identity refers to the degree of students' satisfaction of their major, which indicates that students with a high satisfaction on their major are more likely to make improvement in their general competency. The basic findings are consistent with researches showing that high professional identity affects students' learning, and will also have better learning initiative, for example, student with high professional identity may follow the hot spots or cutting-edge trends of their major and learn the professional knowledge more actively, which may help to improve their competency [22].

6   Conclusion

6.1   Teaching Content and Teachers' Teaching Behavior Are the Key Factors on Improving Students' General Ability

Compared to the factor of students' learning behavior and professional identity, teachers' teaching behavior and evaluation of teaching content have a significant positive impact on the improvement of students' general competency for freshmen, sophomores and juniors. In line with previous studies, the effective way for teachers to gain the recognition of students is to improve the teaching content and methods [23]. These data show that basic, practical, in-depth and extensive teaching content and diversified teaching methods, strategies and other teaching behaviors can help students improve their general competency, which requires universities to pay more attention to the development of curriculum construction and provide students with better learning resources. At the same time, we should give emphasis to the guidance of teachers' teaching behavior by adopting diversified teaching methods and teaching strategies into classroom, such as seminar and group cooperative learning.

6.2   Teaching Content and Teachers' Teaching Behavior Have the Greatest Impact on the Improvement of Students' Competency, Which May Hinder the School's Guidance on Students' Learning Behavior

As mentioned above, teachers' teaching behavior and the evaluation of teaching content have a significant positive impact on the improvement of students' general competency, and students' learning behaviors also have impact. For example, for freshmen, students' learning behaviors in class and after class, interaction with classmates can help students improve their general competency; for sophomores, students' learning behavior after class and interaction with classmates can help students improve their general competency; for junior students, their learning behavior after class can help them improve their general competency. Students' learning behavior has little impact regardless of their grades, which means students' learning behavior does not play a key role in improving general competency. In this case, universities may neglect the guidance of students' learning behavior. One possible reason is that students in application-oriented universities do not have strong learning ability and good learning habits. Most of them cannot learn independently and often rely on teachers' teaching behavior or only pay attention to the school's teaching contents. Therefore, their own learning behavior does not play a greater role than teachers' teaching behavior and the evaluation factors of curriculum content, which may hinder colleges and universities from paying attention to the effective guidance of students' learning behavior and the cultivation of learning ability. Findings from this research show that the role of students' learning behavior is not as important as the factor of teachers' teaching behavior and evaluation of teaching content, which may hinder universities from paying attention to the effective guidance of students' learning behavior and the cultivation of learning ability.

Acknowledgment. This paper is the research result of Guangxi's 2021 private higher education research project "Research on Improving the Teaching Ability of Teachers in Private Applied Colleges and Universities" (No.2021ZJY667).

References

1. Soumi, M., Soumalya, C., Sayan, C.: Interactive web-interface for competency-based classroom assessment. Int. J. Educ. Manage. Eng. (IJEME) **13**(1), 18–28 (2023)
2. Yuan, G.: Cultivation of application-oriented undergraduate general professional skills: value and approach. Mod. Educ. Manage. **377**(08), 105–111 (2021). (in Chinese)
3. Nadler, D.A., Tushman, M.: The organization of the future: strategic imperatives and core competencies for the 21st century. Organ. Dyn. **27**(1), 45 (1999)
4. Le Deist, F.D., Winterton, J.: What is competence? Hum. Resour. Dev. Int. **8**(1), 27–46 (2005)
5. Zuo, J., Zhang, G.E., Huang, F.: Construction of ability and quality model of engineering talents based on analytic hierarchy process. In: Hu, Z., Zhang, Q., Petoukhov, S., He, M. (eds.) Advances in Artificial Systems for Logistics Engineering, pp. 356–366. Springer International Publishing, Cham (2022).https://doi.org/10.1007/978-3-031-04809-8_32
6. Tarimoradi, M., Zarandi, M.F., Türksen, I.B.: Hybrid intelligent agent-based internal analysis architecture for CRM strategy planning. Int. J. Intell. Syst. Appl. (IJISA) **6**(6), 1–20 (2014)
7. Young, J., Chapman, E.: Generic competency frameworks: a brief historical overview. Educ. Res. Perspect. **37**(1), 1–24 (2010)
8. Ministry of Education of the People's Republic of China: Chinese Students Develop Core Literacy" was released [EB/OL] [2022–12–26]. http://www.scio.gov.cn/zhzc/8/4/Document/1491185/1491185.htm
9. EU publications: Key competences for lifelong learning [EB/OL] [2022–12–26]. https://op.europa.eu/en/publication-detail/-/publication/297a33c8-a1f3-11e9-9d01-01aa75ed71a1/language-en
10. Secretary's Commission on Achieving Necessary Skills. Background [EB/OL] [2022–12–26]. https://wdr.doleta.gov/SCANS/
11. Wan, Z.: The classification, cultivation and inspiration of competence in American – taking SCANS as an example. Educ. Res. Exp. **160**(05), 36–39 (2014). (in Chinese)
12. Wu, T., Lou, L., Zhou, H., et al.: The current status and influencing factors of intercultural communication competence of nursing students. Chin. J. Nurs. Educ. **19**(07), 626–630 (2022). (in Chinese)
13. Zhuang, Y., Liu, Y.: The Influencing factors of lower grade undergraduates' innovation capability in science and technology based on structural equation model. China High. Educ. Res. **4**, 51–56 (2022). (in Chinese)
14. Zhang, Z., Zhao, M., Jie, W., et al.: Analysis of influencing factors of problem analysis ability of critical thinking among undergraduates in a medical college in Wuhan. Med. Soc. **35**(04), 100–105 (2022). (in Chinese)
15. Xiao, X., Yuan, H., Liao, H., et al.: Study on influencing factors of autonomous learning ability of medical students in Guangzhou city during COVID-19 epidemic. Med. Soc. **35**(02), 101–104+111 (2022). (in Chinese)
16. Yu, T., Gu, X.: The generic competency mismatch of engineering graduates and its influencing factors. Res. High. Educ. Eng. **05**, 43–49 (2022). (in Chinese)
17. Rudenko, I.V., Fedorova, N.I., Muskhanova, I.V., et al.: Organizational and pedagogical conditions development of general competences formation in the educational activities of the university. J. Sustain. Dev. **8**(6), 45 (2015)
18. Wang, Y.: Talent training and curriculum innovation of drama, film and television literature under the background of "New Liberal Arts." Drama **198**(04), 163–173 (2021). (in Chinese)
19. Wang, X., Wu, Y.: Exploration and practice for evaluation of teaching methods. Int. J. Educ. Manage. Eng. (IJEME) **2**(3), 39–45 (2012)
20. Gull, F., Shehzad, S.: Effects of cooperative learning on students' academic achievement. J. Educ. Learn. **9**(3), 246–255 (2015)

21. Meeks, M.D., et al.: Deep vs. surface learning: an empirical test of generational differences. Int. J. Educ. Res. **1**(8), 1–16 (2013)
22. Hou, Z., He, W., Wang, Z.: Research on the influence of tutor's guidance style on graduate students' knowledge sharing and innovation. Academic Degree & Graduate Education **279**(02), 62–67 (2016). (in Chinese)
23. Zuo, J., Wu, X., Zhang, G., et al.: Statistical analysis based on students' evaluation of teaching data in an application-oriented university. In: 2022 International Conference on Educational Innovation and Multimedia Technology (EIMT 2022), pp. 490–498. Atlantis Press (2022)

Dynamic Adjustment of Talent Training Scenario for Electrical Engineering and Automation Majors Based on Professional Investigation

Haoliang Zhu[✉]

School of Intelligent Manufacturing, Nanning University, Nanning 530200, Guangxi, China
412363055@qq.com

Abstract. For the purpose of realizing the effective connection between the professional talent training program and the social needs, this paper has carried out a survey of previous graduates and related enterprises and industries for the electrical engineering and automation specialty of Nanning University. Starting from the needs of enterprises, this paper analyzes the scientificity and applicability of the development of talent training scenarios for electrical engineering and automation majors, invites enterprises to participate in the whole process of specialty setting, curriculum development and teaching reform, timely discovers the deficiencies of this major in school running ideas, specialty setting, curriculum system, student ability and quality training, student and other aspects, and dynamically adjusts the professional talent training plans and programs. While improving the talent training program, it is also necessary to constantly expand the channels of enrollment and employment. The research ideas and methods of this paper can provide some reference and reference for other colleges and universities.

Keywords: Talent training · Major in electrical engineering and automation · Professional investigation · Talent training scheme · Dynamic adjustment

1 Introduction

Electrical automation technology mainly refers to a multi-disciplinary comprehensive technology that uses electrical and electronic technology and devices to realize automatic control of production process [1]. Its emergence, development and increasingly extensive application have greatly improved productivity, and have penetrated into all aspects of industrial production, becoming an important basis and leading role in the development of modern industry [2–4]. The specialty of electrical engineering and automation is set up to train such specialized talents.

Talent is the core competitiveness of enterprises and the cornerstone of industrial development and social progress [5, 6]. With the vigorous development of the equipment manufacturing industry and modern service industry in the Beibu Gulf urban agglomeration with Nanning as the core, relying on the developed industrial base of electrical and electronic industries in the Pearl River Delta, the demand for electrical automation technology professionals in regions including the Pearl River Delta has been strong in recent years, and the trend is increasing year by year, which requires colleges and universities to cultivate and cultivate a large number of professional theoretical knowledge that both adapt to the characteristics of the times, High-quality application-oriented talents with professional operating skills [7, 8]. In response to the new requirements of relevant industries for talents, colleges and universities should maintain real-time contact and effective communication with enterprises and industries at all times, and take the survey results and sampling data as feedback information in the formulation of talent training programs, and constantly make "dynamic" adjustments.

Nanning University, as one of the first pilot universities for the reform of applied science and technology universities of the Ministry of Education, focuses on the development of professional clusters of applied disciplines, builds a disciplinary and professional system with engineering as the main body and multi-disciplinary and coordinated development of disciplines, and actively promotes the reform of the training mode of applied talents with the cultivation of applied talents as the main line [9]. In recent years, teaching reform has been continuously deepened around the cultivation of "application-oriented talents", and a series of effective measures have been formulated, especially in the formulation of talent training programs, and outstanding achievements have been achieved. In the construction of education and teaching system, professional research is an effective means, providing an important basis for the formulation and reform of talent training scheme [10].

The specialty of electrical engineering and automation is the dominant specialty of the school, serving the development needs of Guangxi's "14 + 10" industrial cluster of 100 billion yuan [11]. At present, there are more than 700 students in the university, and there are more than 150 graduates of the first undergraduate course of this major in 2022. It is necessary to conduct comprehensive and in-depth research on graduates and conduct interviews with relevant enterprises to guide the revision of talent training plans.

2 Significance and Method of Investigation and Research

Investigation and research is a scientific method for people to understand the real situation of things in a planned and purposeful way through various ways and methods, so as to understand and transform society. Through thinking and processing the data and materials obtained from the survey, we can gain the understanding of the nature and laws of objective things [12, 13].

2.1 Significance of Professional Investigation and Research

The professional research discussed in this paper includes both the research on graduates and the research on enterprises. The enterprise survey is mainly carried out for those enterprises that have a certain degree of relevance to the specialty. From the actual production of the enterprise, understand the job responsibilities and requirements for talent quality [14–16].

The significance of conducting professional research can be summarized into four aspects.

(1) Provide reference for the development of professional talent training programs;
(2) Provide reference for curriculum construction;
(3) Open up a path for school-enterprise cooperation;
(4) Provide learning opportunities for the construction of teaching staff.

Through in-depth communication, the school understands the characteristics, culture and talent needs of the enterprise. At the same time, it invites the technical personnel and managers of the enterprise to effectively participate in the school's professional setting, curriculum development and teaching reform, which can improve the teaching quality of electrical engineering and its automation specialty, and timely find out the teaching ideas, professional setting, curriculum system, and student's ability and quality training of the specialty Deficiencies in student management and other aspects and suggestions for modification [17–20].

2.2 Research Methods

Field survey, interview survey, conference survey, questionnaire survey, expert survey and statistical survey have all been proved to be effective methods of investigation and research [21, 22]. The two types of objects discussed in this paper - graduates and enterprises - can be investigated in different ways.

(1) Graduate survey

For graduates who have taken up their jobs, the most convenient way is to use questionnaire stars for questionnaire survey. Interviews, meetings and field surveys are also common methods [23, 24]. The main contents of the survey are: whether the graduates' employment distribution, post treatment, professional knowledge, job distribution and interpersonal relationship meet the requirements of employers and posts; Check whether the students trained by the school are "satisfactory" to the enterprise; Analize whether the students are "satisfied" with the school's professional setting, talent training objectives and procedures.

(2) Research on enterprises

The above-mentioned research methods are effective and often used for enterprise research. The most important purpose of enterprise research is to go deep into enterprises and industries and understand their requirements for talents [25].

In order to make the survey results broadly representative and scientific, the survey objects are divided into electronic and electrical enterprises in Nanning and its surrounding cities and the Pearl River Delta according to their geographical characteristics. According to the methods and steps of developing talent training programs, the contents of the survey are divided into two major aspects: first, the survey of the application of electrical automation technology in enterprises, the demand and source of talents, the law of professional growth of technicians, the professional positions and responsibilities of the corresponding students, the knowledge, ability and quality they have, and the key courses they should study. The second is to investigate the actual employment positions and career growth process of graduates of electrical automation technology in recent years. The survey needs to be based on the growth and development of students [26, 27].

3   Questionnaire Survey on Graduates

A questionnaire survey was conducted using a questionnaire star. The collected questionnaires were mainly aimed at graduates engaged in electrical and electronic industries. A total of 57 valid questionnaires were collected.

3.1   Employment Situation of Graduates

Figure 1 shows that 61% of graduates are employed in Guangxi and 28% of graduates are employed in Guangdong, which shows that the talent training of this major is basically in line with the service orientation of the school, which is "based on the Beibu Gulf urban agglomeration with Nanning as the core, to cultivate high-quality application-oriented talents with" character, employment ability, entrepreneurship ability, foundation for further study, and development potential", Therefore, the orientation of the talent training service of this specialty is basically accurate. In the future, the talent training orientation of this specialty should continue to adhere to "local" and "application-oriented", and adhere to the implementation of talent training oriented to the training of talents serving local economic construction.

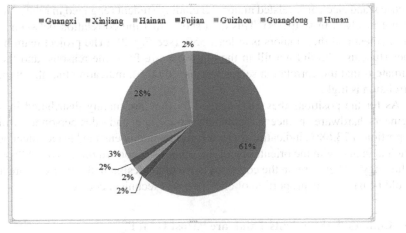

Fig. 1. Employment distribution of graduates

3.2   Salary of Graduates

It can be seen from Table 1 that the average salary of graduates is 4850 yuan, slightly lower than the national statistics, but higher than the average level in Guangxi. From the student feedback, students are basically satisfied with talent training. Basically, it can also be reflected from the side that the quality of talent training in the electrical engineering and automation specialty of our university is affirmed, but the per-capita treatment of the graduates of this specialty is generally lower than the national level (5700 yuan per capita), which indicates that the lack of local industrial resources is also an important condition to constrain the professional development.

Table 1. Wages and benefits of graduates

Initial employment treatment (RMB)	Subtotal	Proportion
> = 20000	1	1.75%
10000–20000	6	10.53%
5000–10000	17	29.82
<5000	33	57.89%

3.3 Industry of Graduates

A total of 162 people who graduated in 2022 were sent questionnaires through QQ group, and 57 valid questionnaires were collected. They were mainly from students engaged in electrical and electronic related majors (excluding students engaged in related industries but not involved in statistics). This data shows that the correlation between students' employment and their majors is at least 35% (see Fig. 2). (The project team has asked other students who did not fill in the questionnaire for some reasons, and the overall estimate is that the correlation seems to close to 60%), indicating that the employment correlation is high.

As for the positions they are engaged in, they are mainly distributed in software engineers, hardware engineers, maintenance personnel and sales personnel, with a total proportion of 73.68%, indicating that students are mainly engaged in technical positions, which is in line with the orientation of the school's school-running type of "application technology". Therefore, in the course of curriculum design in the future, more attention should be paid to the proportion of class hours of technical courses.

3.4 Courses that Students Think are "Most Useful"

From the survey results (Fig. 3), it can be found that the "most useful" in students' minds mainly include circuit foundation, electrical CAD, analog electronics, C language, PLC and electrical control technology, relay protection principle, etc. This shows a problem: they generally believe that basic courses are the most important. Therefore, in the process of talent training in schools, we should "lay stress on the foundation and strengthen the technology".

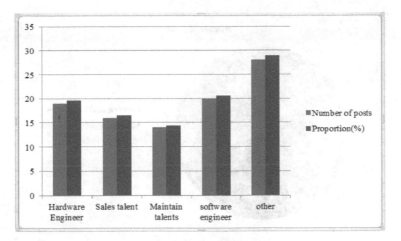

Fig. 2. Graduates' industries

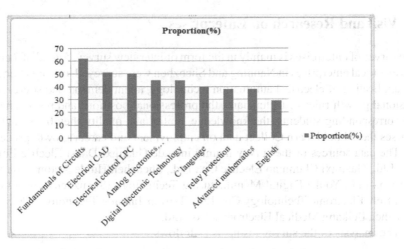

Fig. 3. "Most useful" courses in students' minds.

3.5 About the So-Called "Most Useful" Knowledge and Ability

The questionnaire designed a two-choice question "the most useful knowledge and ability learned in college", shown in Fig. 4.

Most of the students who participated in the survey believe that learning ability, professional knowledge and interpersonal communication are the core abilities of graduation design. Therefore, in the formulation of the school's talent training scheme, in addition to the training of professional knowledge, we should also pay attention to the cultivation of students' autonomous learning ability and interpersonal skills.

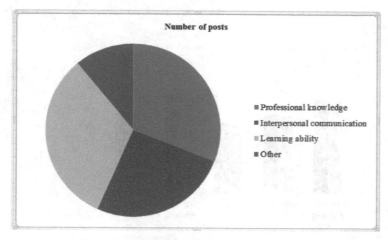

Fig. 4. What students think is the "most useful" knowledge in university

4 Visit and Research on Enterprises

The survey of enterprises is mainly in the form of interview survey. A total of 7 electronic and electrical enterprises in Nanning and Shenzhen were surveyed, mainly to understand the application of electrical automation technology, talent demand and source, the professional growth rules of technicians, the professional positions and responsibilities of the corresponding students, the knowledge, ability and quality they have, and the key courses they should learn Collect data on employment and career growth process.

The data sources of the survey mainly include Guangxi Dansi Electric Equipment Co., Ltd., Guangxi Chunmao Electric Co., Ltd., Shenzhen Huide Automation Preparation Co., Ltd., Youke Digital Manufacturing Technology (Shenzhen) Co., Ltd., Shenzhen Kelu Electronic Technology Co., Ltd., Taiwan Taikang Technology Co., Ltd., and Shenzhen Beikang Medical Electronics Co., Ltd.

The analysis of the survey results is as follows.

4.1 Professional Setting

Employers generally believe that this major is relatively reasonable, can meet the needs of social development, and the employment rate of students is relatively high. According to the questionnaire survey, based on the requirements of enterprises for students to master professional knowledge, it is believed that electrical principle, electronic technology, electrical engineering, electrical measurement technology, power electronic technology, electronic CAD, PLC principle and application, power transmission and automatic control system, single-chip microcomputer principle and application, sensor and application, fault diagnosis and maintenance of CNC machine tools, maintenance electrician skill training Graduation comprehensive practice is a very important course.

4.2 Problems of Graduates

Employers have the following opinions on the problems of new graduates.

(1) Lack of clear and long-term goals. After entering the enterprise, there is no long-term plan. In the absence of goals, the constant job-hopping also causes employers to lose confidence in the training of students, and suggests that "job-hopping is necessary, but it is better not to be too frequent";
(2) The foundation is not solid enough. The foundation of professional knowledge learned in school is not solid enough and the self-positioning is high, but there is a gap between the actual application ability and the actual situation of the enterprise.

4.3 Concerns and Expectations of Enterprises

The quality that enterprises pay most attention to talents mainly includes:

(1) Have the spirit of hard work and dedication;
(2) Have good moral quality and the spirit of unity and cooperation;
(3) Solid basic knowledge, theory and basic skills;
(4) Compound talents have comprehensive knowledge (especially small companies) and have some "advantages";
(5) Be good at handling interpersonal relationships, especially with superiors and subordinates. This also shows that in addition to the basic professional ability, the cultivation of comprehensive quality can not be ignored, especially in terms of team spirit and professional quality.

5 "Dynamic" Adjustment of Talent Training Scheme

According to the situation of graduate research and enterprise research, and in combination with our school's "application-oriented and local" orientation, we have adjusted the talent training program for the major of electronic engineering and automation. Fine adjustments have been made in the curriculum and class hour arrangement, and new requirements have been put forward in the teaching methods and means.

5.1 "Dynamic" Adjustment of Curriculum and Class Hours

On the basis of research, the teaching team has fine-tuned the talent training plan for many times to meet the requirements of paying equal attention to the stability and dynamics of the teaching plan.

In the latest adjustment, in order to cultivate students' hard-working and professional dedication, a 24-h "labor education" course was added, which is a compulsory course, with the purpose of helping students establish the concept of love of labor and cultivate their sense of social responsibility; Three optional courses, "Etiquette", "Speech and Eloquence" and "Aesthetic Education", have been added. Each course has 24 class hours, and students can take it independently.

Through these three courses, students' language communication ability is developed, and students' aesthetic and creative ability is improved. In addition, the proportion of class hours and practical teaching has been adjusted for some courses, mainly increasing the proportion of practical teaching and appropriately reducing the class hours of theoretical teaching, so as to highlight and improve the applicability of the course.

The latest course adjustment is shown in Table 2.

It should be noted that with the change of the situation and the progress of technical means, the requirements of employers for talent training will also change. Therefore, the professional investigation adopted in this paper should also form a long-term mechanism. The survey of relevant enterprises, industries and graduates should be held regularly or irregularly to obtain "dynamic feedback information" in a timely manner. At the same time, "dynamic adjustment" should be made to the talent training program, or, curriculum syllabus and teaching content in a timely manner. Only in this way can the university ensure that the talents we cultivate are "useful talents" for the society.

Table 2. Adjustment of course hours

Course name	Course nature	Class hour(Theory + Practice)	
		Original	After adjustment
Labor education	Compulsory	0 + 0	0 + 24
Aesthetic education	Elective	0 + 0	0 + 24
Ceremony	Elective	0 + 0	0 + 24
Speech and Eloquence	Elective	0 + 0	0 + 24
Fundamentals of Circuits	Compulsory	48 + 8	48 + 16
Electrical CAD	Compulsory	32 + 16	24 + 24
Electrical control LPC	Compulsory	48 + 8	40 + 16
Analog Electronics	Compulsory	48 + 8	40 + 16

5.2 "Dynamic" Reform of Teaching Methods and Means

Some adjustments and reforms have been made in the teaching methods and means of some courses. Some practical courses, such as "C Language", "Principles and Applications of Single-chip Computers", "PLC Control Technology", "Electronic Circuit CAD", encourage teachers to adopt integrated project teaching methods, introduce practical links, and let students achieve "learning by doing"; At the same time, the assessment methods of these courses have also been adjusted, changing the original "usual performance + final test" into the project assessment method, paying more attention to the evaluation of the learning process and learning results, and changing the examination-oriented assessment into the ability assessment, so as to better guide students to learn, so that they can adapt to the requirements of enterprises more quickly in the future, and better fulfill the mission of serving the society.

6 Conclusion

Professional research is an effective means to realize the dynamic matching between talent training and enterprise industry, and also an important basis for the formulation, adjustment and continuous improvement of talent training programs.

The targeted research on the revision of the talent training plan for the electrical engineering and automation majors of Nanning University is to take "graduate research" and "enterprise visit research" as the main research plan. Through real and effective data sampling and analysis, we can judge the talent training direction of our specialty, and the orientation of specialty construction is therefore scientific, so as to promote the effective achievement of talent training objectives.

There are many micro aspects that need to be improved, such as curriculum implementation, proportion of practice class hours, and cultivation of students' professional quality. More in-depth and extensive research can provide a clear direction for the formulation and reform of talent training programs in the future.

The shortcomings of this survey are mainly reflected in the small sample size of data and the relatively single survey dimension, which is the area that the project team needs to improve.

When professional research becomes normalized, timely and sufficient dynamic feedback can effectively promote the dynamic adjustment and improvement of talent training programs. The talents trained by the school will play a greater role in the development of the enterprise.

Acknowledgment. This research is supported by two Projects:

(1) The undergraduate core curriculum construction project of Nanning University "Construction of Electrical Measurement Technology" (2021BKHXK09);

(2) The scientific research platform of Nanning University "Construction of Electronic Design and Research Center of Nanning University" (KYPT202222).

References

1. He, H., Long, Y.: Electrical engineering and automation technology in electrical engineering. J. Phys. Conf. Ser, **1744**(2), 022112 (2021)
2. Lin, Y.: Application analysis of electrical automation technology in power engineering. Chin. J. Electr. Eng. (English) **6**(4), 21–24 (2020)
3. Li, Y.: Application of power system in electrical engineering and automation. J. Phys: Conf. Ser. **1574**, 012128 (2020)
4. Ye, X.: Application analysis of artificial intelligence in electrical engineering automation. Chin. J. Electr. Eng. (English) **6**(4), 29–32 (2020)
5. Girdharwal, N.: Talent on demand: retaining talent in manufacturing industry. Int. J. Recent Technol. Eng. **8**(3), 3367–3370 (2019)
6. Chen, J., Wu, J., Jiang, W., et al.: Demand analysis of employment talents based on deep learning. IOP Conf. Ser. Earth Environ. Sci. **692**, 042017 (2021)
7. Chen, H., Li, Z., Li, W., Mao, H.: Discussion on teaching pattern of cultivating engineering application talent of automation specialty. Int. J. Educ. Manage. Eng. (IJEME) **2**(11), 30–34 (2012)

8. Caunedo, J., Keller, E.: Technical change and the demand for talent. J. Monetary Econ. **129**, 65–88 (2022)

9. Zhang, Y.: Research on the cultivation mode of intelligent financial talents in the era of big data - based on the research of big data and financial management specialty in a vocational college. Mod. Rural Sci. Technol. **04**, 109–110 (2022). (in Chinese)

10. Chen, X., Liu, Z., Gao, D., et al.: Carry out professional research to build "empowerment" for professional connotation. Pig Science **39**(08), 62–65 (2022). (in Chinese)

11. Chen, Q.-W., Gong, P.: Design of virtual simulation experiment teaching system for electrical engineering and automation specialty. In: Fu, W., Liu, S., Dai, J. (eds.) eLEOT 2021. LNICSSITE, vol. 389, pp. 381–394. Springer, Cham (2021). https://doi.org/10.1007/978-3-030-84383-0_33

12. Wei, J., Chen, K., Yang, J.: Reflections on the teaching of urban and rural social comprehensive survey curriculum – from the perspective of the quality of student research report. Educ. Teach. Forum **30**, 63–64 (2018). (in Chinese)

13. Wang, C.: An Investigation and structure model study on college students' studying-interest. Int. J. Mod. Educ. Comput. Sci. (IJMECS) **3**(3), 33–39 (2011)

14. Liang, X.: Research and analysis of the university-enterprise cooperation in the construction of pharmaceutical specialty. J. Shijiazhuang Vocat. Tech. Coll. **34**(4), 52–55 (2022). (in Chinese)

15. Liu, Y., Tang, F., Chen, S.: Research and development suggestions of solar photothermal technology and application specialty in higher vocational education. Hunan Education (Version C) **04**, 52–54 (2022). (in Chinese)

16. Wang, Y.: New changes in higher vocational education based on modern agricultural research. Mod. Agric. Sci. Technol. **06**, 251–253 (2021). (in Chinese)

17. Han, P.: Research and analysis on the demand for new energy vehicle technology professionals in Shanghai. Auto Maintenance and Repair **20**, 42–45 (2020). (in Chinese)

18. Long, F.: Logistics professional research based on the post demand of travel enterprises. Mod. Vocat. Educ. **16**, 162–163 (2020). (in Chinese)

19. Gan, Z.: Learn from experts and do professional research. Appl. Writ. **03**, 59–60 (2020). (in Chinese)

20. Zhang, B.: Research and exploration of teaching standards research methods for software technology majors in the new era. J. Tianjin Vocat. Coll.**22**(01): 38–42+53 (2020). (in Chinese)

21. Le, Z., Zhang, D., Duan, H., et al.: Analysis of the correction effect of professional research activities on engineering talent training program under the background of new engineering. Guangzhou Chem. Ind. **47**(21), 165–167 (2019). (in Chinese)

22. Chen, W.: Design and practice of professional research conducted by electrical majors in the context of the integration of secondary and higher vocational education. Science Park of Middle School **14**(05), 33–35 (2018). (in Chinese)

23. Zhou, Y.: Research and analysis report on business English undergraduate majors oriented by the demands of compound talents. Campus English **40**, 84–85 (2018). (in Chinese)

24. Yan, S.: The necessity and practical significance of professional research on agricultural product marketing, storage and transportation in secondary vocational schools. Global Market Inf. Guide **38**, 23 (2017). (in Chinese)

25. Yanqing, S.: Research on social sports and market demand of Shandong university of traditional Chinese medicine. Sci. Technol. Inf. **15**(20), 211–213 (2017). (in Chinese)

26. Al-Hagery, M.A., Alzaid, M.A., Alharbi, T.S., Alhanaya, M.A.: Data mining methods for detecting the most significant factors affecting students' performance. Int. J. Inf. Technol. Comput. Sci. (IJITCS) **5**, 1–13 (2020)

27. Liu, Z., Fei, J., Wang, F., Deng, X.: Study on higher education service quality based on student perception. Int. J. Educ. Manage. Eng. (IJEME) **2**(4), 22–27 (2012)

The Effect of Integrating Higher Vocational Aesthetic and Ideological and Political Education Based on FCE

Lili Li[✉] and Feifei Hu

Wuhan Railway Vocational College of Technology, Wuhan 430052, China
642320327@qq.com

Abstract. As an important carrier of moral education, the integration of aesthetic education and ideological and political education is more conducive to realizing the basic task of moral education, and has the significance of the contemporary value. Based on the FCE, the effect of the integration of higher vocational aesthetic education and ideological and political education is as follows: (1) The integration effect of higher vocational aesthetic education and ideological and political education is excellent. (2) Students' enthusiasm for learning is still very high, but interest and participation in activities need to be improved. (3) Ideological and political teachers have good performance in knowledge structure, teaching ability and scientific research ability, but there are still deficiencies and need to be further improved. (4) The system and platform construction of integrating aesthetic education and ideological and political education are still quite perfect, but be strengthened in terms of capital investment, atmosphere construction and media publicity. (5) The design of educational content organically integrates the cultural vision, artistic accomplishment, aesthetic needs and mood of the educated, and deepens the understanding and experience of the beauty of ideological and political education through ideological guidance, political orientation and educational promotion.

Keywords: Aesthetic education · Ideological and political education · Fuzzy comprehensive evaluation (FCE) · Fusion effect

1 Introduction

The integration of aesthetic education and ideological and political education is actually the recognition, experience and judgment of the beauty of ideological and political education. And in a beautiful way to expand, enhance and create the beauty of ideological and political education and its value, unifying ideological and political education in the creation of human self-transcendence and development, so as to manifest the essential power of human beings and access to the realm of human performance. Since the Spring and Autumn Period and the Warring States period, China has been using aesthetic education to strengthen ideological consciousness, such as "poetry to express the will" advocated by the Book of Songs. At the beginning of the last century, Tao

© The Author(s), under exclusive license to Springer Nature Switzerland AG 2023
Z. Hu et al. (Eds.): ICCSEEA 2023, LNDECT 181, pp. 1022–1032, 2023.
https://doi.org/10.1007/978-3-031-36118-0_87

Xingzhi proposed the concept of "poetic" moral education, and Cai Yuanpei pointed out the use of artistic means in education. In the 1920s, the early pioneers of Marxist thought realized the effectiveness of literature and art as forms of aesthetic education for the spread of ideas among the people, and Li Dazhao proposed that "beautiful literature and art embody the spirit of the times and use popular literary forms to make them understandable to the general labor force." Mao Zedong, Qu Qubai, Yun Daying, and others initiated the experimentation and practice of beauty education activities in the field of ideological and political education. In the 1980s, scholars interpreted beauty at the theoretical level with a Marxist perspective, combining reasoning and empathy, rationality and sensibility, and using forms and contents of beauty such as sound, form, color, and meaning in political science classes to strengthen the teaching effect. After the new century, research on the aesthetic characteristics of ideological and political education, research on theories and methods of aesthetic education in ideological and political education, and research on the integration of aesthetic education and ideological and political education have been conducted to achieve the unity of "truth, goodness and beauty" and a high degree of conformity between ideological and political education and aesthetic experience [1], so as to better achieve educational goals and promote the cultivation of ideological and political qualities of all people.

2 Materials and Methods

2.1 Fuzzy Comprehensive Evaluation (FCE)

Fuzzy Comprehensive Evaluation (FCE) is a comprehensive evaluation method based on fuzzy mathematics [2], which transforms qualitative evaluation into quantitative evaluation based on the affiliation theory of fuzzy mathematics that is to make an overall evaluation of things or objects subject to multiple factors using fuzzy mathematics. It has the characteristics of clear results and strong system, which can better solve the fuzzy and difficult to quantify the problems, and is suitable for solving various non-deterministic problems. In the fuzzy comprehensive evaluation method, it contains six basic elements: First, the evaluation factor theory domain U, by an attribute, divide it into multiple subsets; The second is the domain V, it means that in the fuzzy comprehensive judgment a collection of the comments that is a collection of possible experts to evaluate, this paper uses excellent, good, qualified, unqualified as the evaluation collection; Third, the fuzzy relationship matrix R, it is the single-factor evaluation result of fuzzy comprehensive evaluation; Fourth, the right vector of judging factors H, it is to measure the importance of each evaluation factor to the evaluation object, used to weight the processing of H in the fuzzy comprehensive evaluation; Fifth, the synthetic operator, it refers to the calculation method used to synthesize H and R; Sixth, the evaluation result vector B, it is the comprehensive evaluation results of all the judging factors, generally is in the form of a matrix. According to the six basic factors of the above fuzzy comprehensive evaluation method, the basic process of evaluation can be divided into the following steps:

(1). Established the Factor Set

Factor refers to the elements or causes that affect things. Systematic analysis needs to judge the object, that with the least factors to describe it, grasp the key points, conducive to simplify the operation process and reduce the overall evaluation between.

Let the evaluation factor set of the first level index be U.

$$U = \{u_1, u_2, \ldots, u_m\} \tag{1}$$

(2) Establish the Weight Set

According to the importance of the factors given the corresponding weight. Should be the evaluation of each level of index. The weight set of evaluation factors for first level indicators is H, and the weight set of second level indicators is as follows.

$$H_i = \{h_{i1}, h_{i2}, \ldots, h_{in}\} \tag{3}$$

$$\sum_{j=1}^{n} h_{ij} = 1, h_{ij} > 0, i = 1, 2, \ldots, m \tag{4}$$

(3) Establish the Weight Set

The evaluation set is a set of various possible results of total evaluation. There arc p possible results of total evaluation. The evaluation set can be expressed as V.

$$V = \{v_1, v_2, \ldots, v_p\} \tag{5}$$

(4) Single-factor Fuzzy Evaluation

The evaluation is conducted from a factor alone to determine the membership degree of the evaluation object to the evaluation set elements, which is called single-factor fuzzy evaluation. The evaluation object is evaluated according to the i factor u_i, and the membership degree of the k element v_k in the evaluation concentration is r_{ik}, then the evaluation result according to the i factor u_i is R_i.

$$R_i = \{r_{i1}, r_{i2}, \ldots, r_{ip}\}, i = 1, 2, \ldots, n \tag{6}$$

The results of each single factor as a matrix of rows R. As the one-factor evaluation matrix.

$$R = (r_{ik}) = \begin{bmatrix} r_{11} & r_{12} & \cdots & r_{1p} \\ r_{21} & r_{22} & \cdots & r_{2p} \\ \vdots & \vdots & \ddots & \vdots \\ r_{n1} & r_{n2} & \cdots & r_{np} \end{bmatrix} \tag{7}$$

(5) Fuzzy Comprehensive Evaluation

    Fuzzy comprehensive evaluation is to consider the influence of all factors and get the correct evaluation results. If the one-factor evaluation matrix of the fuzzy comprehensive evaluation of the first-level indicators is set, the comprehensive evaluation set is B.

$$B = H \cdot R = \left(b_1, b_2, \ldots, b_p\right) \tag{8}$$

    The one-factor evaluation matrix of the fuzzy comprehensive evaluation of the second-level index is R_i, then the fuzzy comprehensive evaluation set of the second-level index is B_i.

$$B_i = H_i \cdot R_i, i = 1, 2, \ldots, m \tag{9}$$

    The fuzzy comprehensive evaluation results of second-level indicators are taken as the single factor evaluation of the fuzzy comprehensive evaluation of first-level indicators.

$$R = \begin{pmatrix} B_1 \\ B_2 \\ \vdots \\ B_m \end{pmatrix}_{mxp} \tag{10}$$

According to the maximum membership principle, if

$$b_t = max\{b_j\}, j = 1, 2, \ldots, p \tag{11}$$

    Then the evaluation results are subordinate to v_i

2.2 Effect Evaluation

The effect can be evaluated from the following five aspects.

(1) Establish the Effectiveness Evaluation Index Set

    From 2.1, the evaluation index is divided into first-level indicators and second-level indicators. The primary indicators are U = {educatee, educators, educationalresources, integrationapproach}; the secondary indicators are U_1 = {interest degree, enthusiasm for learning, participation, ability to participate}; U_2 = {knowledge structure, teaching ability, scientific research ability}; U_3 = {system construction, platform construction, capital investment, media publicity and guidance}; U_4 = {design interesting ideological and political classroom to enrich campus aesthetic creation and experience, lead students' better life and social practice, improve teachers' aesthetic quality and cultivate aesthetic feelings}. The comprehensive evaluation index system and weight of the effect of the integration of higher vocational aesthetic education and ideological and political education are shown in Table 1.

(2) Establish a Weight Set

    According to the index weights in Table 1, the weight set of each factor can be drawn as follows.

    Weight set of evaluation factors for first-level indicator H.

$$H = (0.4096, 0.2951, 0.1056, 0.1897) \tag{12}$$

Table 1. Comprehensive evaluation index system and weight of the effect of the integration of higher vocational aesthetic education and ideological and political education

Level 1 Indicators	Weight	Secondary Indicators	Weight	Total Weight
educatee (student)u_1	0.4096	degree of interest u_{11}	0.2911	0.0168
		enthusiasm for learning u_{12}	0.3252	0.0646
		participation in activities u_{13}	0.3409	0.1496
		ability to participate u_{14}	0.0428	0.0232
educators (teacher)u_2	0.2951	knowledge structure u_{21}	0.4326	0.1305
		teaching ability u_{22}	0.4928	0.1568
		ability to research u_{23}	0.0746	0.0035
educational resources (educational institution)u_3	0.1056	system construction u_{31}	0.4547	0.1440
		platform construction u_{32}	0.3081	0.0761
		funding u_{33}	0.1012	0.0054
		media publicity and guidance u_{34}	0.1359	0.0084
Fusion way u_4	0.1897	design an interesting ideological and political class u_{41}	0.5262	0.1874
		enrich the campus aesthetic creation and experience u_{42}	0.2079	0.0112
		lead students to live a better life and social practice u_{43}	0.1503	0.0178
		improve teachers' aesthetic quality and cultivate their aesthetic feelings u_{44}	0.1256	0.0047

The weight set of the factors evaluating the secondary index H_1, H_2, H_3, H_4.

$$H_1 = (0.2911, 0.3252, 0.3409, 0.0428) \tag{13}$$

$$H_2 = (0.4326, 0.4928, 0.0746) \tag{14}$$

$$H_3 = (0.4547, 0.3081, 0.1012, 0.1359) \tag{15}$$

$$H_4 = (0.5262, 0.2079, 0.1503, 0.1256) \tag{16}$$

(3) Establish the Evaluation Set

The effect evaluation of the integration of higher vocational aesthetic education and ideological and political education is set as four levels, namely the evaluation set $V = \{Verygood, good, average, bad\}$. If the percentage-point system is used to represent the evaluation level, $V = |90, 80, 70, 50|$.

Taking higher vocational colleges in Hubei Province as an example, the effect of the integration of higher vocational aesthetic education and ideological and political education is adjusted through the form of questionnaire adjustment. A total of 500 questionnaires were distributed and 469 questionnaires were recovered. After statistics, the evaluation of each factor in the index system is shown as in Table 2.

Table 2. Summary table of the results of the secondary index item questionnaire

Desig-nation	Very good	good	average	bad	Desig-nation	Very good	good	average	bad
u_{11}	56	152	226	35	u_{31}	100	340	27	2
u_{12}	96	308	55	10	u_{32}	99	344	23	3
u_{13}	102	316	43	8	u_{33}	45	353	58	13
u_{14}	52	146	182	89	u_{34}	65	363	37	4
u_{21}	89	353	26	1	u_{41}	132	331	5	1
u_{22}	112	353	3	1	u_{42}	86	350	26	7
u_{23}	112	350	6	1	u_{43}	83	337	34	15
					u_{44}	98	348	22	1

(4) Obtain the Fuzzy Relation Matrix

Data processing in Table 2 obtained the evaluation proportion of each factor in the secondary index system, namely the one-factor evaluation matrix R_1, R_2, R_3, and R_4 of the fuzzy comprehensive evaluation of the secondary index.

$$R_1 = \begin{bmatrix} 0.1192 & 0.3241 & 0.4819 & 0.0763 \\ 0.2047 & 0.6567 & 0.1173 & 0.0213 \\ 0.2175 & 0.6738 & 0.0917 & 0.0171 \\ 0.1109 & 0.3113 & 0.3881 & 0.1898 \end{bmatrix} \quad R_2 = \begin{bmatrix} 0.1898 & 0.7527 & 0.0554 & 0.0021 \\ 0.2388 & 0.7527 & 0.0064 & 0.0021 \\ 0.2388 & 0.7463 & 0.0128 & 0.0021 \end{bmatrix}$$

(17)

$$R_3 = \begin{bmatrix} 0.2132 & 0.7249 & 0.0576 & 0.0043 \\ 0.2111 & 0.7335 & 0.0490 & 0.0064 \\ 0.0959 & 0.7527 & 0.1237 & 0.0277 \\ 0.1386 & 0.7740 & 0.0789 & 0.0085 \end{bmatrix} \quad R_4 = \begin{bmatrix} 0.2814 & 0.7058 & 0.0107 & 0.0021 \\ 0.1833 & 0.7463 & 0.0554 & 0.0149 \\ 0.1770 & 0.7186 & 0.0725 & 0.0320 \\ 0.2090 & 0.7420 & 0.0469 & 0.0021 \end{bmatrix}$$

(18)

$$R = \begin{bmatrix} 0.1192 & 0.3241 & 0.4819 & 0.0763 \\ 0.2047 & 0.6567 & 0.1173 & 0.0213 \\ 0.2175 & 0.6738 & 0.0917 & 0.0171 \\ 0.1109 & 0.3113 & 0.3881 & 0.1898 \\ 0.1898 & 0.7527 & 0.0554 & 0.0021 \\ 0.2388 & 0.7527 & 0.0554 & 0.0021 \\ 0.2388 & 0.7463 & 0.0128 & 0.0021 \\ 0.2132 & 0.7249 & 0.0576 & 0.0043 \\ 0.2111 & 0.7335 & 0.0490 & 0.0064 \\ 0.0959 & 0.7527 & 0.1237 & 0.0277 \\ 0.1386 & 0.7740 & 0.0789 & 0.0085 \\ 0.2814 & 0.7085 & 0.0107 & 0.0021 \\ 0.1833 & 0.7463 & 0.0554 & 0.0149 \\ 0.1770 & 0.7186 & 0.0725 & 0.0320 \\ 0.2090 & 0.7420 & 0.0469 & 0.0021 \end{bmatrix} \quad (19)$$

(5) Calculate the Evaluation Results

According to the above formula, the fuzzy comprehensive evaluation results of the secondary indicators can be obtained B_1, B_2, B_3, B_4.

$$B_1 = H_1 \cdot R_1 = (0.1802, 0.5509, 0.2263, 0.0431) \quad (20)$$

$$B_2 = H_2 \cdot R_2 = (0.2176, 0.7522, 0.0281, 0.0021) \quad (21)$$

$$B_3 = H_3 \cdot R_3 = (0.1905, 0.7370, 0.0645, 0.0079) \quad (22)$$

$$B_4 = H_4 \cdot R_4 = (0.2389, 0.7277, 0.0338, 0.0092) \quad (23)$$

Taking the fuzzy comprehensive evaluation results of second-level index as single-factor evaluation of first-level index get R, and bring into formula above.

$$B = H \cdot R = (0.2035, 0.6635, 0.1142, 0.0209) \quad (24)$$

3 Evaluation Results

From the above results, it can be seen that in the investigation of the effect of the integration of higher vocational aesthetic education and ideological and political education in Hubei Province, 20.35% of teachers and students think that the effect of the integration of higher vocational aesthetic education and ideological and political education is very good, 66.35% think it is good, 11.42% think it is general, and 2.09% of teachers and students think it is poor. According to the principle of maximum subordination, it can be concluded that the effect of integrating higher vocational aesthetic education and ideological and political education in Hubei Province is excellent. However, there are also some deep-rooted reasons worth analyzing, so as to further promote the integration of aesthetic education and ideological and political education [3–5], constantly realize the unity of "true, good and beautiful", realize the high degree of ideological and political education and aesthetic experience, better realize the educational goal, and promote the cultivation of ideological and political quality of the whole people [6].

While in the study, it was found that students were still very enthusiastic about learning, that ideological and political teachers have good performance in knowledge structure, teaching ability and scientific research ability [7], but there is still a deficit of 2.1%, which needs to be further improved. The survey found that the system and platform construction of integrating aesthetic education and ideological and political education is quite perfect, but it needs to be strengthened in the aspects of capital investment, atmosphere construction and media publicity [8]. From the questionnaire survey, it is found that 23.89% of the students and teachers respond very well to the integration of higher vocational aesthetic education and ideological and political education in general, 72.77% of them think it is good, but 3.38% of them still think it is average and 0.92% of them think it is poor. Among them, "designing a tasty Civic Science classroom" and "improving teachers' aesthetic quality and cultivating aesthetic feelings" are well developed [9], but "enriching aesthetic creation and experience on campus and leading students to social practice for a better life" are less satisfactory.

4   Evaluation Analysis

4.1   Improve Students' Interest and Participation in Activities

Except for art students, most students have no professional education or training in aesthetic education [10], and their ability to carry out activities is limited, which to some extent increases students' sense of loss and reduces their interest in learning and enthusiasm for participation. Therefore, in the practice of integrating aesthetic education with ideological and political education, the strong motivation and desire for expression of young students can be explored and brought into play according to their characteristics and strengths, and their initiative and creativity can be actively mobilized, and different types of activities can be carried out to give students a greater sense of access. The "artistic", "aesthetic", "experiential", and "emphatic" methods can be used to stimulate and train the educated to shape beautiful objects [11], actively explore existing cultural and artistic works, create situations with beautiful expressions, expand the emotional experience, enrich the aesthetic experience, and promote the depth of aesthetic understanding. In turn, they are able to receive theoretical inculcation, transform into practical ideological and political qualities, core literacy and key competencies, and internalize them into practical actions in a pleasant way.

4.2   Further Enhance Teacher' Knowledge Reserve and Teaching and Scientific Research Ability

The use of aesthetic education in ideological and political education is to unify and integrate the role of aesthetics in promoting the free and comprehensive development of human beings with the goal of ideological and political education striving to achieve comprehensive human development. Through the emotional experience of beauty, the use of artistic conception, creation [12], expression and appreciation and other levels of thinking and practice of beauty, to mobilize the senses and thinking, the formation of aesthetic understanding, and get a sense of pleasure from the acquisition, and then the occurrence of changes in thought, prompting the adjustment of personal behavior, in order to achieve the unity of knowledge and action. Therefore, the knowledge reserve and teaching and scientific research ability of ideological and political teachers is to solve the problem of "explaining the deep, profound and vivid truth full of philosophical thinking, historical deposits and reality".

4.3   School Funding Investment and Media Publicity Be Strengthened

As the backbone of education, schools should fully play to the role of organization and leadership [13], enrich the construction of the system and platform, and appropriately increase some funds in the actual integration of aesthetic education and ideological and political education, increase the publicity and guidance, and strive to create a stronger atmosphere.

4.4   The Integration Way Be Further Broadened

The national unified teaching materials for Civics reflect the highest level of current Marxist theory research, reflect the requirements of national will and ideology, and provide a strong theoretical foundation. And the rooting of the teaching material system requires the effective transformation of the teaching material system to the teaching system. The use of aesthetic education in ideological and political education means that the design of educational contents, educational methods and educational forms are set around the development of aesthetic activities, educational activities stimulate the leading role of the teacher and the enthusiasm [14], initiative and creativity of the taught in the educational activities with the help of the concept and methods of aesthetic education, the educator and the educated communicate with the intermediary of aesthetic objects, realize interaction through the production of beauty, promote the overall human development and improve the practical effect of thought transformation. Therefore, by expanding the ways of integration of higher vocational aesthetic education and ideological and political education, we can better realize the high degree of fit between ideological and political education and aesthetic experience, better achieve the education goal and promote the ideological and political quality of college students.

5   Conclusion

The cultivation of aesthetic ability plays an important role in ideological and political education. Aesthetics has a unique advantage in the penetration of people's deep consciousness. The emotional experience connects the past and future of the aesthetic subject, and the sensory pleasure as well as the imagination inspired in the process of transition to rationality are more deeply recorded in people's physiological and psychological structures, reaching a depth that is difficult to achieve by purely rational thinking, thus making people's understanding closely connected with people's emotions, establishing emotional in people's behavior system. The mechanism of stimulation, people's judgment is not only utilitarian rational judgment, but also adds more actionable emotional judgment, which promotes the internalization of ideological and political education.

Overall, this paper uses the fuzzy comprehensive evaluation method to assess the effectiveness of the current integration effect of higher vocational aesthetic education and ideological and political education, on the basis of which the following conclusions are drawn.

1) The current integration effect of higher vocational aesthetic education and ideological and political education is excellent, reaching 86.7%. However, there are some deep-seated reasons worth analyzing.
2) Students' enthusiasm for learning is still very high, but most of them have no professional education or training in aesthetic education, and their ability to carry out activities is limited except for art students, which to a certain extent breeds students' sense of loss and reduces their interest in learning and enthusiasm for participation. Therefore, in the practice of integrating aesthetic education with ideological and political education, different types of activities can be carried out according to students' characteristics and specialties, so that students can have more access to them.
3) Teachers of ideology and political science courses perform well in knowledge structure, teaching ability and scientific research ability, but there are still shortcomings that need further improvement.
4) The system and platform construction of the integration of aesthetic education and ideological and political education in higher vocational institutions is still quite perfect, but the financial investment, atmosphere creation and media publicity need to be strengthened.
5) The current ways of integrating aesthetic education with ideological and political education are generally well reflected. Among them, the educational content is designed to organically integrate the cultural vision, artistic literacy, aesthetic needs and state of mind of the educated, and deepen the understanding and experience of the beauty of ideological and political education through ideological leadership, political orientation and educational promotion.

Acknowledgment. This project is supported by Hubei Province education planning 2021 annual key topics (2021GA092).

References

1. Li, L., Wang, Z., Tang, X.: Study on practical reform of ideological and political theory course in colleges and universities from the perspective of cultivating "core competency" and "key ability." Educ. Circles **15**, 66–68 (2019). (in Chinese)
2. Wang, Y.: Study on the effectiveness of ideological and political education in the major tasks of the armed police. J. Higher Educ. **15**, 38–39 (2019). (in Chinese)
3. Sathe, M.T., Adamuthe, A.C.: Comparative study of supervised algorithms for prediction of students' performance. Int. J. Modern Educ. Comput. Sci. (IJMECS) **2**, 1–21 (2021)

4. Rafida, V., Widiyatni, W., Harpad, B., Yulsilviana, E.: Implementation of multi-attribute rating technique simple in selection of acceptance scholarship of PMDK (Case study: STMIK widya cipta dharma). Int. J. Modern Educ. Comput. Sci. (IJMECS) **2**, 22–33 (2021)
5. Sriram, B.: Learner's satisfactions on ICT innovations: omani learners viewpoints. Int. J. Modern Educ. Comput. Sci. (IJMECS) **12**(5), 1–15 (2020)
6. Yunusovich, A.V., Ahmedov, F., Norboyev, K., Zakirov, F.: Analysis of experimental research results focused on improving student psychological health. International Journal of Modern Education and Computer Science (IJMECS) **14**(2) (2022)
7. Xiao, W.: Integrity and innovation of ideological and political teaching in colleges and universities--interview with Lingjun Huang, dean of school of marxism in huazhong university of science and technology. People's Daily, 2019–1–25(9) (in Chinese)
8. Noor, M.B., Ahmed, Z., Nandi, D., Rahman, M.:Investigation of facilities for an m-learning environment. International Journal of Modern Education and Computer Science (IJMECS) **1**, 34-48 (2021)
9. Bouzid, T., Kaddari, F., Darhmaoui, H., Bouzid, E.G.: Enhancing math-class experience throughout digital game-based learning, the case of moroccan elementary public schools. International Journal of Modern Education and Computer Science (IJMECS) **13**(5) (2021)
10. Divayana, D.G.H.: ANEKA-based asynchronous and synchronous learning design and its evaluation as efforts for improving cognitive ability and positive character of students. Int. J. Modern Educ. Comput. Science (IJMECS) **10**, 14–22 (2021)
11. Kyrpychenko, O., Pushchyna, I., Kichuk, Y., Shevchenko, N., Luchaninova, O., Koval, V.: Communicative competence development in teaching professional discourse in educational establishments. Int. J. Modern Educ. Comput. Sci. (IJMECS) **8**, 16–27 (2021)
12. Firdaus, H., Zakiah, A.: Implementation of usability testing methods to measure the usability aspect of management information system mobile application (Case study sukamiskin correctional institution). Int. J. Modern Educ. Comput. Sci. (IJMECS) **13**(5), 58–67 (2021)
13. Perez, J.G., Perez, E.S.: Predicting student program completion using naïve bayes classification algorithm. International Journal of Modern Education and Computer Science (IJMECS), **13**(3), 57–67 (2021)
14. Dolgikh, S.: Categorization in unsupervised generative self-learning systems. Int. J. Modern Educ. Comput. Sci. (IJMECS) **6**, 68–78 (2021)

Teaching Reform and Practice of "Sensor and Detection Technology" Course Based on "Specialty and Innovation Integration" and Project-Based Education

Deshen Lv[✉] and Ying Zhang

School of Intelligent Manufacturing, Nanning University, Nanning 530200, China
876357930@qq.com

Abstract. The course "Sensor and Detection Technology" is a required course for many colleges and universities in electronics, electrical, communication and automatic control and related majors. It is a course combining theory and practice and plays a very important role in the cultivation of innovative talents. Under the traditional teaching method, the course arrangement is mostly limited to the class teaching and corresponding in-school experiments. The course design is not well integrated with enterprise practice. Students are relatively passive in the course learning, and high-quality course resources are relatively scarce. In view of the above problems, the paper discusses the objectives and design ideas of curriculum teaching reform under the guidance of the principle of "integration of specialty and innovation", the specific items of curriculum teaching and the evaluation methods, focuses on the deep integration of "professional education" and "innovation and entrepreneurship education", optimizes the teaching content, aims to cultivate students' innovative thinking, and lays a solid foundation for the society to cultivate innovative talents.

Keywords: Sensor and detection technology · Integration of specialty and innovation · Innovation and entrepreneurship · Teaching reform

1 Introduction

Sensor is the primary link to realize automatic detection and automatic control [1, 2]. As a device that converts external input signals into electrical signals, sensors have become an indispensable tool in today's production and life [3–5]. Industrial production lines, security systems, car navigation systems, aerospace systems, household appliances, etc. are not difficult to see the traces of sensors [6–8]. The sensor and its detection technology have become a key field of research in the industry and academia [9].

Many colleges and universities have taken the course of "Sensor and Detection Technology" as a compulsory basic course for electronics, electrical, communication and automatic control majors [10, 11]. The purpose is to enable students to master sensor technology, understand relevant knowledge in measurement and control, computer

© The Author(s), under exclusive license to Springer Nature Switzerland AG 2023
Z. Hu et al. (Eds.): ICCSEEA 2023, LNDECT 181, pp. 1033–1045, 2023.
https://doi.org/10.1007/978-3-031-36118-0_88

application, electronic information and other fields, and cultivate students' ability of product design and innovative application [12–14]. Graduates of electronics, electrical, communication and automatic control majors have many jobs related to sensors, such as detection and installation of electronic products, intelligent instrument development, robot application development, etc. [15, 16]. It occupies a large proportion in the whole basic curriculum system.

The teaching of this course attaches great importance to the combination of "specialty" and "innovation". The former focuses on the vertical development of professional technology to understand the essence of knowledge and skills; The latter - innovation and entrepreneurship education needs to pay attention to the innovation and application of knowledge, which is conducive to the horizontal development of professional technology [17–19]. In recent years, the integration of specialty and innovation has been the focus of curriculum teaching reform in colleges and universities. It attaches importance to both professional education and innovation and entrepreneurship education [20, 21]. Traditional teaching only pays attention to professional education, which cannot effectively help students grasp and understand the content of "sensor and detection technology" and apply it to future work [22, 23]. In the context of "mass entrepreneurship and innovation", it is required to closely link professional knowledge with innovation and entrepreneurship education to carry out curriculum reform [24, 25]. Therefore, it is urgent to link sensor technology with innovation and entrepreneurship education, and it is very urgent to develop a curriculum reform plan of "sensor and detection technology" based on the integration of specialty and innovation [26]. How to integrate innovation and entrepreneurship education into the teaching of "sensor and detection technology" is the key to reform [27, 28].

2   Teaching Status and Reform Content of the Course "Sensor and Detection Technology"

2.1   Current Situation of Course Teaching

The teaching of specialized and creative integration course is a teaching reform method that has emerged in recent years. Many colleges and universities have not reformed the teaching of "sensor and detection technology" course, and still use the traditional teaching method. In traditional teaching methods, teachers regard themselves as the controller and dominator of the classroom, and students' learning state is always particularly passive. Students need to listen to what the teacher says. Some teachers put too much emphasis on the discussion of principles, but did not guide students to carry out diversified practice, relied too much on the final written test results, and did not allow students to participate in the operation by themselves, resulting in very poor practical results of students. At the same time, the content of the textbook lags behind the development of the discipline and does not show the cutting-edge dynamics of sensor technology. In addition, most teachers use the method of classroom teaching to guide students to follow their own ideas and learn theoretical knowledge mechanically, without leaving enough room for students to think. When a question is thrown out, students will directly publish the correct answer before they have time to think. Over time, students' initiative and enthusiasm will be greatly weakened.

Some colleges and universities have carried out the curriculum reform of "sensor and detection technology" specialized innovation integration, but there are still some problems, such as the curriculum of specialized innovation integration is not reasonable enough, and the high-quality curriculum resources of specialized innovation integration are lacking.

2.2 The Curriculum is not Reasonable Enough

In the process of "sensor and detection technology" innovation and entrepreneurship integration reform, many colleges and universities still separate sensor knowledge from innovation and entrepreneurship, and have not designed an organic curriculum integration mode, which makes learners mechanically contact with innovation and entrepreneurship content after the completion of professional learning in the course of curriculum learning, and does not have a connection between the two in their minds. Such curriculum settings have lost the original significance of innovation and entrepreneurship integration, It deviates from the original intention of innovation and entrepreneurship based on the advantages of sensor knowledge. The sensor curriculum ignored innovation and entrepreneurship practice at the beginning of the integration design, or the school lacked innovation and entrepreneurship practice resources, and the curriculum theory and practice system design was not perfect.

2.2.1 Lack of High-Quality Curriculum Resources

In the specific practice of the integration of specialty and innovation, the construction of curriculum resources is particularly important, which is the implementation of details in practice. Most colleges and universities lack high-quality curriculum resources for "sensor and detection technology" courses. Professional education continues to be carried out in the traditional professional education mode, and innovation and entrepreneurship education is also carried out in the traditional way. The current development status of the industry is not included in the construction of curriculum resources, and the existing problems of the industry are analyzed. For the specific key and difficult points of the course, no in-depth analysis and sorting has been carried out, and the organic relationship between the key and difficult points and the innovation and entrepreneurship knowledge points has been sorted out, so as to achieve the one-to-one mapping between the course content and innovation. In the resource construction, real cases should be added, and the case rational analysis of sensor knowledge should be used while cultivating innovation literacy.

The lack of strong participation of enterprises is also an important reason for the lack of resources in the construction of specialized creative courses. The enterprise has rich project cases, which can help the school promote the teaching of experiment and practice. At the same time, the rich technology and management advantages of enterprises can also inspire the teachers and students of the school to pursue innovation and carry out innovation and entrepreneurship. Increasing the participation of enterprises in the integration of specialty and innovation in colleges and universities can provide strong support in the construction of resources, help schools truly realize the integration of specialty and innovation, and promote talent cultivation.

2.3 Ideas and Contents of Curriculum Reform

Starting from the principle of "integration of specialty and innovation", in order to make the curriculum design of "sensor and detection technology" more reasonable, provide high-quality and realistic projects, and promote the collaborative development of students' professional technology and innovation and entrepreneurship. As shown in Fig. 1, the curriculum reform mainly includes three modules - the design of curriculum objectives, the implementation plan of project teaching method, and diversified assessment methods.

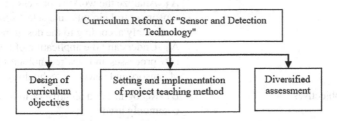

Fig. 1. Contents of the curriculum reform of "sensor and detection technology"

Module 1 "Design of teaching objectives" is to reset the old teaching objectives, adjust and deepen the knowledge objectives, ability objectives and quality objectives, and increase the goal of innovation and entrepreneurship; Module 2 "Project teaching method implementation plan" is to establish the idea and steps of project-based teaching method, deconstruct the old knowledge points, adopt project teaching method, reconstruct the items that meet the curriculum, and give the teaching items of "sensor and detection technology" curriculum; Module 3 "diversified assessment methods" is to establish a diversified process assessment and evaluation system, more objectively and scientifically reflect the learning effect of students, and help to cultivate diversified innovative talents.

3   Curriculum Reform of "Sensor and Detection Technology" Based on the Integration of Specialty and Innovation

3.1   Design of Teaching Objectives

By learning "sensor and detection technology", students can master the basic working principles of temperature sensor, humidity sensor, photoelectric sensor, mechanical sensor, etc., master the conversion circuit, error analysis and correction circuit of various sensors, and master the circuit analysis and detection circuit design methods of various sensors. Adhering to the comprehensive talent training objectives of specialty and innovation and entrepreneurship, the design of teaching objectives in the four directions of knowledge, ability, literacy and innovation and entrepreneurship is determined. The course objectives of "Sensor and Detection Technology" are shown in Table 1.

The conceptual model of curriculum objectives can be expressed as:

$$T_{kcqi} = \sum_{i=1}^{4} A_i + \sum_{j=1}^{4} B_j + \sum_{k=1}^{3} C_k + \sum_{m=1}^{3} D_m \tag{1}$$

Table 1. Course objectives of "sensor and detection technology"

Teaching objectives	Concrete content
Knowledge objectives A	A_1 -Master the basic method of measurement, the error representation method of sensor measurement data and the data processing method A_2 -Familiar with the types, basic structures, working characteristics, working parameters and applications of various sensors A_3 -Analyze the working process of various sensor control circuits, and select sensors reasonably according to the design needs A_4 -Understand the application of new materials, new processes and new technologies in the field of sensors and detection technology
Capability objectives B	B_1 -Recognize and identify various sensors commonly used B_2 -Be able to use multi-meter, oscilloscope, signal generator and other instruments to detect the quality of various sensors B_3 -Be able to select and design the detection circuit that actually meets the requirements according to the needs B_4 -Be able to analyze the typical application circuits and working principles of various sensors
Quality objectives C	C_1 -Cultivates students' patriotic feelings and dialectical thinking ability, and has a rigorous and realistic attitude of seeking knowledge and a craftsman's spirit of keeping improving C_2 -Improves students' ability to analyze and solve problems, as well as team cooperation C_3 -Has the awareness of quality and serious work attitude, establishes a correct outlook on life, and strengthens the sense of social responsibility
Innovation and entrepreneurship objectives D	D_1 -Professional knowledge and innovation capabilities are integrated. Introduces the case of competitions into the learning of professional knowledge D_2 -Has the ability of market research and scheme design, team cooperation and division of labor, unity and cooperation, as well as a certain degree of leadership D_3 Cultivate the spirit of innovation and practice, stimulate students' professional honor and industry pride, and lay a good foundation for work after graduation

3.2   Course Design Ideas

3.2.1   The Course Adopts the Project-Based Teaching Method

The project-based teaching method used in the teaching of "sensor and detection technology" can not only make students pay attention to the relationship between their living environment and professional knowledge, but also use the knowledge learned to find and solve practical problems, effectively improve their comprehensive qualities such as communication, cooperation and display, and make them become lifelong learners, which is a significant advantage of project-based teaching.

The course teaching is guided by engineering practice problems, guided by the teacher's explanation and analysis, and guided by the theory. The teaching with students' hands as the main body is a modern and scientific teaching method that focuses on practical skills training and develops specific teaching contents. Under the guidance of the teacher, students can collect knowledge one by one according to the outline requirements designed for the content of each specific experimental project, consult various relevant books and materials, and then analyze and solve the problems existing in the actual problems through in-depth study and cooperation and discussion of research topics among the team members, until all the project tasks are completed, so as to fully meet the requirements of the actual teaching objectives of the course.

3.2.2   Design Idea of Project-Based Teaching

In project-based teaching, teaching is generally carried out according to five steps: setting tasks, grouping, dividing tasks, assessing, reviewing and summarizing. The details are as follows:

(1) Task setting
    According to the characteristics of the talent training teaching program and the content requirements of the syllabus, teachers closely combine the teaching content of the theoretical knowledge of the course with the tasks of professional practice to design the teaching tasks, intersperse innovation and entrepreneurship knowledge, and try to integrate, optimize and adjust multiple knowledge points, so as to achieve the dual purpose of strengthening the effective training of students' skill quality and comprehensive practical application. In the process of independently completing the project tasks, Continuously gain a sense of experience and achievement, and stimulate their learning initiative and independent innovation consciousness.

(2) Grouping
    Teachers can group students according to their wishes and the degree of mastery of course learning, and try to achieve the overall level balance of students in each group. A group leader is selected from each group to be responsible for the implementation process and reporting of the project, so as to ensure the smooth completion of the teaching work.

(3) Tasks' assigning
    Determine the specific projects of each group and their organization and distribution work, communicate and discuss within the group, propose project solutions based on division of labor and cooperation, and constantly adjust and improve them to obtain specific and effective solutions and complete the group tasks.

(4) Communication and reporting

The plans of each group should be communicated and reported in the whole class, so that they can inspire, learn from each other, comment on and discuss with each other, which is obviously beneficial to the common improvement of all students.

(5) Review and summary

When summarizing and commenting on the project results of each group of students, the teacher can clarify the relationship between the solution and the knowledge points, inspire students to make horizontal comparison of each group of projects and vertical comparison of the improvement effect of the plan, thus triggering in-depth thinking and training students' thinking and expression abilities.

3.3 "Sensor and Detection Technology" Teaching Project

The "sensor and detection technology" teaching project mainly includes the basic sensor detection technology, temperature detection, pressure detection, material level detection, wave detection, environmental detection and other projects, with a total of 64 class hours. Its main content and class hour distribution are shown in Table 2.

The course project of "Sensor and Detection Technology" is based on the familiar projects of students, integrates the knowledge points and experimental contents of traditional theoretical teaching, combines theory with practice, implements students' innovation and entrepreneurship in place, and trains students' innovation ability to achieve twice the result with half the effort.

3.4 Assessment of Students' Learning Effect

Before the reform of "sensor and detection technology", the total score = attendance score (5%) + classroom score (25%) + final exam score (70%). After the reform, the project is based on the project, and each project is designed based on the work process. The assessment method is shown in Table 3.

In the previous teaching class, the subjective score of teachers is relatively high, and the proportion of final score is high, which is easy to lead to poor final exam scores of students who usually learn well, so the total score obtained deviates greatly from the actual learning effect of students. "The course of sensor and detection technology" is specially created for integrated teaching, with students occupying the dominant position and playing a leading role, focusing on the process assessment of project teaching, and viewing students' learning achievements from multiple perspectives, which has great advantages.

Table 2. "Sensor and detection technology" teaching items and class hour allocation table

No	Project	Main contents	Hours
1	Item 1 Fundamentals of sensor detection technology	-Fundamentals of sensor and detection technology -Measurement error and data processing -Data processing cases	8
2	Item 2 Temperature detection	-Thermocouple principle and detection circuit analysis -Thermal resistance principle and detection circuit analysis -Principle and detection of thermistor -Design of other temperature sensors and temperature detection circuits	12
3	Item 3 Pressure detection	-Principle and detection of strain pressure sensor -Principle and detection of piezoelectric pressure sensor -Principle and detection of capacitive pressure sensor -Principle and detection of inductive pressure sensor -Design of pressure detection circuit	12
4	Item 4 Speed detection	-Magnetoelectric induction sensor -Eddy current sensor -Hall sensor -Semiconductor magnetic sensor -Overview of photoelectric sensors -Photoelectric effect and typical devices -Analysis and Design of Speed Detection Circuit	12
5	Item 5 Detection of sound waves	-Overview of wave sensor -Ultrasonic testing application -Infrasound detection application -Design of ultrasonic testing circuit	10
6	Item 6 Environmental detection	-Gas sensor and its application -Humidity sensor and its application -Environment detection circuit design	10

Table 3. Post-reform evaluation table of "sensor and detection technology" course

Evaluation content	Evaluation items	Evaluation method	Proportion
Process assessment	Attendance and discipline	"Learning Pass" sign-in	5%
	Team cooperation	Self and inter-group evaluation	5%
	Project research and scheme design	Scheme evaluation	10%
	Leadership and communication	Self and inter-group evaluation	10%
	Basic professional knowledge	Quiz	10%
	Knowledge application and creation	Quiz	10%
	Individual contribution	Self and inter-group evaluation	5%
	Language expression and project analysis	PPT, speech, evaluated by group leaders	10%
	Project results	Project results judged by team leaders	15%
Result assessment	Final exam	Teachers evaluate according to the test	20%

4 Achievements of the Current Stage of Innovation and Innovation Integration Reform

4.1 Significant Improvement of Students' Course Performance

Since the implementation of the innovation and innovation integration curriculum reform in the second semester of 2020, the curriculum content of "sensor and detection technology" has been restructured, the project-based teaching method has been adopted, the innovation and entrepreneurship content has been integrated, and the process assessment has been emphasized. The students' learning achievements have been significantly improved. Figure 2 shows the total results of electrical engineering and its automation, mechanical design and manufacturing and its automation in the past three years. It can be seen that after the reform, the achievements of the two majors have generally improved by more than 5 points compared with the previous ones, with obvious progress.

4.2 Achievements in Innovation and Entrepreneurship Competitions have Improved Significantly

Nanning University attaches great importance to the teaching reform of the integration of specialty and innovation. In the past three years, it has approved 60 projects of the

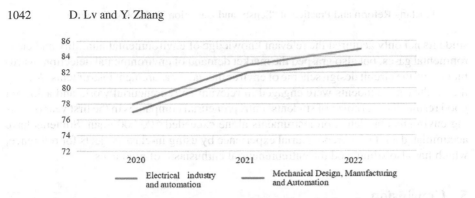

Fig. 2. Three years' scores in "sensor and detection technology" of two majors

integration of specialty and innovation, which has effectively improved the quality of talent training and achieved excellent results. In particular, outstanding achievements have been made in the establishment of Innovation and entrepreneurship training program for college students (abbreviated as IET) and the International "Internet + " Undergraduate Innovation and Entrepreneurship Competition Guangxi Division (referred to as "INT + IEP" for short). The project approval and awards are shown in Table 4.

Table 4. Awards of IET and "INT + IEP"

Year	IET State level	IET Provincial level	"INT + IEP" Guangxi Provincial Competition		
			Gold award	Silver award	Bronze award
2020	38	93	3	8	30
2021	39	103	14	26	31
2022	44	122	10	17	68

As can be seen from Table 4, the number of major founding projects and Internet plus awards are increasing year by year, as well as national level projects. Through the implementation of the integration of specialty and innovation teaching, the students' professional level and innovation and entrepreneurship literacy have been continuously improved, and the new type of entrepreneurship and innovation talents have been growing. The course teaching team guided the students to achieve good results in the innovation and entrepreneurship competition, and the students obtained dozens of achievements and awards in the national and regional competitions of the above two major events.

4.3 Stimulate Graduates' Entrepreneurial Enthusiasm

Through the teacher's classroom teaching and related content experiments, as well as the promotion of project teaching method, students have accumulated a lot of entrepreneurial experience while learning knowledge, which provides very effective help for their independent entrepreneurship. For example, through the environmental detection project, students not only acquired the relevant knowledge of environmental humidity and environmental gases, but also grasped the market demand of environmental detection instruments and the circuit design scheme of environmental measurement instruments. Among the graduates, 2 students were engaged in relevant entrepreneurial work, and achieved good results. The turnover of students' entrepreneurial companies in the first year of selling environmental detection instruments alone exceeded 500,000 yuan. Students have accumulated a lot of entrepreneurial experience by using in-class projects for reference, which has also stimulated the entrepreneurial enthusiasm of graduates.

5 Conclusion

The teaching reform of "sensor and detection technology" based on the integration of specialty and innovation, taking students' project practice ability and innovation consciousness as the foothold of talent cultivation, has not only improved students' theoretical knowledge and practical ability, but also improved students' innovation ability. By adopting the project-based teaching method, the transformation of the main role of classroom teaching has been realized, so that each student has always played the main and leading role, and the rapid improvement of their comprehensive quality has been continuously realized, and the students' internal enthusiasm for course learning, experiment and practice has been enhanced, and the expected teaching purpose has been achieved. At the same time, the overall teaching level of the school has been rapidly, effectively and comprehensively improved.

The teaching practice has proved that the teaching reform of the "sensor and detection technology" course based on the integration of specialty and innovation has optimized the teaching content around the requirements of the deep integration of "professional education" and "innovation and entrepreneurship education", which is more in line with the requirements of education and teaching for professional courses, at the same time, it is conducive to students mastering the application of the curriculum in the specialty and the frontier of professional development, and cultivating students' innovative thinking.

Acknowledgment. This project is supported by the third batch of curriculum reform project of the integration of specialty and innovation of Nanning University "Teaching reform and practice of 'sensor and detection technology' based on the integration of specialty and innovation" (2021XJZC04).

References

1. Anayat, S., Butt, S., Zulfiqar, I., Butt, S.: A deep analysis of applications and challenges of wireless sensor network. International Journal of Wireless and Microwave Technologies (IJWMT), **10**(3), 32–44 (2020)
2. Kori, G.S., Kakkasageri, M.S.: Game theory based resource identification scheme for wireless sensor networks. International Journal of Intelligent Systems and Applications (IJISA) **14**(2), 54–73 (2022)
3. Altay, A., Learney, R., Güder, F., et al.: Sensors in blockchain. Trends in Biotechnology **40**(2), 141–144 (2022)

4. Ashraf, S., Saleem, S., Ahmed, T.: Sagacious communication link selection mechanism for underwater wireless sensors network. International Journal of Wireless and Microwave Technologies (IJWMT) **10**(4), 22–33 (2020)
5. Kawamoto, K., Shiomi, H., Ito, T., et al.: Vector sensor imaging. Optics and Lasers in Eng. **162**, 107439 (2023)
6. Nair, T., Singh, A., Venkateswarlu, E., et al.: Generation of analysis ready data for indian resourcesat sensors and its implementation in cloud platform. International Journal of Image, Graphics and Signal Processing (IJIGSP), **11**(6), 9–17 (2019)
7. Shia, H.H., Tawfeeq, M.A., Mahmoud, S.M.: High rate outlier detection in wireless sensor networks: a comparative study. International Journal of Modern Education and Computer Science (IJMECS), **11**(4), 13–22 (2019)
8. Puranikmath, V.I., Harakannanavar, S.S., Kumar, S.: Dattaprasad Torse. Comprehensive study of data aggregation models, challenges and security issues in wireless sensor networks. International Journal of Computer Network and Information Security (IJCNIS), **11**(3), 30–39 (2019)
9. Rambabu, C., Prasad, V.V.K.D.V., Prasad, K.S.: Multipath cluster-based hybrid mac protocol for wireless sensor networks. International Journal of Wireless and Microwave Technologies (IJWMT), **10**(1), 1–16 (2020)
10. Chunjiao, Z.: Teaching design of sensor and detection technology course based on OBE concept. China Modern Education Equipment **11**, 88–90 (2022). (in Chinese)
11. Chen, J.: Construction of practical training course assessment system for applied electronic technology specialty based on OBE concept -- taking the Comprehensive Practical Training Course for Sensors as an example. Computer Knowledge and Technology, **18**(33): 115–118 (2022). (in Chinese)
12. Zebin, L.: Thinking on the teaching reform of sensor principle and application. Science and Technology Vision **34**, 190–191 (2021). (in Chinese)
13. Siqi, W.: Teaching design of "sensor working principle and application" based on STEM theory. Physics Teaching Discussion **40**(08), 30–35 (2022). (in Chinese)
14. Guo, H.: Research on the construction mode of open courses based on OBE concept -- take the sensor course of photovoltaic specialty as an example. Nanfang Agricultural Machinery, **53**(16), 195–198 (2022). (in Chinese)
15. Cao, X., Chen, Y.: Practical research on the course "Fundamentals of Sensor and Engineering Test Technology" based on "project-based teaching". Theoretical Research and Practice of Innovation and Entrepreneurship, **5**(23), 162–165 (2022). (in Chinese)
16. Bai, Y., Liu, G., Chuai, Y., et al.: First-class course construction of detection technology and sensors under OBE concept. Metallurgical Manage. (19), 75–76 (2021). (in Chinese)
17. Sun, L., Wan, L., Wang, Y., Wang, X.: Research on BOPPPS teaching mode under the background of integration of specialty and innovation - taking the analogical innovation method in innovative thinking and innovation method as an example. Science and Education, (24), 79–83 (2022). (in Chinese)
18. Yang, K.: Exploration and practice of the integration of innovation and entrepreneurship education and professional education in the context of "special innovation". Heilongjiang Education (Higher Education Research and Evaluation), (12), 42–44 (2022). (in Chinese)
19. Wu, J., Yang, B.: Analysis on the teaching reform of "product design" course based on the integration of specialty and innovation. Intelligence (35), 60–63 (2022). (in Chinese)
20. Geng, Y., Zhang, R.: Construction and implementation of specialized and creative chemistry curriculum system under the mechanism of "three complete education". Science and Education Guide, (17), 33–35 (2022). (in Chinese)

21. Liu, S.: Research on the teaching reform of Research Travel Course Design based on the integration of specialty and innovation. In: Proceedings of 2022 the 6th International Conference on Scientific and Technological Innovation and Educational Development, pp. 568–570 (2022)
22. Zhou, H., Zhang, W., Qin, H.: Research and practice on the "student-centered" teaching reform of specialized and creative integration in higher vocational colleges. Modern Vocational Education, (32), 121–123 (2022). (in Chinese)
23. Wu, C., Fei, F., Yao, E., et al.: Course reform practice of sensor principle for engineering education professional certification and training students' ability to solve complex engineering problems. China Modern Education Equipment, (21), 70–72 (2021). (in Chinese)
24. Han, Y.: Reflection and reconstruction of the curriculum system of "integration of specialty and innovation" in higher vocational colleges. Jiangsu Higher Education, (12), 122–127 (2022). (in Chinese)
25. Hu, D.: Exploration of teaching reform in split class in the course of "Sensor Principle and Application". Science and technology wind, (20), 100–102 (2022). (in Chinese)
26. Zhao, C., Wu, Y., Fan, K., et al.: Optimization of the education model of application-oriented universities from the perspective of "integration of specialty and innovation". Journal of Texas University, **38**(06), 107–110 (2022). (in Chinese)
27. Li, X., Zhang, J., Zhang, H., et al.: Teaching reform of "sensor" course under the new situation. Industry and Information Education, (12), 69–72 (2022). (in Chinese)
28. Wang, H.: Survey and thinking on entrepreneurship education of local college students. International Journal of Education and Management Engineering (IJEME), **2**(7), 25–30 (2012)

Improved Method for Grading Bilingual Mathematics Exams Based on Computing with Words

Danylo Tavrov[✉] and Liudmyla Kovalchuk-Khymiuk

National Technical University of Ukraine "Igor Sikorsky Kyiv Polytechnic Institute",
37 Peremohy Avenue, Kyiv 03056, Ukraine
d.tavrov@kpi.ua

Abstract. As the Content and Language Integrated Learning approach to studying subjects such as mathematics through foreign languages gains more recognition, the problem of assessing students' performance in bilingual settings remains largely understudied. In this work, we propose an improved method for grading mathematics exams using words that takes into account correctness, optimality, thoroughness, and tidiness of solutions written by students.

We outline the general multilayer framework for grading bilingual mathematics exams, where assessment at each layer is computed as a fuzzy set. Our approach is an improvement upon previously published methods and considers four main competences that a student should obtain. Applicability of our method is illustrated using model data.

Keywords: bilingual education · CLIL · computing with words · perceptual computing · type-2 fuzzy set

1 Introduction

In a modern globalized world, the idea that students should learn at least one foreign language as early as possible has gained wide popularity. One of the forms of such learning is known by the name of Content and Language Integrated Learning (CLIL), broadly defined as [1] the educational approach, whereby learners study content subjects (computer science [2], mathematics, history, biology, and so on) in a foreign language. The idea is that students acquire knowledge of a content subject and at the same time master language competences. This idea has proven to be very effective, which is why the European Council recommended [3] that, among other provisions, CLIL practices be further implemented. Some studies indicate [4, 5] that learners in bilingual classes were more motivated than those who studied through their primary language, and that CLIL has a positive impact on mathematical performance. In general, learning in a foreign language is considered an advantage [6] to students.

In this work, we discuss the problem of assessing students' performance in bilingual setting. CLIL is not equivalent to learning a foreign language. Subjects such as mathematics are taught not *in*, but rather *with the help of* a foreign language. This means that,

Z. Hu et al. (Eds.): ICCSEEA 2023, LNDECT 181, pp. 1046–1056, 2023.
https://doi.org/10.1007/978-3-031-36118-0_89

in general, such assessment should [1] be carried out in a foreign language, but with the intent of testing knowledge gained in the subject being studied.

At the same time, importance of the foreign language should not be downgraded. Both the content subject and the foreign language are involved equally [7] in defining learning goals. Proper assessment in such classes should thus take into account linguistic competences, subject competences, sociolinguistic competences, and what can be called soft skills competences (work in a team, debates, and so on).

In this work, given expertise of the second author, we focus on teaching mathematics in French for students whose primary language is Ukrainian. Therefore, emphasis is put on linguistic and subject (mathematics) competences.

In [8], the authors discuss a framework for assessing both subject and language dimensions of students' work. The authors split subject questions into recall, application, and analysis tiers depending on their cognitive requirements. Linguistic requirements of a given task are treated at three levels—vocabulary, sentence, and text. However, teaching mathematics in a bilingual setting is complicated by the fact that [9] students perform code-switching, i.e. they switch languages during arithmetic computation and conversation. As a result, there do not currently exist assessments for language proficiency in the context of mathematical competence, for different age groups and topics.

Further complications arise from the fact that the overall grade cannot be computed as an arithmetic sum of mathematical and linguistic parts. Even though some authors claim otherwise [10], it is our conviction that every problem on the exam touches upon several competences at once. Moreover, assigning specific numeric grades to problems on an exam is highly subjective, not in the sense that the teacher is not impartial, but in the sense that she is typically assigning grades based on her experience. Even when rough guidelines exist, the final call is upon the teacher who needs to consider not only correctness of a solution but also how optimal it is, how thorough the explanation is, how neat and tidy a student's work is, and so on.

In this paper, we propose to grade exams using fuzzy sets to capture uncertainty and subjectivity involved. Fuzzy sets have been applied in other works [11, 12], but we propose an improvement of our previous assessment framework [13] tailored to grading mathematical exams for Ukrainian students studying mathematics in French. We use a computing with words (CWW) [14] approach that enables the teacher to use words (expressed as fuzzy sets), making her work user-friendly and allowing to minimize subjectiveness by treating imprecision related to linguistic concepts in a formal way.

The paper is structured as follows. In Sect. 2, we describe our framework for grading bilingual mathematics tests and discuss modelling linguistic concepts using fuzzy sets. Section 3 gives a version of an exam that can be graded using our approach, and results for synthetic data are provided to illustrate its validity. Section 4 concludes.

2 CWW Based Method for Grading Mathematics Exams

2.1 Grading Process Outline

Let us consider a mathematics exam of N problems that need to be graded (a concrete example will be shown in Sect. 3). In accordance with the above discussion, we propose to obtain an overall grade according to a decision-making process shown in Fig. 1.

Such decision making is called hierarchical and distributed [15], because it involves aggregating independent grades for individual problems in an hierarchical fashion.

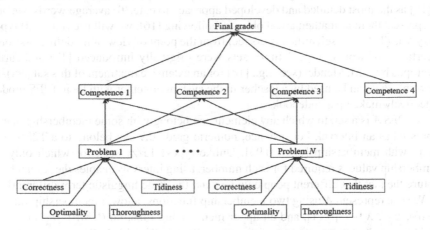

Fig. 1. Hierarchical and distributed grading of mathematical exams

A teacher evaluates each problem according to four criteria:

- *correctness* (C1). In some cases, a solution can be unequivocally correct, and in others, it can be "mostly" correct (e.g., when all the steps of the solution are correct save for some minor arithmetic mistakes);
- *optimality* (C2). For problems, where it is appropriate, a chosen method is typically either optimal or not at all. We suggest grading this criterion in a binary way;
- *thoroughness* (C3). To assess linguistic competences of a student, it is important to consider explanations they provide, which can only be assessed subjectively;
- *tidiness* (C4) of a solution, including its legibility and structure.

In general, these criteria are evaluated using words, as they are mostly subjective and hard to formalize. In some cases, it is acceptable to use numbers 0 (if an answer is absent or it is not optimal) or 1 (if an answer is absolutely correct).

At each node in Fig. 1, inputs are incoming from the lower-level nodes connected to it, and the output is computed as a linguistic weighted average (LWA) discussed below.

Improving upon our previous findings in [11], we consider the following competences: solving a problem (competence 1), understanding a problem (competence 2), explaining a chosen solution (competence 3), and ability to integrate language and mathematics (competence 4). The last competence is evaluated directly by the teacher based on the overall impression from the exam being graded.

Arcs connecting nodes are weighted, also with words (or in some cases with numbers 0 or 1), to allow for different characteristics to have different effects on the overall grade. Weights can be assigned by independent experts, which makes the process of grading more objective. The teacher does not know which criteria and competences are more important, and thus cannot influence the overall result, even subconsciously.

2.2 Aggregating Grades Using Linguistic Weighted Average

To calculate a weighted average of several words, we use perceptual computing described in [15] as the most detailed and developed approach to date. To average words, we need to represent them as mathematical objects. Following [16], we will use (interval) type-2 fuzzy sets (T2FS) as scientifically correct from the point of view of modeling first-order uncertainty of words. Type-2 fuzzy sets were originally introduced [17] and further developed by J. M. Mendel (see, e.g., [18] for an extensive treatment of this subject). All the words that can be used by a teacher in the system, together with their T2FS models, collectively make up a codebook.

A T2FS A is a set, to which any element can belong with some membership degree, expressed as an interval. For instance, numeric grade 0.9 can belong to a T2FS "very good" with membership [0.85; 0.95]. Unlike type-1 fuzzy sets, for which only one membership value is required for each number, using intervals enables the researcher to capture the fact that different people understand the same linguistic term differently.

We can represent A using two membership functions: a lower membership function (LMF), $\underline{\mu}_{\tilde{A}} : X \rightarrow [0; 1]$, and an upper membership function (UMF), $\overline{\mu}_{\tilde{A}}(x) : X \rightarrow [0; 1]$, where X is some universe of discourse. We take $X = [0; 1]$. Then we can say that any value $x \in X$ belongs to A with an interval degree expressed as $[\underline{\mu}_{\tilde{A}}(x); \overline{\mu}_{\tilde{A}}(x)]$.

In applications, it is customary to use trapezoidal membership functions:

$$\mu(x; a, b, c, d) = \begin{cases} \frac{x-a}{b-a}, & a \leq x \leq b \\ 1, & b \leq x \leq c \\ \frac{d-x}{d-c}, & c \leq x \leq d \\ 0, & \text{otherwise} \end{cases} \tag{1}$$

If LMF and UMF are trapezoidal, we obtain a T2FS from Fig. 2. Numbers 0 (or 1) are represented as T2FSs by setting $a = b = c = d = 0$ (or 1) for both LMF and UMF.

To compute weighted average of several real numbers $x_1, \ldots, x_n$, with associated real positive weights $w_1, \ldots, w_n$, one would compute

$$y = \frac{\sum_{i=1}^{n} x_i w_i}{\sum_{i=1}^{n} w_i} \tag{2}$$

When values $X_1, \ldots, X_n$ and their associated weights $W_1, \ldots, W_n$ are expressed as T2FSs, (2) can be technically rewritten as

$$Y = \frac{\sum_{i=1}^{n} X_i W_i}{\sum_{i=1}^{n} W_i} \tag{3}$$

This expression is computed as a series of interval weighted averages [15].

2.3   Assigning a Final Grade

After the words are assigned by a teacher for each criterion of each problem, and they are averaged according to the connections depicted in Fig. 1, one obtains a T2FS G at the "Final grade" node. There are several options that the teacher has at this point:

- she can select a word from the codebook that has the most similar T2FS;
- she can rank the students according to their T2FSs;
- she can assign G to one of the predefined classes corresponding to different grades.

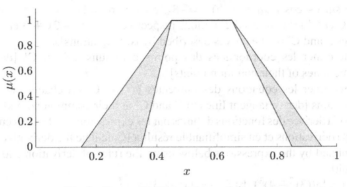

Fig. 2. Trapezoidal interval type-2 fuzzy set

Following [11], we proceed according to the third option and assign one of the four grades expressed as intervals: $A = [0.75; 1]$, $B = [0.5; 0.75]$, $C = [0.25; 0.5]$, or $D = [0; 0.25]$. We do it using a well-known Vlachos and Sergiadis's T2FS subsethood measure [19] adapted to our case as follows:

$$ss_{VS}(G, I) = \frac{\sum\limits_{i=1}^{N} \min(\underline{\mu}_G(x_i), 1_I(x_i)) + \sum\limits_{i=1}^{N} \min(\overline{\mu}_G(x_i), 1_I(x_i))}{\sum\limits_{i=1}^{N} \underline{\mu}_G(x_i) + \sum\limits_{i=1}^{N} \overline{\mu}_G(x_i)} \tag{4}$$

where $x_i \in X$, $i = 1, \ldots, N$, are equally spaced points of the support of G, $I \in \{A, B, C, D\}$, and 1_I is the indicator function. We select the class that yields the highest value of (4). Classes can be defined differently depending on a specific application.

3 Application of CWW to Grading Mathematical Exam

3.1 Structure of the Exam

In this section, we present an example of a mathematical exam that enables us to illustrate the strongest features of the CWW applied to grading. This exam was created by the second author of this paper as a practicing educator in the field of bilingual mathematical education in French and Ukrainian.

1. En utilisant le tableau des dérivées des fonctions usuelles, calculer [Using the table of derivatives of the common functions, calculate]:
   (a) $y = \sin x - \cos x$, $x_0 = \pi$ (b) $y = -8\sqrt{x} - x^{-1}$, $x_0 = 1$
2. Soient C et C' les courbes d'équations respectives, $y = x^3 - 2x + 3$ et $y = 2x^2 - 3x + 3$ [Let C and C' be the curves described by two equations]:
   (a) Déterminer les coordonnées des points communs à C et C' [determine the coordinates of their common points]
   (b) Déterminer les équations des tangentes à C et C' en chacun de leurs points communs [derive tangent lines to C and C' at their common points]
3. Calculer la dérivée des fonctions définies par les expressions ci-dessous en utilisant les règles de dérivations et en simplifiant le résultat [Calculate the derivative of the functions defined by the expressions below using the rules of derivation and simplifying the result]:
   (a) $f(x) = \ln((3x^2 - 4)^3)$ (b) $f(x) = (x^2 + 3)e^{2x-5}$
4. Mettre les signes convenables [Insert correct sign]:

   Soit une fonction f définie et dérivable sur un intervalle I [Let f be a function defined and differentiable on an interval I].
   Si [If] $f'(x)$? 0, alors f est décroissante sur I. [f is decreasing on I]
   Si $f'(x)$? 0, alors f est croissante sur I. [The same for "increasing"]
   Si $f'(x)$? 0, alors f est constante sur I. [The same for "constant"]

5. Étudier complètement la fonction $f(x) = (-x^2 + 3) / (x + 2)$. Pour cela parcourir les étapes ci-dessous [Completely analyze the function f. Follow the steps]:
   (a) Déterminer le domaine de définition de f [Determine its domain]
   (b) Déterminer les points d'intersections de f avec les axes [Determine points of intersection with the axes]
   (c) Déterminer les équations de ses asymptotes [Derive equations of its asymptotes]
   (d) Déterminer la dérivée de f. Puis déterminer les coordonnées des extremums de f. [Find its derivative and points of extrema]
6. Esquisser soigneusement le graphique de la fonction $f(x) = (-x^2 + 3) / (x + 2)$ [Sketch the graph of f]
7. Déterminer une primitive de la fonction f lorsque celle-ci est définie par [Find antiderivative]:
   (a) $f(x) = (x - 1) / (x^2 - 2x)^3$ (b) $f(x) = x^2\sqrt{2x^3 - 3}$ (c) $f(x) = 6\cos^2(4x)\sin x$
8. Dire si les affirmations suivantes sont vraies ou fausses. Justifier chaque réponse [True of False, justify]:
   (a) Toute fonction intégrable possède une unique primitive [Antiderivatives are unique]

(b) Le résultat du calcul d'une intégrale définie est toujours positif [Definite integral is always positive]

9. Calculer les intégrales suivantes [Compute the following integrals]:

(a) $\int_1^3 \frac{3x+1}{3x^2+2x+2}dx$ (b) $\int_{-1}^2 5xe^{x^2-1}dx$ (c) $\int_{\pi/6}^{\pi/2} \frac{3\cos x\, dx}{\sin x}$

10. On considère une fonction f affine par morceaux. Calculer $\int f(t)\, dt$ en utilisant deux méthodes [Compute an integral of a given piecewise affine function using two methods]

11. Complete le Théorème (Cauchy): Si la fonction f est—sur $[a; b]$ alors est—sur $[a; b]$ [Complete the Cauchy theorem: of a function f is [blank] on $[a; b]$, it is [blank] on $[a; b]$]]

12. Dans un ZOO, on veut construire 3 enclos rectangulaires de même taille délimités par les clôtures dégrillages, selon du plan ci-dessous. L'aire totale des enclos doit être de 288 m². Pour minimiser les frais de clôture, quelles doivent être les dimensions de chaque enclos pour la longueur totale des clôtures utilisées soit minimale ? [In a zoo, we want to build 3 rectangular enclosures of the same size delimited by the fences, according to a given plan. The total area should be 288 m². Find the dimensions of each enclosure, so that the total length of fencing is minimal.]

13. Soit f et g deux fonctions définies respectivement par $f(x) = x^3 - x + 3$ et $g(x) = 2x + 1$ [Consider two functions]:

(a) Calculer les coordonnées des points d'intersection du graphe de f avec celui de g [Find points of intersection]

(b) Calculer l'aire de la surface plane hachurée délimitée par les graphes de f et g [Compute the area between two graphs]

3.2 Construction of Codebook and Assignment of Weights

For the method discussed here, the codebook contains words of two types:

– words used to evaluate problems. Following [11, 13], we use eight words: excellent (EX) very good (VG), good (G), sufficient (S), satisfactory (SA);
– words used as weights on connecting arcs in Fig. 1. Following [11, 13], we use five words: highly influential (HI), influential (I), moderately influential (MI), weakly influential (WI), not influential (NI).

All the words are represented as T2FSs with support in [0; 1]. Each T2FS was created using a person FOU approach [20] that requires knowledge elicitation from one expert. The second author of this paper provided appropriate data for this method based on her extensive expertise in this domain. Resulting T2FSs are given in Table 1 (rounded to two significant digits) in reference to (1).

All weights on the arcs from Fig. 1 were also assigned by the second author of this paper. Weights on the arcs connecting criteria, problems, and competences are given in Table 2. Competences are connected with the final grade node as follows: competence 1 has weight I, and other competences have weight HI.

3.3 Discussion of the Results for Synthetic Students

Validity of the proposed approach can be illustrated using model data corresponding to three synthetic students. Grades for each problem are given in Table 3. Grades for competence 4 were assigned as follows: student 1 got S, and students 2 and 3 got 1.

Final grades for each student are given in Table 4 as T2FSs. Also, we show subsethood (4) for each of the four classes (A, B, C, D). Based on these results, a teacher would give a B to the first student, and A to students 2 and 3.

Table 1. Parameters of T2FSs used to represent words from the codebook

Word	LMF				UMF			
	a	b	c	d	a	b	c	d
EX	0.96	0.97	0.99	1.00	0.96	0.97	0.99	1.00
VG	0.84	0.85	0.91	0.93	0.83	0.85	0.91	0.94
G	0.76	0.77	0.80	0.81	0.75	0.77	0.80	0.81
S	0.66	0.67	0.71	0.72	0.65	0.67	0.71	0.73
SA	0.56	0.57	0.61	0.62	0.55	0.57	0.61	0.63
BA	0.47	0.48	0.52	0.53	0.46	0.48	0.52	0.53
U	0.37	0.38	0.42	0.43	0.37	0.38	0.42	0.44
VB	0.12	0.23	0.32	0.33	0.05	0.23	0.32	0.34
HI	0.94	0.95	0.99	0.99	0.94	0.95	0.99	1.00
I	0.83	0.85	0.90	0.91	0.82	0.85	0.90	0.92
MI	0.71	0.73	0.78	0.79	0.70	0.73	0.78	0.80
WI	0.59	0.61	0.66	0.67	0.58	0.61	0.66	0.69
NI	0.26	0.41	0.54	0.55	0.21	0.41	0.54	0.57

As can be seen from these model results, a teacher is able to arrive at final grades using words and binary numbers to grade criteria of each problem on the exam. The results adhere to general expectations, in the sense that the students who get higher grades for individual problems get higher overall grades as well.

The case of Student 1 is important, as it illustrates that even getting very substandard grades for tidiness (criterion 4) does not lower the overall grade very much. This is explained by the fact that criterion 4 is considered not very influential according to Table 2 for most problems. Weights of each criterion can be assigned not by a teacher, but by her supervisor, which makes the process of grading more objective, as the teacher does not have access to the inner workings of the system.

Table 2. Weights on arcs connecting nodes in the framework

Problem	C1	C2	C3	C4	Comp. 1	Comp. 2	Comp. 3
Problem 1	HI	MI	MI	WI	HI	I	NI
Problem 2	HI	I	I	NI	HI	HI	I
Problem 3	HI	MI	MI	WI	HI	I	I
Problem 4	HI	0	0	0	NI	HI	NI
Problem 5	HI	I	HI	I	HI	HI	I
Problem 6	HI	I	I	HI	HI	HI	I
Problem 7	HI	MI	MI	WI	HI	I	NI
Problem 8	HI	I	HI	I	I	HI	HI
Problem 9	HI	MI	MI	WI	HI	HI	NI
Problem 10	HI	MI	I	MI	HI	I	WI
Problem 11	HI	0	0	0	NI	HI	NI
Problem 12	HI	HI	HI	WI	HI	HI	HI
Problem 13	HI	I	I	I	HI	HI	MI

Table 3. Grades for synthetic students

Problem	Student 1				Student 2				Student 3			
	C1	C2	C3	C4	C1	C2	C3	C4	C1	C2	C3	C4
1	EX	1	G	U	EX	1	VG	G	G	1	S	S
2	VG	1	G	U	EX	1	G	G	G	1	G	G
3	G	0	VB	U	G	0	G	G	EX	1	EX	G
4	0	0	VB	U	1	0	G	G	1	0	G	G
5	G	1	VB	VB	VG	1	VG	VG	G	1	VG	VG
6	S	1	VB	VB	EX	1	EX	VG	EX	1	EX	VG
7	G	1	G	U	G	1	G	SA	G	1	G	G
8	G	1	VB	VB	EX	1	VG	VG	VG	1	VG	VG
9	0	0	BA	U	1	0	G	G	1	0	G	G
10	VG	1	G	U	VG	1	G	G	EX	1	G	G
11	0	0	G	U	1	0	G	G	1	0	G	G
12	S	0	G	U	EX	1	VG	G	EX	1	VG	G
13	G	0	VB	VB	G	0	G	G	EX	1	EX	G

Table 4. Parameters of T2FSs for final grades for synthetic students and degrees of subsethood

Student	Grade LMF				Grade UMF				Subsethood			
	a	b	c	d	a	b	c	d	A	B	C	D
1	0.49	0.52	0.58	0.61	0,47	0.52	0.58	0.62	0	0.95	0.05	0
2	0.86	0.87	0.90	0.91	0.85	0.87	0.90	0.92	1	0	0	0
3	0.90	0.91	0.93	0.94	0.89	0.91	0.93	0.94	1	0	0	0

4 Conclusions and Further Research

In this paper, we presented an improved method for grading bilingual mathematics exams using computing with words. The field of assessing students' performance in bilingual settings remains largely understudied, and we aimed to fill the gap by proposing a method that would allow teachers to grade bilingual mathematics exams in a friendly and coherent way. Our method differs from existing ones in two respects.

First, it enables a teacher to use words in place of arbitrary and hard-to-determine numeric grades. Grading is very subjective, therefore computing with words helps qualify subjectiveness by formalizing uncertainty using type-2 fuzzy sets.

Second, grading is organized in several layers. The teacher is responsible for evaluating correctness, optimality, thoroughness, and tidiness of solutions. These evaluations are aggregated using relative influence of each problem on each of the four core competences, and of each competence on the overall grade. By specifying relative weights of each problem and competence, it is possible to separate the grading process from the determination of the outcome, which makes assessment more objective.

Our approach can be developed by adding more words to the codebook to make it more nuanced. Field experiments involving teachers and students should be conducted to investigate the benefits and improvements in grading that our approach can offer.

References

1. Eurydice Report. Content and Language Integrated Learning (CLIL) at School in Europe. Brussels (2006)
2. Xingle, F., Zhaoyun, S., Yan, C., Yupu, B.: The exploration and research on bilingual education of computer discipline. IJEME **2**(10), 52–58 (2012)
3. Council of the European Union: Council Recommendation of 22 May 2019 on a comprehensive approach to the teaching and learning of languages. Off. J. Eur. Union **62**, 15–22 (2019)
4. Mearns, T., de Graaff, R., Coyle, D.: Motivation for or from bilingual education? A comparative study of learner views in the Netherlands. Int. J. Biling. Educ. Biling. **23**(6), 724–737 (2020)
5. Surmont, J., Struys, E., Van Den Noort, M., Van de Craen, P.: The effects of CLIL on mathematical content learning: a longitudinal study. Stud. Second Lang. Learn. Teach. **6**(2), 319–337 (2016)
6. Wang, L., Han, X., Li, M.: Research on bilingual teaching of graduates for computer specialty in financial and economical colleges. IJMECS **1**(1), 53–59 (2009)

7. Tardieu, C., Dolitsky, M.: Impact of the CEFT on CLIL. Integrating the task-based approach to CLIL teaching. In: Agudo, J., de, D.M. (ed.) Teaching and Learning English Through Bilingual Education. Cambridge Scholars (2012)
8. Lo, Y.Y., Fung, D.: Assessments in CLIL: the interplay between cognitive and linguistic demands and their progression in secondary education. Int. J. Biling. Educ. Biling. **23**(10), 1192–1210 (2020)
9. Moschkovich, J.N.: Bilingual/multilingual issues in learning mathematics. In: Lerman, S. (ed.) Encyclopedia of Mathematics Education, pp. 75–79. Springer, Dordrecht (2020)
10. Massler, U., Stotz, D., Queisser, C.: Assessment instruments for primary CLIL: the conceptualisation and evaluation of test tasks. LLJ **42**(2), 137–150 (2014)
11. Tavrov, D., Kovalchuk-Khymiuk, L., Temnikova, O., Kaminskyi, N.-M.: Perceptual computer for grading mathematics tests within bilingual education program. In: Hu, Z., Petoukhov, S., Dychka, I., He, M. (eds.) ICCSEEA 2018. AISC, vol. 754, pp. 724–734. Springer, Cham (2019). https://doi.org/10.1007/978-3-319-91008-6_71
12. Liu, Y., Zhang, X.: Evaluating the undergraduate course based on a fuzzy AHP-FIS model. IJMECS **12**(6), 55–66 (2020)
13. Tavrov, D., Kovalchuk-Khymiuk, L., Temnikova, O., Kaminskyi, N.-M.: Evolutionary algorithm for fine-tuning perceptual computer for grading mathematics tests within bilingual education program. In: Shahbazova, S.N., Kacprzyk, J., Balas, V.E., Kreinovich, V. (eds.) Recent Developments and the New Direction in Soft-Computing Foundations and Applications. SFSC, vol. 393, pp. 173–187. Springer, Cham (2021). https://doi.org/10.1007/978-3-030-47124-8_15
14. Zadeh, L.A.: Fuzzy logic = computing with words. IEEE Trans. Fuzzy Syst. **4**(2), 103–111 (1996)
15. Mendel, J.M., Wu, D.: Perceptual Computing: Aiding People in Making Subjective Judgments. John Wiley & Sons, Inc., Hoboken, New Jersey (2010)
16. Mendel, J.M.: Computing with words: zadeh, turing, popper and occam. IEEE Comput. Intell. Mag. **2**(4), 10–17 (2007)
17. Zadeh, L.A.: The concept of a linguistic variable and its application to approximate reasoning-1. Inf. Sci. **8**, 199–249 (1975)
18. Mendel, J.M.: Type-2 fuzzy sets and systems: an overview. IEEE Comput. Intell. Mag. **2**(1), 20–29 (2007)
19. Vlachos, I., Sergiadis, G.: Subsethood, entropy, and cardinality for interval-valued fuzzy sets—an algebraic derivation. Fuzzy Sets Syst. **158**, 1384–1396 (2007)
20. Mendel, J.M., Wu, D.: Determining interval type-2 fuzzy set models for words using data collected from one subject: person FOUs. In: 2014 IEEE International Conference on Fuzzy Systems (FUZZ-IEEE), pp. 768–775 (2014)

Education System Transition to Fully Online Mode: Possibilities and Opportunities

Md. Asraful Haque[1](✉), Tauseef Ahmad[2], and Shoaib Mohd[1]

[1] Department of Computer Engineering, Aligarh Muslim University, Aligarh 202002, India
md_asraf@zhcet.ac.in
[2] Department of Information Technology, Rajkiya Engineering College, Azamgarh 276201, India

Abstract. Digital transformation is one of the main objectives of educational institutions nowadays. Institutions may gain from the digital revolution by enhancing teaching and learning and their ability to run their operations more effectively. Both of these factors are crucial for better serving students. Online education became necessary due to a number of factors, including the increased accessibility and affordability of technology, the need for more flexible and convenient learning options, and the COVID-19 pandemic which forced schools and universities to close their physical classrooms and move instruction online. People who may not be able to attend traditional in-person classes owing to job, family, or other commitments will find that online education is an excellent alternative. Technical difficulties, such as slow internet connectivity or software glitches, can interfere with online learning and make it challenging for students to access instructional resources. Addressing these challenges and improving the online education experience requires a systematic and ongoing review of the online education system, with a focus on ensuring access to quality education for all students, addressing equity and inclusiveness concerns, and fostering innovation and improvement in the delivery of online education. The online education system has many benefits, as well as some drawbacks, and is likely to continue growing in popularity in the future. This article provides a general overview of online education systems, the technological requirements for their implementation, some potential future issues, and some approaches to make online education effective.

Keywords: Online education · e-Learning · Information technology · IPR

1 Introduction

Academic institutions, such as schools, colleges, and universities, have long served as the hubs of knowledge creation and exchange. Advanced digital technologies transformed and enriched so many other industries but education sector was still in the early stages of adoption until a few years ago. The health emergency due to COVID-19 pandemic forced the educational institutes towards the use of online mode or digital platforms [1]. Knowledge and information are no longer tied to the physical locations of educational institutions. Instead, a variety of platforms, software, encyclopedias, open-source databases, and web browsers may be used to collect information and expertise on a wide range of topics, allowing users to supplement their learning. Online learning can be less expensive than traditional brick-and-mortar institutions, as it eliminates the need for physical classrooms, campus infrastructure, and other expenses [2]. Online learning can give students more control over their learning experience, which can boost motivation and engagement. It allows students to learn on their own schedule and can reduce the stress of commuting or trying to fit classes into a busy schedule. Educators are adopting technology in classrooms today, from massive open online courses (MOOCs) to flipped classrooms, there are new approaches and strategies to improve student learning. The drastic changes in education are generating news because they are having a significant influence on student learning. The ability to access educational resources from any location with an internet connection has led to an increase in the popularity of online education in recent years. According to the Impact Report of the renowned online learning platform Coursera, which was published in 2021, the overall number of enrollment more than twice in 2020 and jumped by 32% the following year, reaching at 189 million (Fig. 1) [3].

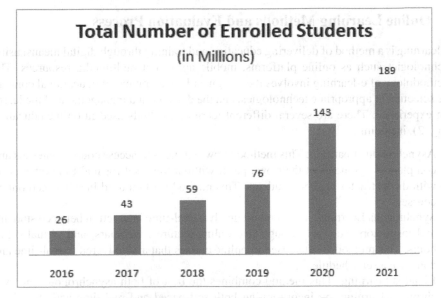

Fig. 1. Increasing trend of enrolled students in Coursera

Although digital education offers many advantages and gives unique access to high-quality education, it also has several drawbacks that might make it difficult to succeed [4–8]. For example, the lack of digital infrastructure disrupts online education. Not all students have equal access to technology and the internet, which can create disparities in educational opportunities and outcomes. The students should have a minimum degree of computer literacy to learn effectively in an online context. The faculty members to be equipped with digital skills and knowledge to effectively design, deliver, and evaluate online programs, which can be challenging for some institutions and educators.

Certain fields such as Science, Technology, Engineering and Medical require hands-on experiences which are not always possible through online education. The effectiveness of online education can depend on factors such as the specific course or program, the teaching methods used, the technology infrastructure and support provided, and the individual student's learning style and needs [9]. Despite the challenges this new trend brings, it has many opportunities as well. It is necessary to review and analysis the different facets of online learning techniques for the following reasons:

- Rapid increase in online education due to COVID-19 pandemic.
- Need to assess effectiveness and efficiency of online education.
- Address challenges and improve student experience.
- Ensure access to quality education for all students.
- Keep up with advancements in technology and education.
- Address equity and inclusiveness concerns in online education.
- Ensure compliance with regulations and accreditation standards.
- Evaluate the impact on employment and career opportunities.
- Prepare for potential future disruptions and emergencies.
- Foster continuous improvement and innovation in online education.

2   Online Learning Methods and Evaluation Process

E-learning is a method of delivering education and training through digital means, using technologies such as online platforms, mobile apps, and multimedia resources. The methodology of e-learning involves the design and development of educational content, the selection of appropriate technologies, and the delivery and management of the learning experience. There are several different learning methods used in online education (Fig. 2), including:

1) Asynchronous Learning: This method allows students to access course materials and complete assignments at their own pace, without the need for real-time interaction with the instructor or other students. This method is often used in self-paced online courses.
2) Synchronous Learning: This method involves real-time interaction between students and instructors, such as through live online lectures, webinars, and virtual office hours. This method is often used in online classes that are delivered in real-time and that have a set schedule.
3) Blended Learning: This method combines the best of both asynchronous and synchronous learning, by incorporating both self-paced and real-time activities. This method is often used in online classes that have some scheduled meetings and some self-paced activities.
4) Self-directed Learning: This method allows students to take responsibility for their own learning by giving them the autonomy to choose their learning activities and assessments. This method is often used in MOOCs (Massive Open Online Courses) and other self-paced online programs.
5) Gamified Learning: This method uses game elements such as points, badges, and leaderboards to make the learning experience more engaging and motivating. This method is often used in online courses that are targeted to children and younger students.

6) Project-Based Learning: This method allows students to apply their knowledge and skills to real-world problems or projects. This method is often used in online courses that focus on applied skills, such as computer programming or design.

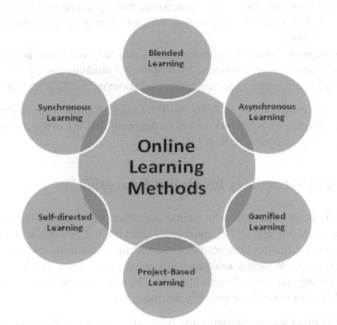

Fig. 2. Online Learning Methods

Each approach has benefits and drawbacks of its own, and the best approach to adopt will depend on the course-objectives, the students' background, and the available resources. In general, a combination of different methods is often used to create a more engaging and effective learning experience for students. The evaluation process in e-learning assesses the effectiveness and impact of the learning program. It involves collecting data on various aspects of the learning experience, such as student engagement, content effectiveness, and overall satisfaction. Some of the most common evaluation methods include:

1) Online quizzes and exams: These can be used to assess students' knowledge and understanding of course content, and can include multiple-choice, short-answer, and essay questions.
2) Discussions and forums: Online discussions and forums can be used to assess students' critical thinking, communication, and collaboration skills, as well as their understanding of course content.
3) Projects and presentations: Online students can submit projects, research papers, and presentations that demonstrate their understanding of course content and their ability to apply what they have learned.

4) Self-reflection and peer evaluations: Students can be asked to reflect on their own learning and to evaluate the work of their peers. This can provide valuable feedback to both the students and the instructor.

5) Analytics: Data analytics can be used to track students' engagement with course content and activities, such as how much time they spend on the course website, how many times they access the course materials, and how many times they participate in online discussions.

6) Online proctoring: Remote proctoring tools can help to ensure academic integrity of online exams and assessments, by monitoring students during the test, flagging suspicious behavior and capturing video and audio of the student during the test.

An effective evaluation of students' performance in online courses should take into account the unique characteristics of online learning and the different ways in which students engage with course content [10–12].

3 Tools and Technology for Online Education

The use of modern technologies in online education can enhance the interactive and engaging nature of online learning, making it more effective and efficient. However, they require adequate infrastructure, resources and training for both teachers and students [13]. A wide range of digital tools and technology have been developed to meet the students' evolving educational needs [14, 15]. The tools and technologies needed for an educational system to function virtually are listed below:

1) Reliable internet access: Students and educators need a stable internet connection to access online resources and participate in virtual classes.

2) Hardware and software: Students and educators will need devices such as smartphones, laptops or tablets, as well as appropriate software for video conferencing, online learning management systems, and other tools.

3) Virtual Classrooms: These are interactive online environments where students and teachers can meet, collaborate, and engage in real-time learning. Examples include Zoom, Google Meet, and Webex.

4) Digital content and resources: Educators need access to digital resources such as e-books, PPTs, videos, and other materials to support online instruction.

5) Professional development: Educators will need training and support in using digital tools and best practices for online instruction.

6) Learning Management System (LMS): A LMS is a software application for the administration, documentation, tracking, reporting, and delivery of educational courses, training programs, or learning and development programs. Examples include Blackboard, Canvas, and Moodle.

7) Artificial Intelligence (AI) and Gamification: AI and ML are increasingly being used to create personalized and interactive learning experiences, as well as to assist with assessment and feedback. The gamification approach uses game design elements, such as points, badges, and leaderboards, to make learning more engaging and interactive.

8) Social Learning Platforms: These are online platforms that allow students to collaborate, share resources, and build networks with their peers, such as Edmodo, Schoology, and Classcraft.
9) Technical support and data security: A dedicated technical support team is required to troubleshoot and solve any technical issues that may arise with the infrastructure. Implementing security measures and protocols to protect student and faculty data is crucial.

Implementing digital infrastructure for e-learning system requires a comprehensive approach, involving not just technology but also the development of policies, training, and support to ensure a successful transition to online education [16].

4   Possible Consequences in Online Education System

Modern technologies, such as artificial intelligence, virtual and augmented reality, and online learning platforms, are likely to have a significant impact on the way teaching is conducted in the future. These technologies have the potential to personalize learning, provide more interactive and immersive experiences, and make education more accessible to people all over the world. Additionally, the use of data analytics and machine learning can help teachers to better understand how their students are learning, and make adjustments to their teaching methods accordingly. However, online learning can lack the face-to-face interaction and engagement that is present in traditional classroom settings, which can make it more difficult for some students to stay motivated and engaged. Ensuring the quality of online education can be a challenge. The future of online education is likely to be shaped by a number of factors, including advances in technology, changing student demographics, and shifting societal and economic trends.

We conducted a survey to determine the direction and future of online education. We have gathered feedback from 200 teachers and 500 students in the 18 to 25 year old age range. All are from India and are proficient with computers and the Internet. Six questions made up the survey's foundation. The participants were asked to respond with (Yes/No) when appropriate and to offer any comments they might have (optional).

The survey findings are presented in Table 1. 69.14% of participants think that offering education entirely online will lower academic standards. However, the majority of them support blended learning. Blended learning, which combines online learning with face-to-face instruction, will become more prevalent, providing students with the best of both worlds. According to 84.86% of respondents, online learning has a negative impact on students' mental health. 75.86% believe that e-learning will increase socioeconomic inequality. 63.71% of participants believe that cyber-attacks and hacking make online education susceptible to loss or theft of important data and intellectual property, including educational assets like video lectures, eBooks, and other interactive content. However, the majority of participants disregard the problems with discipline and gender inequality. We prefer to talk about four potential issues with a future online education system based on the survey findings.

Table 1. A survey on future of online education

Significant issues with online education	Students opinion (Out of 500)		Teachers opinion (Out of 200)		Results (agreed-upon percentage)
	Yes	No	Yes	No	
Will e-learning lead to a drop in educational quality?	329	171	155	45	69.14%
Will students face a lot of psychological problems?	412	88	182	18	84.86%
Will e-learning increase socioeconomic divide?	367	133	164	36	75.86%
Will e-learning lead to further gender inequality?	44	456	28	172	10.29
Will the flexibility of e-learning lead to a lack of discipline among students?	118	382	69	131	26.71%
Will the violations of IPR increase?	305	195	141	59	63.71%

4.1 Lack of Quality Education

The shift to online education has led to concerns about the quality of education being delivered. One factor affecting the quality of online education is the lack of in-person interaction between students and teachers. Online education relies heavily on technology and asynchronous communication, which can make it difficult for teachers to engage students and provide immediate feedback. The absence of in-person interaction also means that students may miss out on the social and emotional benefits of traditional in-person education, such as building relationships with classmates and teachers. In the future, the education system is likely to become less human-driven and more technology-driven as advancements in technology continue to shape the way we learn. The potential over-reliance on modern tools like ChatGPT may lead to a decrease in critical thinking skills. Moreover, the content generated by these AI-systems like ChatGPT may not be accurate or relevant [17]. This could result in students receiving incomplete or incorrect information, leading to misunderstandings or misinformation [18].

4.2 Psychological Issues

Online education has become a prevalent form of learning in recent years; however it is not without its psychological challenges. The lack of social interaction and face-to-face interaction can lead to feelings of isolation and disconnection, which can negatively impact a student's mental health and motivation [19]. Technical difficulties and frustration can also disrupt the learning experience and cause added stress and anxiety. Online education requires a high level of self-discipline and motivation to stay on task and complete assignments, which can be difficult for some students. The virtual learning environment can reduce the sense of community and belonging for online students, leading to feelings of loneliness and a lack of connection. Additionally, online education can limit the amount of feedback and support students receive from instructors and peers, leading to feelings of uncertainty and a lack of confidence in their ability to succeed.

4.3 Socioeconomic Divide

The increased reliance on online education as a result of the COVID-19 pandemic has raised concerns that it may exacerbate socioeconomic divides. On one hand, online education offers more flexible and accessible options for students, especially for those who live in rural or remote areas or have disabilities. This can level the playing field for students who previously faced barriers to accessing traditional in-person education. However, on the other hand, online education also requires access to technology and a stable internet connection, which many students from low-income families may not have. These students may therefore fall behind their peers and face further disadvantage. Additionally, the lack of in-person interaction and support can also negatively impact students who come from disadvantaged backgrounds, as they may not have access to the same resources and support systems as their more privileged peers. Overall, while online education has the potential to bridge some of the gaps in access to education, it can also deepen existing inequalities if steps are not taken to address the technology and resource disparities among students.

4.4 Increased Violations of Intellectual Property Rights (IPR)

The breach of intellectual property rights for e-learning materials is one of the major concerns with online education systems [20]. Content utilized in online learning environments, such as instructional software, e-books assessments, videos, audio recordings, photographs, and other multimedia components, is considered intellectual property. Depending on how they are used and their unique nature, these resources may be covered by intellectual property regulations i.e. copyright, trademark, or patent laws. The sharing of educational resources is actually a good thing. Students, scholars and even other teachers get exposure to you. But when they go too far, stealing, plagiarism and piracy are all unethical practices that can be a risk, as it can lead to loss of revenue and reputation. Intellectual property rights in e-learning can be threatened by various factors, including: copyright infringement, plagiarism of course content, unauthorized distribution of course materials, misuse of trademarks and branding, theft of trade secrets and proprietary information, piracy of online courses and exams, unauthorized use of patented technology, and more [21]. E-learning is becoming more susceptible to hacking and cyber-attacks, which can lead to the loss or theft of valuable data and intellectual property. E-learning materials must be managed in compliance with the applicable privacy laws and regulations.

5 Suggestions for Effective Online Education

With the increasing use of digital tools and resources, students will have access to a wealth of information and resources at their fingertips, enabling them to explore and learn in new and innovative ways. While online education has made education more accessible for many students, it has also created new challenges and limitations that can affect the quality of education being delivered. Effective online learning requires careful planning and strategies to ensure that the learning experience is engaging, interactive, and tailored to the needs of the students [22]. The following are some planning strategies for effective online learning:

5.1 Defining Learning Objectives

Clearly defining the learning objectives and outcomes for the course will help to ensure that the content and activities are aligned with the goals of the students. Tailoring the course to the needs of the students, such as providing flexible scheduling, accommodating the difficulty level of the content, and accommodating the teaching methodologies, will help to accommodate different learning styles and preferences. To ensure the quality of online programs, accreditation is necessary. Typically, accreditation is carried out by independent, nationally recognized accrediting bodies. These bodies have specific criteria and standards that institutions must meet in order to be accredited. The question of accreditation, which is a process of evaluating institutions and programs to ensure that they meet certain standards, is still a concern in online higher education.

5.2 Developing Proper Counseling System

Counseling is important in an online education system as it provides critical support and guidance to students. Online education can pose unique challenges such as isolation, technology difficulties, and time management issues. Counselors can help students navigate these challenges and succeed in their studies. Encouraging self-paced learning and providing flexibility in terms of when and how students complete assignments will help to accommodate different schedules and learning styles.

5.3 Facilitating Communication and Collaboration

Providing opportunities for communication and collaboration, such as virtual office hours, discussion forums, and group projects, will help to create a sense of community and support among students. Creating a sense of community and connection, such as through virtual meetings, student clubs, and group projects, will help students to feel connected to the class and to the instructor, even though they are learning remotely.

5.4 Designing Interactive and Engaging Content

Designing interactive and engaging content that makes use of multimedia, simulations and other interactive elements will help to keep students engaged and motivated. Incorporating technology, such as learning management systems, virtual reality, and artificial intelligence, to enhance the learning experience can increase engagement and improve the students' learning outcomes.

5.5 Developing Effective Feedback Mechanism and Support System

Providing clear and timely feedback on assignments and assessments will help students to understand what is expected of them and to make progress in their learning. Regularly review and assess the online learning system, and make changes as necessary to improve the experience of students and instructors. Providing support and resources, such as tutoring, office hours, or academic advising, will help students to stay on track and to manage their time more effectively.

5.6 Implementing an IPR Management Policy

It is important for online education institutions and educators to have a clear understanding of IPR laws and to have policies in place to prevent and address these threats. Figure 3 shows some standard practices that combine both technical and legal safeguards for Intellectual Property Rights (IPR).

- Copyright registration: registering the e-learning materials with the appropriate government copyright office provides legal protection against unauthorized reproduction and distribution.
- Watermarking: adding a visible or invisible mark to the e-learning materials that identifies the owner can help deter unauthorized use and aid in enforcement of IPR.
- Digital rights management (DRM) technology: this technology is used to control access to and usage of the e-learning materials, such as by requiring a login or preventing the materials from being downloaded or printed.
- Licensing: using a license agreement to clearly set out the terms and conditions under which the e-learning materials can be used, such as specifying that the materials can only be used for personal or non-commercial use.
- Legal action: taking legal action against individuals or organizations that are found to be infringing on the IPR of the e-learning materials.

The architecture of IPR protection will vary depending on the nature of the e-learning materials, the intended audience, and the laws of the country where the materials are being used.

It is expected that online education will become more widely recognized and accepted, and will be seen as a viable alternative to traditional, on-campus education. Online education should become more affordable and accessible, as more institutions and organizations invest in the infrastructure and resources needed to deliver high-quality online education. It's also important to provide feedback and support to the students throughout the course to help them achieve their goals.

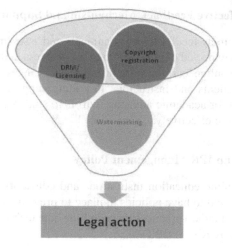

Fig. 3. IPR Protection Methods

6 Summary and Conclusion

The trend of online education has been on the rise in recent years, and it has accelerated significantly due to the COVID-19 pandemic. The future of education in an online environment is likely to be characterized by a continued growth in the number of online courses and degree programs offered, as well as an increase in the use of technology to enhance the learning experience. Online education has the potential to make education more accessible to a wider range of students, including those who live in remote or underserved areas, those who have work or family obligations that make it difficult to attend traditional on-campus classes, and those who are seeking to further their education while balancing other responsibilities. The use of virtual and augmented reality technologies, artificial intelligence, and other advanced technologies is also likely to become more prevalent in online education in the future. These technologies can enhance the interactive and immersive nature of online learning, making it more engaging and effective for students. Additionally, digital transformation offers institutions exciting opportunities to enhance their teaching and learning and the ability to effectively manage their operations – all of which are key to better serving students. However, according to the survey results, many people believe that maintaining the quality of education and ensuring that online courses are rigorous and aligned with standards. An education system that is overly reliant on technology will not help students develop ethical values. We believe that while technology will play a central role in shaping the future of education, the human touch will still be critical. Teachers and instructors will continue to play a crucial role in guiding students, providing emotional support, and fostering critical thinking skills. The future of education will be a balance between technology and human interaction, with each playing a complementary role in promoting learning and growth. Cyber security to protect intellectual property and student data in a completely online environment is also a major concern. It is important to note that the survey was conducted on a small sample data collected from a specific geographical region. In the future, we will address these issues through a large-scale survey because it is critical to ensuring a successful and high-quality online education experience for students.

References

1. Ahmed, N., Nandi, D., Zaman, A.G.M.: Analyzing student evaluations of teaching in a completely online environment. Int. J. Mod. Educ. Comput. Sci. **14**(6), 13–24 (2022)
2. Nguyen, T.: The effectiveness of online learning: beyond no significant difference and future horizons **11**(2), 309–319 (2015)
3. 2021 Impact Report, Serving the world through learning, Coursera (2021). https://about.cou rsera.org/press/wp-content/uploads/2021/11/2021-Coursera-Impact-Report.pdf
4. Dalgaly, T.: Benefits and drawbacks of online education. In: International Scientific Conference on Innovative Approaches to the Application of Digital Technologies in Education, Stavropol, Russia (2020)
5. Basar, Z.M., et al.: The effectiveness and challenges of online learning for secondary school students – a case study. Asian J. Univ. Educ. (AJUE) **17**(3), 119–129 (2021)
6. Hassan, M.M., Mirza, T., Hussain, M.W.: A critical review by teachers on the online teaching-learning during the COVID-19. Int. J. Educ. Manage. Eng. **10**(6), 17–27 (2020). https://doi.org/10.5815/ijeme.2020.05.03
7. Hassan, M.M., Mirza, T.: The digital literacy in teachers of the schools of Rajouri (J&K)-India: teachers perspective. Int. J. Educ. Manage. Eng. **11**(1), 28–40 (2021)
8. Arkorful, V., Abaidoo, N.: The role of e-learning, the advantages and disadvantages of its adoption in higher education. Int. J. Educ. Res. **2**(12) (2014)
9. Nawaz, A., Khan, M.Z.: Issues of technical support for e-learning systems in higher education institutions. Int. J. Mod. Educ. Comput. Sci. **4**(2), 38–44 (2012)
10. Dixson, M.D.: Measuring student engagement in the online course: the online student engagement scale (OSE). Online Learn. **19**(4) (2015)
11. Abbasi, S., Ayoob, T., Malik, A., Memon, S.I.: Perceptions of students regarding e-learning during Covid-19 at a private medical college. Pak. J. Med. Sci. **36**, 57–61 (2020)
12. Zheng, M., Bender, D., Lyon, C.: Online learning during COVID-19 produced equivalent or better student course performance as compared with pre-pandemic: empirical evidence from a school-wide comparative study. BMC Med. Educ. **21**, 495 (2021). https://doi.org/10.1186/s12909-021-02909-z
13. Andrews, R.: Does e-learning require a new theory of learning? Some initial thoughts. J. Educ. Res. Online **3**(1), 104–121 (2011)
14. Nermend, M., Singh, S., Singh, U.S.: An evaluation of decision on paradigm shift in higher education by digital transformation. Procedia Comput. Sci. **207**, 1959–1969 (2022)
15. Saykili, A.: Higher education in the digital age: The impact of digital connective technologies. J. Educ. Technol. Online Learn. **2**(1), 1–15 (2019)
16. Huang, R.H., Liu, D.J., Tlili, A., Yang, J.F., Wang, H.H.: Handbook on facilitating flexible learning during educational disruption: the Chinese experience in maintaining undisrupted learning in COVID-19 outbreak. Smart Learn. Inst. Beijing Norm. Univ. UNESCO, 1–54 (2020)
17. Haque, M.A.: A brief analysis of 'ChatGPT' – a revolutionary tool designed by OpenAI. EAI Endorsed Trans. AI Robotics **1**(1), e15 (2023)
18. Mintz, S.: ChatGPT: threat or menace? Are fears about generative AI warranted? Inside Higher Ed. (2023). https://www.insidehighered.com/blogs/higher-ed-gamma/chatgpt-threat-or-menace

19. Sakpere, A.B., Oluwadebi, A.G., Ajilore, O.H., Malaka, L.E.: The impact of COVID-19 on the academic performance of students: a psychosocial study using association and regression model. Int. J. Educ. Manage. Eng. **11**(5), 32–45 (2021)
20. Lemoine, A.P.: The impact of online learning on global intellectual property issues. In: Munigal, A. (ed.) Scholarly Communication and the Publish or Perish Pressures of Academia, pp. 279–311. IGI Global (2017)
21. Casey, J., Proven, J., Dripps, D.: Managing intellectual property rights in digital learning materials. TrustDR (2007)
22. Alonso, F., López, G., Manrique, D., Viñes, J.M.: An instructional model for web-based e-learning education with a blended learning process approach. Br. J. Edu. Technol. **36**(2), 217–235 (2005)

Collaborative Education Mechanism of In-Depth Integration Between Schools and Enterprises Under the Background of the Construction of EEE -- Taking the Major of Data Science and Big Data Technology of Nanning University as an Example

Zhixiang Lu[✉] and Suzhen Qiu

College of Artificial Intelligence, Nanning University, Guangxi 530200, China
Luzhixiang2019@126.com

Abstract. School-enterprise cooperative education is an important starting point for the construction of new engineering. How to cultivate "Emerging Engineering Education (EEE)" talents in line with the requirements of the times has become the focus of the construction of application-oriented universities. Taking the construction of the information technology specialty of Nanning University as an example, this paper analyzes the background of the construction of emerging engineering education, and forms a new talent cultivation mode of "dual mainline of integration of industry and education, and dual mainline of integration of professional education and vocational education", which opens up a new path for the cultivation of applied talents, including the joint development of industry-education cooperation projects, joint design of courses, joint preparation of textbooks, joint training of teachers, and sharing of resources Jointly determine the project, jointly teach, jointly manage, jointly evaluate the quality and jointly assist in employment. The practice shows that the deep integration and collaborative education mechanism between Nanning University and enterprises is conducive to improving the quality of applied and innovative talents.

Keywords: Emerging Engineering Education (EEE) · School-enterprise integration · Collaborative education

1 Introduction

The role of higher education in talent training and social progress is huge and cannot be ignored [1–3]. The fourth industrial revolution and industrial transformation are emerging, affecting the form and form of higher education. From "Reindustrialization" in the United States to "Industry 4.0" in Germany, all countries around the world seize the opportunity of the times, vigorously develop engineering education, and have introduced relevant reform policies and measures [4–6]. Under such a background, China's

© The Author(s), under exclusive license to Springer Nature Switzerland AG 2023
Z. Hu et al. (Eds.): ICCSEEA 2023, LNDECT 181, pp. 1070–1084, 2023.
https://doi.org/10.1007/978-3-031-36118-0_91

engineering education has entered a new stage of innovation and reform, and higher education has turned into a connotative development. The "Made in China 2025" strategy emphasizes that as China's economic development evolves to a new stage with more advanced form, more complex division of labor and more reasonable structure, engineering talents should not only have solid technical capabilities, but also have comprehensive qualities such as globalization perspective, cultural awareness, innovation and entrepreneurship [7]. National leaders have repeatedly stressed that "we should promote collaborative innovation between industry, university and research, actively participate in the implementation of innovation-driven development strategy, and focus on cultivating innovative, compound and application-oriented talents". In October 2017, the 19th National Congress of the Communist Party of China clearly put forward the major reform task of deepening the integration of industry and education, and promotes the co-construction and sharing of resources in the collaborative interaction between industry and education [8]. In the same year, the Ministry of Education launched the construction of "Emerging Engineering Education", also abbreviated as EEE, and accelerated the training of high-level engineering and technological innovation talents in emerging fields [9]. The construction of EEE science is proposed in response to the needs of industrial transformation and upgrading, national strategy and the transformation of engineering education paradigm. It emphasizes the orientation of industrial demand and requires the integration of multiple resources and advantages [10]. How to coordinate the interests and responsibilities of both schools and enterprises and achieve benign interaction is the core issue of building a 3-E and middle-class education integration and collaborative education mechanism.

2 The Necessity and Problems of School-Enterprise Collaborative Education

2.1 The Necessity of College-Enterprise Collaborative Education

Under the current situation, it is very necessary for colleges and universities to vigorously promote the school-enterprise cooperation in educating people.

2.1.1 Integration of Production and Education is an Important Way to Cultivate High-Quality Application-Oriented Talents

In October 2019, the National Pilot Implementation Plan for the Integration of Industry and Education emphasized that the integration of industry and education is an important basis for the reform of human resources in China, and it is to solve the mismatch between talent cultivation and industrial demand in China's economic and social development. It is necessary to give full play to the promotion role of enterprises and industries in talent cultivation [11]. In engineering practice, talent training can further strengthen the understanding of professional knowledge and improve professional quality by combining practical production with theoretical knowledge [6, 12]. As an important mode of collaborative education in the context of industrial transformation and upgrading, industry-education integration can effectively promote the integrated development of education and industry, and is an important way to cultivate high-quality application-oriented talents [13].

2.1.2 The Needs of Local Regional Economic Development

In order to implement the spirit of the Notice of the State Council on Printing and Distributing the Action Plan for Promoting the Development of Big Data, promote industrial innovation and development, and speed up the construction of the China-ASEAN Information Port, the People's Government of Guangxi Zhuang Autonomous Region issued a series of documents, including the Development Plan for the Information and Innovation Industry of Guangxi Zhuang Autonomous Region (2021–2025), the Development Plan for the Digital Economy of Guangxi Zhuang Autonomous Region (2018–2025), and clearly proposed the construction of big data infrastructure, The construction of digital Guangxi was fully launched [14–16].

Nanning University adheres to the localization and application-oriented orientation of running a school, serving the local regional economy and implementing the integration of production and education is the fundamental to the development of the school. The development of regional economy and society requires schools to cultivate high-quality applied talents, and local applied schools should determine the goal of talent cultivation by serving the regional economy. Therefore, only through the integration of industry and education and school-enterprise cooperation to improve the ability to serve local industry enterprises can we better serve the local regional economic development. Guangxi's information technology application and innovation industry has just started. The professional construction should be closely combined with the information and innovation industry, and the orientation of professional talent training should adapt to local industries. Through in-depth cooperation between schools and enterprises, integration of industry and education, and joint training of application-oriented talents, promote regional economic development.

2.1.3 The Needs of Professional Construction

The construction of the EEE, specialty should be in line with the market demand from the aspects of specialty orientation, talent training mode, teaching resource construction, teaching process, teaching assessment, etc. [17–19]. Nanning University is a resource-constrained local newly-built application-oriented university, with relatively weak resources such as teaching staff, teaching software and hardware, which is difficult to meet the training needs of information technology professionals. However, information technology-related enterprises' advantageous resources in technology development, application, management, and software and hardware environment are exactly the resources needed for professional construction. The specialty should actively connect with the market, carry out in-depth cooperation with the industry, complement each other's advantages, jointly establish a market-oriented new data science and big data technology talent training mode, and cultivate high-quality application-oriented talents serving regional economic development [20, 21]. Through the in-depth cooperation between schools and enterprises, the educational logic and industrial logic are connected to realize the connection between professional education and industrial production, and the connection between curriculum teaching and professional positions. At the same time, the problem of insufficient teachers and teaching resources in the school is solved, and the problem of the disconnection between theoretical teaching and actual production is solved.

2.2 Main Problems Faced by School-Enterprise Collaborative Education

2.2.1 There is no Mature and Stable Education Mode for the Construction of EEE

China's Emerging Engineering Education construction is still in its infancy, and there are many deficiencies in the training mode of new engineering talents. In the context of the construction of 3-E, the training objectives of new engineering talents have gradually changed. The industry-university-research partners have not formed a mature collaborative education model in this field. The existing education model is relatively simple, mostly from the perspective of colleges and universities to talk about cooperation, ignoring the need for high-quality skilled talents in the construction of new engineering talents. In the construction of the EEE talents training model, the synergy of the various subjects of industry-university-research is insufficient, As a result, colleges and universities, enterprises, research institutes and other resource groups have not formed a mature collaborative education model.

2.2.2 Incomplete Multi-party Responsibility and Authority Mechanism

The imperfection of the responsibility and right mechanism of the integration of industry and education in China's engineering education is mainly due to the lack of incentive and restraint mechanism, which leads to the low participation enthusiasm of all subjects; On the other hand, the division of responsibilities and rights of each subject is not clear, which leads to hindered coordination and low efficiency. The specific performance is as follows: First, there are few incentive and restraint mechanisms in the collaborative education mechanism of industrial and educational integration of engineering education in China, and few policies mention incentive and restraint mechanisms. The collaborative education of industrial and educational integration mainly depends on the cooperation oriented by the industrial demand of universities and enterprises. Second, the division of responsibilities and rights of each subject is not clear, which leads to many obstacles to synergy and poor synergy effect. For universities, the integration of production and education can promote students to combine theoretical knowledge and practice, which is conducive to the cultivation of talents required by the industry. For enterprise employers, participating in collaborative education can make it easier to obtain talents, enhance the publicity effect of enterprises, and reduce the cost of employment. In the actual operation process, there are still some problems, such as the university's education responsibility and the low participation of enterprises.

2.2.3 Lack of Scientific Evaluation System

In the operation process of industry-education integration and collaborative education mechanism, both schools and enterprises assume the responsibility of education, and the evaluation subject should be carried out jointly by multiple parties. The research history of the collaborative education evaluation of the integration of industry and education in engineering education is relatively short, and the relevant rules and regulations are not perfect, showing the current situation that the evaluation subject is lack of diversification, the evaluation content is not systematic and comprehensive, and the evaluation method is not scientific enough.

3 Construction of a Collaborative Education Model with Deep Integration of Schools and Enterprises

3.1 College and Enterprise In-Depth Integration and Joint Construction of Professional Ideas

With the reform concept of "professional construction and industrial development in the same direction and in the same direction as the same industry", the demand side promoting the structural reform and construction of the professional supply side as the reform direction, and comprehensively implementing the reform task of "cultivating morality and educating people", adhere to the "open" path of professional construction and development, take the integration of industry and education as the starting point, with the help of external forces of emerging industries, and through school-enterprise cooperation to build, change, improve and strengthen majors, Constantly strengthen the construction of professional connotation, promote the application-oriented transformation of professional, improve the level of professional construction, and cultivate professional characteristics.

3.2 Deepen the Reform of the Cooperative Education Mechanism Between Schools and Enterprises, and Continue the Reform and Innovation of the Cultivation Mode of Application-Oriented Talents

In the training and model reform of applied talents, school-enterprise cooperation is a very effective way.

(1) Professional ecological construction. With "resource integration, team integration, management integration, teaching integration, and cultural integration" as the starting point of reform, deepen the reform of industry-education integration mechanism based on the community of interests, deepen cooperative development, create a new education community that integrates talent training, scientific research, technological innovation, enterprise services, student entrepreneurship and other functions, and vigorously promote the sound development of professional ecology. The construction of the new mechanism for the construction of the "five integration" industry-education in-depth integration specialty is shown in Fig. 1.

(2) The continuous improvement and innovation of the talent training mode of the cooperation between industry and education. 1) Based on the path of "reverse design and positive implementation", according to the professional certification standards of engineering education and the needs of the school's orientation, industry chain and innovation chain, put forward the objectives and requirements of talent cultivation, reconstruct the curriculum system according to the post ability and career development, and develop a new plan for the cultivation of applied talents. 2) Through the "ten joint" construction of the industrial college's industry-education cooperation

plan, curriculum, teaching materials, teacher training, resource sharing, project setting, teaching, management, quality evaluation and employment assistance, the new
mode and path of industry-education cooperation and application-oriented talent cultivation in line with application-oriented universities will be reconstructed, and the
role of the two education subjects in the process and each link of application-oriented
talent cultivation will be given full play.

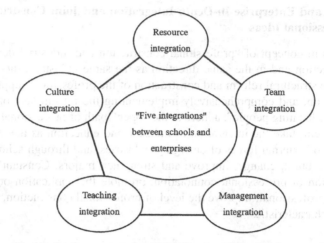

Fig. 1. New mechanism for the construction of the "five integration" industry and education deep
integration specialty

Through the implementation of the "ten joint" of industry and education, explore
the talent cultivation path of "professional and industrial docking, curriculum and real
job docking, classroom and work situation docking", and realize the docking of talents'
professional skills and work fields. 3) Innovate the training process of "double mainline education", implement new collaborative education measures such as "production
and education, double teachers and double abilities" teaching, "production and education, double bases" practice, "production and education, double cooperation" training,
and "production and education, double standards" evaluation, so that the educational
integration of production and education synergy runs through the whole process of
talent training, and achieve a good connection between the talent supply side and the
industry demand side. Figure 2 shows the framework of the talent cultivation model of
industry-education cooperation.

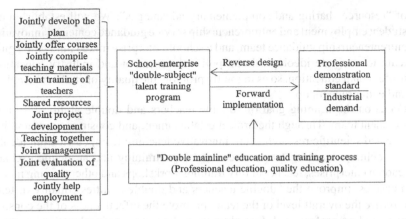

Fig. 2. Talents cultivation mode of industry-education cooperation

3.3 Establish a Teacher Development and Training Mechanism, Focus on the Training Goal of "Double Teachers and Double Abilities", and Build an Excellent Team

The building of an excellent teaching team should focus on the goal of "double teachers and double abilities".

(1) Innovate the training mechanism and promote the transformation and construction of teachers. Establish a teacher development center, focus on the goal of team construction of "structural transformation, capacity transformation and role transformation", implement a series of measures such as the "dual ability promotion drive" mechanism of teachers, "innovative teaching team cultivation", strengthen the reform of the construction of teachers and grass-roots teaching organizations, optimize the team structure, improve the team level, and promote the overall transformation of teachers' role to application-oriented undergraduate education, Cultivate a team of young teachers who grow together with the school.

(2) Based on the integration mechanism of industry and education, promote the structural reform of the team. Focusing on the structural objectives of the teaching staff under the mode of industrial college, giving full play to the institutional advantages of the cooperation with *iFLYTEK* to jointly build majors, promoting the structural reform of the team based on the concept of "team integration", establishing high-level talent introduction and training measures, and building a high-level mixed teaching staff with "school-enterprise connectivity, full-time and part-time combination, and complementary advantages".

(3) Implement the "double leadership" system of schools and enterprises, and innovate the construction mode of teaching grass-roots organizations. Establish a professional teaching and research office and an enterprise teaching department to implement the grass-roots teaching organization construction of "double leadership system" of schools and enterprises; Establish the curriculum responsibility system, promote the construction of "one course, two teachers" embedded teaching organization, and form a working environment, teaching and research pattern and management mode

of "resource sharing and complementary advantages"; We will establish a college student employment and entrepreneurship service guidance center, an innovation and entrepreneurship guidance team, and a school-enterprise mixed innovation guidance team to integrate ideological and political education and industry education into professional education, so as to jointly promote the quality of students' employment and entrepreneurship.

(4) Focus on the training goal of "double teachers and double abilities" to build an excellent team. Through the project establishment and construction of application-oriented scientific research team, innovative teaching team, curriculum ideological and political teaching team, core technology training and academic exchange of leading enterprises, teacher ability training workshops and other training modes and measures, improve the "double teachers and double abilities" quality of teachers, improve the overall level of the team, promote the effectiveness of the construction of first-class professional, first-class team, first-class curriculum, and promote the ability of professional leaders and the comprehensive development of teachers.

3.4 Promote Curriculum Construction and Reform Around Application-Oriented Characteristics

Curriculum construction and reform should highlight the characteristics of application-oriented schools and specialties.

(1) Curriculum system reform. Based on the needs of the application "type", application "positioning" and application "ability" of the curriculum, establish the curriculum logic system and curriculum construction objectives, deconstruct and reconstruct the traditional curriculum system according to the technology application logic and work action logic, and form a new application-oriented curriculum system.

(2) Curriculum content reform. Based on the needs of "change" in technology, "cross-border" in production and teaching, "reality" in scenes, and "effectiveness" in application of the curriculum implementation, and based on the reform strategy of "circularization of work tasks, work-orientation of teaching tasks, and systematization of work processes", the curriculum content is reconstructed, the curriculum implementation carrier is constructed, the curriculum implementation situation is optimized, and the curriculum quality evaluation is reformed.

(3) First-class curriculum building. Take the first-class curriculum standard as the benchmark, build the golden curriculum, promote the full coverage of the ideological and political construction of the professional curriculum, and improve the overall level of the professional core curriculum group; Clearly standardize the process and objectives of each course construction, and implement the reform to achieve results through the establishment of course archives.

3.5 Promote the Construction and Reform of Teaching Mode Around Application-Oriented Characteristics

The reform of teaching mode should be carried out around the characteristics of application type.

(1) Classroom teaching reform. Based on the ability required by a position working in the field of big data, we will position the training objectives, design the teaching

content, determine the teaching methods, manage the teaching process, evaluate the teaching effect, and form a new teaching model of output-oriented ability. We launched the "project-based task-driven" teaching reform demonstration, promoted the innovation of teaching methods and technical means such as flipped teaching, design experiment and innovative experiment, and formed a new model and method suitable for classroom teaching in applied technology universities.

(2) Practice teaching reform. 1) Correctly understand the goal and important position of practical teaching, innovate the professional practical teaching system, form the curriculum system structure of "double system" of theory and practice, and reflect the applicability, professionalism, technology and innovation of practical teaching; 2) Give full play to the advantages of the construction mechanism of the Industrial College, build and develop the experimental training center, build and continuously optimize the online experimental platform, standardize and strengthen the construction of the joint off-campus training base with strong complementarity and relatively stable with the campus, and improve the utilization rate and effect of the in-campus and off-campus experimental training teaching base; 3) In the reform of practical teaching mode, we should strive to achieve "four changes", that is, from demonstration to actual combat, from verification to design, from restriction to openness, and from singleness to comprehensiveness, and focus on cultivating students' engineering practice ability and innovation ability; In the implementation of practical teaching innovation, actively expand the practical teaching path of school-enterprise cooperation, industry-teaching integration, and work-study alternation, so that students can truly experience practical training; In the reform of practical teaching methods, give full play to the main role of students, and strive to promote "teaching as one" and "cooperation between production, teaching and research"; In the construction of practical teaching carriers, we should enrich the teaching carriers, widely carry out innovation and entrepreneurship practice, skill training and certification, discipline competition, campus cultural activities, and various social practice activities to guide students' personality development.

3.6 Cooperation Between Production and Teaching, Technology Guidance, and Strengthening the Construction of Application-Oriented Teaching Resources

The construction of teaching resources can be started from the two aspects mainly.

(1) Construction of application-oriented teaching materials. The joint venture develops professional core courses, selects and compiles applied course textbooks, and focuses on highlighting the application and progressiveness of textbooks.

(2) Construction of digital technology teaching resources. 1) Relying on new technology, develop and enrich technical engineering training programs and work scenario experience; Guided by the basic concepts and standards of engineering education professional certification, establish a professional knowledge map and skill map based on the big data talent training process, and help form a new intelligent learning model of "precision and intelligence"; 2) Industry-teaching collaboration continuously improves and optimizes the teaching platform, continuously improves the level of closed-loop management of "teaching, learning, training, management and

testing" of the platform in the talent training process, improves the scientificity and accuracy of teaching management, and improves the organizational efficiency and teaching quality of teaching.

3.7 Promote the Comprehensive Reform of Professional Education and Teaching with Advanced Quality Assurance Concept

Adhering to the quality concept can guarantee the comprehensive reform of professional education and teaching.

(1) Adhere to the student-centered concept and promote the reform of the main function of quality. 1) Reform the system and mechanism, and strengthen the main function of school quality: adhere to the "people-oriented", and strengthen the central position of undergraduate teaching; 2) Reform the teaching mode and strengthen the quality subject function of teachers and students: first, change the education and teaching concept and promote the reform of student-centered teaching mode; The second is to use modern information technology to transform traditional teaching and drive the modernization of education.

(2) Adhere to the output-oriented concept and promote the reform of quality factor allocation. 1) Reform the allocation of process elements to produce excellent quality through excellent education and teaching process: actively respond to the requirements of the times, and gradually promote the implementation of the strategy of cultivating outstanding talents through the selection and training of discipline and professional competitions, the further study and the cultivation of innovative achievements; The second is to vigorously promote innovation and entrepreneurship education and cultivate innovative, application-oriented and compound talents by using the "Create the Future" college students' innovation and entrepreneurship practice education platform, the brand of college students' career planning competition, the construction of specialized and innovative integrated courses, and the innovation and entrepreneurship competition; 2) Reform the allocation of resource elements and support the excellent education and teaching process with high-quality education and teaching resources: deepen the construction of "resource integration and team integration" of school-enterprise cooperation. First, build a mixed teaching team with noble teachers' ethics, strong teaching ability and willingness to invest, and be a good guide for students in the workplace; The second is to rely on the introduction of advanced technology to improve the modernization level of teaching resources and improve teaching quality.

(3) Adhere to the concept of continuous improvement and promote the reform of quality assurance methods. Establish advanced quality assurance concepts to provide effective guidance for quality assurance work; Improve quality standards and play an important role in talent training; Improve the quality assurance mechanism and enhance the quality assurance effect; Create a strong quality culture and improve the quality assurance level.

Take the effectiveness of professional service for economic and social development, the combination of professional and industrial enterprises, the connection of training process and production practice, and the matching of training quality and industrial demand as the observation points of quality evaluation, and establish a new multiple evaluation mechanism through the integration of "school and enterprise double standards".

4 Achievements of School-Enterprise Integration and Collaborative Education Mechanism Construction

Significant results have been achieved through the construction of the collaborative education mechanism for the in-depth integration of schools and enterprises.

4.1 Course Effectiveness

The evaluation of the quality of talent training mode can establish the evaluation indicators of talent training quality from the aspects of curriculum effectiveness, employment quality, basic work ability, etc. Statistical analysis is made on the survey data of graduates of Data Science and Big Data Technology in our university after graduation in recent two years through questionnaires. The employment of students half a year after graduation is relatively stable, and they can correctly evaluate the importance of their ability requirements in work. Conduct data analysis on the importance and satisfaction of the curriculum surveyed by the graduating students to analyze the effectiveness of the curriculum. The impact of curriculum importance and satisfaction on curriculum effectiveness is shown in Table 1.

Table 1. Relationship between satisfaction of curriculum importance and curriculum effectiveness

Course importance	Course satisfaction	Course effectiveness
high	high	high
high	low	low
low	/	general

According to the survey and analysis of 45 courses in the talent training plan, the number and proportion of various courses in 2021 and 2022 are shown in Table 2.

Table 2. Number and percentage of professional courses in 2021 and 2022

Class label	2021		2022	
Important but not satisfied	24	53.33%	13	28.89%
Important and satisfied	6	13.33%	24	53.33%
unimportance	15	33.33%	8	17.78%

Assume that the coefficient of important and satisfying class courses is 1, the quantity is n_1, the coefficient of important but not satisfying class courses is 0.5, the quantity is n_2, and the coefficient of unimportant class courses is 0, the quantity is n_3. Then the course effectiveness index is:

$$C = \frac{n_1 \times 1 + n_2 \times 0.5 + n_3 \times 0}{n_1 + n_2 + n_3} \tag{1}$$

The value of C is between [0, 1]. The lower the value, the lower the importance and satisfaction. The higher the value, the higher the importance and satisfaction. Table 2 shows that the proportion of important and satisfying courses will increase in 2022. According to the above formula, the curriculum effectiveness index of this major in 2021 and 2022 is *0.40* and *0.68*, respectively. It can be seen that the school-enterprise in-depth industry-university integration collaborative education model can dynamically adjust the curriculum according to the market demand to meet the actual work needs of talents.

4.2 Employment Quality

The quality of employment is an important indicator to evaluate the quality of talent training. The factors that affect the quality of employment include salary level, work status, resignation factors, etc. The evaluation of salary level mainly depends on whether the actual salary is higher than the bottom line of salary expectation, the working status mainly depends on whether it meets the value of career expectation, and the reason for resignation depends on the number of employers who have graduated to the current position. The quality of employment is divided into three categories, good, average and poor, and the comprehensive judgment is made based on the factors such as graduation direction, income, work place, number of employers, and reasons for leaving. We collected *443* survey data of graduates of data science and big data technology in 2021 and *304* in 2022. The statistical analysis is shown in Table 3.

Table 3. Classified statistics of employment quality

Class label	2021		2022	
Good employment quality	211	47.63%	197	64.80%
Average employment quality	147	33.18%	58	19.08%
Poor employment quality	85	19.19%	49	16.12%

Suppose that the coefficient of "good" employment quality is 1, the records is J_1, the coefficient of "average" employment quality is 0.5, the number of records is J_2, the coefficient of "poor" employment quality is 0, and the number of records is J_3, the employment quality index is as shown in formula (2):

$$J = \frac{j_1 \times 1 + j_2 \times 0.5 + j_3 \times 0}{j_1 + j_2 + j_3} \tag{2}$$

The value of J is between [0,1]. The lower the value, the worse the quality of employment. The higher the value, the better the quality of employment. It can be seen from Table 3 that the proportion of good employment quality will increase in 2022. According to the above formula, the employment quality index of this major in 2021 and 2022 is *0.66* and *0.74*, respectively. It can be seen that the school-enterprise deep industry-university integration collaborative education model flexibly adjusts the talent training program, and the employment quality is getting better and better.

4.3 Application and Promotion

Adhering to the "open" path of professional construction and development, taking the integration of industry and education as the starting point, with the help of external forces of emerging industries, and through school-enterprise cooperation in building, changing, optimizing, and strengthening majors, we have built an industry-education integration mechanism of "resource integration, team integration, management integration, teaching integration, and cultural integration", and formed a co-ordering of industry-education cooperation programs, co-setting of courses, co-compiling of textbooks, co-training of teachers, resource sharing, and co-setting of projects The "ten joint", also called "ten common" specialty construction path of teaching, management, quality evaluation and employment assistance, the innovation of the "double main line education" training process, the implementation of the "production and education, double teachers, double abilities" teaching, "production and education, double bases" practice, "production and education, double cooperation" training, "production and education, double standards" evaluation and other collaborative education measures, so that the education integration of production and education cooperation runs through the entire process of talent training, Realize a good connection between the talent supply side and the industry demand side.

Over the past five years of construction, more than 20 colleges and universities from inside and outside the province have come to our university every year to investigate the construction of the collaborative education mode of industry-teaching integration, and the reform results have been used for reference and application by 15 similar colleges and universities. He has been invited to share experiences at the National Computer Education Conference, the International Forum on the Integration of Industry and Education and other conferences and universities for more than 30 times.

5 Conclusion

Nanning University and enterprises have deeply integrated production and education, and jointly built a EEE specialty under the mode of artificial intelligence industrial college, aiming at the cultivation of application-oriented talents, seeking development based on the integration of production and education, adhering to the theoretical guidance of OBE professional certification of "student center, ability output, and continuous improvement", and exploring the effective path of production and education integration based on the "five integration" and "ten-joint" mechanism innovation of schools and enterprises The construction of the new model of application-oriented talent cultivation of "production and education cooperation with two main bodies and two main lines" has achieved remarkable results, showing strong development vitality.

The 3-E construction puts forward new requirements for the training specifications of talents, and also brings new challenges to colleges and universities and enterprises. It is the need of the new engineering construction, the need of industrial transformation, and the need of national development to jointly build the education model that meets the needs of economic development. Due to the relatively new theme of "new engineering" construction, the systematic analysis of Emerging Engineering Education and "industry-university-research collaborative education" is not thorough enough, the school-enterprise collaboration lacks the participation of the government and industry, and there is no quality monitoring system that restricts all parties, and the cultivation of high-quality applied talents is not guaranteed. Therefore, it is necessary to increase the strength of the government and industry and establish a quality monitoring system for collaborative education of schools, governments, banks and enterprises. Through the multi-party collaborative quality monitoring system, improve the enthusiasm of enterprises to participate and cultivate high-quality application-oriented talents that meet the needs of regional economy.

Acknowledgment. This research is supported by the Collaborative Education Project of the Ministry of Education in 2020: Research on The Curriculum Reform of The School-Enterprise Collaborative Education Model (202002273034).

References

1. Waha, L.W., Yeeb, W.S., Soona, W.K., et al.: Collaborative learning experience in higher education: students' perspective. Malaysian Constr. Res. J. **15**(Special 1), 273–286 (2022)
2. Skedsmo, G., Huber, S.G.: Measuring teaching quality, designing tests, and transforming feedback targeting various education actors. Educ. Assess. Eval. Accountability **32**(3), 271–273 (2020). https://doi.org/10.1007/s11092-020-09333-9
3. Hamada, H.: Action research to enhance quality teaching. Arab World English Journal, 2019, Conference Proceedings (1), 4–12 (2019)
4. Nawaz, R., Sun, Q., Shardlow, M., et al.: Leveraging AI and machine learning for national student survey: actionable insights from textual feedback to enhance quality of teaching and learning in UK's higher education. Appl. Sci. **12**(1), 514 (2022)
5. Mantle, M.J.H.: How different reflective learning activities introduced into a postgraduate teacher training programme in England promote reflection and increase the capacity to learn. Res. Educ. **105**(1), 60–73 (2019)
6. Alkhathlan, A.A., Al-Daraiseh, A.: An analytical study of the use of social networks for collaborative learning in higher education. Int. J. Mod. Educ. Comput. Sci. (IJMECS) **9**(2), 1–13 (2019)
7. Li, T., Li, F., Lu, G.: Analysis of the way to improve the quality of engineering talent training for "Made in China 2025." High. Eng. Educ. Res. **06**, 17–23 (2015). (in Chinese)
8. Chen, J.: Put the deep integration of industry and education in an important position. People's Daily/2018–04–03 (7). (in Chinese)
9. Zhong, D.: The connotation and action of new engineering construction. High. Eng. Educ. Res. **3**, 1–6 (2017). (in Chinese)
10. Li, Z.: Key points for design and implementation of the new round of audit and evaluation scheme. High. Eng. Educ. Res. (03), 9–15 (2021). (in Chinese)

11. Central People's Government of the People's Republic of China. National Pilot Implementation Plan for the Integration of Industry and Education [EB/OL] 2. http://www.Gov.Cn/xinwen/2019-10/11/content_5438226.htm. 2019-10-11. (in Chinese)

12. Inayat, I., Amin, R., Inayat, Z., Badshah, K.: A collaborative framework for web based vocational education and training (VET); findings from a case study. Int. J. Mod. Educ. Comput. Sci. (IJMECS) **5**(12), 54–60 (2013)

13. Ma, E.: Transforming education requires collaboration at every level. Child. Educ. **98**(4), 36–39 (2022)

14. The State Council: Notice of the state council on printing and distributing the action outline for promoting the development of big data. Chinese Government Procurement (9), 15–21 (2015). (in Chinese)

15. Peng, X.: Implementation of the big data strategy "digital construction" in Hezhou is accelerated. Guangxi Economy **10**, 23–24 (2018). (in Chinese)

16. Lu, D., Yang, C.: Guangxi power grid corporation issued the "fourteenth five-year plan" development plan. Guangxi Electric Power **5**, 7–8 (2022). (in Chinese)

17. Meng, Q., Li, W.: Consideration and mechanism innovation of the cultivation of legislative talents in China under the mode of cooperation between government, industry, university and research. Hebei Law **40**(10), 76–96 (2022). (in Chinese)

18. Liu, X.: Research on college students' ideological and political education in the process of industry-education integration. Northeast Forestry University, Harbin (2021). (in Chinese)

19. Wei, W.: Research on the cultivation of application-oriented undergraduate talents from the perspective of capability-based education. Jiangsu Higher Education (02), 44–48 (2017). (in Chinese)

20. Li, K., Liang, S., Zhao, Y.: Exploration and practice of the construction of new engineering courses in cooperation between application-oriented undergraduate universities and enterprises – taking Shenyang Institute of Technology as an example. J. Southwest Jiaotong Univ. (Social Science Edition) **22**(01), 134–141 (2021). (in Chinese)

21. Salas-Tarin, C., Gunderman, R.B.: Educating learners in academic-industrial collaboration. Acad. Radiol. **28**(9), 1321–1322 (2021)

Formal and Non-formal Education of Ukraine: Analysis of the Current State and the Role of Digitalization

Maryna Nehrey[(✉)], Nataliia Klymenko, and Inna Kostenko

National University of Life and Environmental Sciences of Ukraine, 15 Heroyiv Oborony Street, Kyiv 03041, Ukraine
marina.nehrey@gmail.com

Abstract. The paper offers an analysis of formal and non-formal components of education in the context of digitalization processes, focusing on indicators in the wartime period. An analysis of formal and non-formal learning in Ukraine under martial law was carried out. The connection between the education of the population and the indicators of the country's innovative development is proven. The influence of the educational level of the population on the level of development of the digital economy was studied. The factors of demand for formal education through the dynamics of educational institutions are analyzed. It has been established that the spread of non-formal education in Ukraine is accelerating due to the introduction of open educational platforms and projects, and education is acquiring signs of digitization.

Keywords: education system · digitalization · formal learning · non-formal learning · regression

1 Introduction

Digitalization is one of the up-to-date trends in education and science, as well as in the economy and society. The Cabinet of Ministers of Ukraine has approved the Concept for the Development of the Digital Economy and Society of Ukraine for 2018–2020 and the Concept for the Development of Digital Competencies until 2025 [1]. This gave a significant boost to the development of the digitalization of education. It has been proven that the level of digitalization and the level of education are highly correlated [2, 3]. The new ranking goals of the Concept for the Development of the Digital Economy and Society of Ukraine are to increase positions in the Networked Readiness Index, Global Innovation Index, ICT Development Index, and Global Competitiveness Index. In general, Ukraine has an unstable position in this indicator and is constantly in the third quantile [4].

The continuous development of information and communication technologies, as well as changes in the structure of in-demand professions during the pandemic and wartime, are leading to a shift in the balance between formal and non-formal education.

In the late 1960s, the concept of the global education crisis was introduced [5]. Its definition was based on the idea that the formal education system was adapting rather slowly to socioeconomic changes in society. The importance and potential of education that can be obtained outside of traditional educational institutions, which is called formal education, gained the scientists' focus. In our view, the problem of education's inconsistency with modern requirements is in the balance between formal and non-formal components, and the underestimation of the role of non-formal education.

Therefore, the paper proposes an analysis of the impact of the level of education on the innovative development of the state and investigates the transformational processes of the structure of formal and non-formal education under the influence of digitalization and wartime challenges.

2 Literature Review

Scientific and technological progress, economic and political dynamics in the world in recent years, and now the challenges of full-scale military actions, have led to the necessity of finding new approaches and models of education that create conditions for a continuous learning process and understanding the world. The lifelong learning model and non-formal education have become significant components of the modern educational environment.

Non-formal education of youth and adults is becoming increasingly important in the context of social adaptation, ensuring sustainable and balanced development of society. Adult education is considered a social indicator of the human dimension of the state policy, one of the ways to achieve socio-economic well-being, and a tool for promoting the ideas of the information society.

Therefore,the demand for non-formal education is growing [6]. Scholars and politicians are discussing lifelong learning. On the one hand, the authors declare that it can be a combination of formal and non-formal education, on the other hand, this term is used to explain any form of systematic learning that takes place outside a formal educational institution [7–12]. Non-formal education programs are growing in many countries [13]. Non-formal learning is broadly aligned with organized institutional learning models (such as schooling, universities), while informal learning describes everyday learning that people do throughout their lives. Informal learning is less well understood, despite its specific use in different contexts of international policy.

Formal learning is relatively well defined in the research literature, which makes it easier to study. It fits well into narrow curriculum models (i.e., models that focus particularly on organized learning). Informal and non-formal learning do not fit well within the narrow confines of a curriculum model and require us to use a broader conceptualization of curriculum. Informal and non-formal learning are complex but powerful concepts, and they pose challenges to curriculum design. Non-formal learning is a hybrid of other forms of learning, which means that it occurs when formal and non-formal elements are combined. Learning in formal systems is simple in terms of all participants in the educational process [7, 14]. Formal learning often ends with certification and recognition of diplomas or qualifications. Formal learning can also be part of the assessment function, as this is required for formal systems [8, 15].

Performance-based budgeting and funding in higher education is an offshoot of the public administration and service reform often associated with the term new public administration. This reform began to spread in the 1980s. Today, many European countries have introduced some form of performance-based funding in higher education [16–19].

It is obvious that the educational infrastructure of Ukraine has become a strategic target for the Russian invaders, as more than 2,600 educational institutions have been damaged by bombing and shelling, more than 330 of them have been completely destroyed. Educational institutions located on captured Ukrainian lands are unable to fulfill their primary function of providing a favorable and safe educational environment and providing support to participants in the educational process [20].

3 Methodology

The paper aims to study the potential of educational services in the context of formal and non-formal education in Ukraine during the martial law period. Dynamic changes in the digitalization of the country's economy tend to affect structural indicators in education. The paper highlights the development of formal education in terms of ownership, focuses on the increasing demand for non-formal education, and presents audience metrics for the recommended platforms by the Ministry of Education and Science.

The following hypotheses were proposed to study the potential of formal and non-formal education in Ukraine in the context of economic digitalization and war:

Hypothesis 1. There is a relationship between the Global Innovation Index and educational level. The level of development of the digital economy (innovative development) depends on the educational level of the population. To confirm this, we used econometric modeling and linear regression. To test the significance of linear regression models, the F-criterion was used, and then the calculated value was compared with the critical value of the corresponding Fisher distribution at a given level of significance.

Hypothesis 2. There are structural changes in the supply of formal and non-formal education in Ukraine under the influence of institutional changes. To confirm Hypothesis 2, a comparative analysis was used, and the analytical indicator used was the growth (decline) coefficient. This is a relative indicator that characterizes the ratio of a given level to the level taken as a basis for comparison.

Absolute indicators focus on the closure of institutions by form of ownership and time periods, the introduction of external independent evaluation, the establishment of an admission system based on open lists by educational levels and specialty through the electronic office of the applicant, reforms to eliminate low-competitive HEIs (2014–2019), the impact of the pandemic and the digitalization of the educational process.

4 Data

For the empirical study, we chose indicators of supply and demand for formal and non-formal education in Ukraine for the period of 10–12 years. The study uses data from the World Bank and the State Statistics Service of Ukraine.

We examine the relationship between educational level and economic development indicators to substantiate the main principles of the Concept of Development of the Digital Economy and Society of Ukraine. The analysis is based on data from the World Bank and the State Statistics Service of Ukraine. The Global Innovation Index was analyzed to investigate the correlation between educational level and the digitalization of the economy. The sample volume is 114 values.

The share of the population with a complete higher education in the total population aged 25 years and older (educational level) is used as an indicator of educational attainment. To represent the innovative development of countries, a weighted average indicator is taken - the country's score on the Global Innovation Index, which includes 81 variables that characterize in detail the innovative development of countries at different levels of economic development (hereinafter referred to as the GII). The sample includes the years 2017–2018, which were chosen to substantiate the main provisions of the Concept of Development of the Digital Economy and Society of Ukraine, the identified factors of influence, and assessments of the results.

The study of formal education in Ukraine used data on the number of applicants by educational level over the past 10 years from the Unified State Database on Education, as well as the dynamics of changes in the number of educational institutions, considering the form of ownership. To study non-formal education in Ukraine, we used the recommended list of educational platforms from the Ministry of Education and Science of Ukraine, and the Similarweb service for competitive analysis and classroom metrics. The data sample includes data for September-December 2022.

5   Empirical Results

5.1   The Relationship Between Education and Indicators of the Country's Competitiveness

Education and human capital development are of course crucial for rapid, sustainable economic growth. Scientists, politicians, and experts are actively discussing the development of formal and non-formal education using various technologies [21–31]. However, the real share of human capital in national wealth remains small. In support of this, we examine gross value added by economic activity (GVA), which is calculated as the difference between output and intermediate consumption of each economic activity, reduced by the amount of financial intermediary fees. It includes the primary income generated by production participants. The share of GVA of educational services in Ukraine in 2000 amounted to 4.35%, then in 2010 - 5.6%; since 2013, it began to decline to the level of 4.4% in 2016, in 2020 it amounted to 5.0% [32]. In the OECD countries, according to Eurostat, the gross value added of education is accounted for in a separate group along with public administration, defense, health, and social services, but this indicator is not available for Ukraine. Traditionally, the UK has one of the highest indicators, its figure in 2020 was 20.94%, Germany - 19.43%, countries bordering Ukraine - Poland - 15.28%, and Hungary - 17.47%. [33]. Previous studies have determined the impact of education on basic indicators of economic development. According to the regression model, with the current level of education, Ukraine should have 95% confidence intervals of GDP per capita ranging from 36 to 48 thousand dollars, with actual GDP per capita not exceeding

3 thousand dollars. According to the regression model, the expected increase in GDP per capita with a 1% increase in educational attainment is approximately 1 thousand dollars. Based on these data, it can also be concluded that Ukrainian education is not sufficiently effective in terms of economic development, despite having a high potential that needs to be realized.

Education has a significant impact on the competitiveness and digitalization of the country's economy. According to the approved Concept for the Development of the Digital Economy and Society of Ukraine, the key goals of the implementation of the Concept for the Development of the Digital Economy and Society of Ukraine are to increase positions in the Networked Readiness Index, Global Innovation Index, ICT Development Index, and Global Competitiveness Index [4]. Economically developed countries, such as the United States, Canada, and Singapore, have the highest positions in terms of the educational level of the population in the world. In 2019, Ukraine ranked 85th out of 141 in The Global Competitiveness Index. It shows the best results in the Skills indicator, which reflects the overall level of qualification of the workforce, as well as the quantitative and qualitative indicators of secondary and higher education (44th position). Ukraine has an unstable ranking in this indicator and is constantly in the third quantile. Ukraine has the best performance in the structure of the critical thinking of the future workforce (26th position). This leads to the conclusion that the level of human capital potential can be increased with appropriate government regulation in the further future. The study used data on the share of the country's population with higher education (%), the country's Global Innovation Index (score) and the country's GDP (current US$).

To test the hypothesis that there is a relationship between the educational level and the digitalization of the economy, it was hypothesized that there is a natural relationship between the global innovation index and the educational level. Highly developed countries are characterized by the relationship between education and the technological process: investments in education contribute to the development of technological innovations, thereby making human capital and labor more productive and generating income growth. As an indicator of educational attainment, we used the share of the population with a complete higher education in the total population aged 25 and older (hereinafter referred to as educational attainment). To represent the innovative development of countries, a weighted average indicator is taken - the country's score on the Global Innovation Index, which includes 81 variables that characterize in detail the innovative development of countries at different levels of economic development (Fig. 1). The dependence of a country's Global Innovation Index on its educational level is increasing. The functional dependence equation indicates that the expected initial level of the Global Innovation Development Index is 26.5 points. A 1% increase in the proportion of the country's population aged 25 and older with a complete higher education will lead to an increase in the Global Innovation Development Index by an average of 1.06 points.

The modeling results show that 54.19% of the variance in the Global Innovation Index is explained by changes in educational attainment.

$$Y = 1.057x + 26.53 \tag{1}$$

The regression standard error is 8.26. If the educational level indicator includes statistical information on the share of the country's population aged 25 and older with

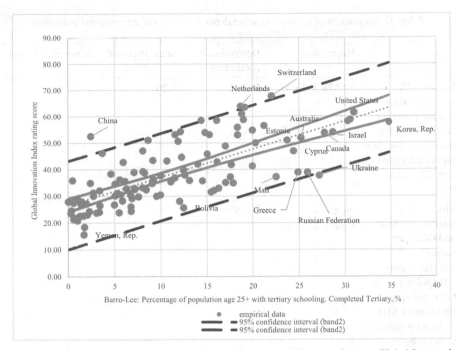

Fig. 1. Global Innovation Index and population level scatter diagram, Source: Global Innovation Index and the World Bank

incomplete higher education, Ukraine will have a value of 39.92%. In this case, the determination index will increase by 0.02%, and the standard error for the value of the Global Innovation Index will decrease by 0.02%.

We tested the hypothesis that there is a correlation between the education level, the level of economic digitalization, and GDP per capita as a baseline indicator of economic development. It is assumed that GDP per capita and educational level influence the development of a country's digital economy (Table 1.).

The analysis showed that 76.84% of the variance of GDP per capita is explained by the variance of the population's educational level and the Global Innovation Index, with their statistical significance equal to 0.05.

5.2 Analysis of the Formal Education Market of Ukraine in the Conditions of Martial Law

Formal education in Ukraine is well developed. It is largely public, both in terms of demand and supply. This trend is typical for both secondary and higher education. Complete secondary education is compulsory in Ukraine, so the largest share of pupils, students, and trainees in Ukrainian educational institutions is made up of students of basic and complete secondary education (62% in the 2020–2021 academic year) [34].

The general demographic situation in the country, the number of places funded by the budget, and the cost of education are the main factors influencing the formation

Table 1. Parameters of regression dependence between the investigated indicators

	Regression	Determination coefficient	Standard error	Significance level
Model "GII - educational level": y is the country's GII (points), x is the population with higher education aged 25 and over (%)	$1.057x + 26.53$	54%	8.26	0.05
Model "GII - educational level and GDP per capita": y is the country's GII (points), x_1 is GDP per capita (thousand USD) x_2 is the population with higher education aged 25 and over (%)	$26.37x + 0.34x_1 + 0.34x_2$	76.84%	5.83	0.05

Source: Global Innovation Index and the World Bank

of demand for education. The demographic situation in the country caused a certain decline in the total number of people, which occurred from 2007 to 2017. In the period from 2016 to 2021, there were trends toward a decrease in the number of secondary school graduates by an average of 4–7% annually, while the percentage of enrolled university students as a share of the total number of secondary school graduates is growing. According to the Unified Database on Education, the demand for bachelor's degrees in 2019 increased by 6.86% compared to 2018. The consequences of military operations in the country - migration processes, mortality, occupation of the territory, destruction of educational institutions and other infrastructure related to the educational process - will have a significant and unpredictable impact.

An analysis of the current structure of Ukrainian higher education institutions by the form of ownership has shown that there have been no significant changes in the education reform since 2014. A slight decrease occurred in 2018–2019 by 5.4% compared to 2015–2016, mainly due to private HEIs. According to the register of educational entities, as of January 01, 2020, there are 151 private higher education institutions in Ukraine, and the number of higher education students enrolled in them is only 10% (72525 people) of the total number of students. At the same time, more than half of these private HEIs (86 HEIs) have less than 200 students, and only 18 private HEIs have a contingent of more than 1000 students. This can be explained by the significant influence of the ranking, popularity, and elite status of the HEIs, and the quality of their educational services, which are generally considered higher in Ukraine for public HEIs than for private ones.

We have analyzed data on the dynamics of the number of vocational pre-university, vocational, and higher education institutions in Ukraine by type of ownership and changes that have occurred in the time periods since the introduction of external independent evaluation (EIE), the establishment of an admission system based on open lists by educational levels and specialty to all HEIs, including the electronic office of the applicant and targeted allocation of budget places based on the specified priorities and EIT, as well as reforms to eliminate low-competitive HEIs (2014–2019), the impact of the pandemic and the digitalization of the educational process (2020–2021), the outbreak of hostilities on the territory as a result of Russia's attack (2022–2023) (Table 2.).

Table 2. Dynamics of the number of institutions of professional pre-higher, professional and higher education in Ukraine

The year of blocking the subject of educational activity in the unified state electronic database on education	In total	Form of ownership of educational institutions		
		State	Private	Communal
Institutions of higher education, amount				
2022-2023	600	350	210	40
change (blocking), total, %	17.92%	19.35%	15.32%	18.37%
change (open), total, %	21.48%	14.06%	37.10%	8.16%
Institutions of higher education: blocking the subject of educational activity in the unified state electronic database on education				
2010-2014	1	1	0	0
2014-2019	51	31	17	3
2020-2021	35	22	11	2
2022-2023	44	30	10	4
Institution of professional (vocational and technical) education and Institutions of vocational pre-higher education, amount				
2022-2023	1506	1029	231	246
change (block), total, %	23.75%	19.86%	39.21%	20.90%
change (open), total, %	8.76%	6.70%	19.74%	3.86%
Institution of professional (vocational and technical) education and Institutions of vocational pre-higher education: blocking the subject of educational activity in the unified state electronic database on education				
2010-2014	0	0	0	0
2014-2019	144	79	43	22
2020-2021	132	67	43	22
2022-2023	193	109	63	21

Source: Unified State Electronic Database on Education

Since the beginning of Russia's full-scale invasion of Ukraine, a significant number of higher education institutions have been under threat of forced relocation or evacuation. In the areas of active hostilities (Kharkiv, Chernihiv, Sumy, Donetsk, Kherson, Zaporizhzhia, Mykolaiv regions and Kyiv), educational buildings have been damaged. Ukraine already has experience in evacuating universities. Some universities in Donetsk, Luhansk, and Crimea were moved to safer cities within these regions or to other regions (20 universities are targeted) after 2014. Based on the comparative analysis, we can conclude that the number of Institutions of professional (vocational and technical) education and Institutions of vocational pre-higher education decreased the most between 2010 and 2023. A significant number of higher education institutions were closed in the period of 2022–2023.

There is currently no data on the number of affected students and teachers who were in the area of active hostilities. There are examples of targeted evacuations of students and teachers from the area of active hostilities (486 people from Kharkiv, 1700 foreign students from Sumy). According to the Ministry of Education and Science, by August 2022, more than 2,000 secondary and higher education institutions were damaged and more than 220 were destroyed, including 131 vocational pre-university and higher education institutions. For various reasons, about 2,000 academic staff was not able to continue teaching and research after February 24, 2022, and 63 educational institutions reported a shortage of teachers. The biggest obstacles to continuing education are the lack of Internet connection (79%), the deteriorating security situation in places of study (46%), and the lack of technical means for online teaching and learning (39%). [35].

5.3 Development of Non-formal Education in Ukraine: Study of Typical Platforms and Projects

Compared to formal education, which has a clear geographical link, in the case of non-formal education, it is very difficult to draw a physical border between markets. The Ministry of Education and Science of Ukraine has recommended a list of online platforms for non-formal education [36]. A comparative analysis of the platforms was conducted based on data on the average monthly number of visitors, their structure, time of visit, and the number of pages viewed after viewing the initial login page. Data received from open web analytics services (Table 3.).

The most popular platforms in terms of audience metrics are www.coursera.org and www.udemy.com, which have a significant number of visitors and page views. The most dynamic platform in terms of the number of users is openuped.eu. Ukrainians prefer international platforms for learning foreign languages. Domestic online platforms were used to study narrowly specialized areas. This can be explained by the monetization of the international non-formal education market. Typical monetization options include selling a program of a set of courses, selling a course, selling a subscription for a period, selling certificates when the course is provided free of charge, selling additional services such as consultations, assignment checking, etc., selling visitor data to advertisers for targeted advertising, processing analytics on the uploaded content and compiling the necessary research for a fee, providing additional paid services such as material storage and organizing discussion platforms.

Table 3. Comparative analysis of educational platforms for non-formal education (audience metrics during October-December 2022)

№	Name of the project (site)	Type	Monthly visits	The structure of visitors by geographical feature (most)	Visit Duration	Bounce Rate	Pages / Visit	The average changes in visitors
Country of origin: USA								
	www.coursera.org	Fixed time	52.83M	USA -21%, India -12%, others – 67%	00:11:06	37.17%	8.16	1.61%
	www.khanacademy.org	Open schedule	43.29M	USA – 47%, India – 6%, others – 47%	00:08:02	49.42%	5.29	−14.95%
	www.edx.org	Fixed time	13.17M	USA – 18%, India –10%, others – 72%	00:06:45	42.36%	6.30	−1.62%
	www.canvas.net	Open schedule	975,017	USA – 51%, others – 49%	00:01:10	58.96%	2.46	−4.71%
	www.udemy.com	Fixed time	101.3M	USA – 12%, India –18%, others – 70%	00:10:09	38%	5.73	1.96%
	online.stanford.edu	Mix	855,722	USA – 16%, India –9%, others – 75%	00:01:27	51.33%	3.66	0.77%
	www.codecademy.com	Open schedule	8.542M	USA – 28%, UK, India –6%, others – 60%	00:09:07	44.18%	5.62	−1.24%
	www.ted.com	Open schedule	10.57M	USA – 18%, India –5%, others – 77%	00:03:32	57.22%	2.80	−11.63%
Country of origin: United Kingdom								
	www.futurelearn.com	Open schedule	3.750M	UK – 20%, India –9%, USA – 8%, others – 63%	00:04:53	58.56%	4.87	−11.08%
	uk.duolingo.com	Fixed time	168,694	Ukraine – 33%, Canada – 15%, others – 52%	00:00:43	41.88%	3.11	4.51%

(continued)

Table 3. (*continued*)

№	Name of the project (site)	Type	Monthly visits	The structure of visitors by geographical feature (most)	Visit Duration	Bounce Rate	Pages / Visit	The average changes in visitors
Country of origin: Germany								
	iversity.org	Open schedule	45,356	Russia, Germany – 18%,, India – 14%, others – 50%	00:03:58	50.76%	3.58	–40.4%
Country of origin: Egypt								
	www.udacity.com	Fixed time	4.454M	Egypt – 21%, USA – 13%, India -8%, others – 58%	00:09:51	36.53%	7.90	–1.9%
Country of origin: Brazil								
	openuped.eu	Mix	8,214	Brazil - 75%, Italy – 6%, others –19 %	00:01:10	66.87%	1.60	75.89%
Country of origin: Ukraine								
	prometheus.org.ua	Mix	782,576	Ukraine –88%, Poland – 3%, others –9 %	00:06:45	49.48%	5.15	–10.86%
	www.ed-era.com	Open schedule	456,178	Ukraine –92%, Poland – 3%, others –5 %	00:05:43	53.72%	4.66	–18.85%
	vumonline.ua	Open schedule	28,469	Ukraine –97%, others –3 %	00:07:30	31.92%	6.92	–9.85%
	https://osvita.diia.gov.ua/	Open schedule	302,623	Ukraine –87%, UK, Germany, Netherlands – 2%, others –7 %	00:02:21	53.65%	4.23	–3.33%

Source: Ministry of Education and Culture of Ukraine and the Similarweb service

One of the initiatives is the Diia. Digital Education platform supported by the Ministry and the Committee for Digital Transformation of Ukraine, which offers free education in an innovative format of educational series. In total, the portal already has more than 75 educational series on digital literacy for lawyers, teachers, doctors, journalists, civil servants, schoolchildren, etc., which increases the level of competitiveness in the labor market.

Since the beginning of 2022, there have been more than 10 million internally displaced persons in Ukraine. Despite the full-scale war, Ukrainian IT companies are maintaining their capacities and developing, so they need qualified employees. At the beginning of the war, there was a shortage of about 30,000 IT specialists to meet the market demand. In 2022, the IT Generation project was launched. Almost 48 thousand Ukrainians took part in the program, submitting more than 211 thousand applications. The high demand for educational retraining programs in Ukraine signals the relevance of non-formal education in the current Ukrainian context. On the other hand, the demand also indicates trends in the labor market, which will necessarily be accompanied by an increase in the level of human potential realization and digitalization of economic processes, and thus the country's economic development.

6 Summary and Conclusion

The study accepts the hypothesis that the digitalization of the economy depends on the level of education. 54% of the variance of the Global Innovation Index is determined by the variance of educational attainment. At first sight, this may not seem like a large enough value. However, Global Innovation Index includes 81 indicators (7 sub-indices) in its calculation, so explaining about 54% of the variation in countries can be considered quite significant. The study of the impact of educational level and GDP per capita on the Global Innovation Index showed that 76.84% of the variance in GDP per capita is explained by the variance in educational level and the Global Innovation Index. We conclude that Ukraine has the potential to have a Global Innovation Index about 30% higher than the actual level, and a GDP per capita 86.21% higher.

The analysis of Ukraine's position in the global rankings of competitiveness and efficiency of education systems showed the need to change the views of the country's population on the effectiveness of knowledge as a source of human capital.

The trends and features of formal education in Ukraine are analyzed in detail. By all quantitative indicators, public higher education institutions significantly outperform private higher education institutions. It is important to note that despite the relatively stable number of HEIs over the past 5 years (until 2020), only 12% of all HEIs are in sufficient demand (70% of students study there). Since the start of the full-scale invasion of Ukraine in 2022, more than 2,000 secondary and higher education institutions have been damaged and more than 220 destroyed, including 131 vocational pre-university and higher education institutions. The biggest obstacles to continuing education are the lack of internet connection (79%), the deteriorating security situation in places of study (46%), and the lack of technical means for online teaching and learning (39%).

Despite the full-scale war, the spread of non-formal education in Ukraine is accelerating through the introduction of open educational platforms and projects, and education is becoming more digitalized. Demand for foreign language learning and skills development for IT professions is growing.

Current research focuses on the potential of non-formal education in Ukraine, which is growing under the influence of the economy's digitalization and the effects of the war. Furthermore, the emphasis is on the demand structure of formal education, which is mainly state-funded and thus subject to strict legislative budget constraints. Further analysis will consider the impact of individual factors on the structural indicators of the education market and the simulation of the main scenarios for the Ukrainian education system, considering the current challenges.

References

1. On the approval of the Concept for the Development of the Digital Economy and Society of Ukraine for 2018–2020 and the Approval of the Plan of Measures for its Implementation. https://zakon.rada.gov.ua/laws/show/67-2018-p
2. Volkova, N.P., Rizun, N.O., Nehrey, MV.: Data science: opportunities to transform education. In: CEUR Workshop Proceedings, vol. 2433, pp. 48–73 (2019). http://ceur-ws.org/
3. Skrypnyk, A., Klimenko, N., Kostenko, I.: The formation of digital competence for the population as a way to economic growth. Inf. Technol. Learn. Tools **78**(4), 278–297 (2020)
4. Global Innovation Index (GII). World Intellectual Property Organization. https://www.wipo.int/global_innovation_index/en/
5. Coombs, P.H.: World Educational Crisis: a Systems Approach. Oxford University Press, New York (1968)
6. Hoppers, W.: Non-formal Education and Basic Education Reform: a Conceptual Review. UNESCO IIEP (2006)
7. Pienimäki, M., Kinnula, M., Iivari, N.: Finding fun in non-formal technology education. Int. J. Child-Comput. Interact. **29**, 100283 (2021)
8. Johnson, M., Majewska, D.: Formal, Non-formal, and Informal Learning: What are they, and how can we Research them?. Cambridge University Press & Assessment Research Report (2022)
9. Romi, S., Schmida, M.: Non-formal education: a major educational force in postmodern era. Cambridge J. Educ. **39**(2), 257–273 (2009)
10. Kucherova, H., Honcharenko, Y., Ocheretin, D., Bilska, O.: Fuzzy logic model of usability of websites of higher education institutions in the context of digitalization of educational services. Neuro-Fuzzy Model. Tech. Econ. **10**, 119–135 (2021). https://doi.org/10.33111/nfmte.2021.119
11. Davydenko, N., Buriak, A., Titenko, Z.: Financial support for the development of innovation activities. Intell. Econ. **13**(2), 144–151 (2019). https://doi.org/10.13165/IE-19-13-2-06
12. Glazunova, O., Morze, N., Golub, B., Burov, O., Voloshyna, T., Parhom, O.: Learning style identification system: design and data analysis. ICT Educ. Res. Indus. Appl. Integr. Harmonization Knowl. Trans. **2732**(2732), 793–807 (2020)
13. Rogers, A.: Non-Formal Education. Springer, New York (2005)
14. Yeasmin, N., Uusiautti, S., Määttä, K.: Is non-formal learning a solution to enhance immigrant children's empowerment in Northern Finnish communities? Migr. Dev. 1–19 (2020)
15. Alnajjar, E.A.M.: The impact of a proposed science informal curriculum on students' achievement and attitudes during the covid-19. Int. J. Early Childhood Spec. Educ. (INT-JECSE), **13**(2), 882–896 (2021)
16. Herbst, M.: Performance-based funding, higher education in Europe. In: Teixeira, P.N., Shin, J.C. (eds.) The International Encyclopedia of Higher Education Systems and Institutions. Springer, Dordrecht (2020)
17. Skrypnyk, A., Klimenko, N., Kostenko, I.: Indicative cost of educational services as a way to optimize Ukrainian higher education. Euro. Cooper. **4**(48) (2020)
18. Rigopoulos, G.: Assessment and feedback as predictors for student satisfaction in UK higher education. Int. J. Mod. Educ. Comput. Sci. (IJMECS) **14**(5), 1–9 (2022)

19. Trianggoro, W.: Tony a the role of curriculum and incubator towards new venture creation in information technology international. J. Educ. Manage. Eng. (IJEME) **9**(5), 39–49 (2019)
20. Education and war in Ukraine Cedos. https://cedos.org.ua/researches/osvita-i-vijna-v-ukr ayini-24-lyutogo-1-kvitnya-2022/#visa_osvita
21. Volkova, N.P., Rizun, N.O., Nehrey, M.V.: Data science: opportunities to transform education. In: Proceedings of the 6th Workshop on Cloud Technologies in Education (CTE 2018), Kryvyi Rih, Ukraine, 21 December 2018, CEURWS.org, pp. 48–73 (2018)
22. Shakhovska, N., Montenegro, S., Kryvenchuk, Y., Zakharchuk, M.: The neurocontroller for satellite rotation. Int. J. Intell. Syst. Appl. **11**(3), 1–10 (2019)
23. Izonin, I., Tkachenko, R., Shakhovska, N., Lotoshynska, N.: The additive input-doubling method based on the SVR with nonlinear kernels: small data approach. Symmetry **13**(4), 612 (2021). https://doi.org/10.3390/sym13040612
24. Mochurad, L., Kotsiumbas, O., Protsyk, I.: A model for weather forecasting based on parallel calculations. In: Advances in Artificial Systems for Medicine and Education VI 21 Jan 2023, pp. 35–46. Cham, Springer Nature Switzerland (2023). https://doi.org/10.1007/978-3-031-24468-1_4
25. Granaturov, V., Kaptur, V., Politova, I.: Determination of tariffs on telecommunication services based on modeling the cost of their providing: methodological and practical aspects of application. Econ. Ann. XXI. **156**(1–2), 83–87 (2016)
26. Matviychuk, A., Lukianenko, O., Miroshnychenko, I.: Neuro-fuzzy model of country's investment potential assessment. Fuzzy Econ. Rev. **24**(2), 65–88 (2019). https://doi.org/10.25102/fer.2019.02.04
27. Kobets, V., Yatsenko, V.: Influence of the fourth industrial revolution on divergence and convergence of economic inequality for various countries. Neuro-Fuzzy Model. Tech. Econ. **8**, 124–146 (2019). https://doi.org/10.33111/nfmte.2019.124
28. Babenko, V., Zomchak, L., Nehrey, M., Salem, A.-B., Nakisko, O.: Agritech startup ecosystem in ukraine: ideas and realization. In: Magdi, D.A., Helmy, Y.K., Mamdouh, M., Joshi, A. (eds.) Digital Transformation Technology. LNNS, vol. 224, pp. 311–322. Springer, Singapore (2022). https://doi.org/10.1007/978-981-16-2275-5_19
29. Munawar, H.S., Khalid, U., Jilani, R., Maqsood, A.: Version management by time based approach in modern era. Int. J. Educ. Manag. Eng. (IJEME), **7**(4), 13–20 (2017)
30. Platforms for improving skills and self-development. https://mon.gov.ua/ua/news/platformi-dlya-vdoskonalennya-navichok-i-samorozvitku
31. Imanov, G., Hajizadeh, M.: The quality assessment of the development of information economy in the republic of Azerbaijan. Neuro-Fuzzy Model. Tech. Econ. **7**, 111–126 (2018). https://doi.org/10.33111/nfmte.2018.111
32. State Statistics Service of Ukraine. http://www.ukrstat.gov.ua/
33. World Bank. http://www.worldbank.org/
34. World Economic Forum. https://www.weforum.org
35. Klymenko, N., Nosovets, O., Sokolenko, L., Hryshchenko, O., Pisochenko, T.: Off-balance accounting in the modern information system of an enterprise. Acad. Acc. Financ. Stud. J. **23** (2019)
36. Karunaratne, N.D.: The globalization-deglobalization policy conundrum. Mod. Econ. **3**(4), 373–383 (2012)

Impact of YouTube Videos in Promoting Learning of Kinematics Graphs in Tanzanian Teachers' Training Colleges

Beni Mbwile[1,2](✉) and Celestin Ntivuguruzwa[1]

[1] African Centre of Excellence for Innovative Teaching and Learning Mathematics and Science (ACEITLMS), University of Rwanda-College of Education (UR-CE), Kayonza, Rwanda
benimbwile1@gmail.com
[2] University of Dar es Salaam-Mkwawa University College of Education, Iringa, Tanzania

Abstract. Tanzanian students have been performing very low on the topic of motion in a straight line which consists of kinematics graphs and formulas for several years. Their low performance is attributed to the dominance of teacher-centred teaching methods such as the lecture method which is contrary to what is being advocated in the current competence-based curriculum. This study investigated the impact of YouTube videos as one of the learner-centred strategies that can promote students learning of kinematics graphs in Tanzanian teachers' training colleges. 150 pre-service Physics teachers from two colleges were distributed into two groups and taught through YouTube videos and lecture methods. Developed and validated physics teachers' kinematics graphs conceptual questions were used to measure pre-service teachers' achievement by administering tests before and after interventions. Simple descriptive and inferential statistics were analysed by using Statistical Package for Social Sciences version 22 and Microsoft Excel. The group of pre-service teachers who were taught kinematics graphs by using YouTube videos performed better than those taught through lectures by achieving average normalised learning gains of 22.87% compared to 1.54% for lecture class on the post-test. Also, there was a statistically significant difference (p-value < 0.001) in performance during the post-test between the lecture method and YouTube video classes while there was no significant difference during the pre-test. The study recommends that Physics educators should incorporate YouTube videos in teaching for the adequate performance of students in kinematics graphs.

Keywords: Kinematics graphs · pre-service teachers · teachers' training colleges · YouTube videos

1 Introduction

Improving the quality of education through the use of technology is becoming increasingly important in educational curricula [1]. Technology always has been part of the teaching and learning process since its inception, whether it was hard copies, writing

Z. Hu et al. (Eds.): ICCSEEA 2023, LNDECT 181, pp. 1099–1109, 2023.
https://doi.org/10.1007/978-3-031-36118-0_93

instruments, or audio-visual media [2]. To cope with the growing technological needs of societies, most countries have employed efforts through curricula reforms [3]. The efforts Tanzania underwent to improve the teaching of science began during the Arusha Declaration in 1967 when the education system had to adopt the philosophy of Education for Self-Reliance [4]. The philosophy emphasized a need for curriculum reform that will enable learners to become creative thinkers, independent, and able to apply theory to real-life [3]. There are four major reforms of curricula from the year 1961 to 2010 which took place in Tanzania [5]. The reforms aimed at improving the teaching and learning of all subjects with more weight being given to Science, Technology, Engineering, and Mathematics (STEM).

Despite all efforts that the government of Tanzania is putting in to improve the way STEM is being taught particularly teaching physics, students' performance in the subject is the lowest compared to other science subjects like chemistry and biology [6, 7]. There are several topics in which students perform very low and hence contribute to the overall poor performance of students in physics, one of them is motion in a Straight line [8]. Concepts of motion in a straight line are presented as kinematics formulas and displayed as kinematics graphs [9]. Kinematics graphs consist of position, velocity, and acceleration as the ordinate and time as the abscissa. There are numerals fundamental errors found when students are plotting, interpreting, and analysing kinematics graphs. For example, in illustrating the position-time graph students often make mistakes in plotting graphs of rest objects [10], difficulty in interpreting area under kinematics graphs [11], and confusion between slope and height [12].

In Tanzania, the topic of motion in a straight line is taught to form two students. Results from the National Examination Council of Tanzania indicate that students are not performing well on the topic in both Form Four and Form Two national examinations [13]. For example, according to [14–18], students had a low performance for consecutively five years in the Form Two nation assessment as shown in Fig. 1.

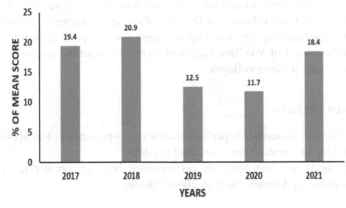

Fig. 1. Students' performance on motion in a straight-line topic

Figure 1 above entails students' highest average performance for consecutively five years was 20.9% and the lowest was 11.7% in 2018 and 2020 respectively. The performance of students is considered to be good, average or weak if their percentage lies in the interval of 65–100, 30–64, and 0–29 respectively [15]. Therefore, for five years students' achievements on the topic were weak.

Several factors are associated with students' poor performance in kinematics such as students' insufficient knowledge of reading graphs, low ability to solve tests in graphical form, and low understanding of the formula used to describe graphs [10]. Other factors not fixed to kinematics include students' negative attitudes toward physics [19] and the dominance of traditional methods of teaching like lecturing and taking notes [20, 21]. Traditional methods make teaching less interactive and more difficult for learners to understand. Thus, deliberate efforts are required to increase students' performance in the topic and Physics in general.

One of the efforts is to use YouTube Videos to teach motion in a straight line. YouTube's website was founded in February 2005 and it enables viewers to upload anything from video clips to full-length videos [22]. Sometimes teachers face the challenge of a shortage of teaching and learning resources while thousands of useful videos are uploaded daily on YouTube [23]. Most of these videos have free access, what is required is an internet connection. The benefit of YouTube videos is that they can be used as a teaching tool (students watch YouTube videos related to the subject) and as a teacher resource (teachers access materials for teaching) [22]. Moreover, YouTube and other social networking sites are being used by schools and university faculties to disseminate information [24].

The recent impact of the COVID-19 outbreak which imposed a lockdown on the entire world forced some education institutions to use online teaching websites, YouTube included [25, 26]. Online mode of teaching and learning was somehow available already but attracted intense attention during the period of COVID-19 [27]. Different ways such as video conferencing and audio were used by educators during online learning. Sometimes educators used to upload videos on learning websites like YouTube for students to use [28]. Little is known in Tanzania about the usefulness of YouTube videos in facilitating the teaching and learning of kinematics graphs. Therefore, this study investigated the impact of YouTube videos in promoting learning kinematics graphs in Tanzanian teachers' training colleges.

1.1 Research Objectives

- Assess significant difference in performance between pre-service teachers in YouTube videos and lecture method classes during pre-test
- Find out significant difference in performance between pre-service teachers in YouTube videos and lecture method classes during post-test

1.2 Research Hypothesis

There is no statistically significant difference in performance between pre-service teachers in YouTube videos and lecture classes

1.3 Significance of the Study

The findings of this study will help tutors in colleges, pre-service teachers and teachers in general to widen the opportunity of using YouTube videos in teaching kinematics graphs as well as other topics in physics subject. In addition, findings will contribute to the existing body of knowledge through publication whereby future researchers can use them as a reference.

2 Literature Review

2.1 YouTube Videos as a Teaching Method

The study by [23] revealed that YouTube videos can play a part in classroom instruction. Through YouTube videos, new concepts can be presented and information can be disseminated. In addition, it was revealed that University faculties use YouTube as a medium for delivering lectures to learners [24]. Also, [29] found that teachers use YouTube to display students' designed art crafts. Furthermore, [30] discovered that teachers use YouTube to keep recorded videos during students' science experiments. Moreover, [31] indicated that students who are far away from normal classrooms would still continue their studies by watching videos uploaded on YouTube.

Meanwhile, [32] mentioned that video lectures tend to reduce loneliness by increasing students' attention towards the subject matter being presented. Similarly, [33] revealed the inclusion of videos in teaching and learning is of paramount benefit in tests and examinations. Likewise, it was argued that investment in YouTube technology can be used for training teachers and saving costs related to face-to-face training [21]. YouTube videos are treasures of resources of materials which can stimulate multiple senses of students and enhance learning [23]. Therefore, learning through YouTube videos can aid the understanding of the subject matter to students.

2.2 Challenges of Kinematics Graphs

Numerous findings have shown students' difficulties when dealing with kinematics graphs. For example, students often make mistakes in plotting graphs of rest objects [10]. The study by [34] revealed two kinds of difficulties including difficulty in connecting graphs to physical concepts and misconceptions in connecting graphs to the real world. [9] revealed students' difficulties with kinematics graphs such as variable confusion, graph transformation, and forming graphs from given kinematics equations. The study further noticed that some students were able to calculate slopes of straight-line graphs which start from the origin but fail when it does not pass through zero. Meanwhile, [12, 35], found that very few students could understand what the slope of a line graph represents. Findings by [11] reported four difficulties facing students when dealing with kinematics graphs; difficulties in calculating slopes from curved graphs, challenges in describing shapes of kinematics graphs, misconceptions in converting between kinematics graphs, and difficulties in interpreting areas under kinematics graphs. Therefore, finding the means of addressing students' challenges on kinematics graphs is very important.

3   Research Methodology

3.1   Approach and Design

This study adopted the quantitative research approach with an experimental research design. In this design, one group of participants (experimental) receives an intervention which is the focus of the researcher and the other (control) receives interventions in normal practices or does not at all [36]. Therefore, the experimental design having control and experimental groups was used in this study.

3.2   Research Participants

A total of 150 respondents comprising 75 pre-service Physics teachers from each of the 2 selected teachers' training colleges in Tanzania participated in the study. Half of the respondents were in year 1 and the other half were in year 2 perusing diploma teachers' programmes by majoring in Physics subject. Males were 118 and females were 32 with ages ranging from 21 to 23. Pre-service Physics teachers were targeted because they are future teachers and each year take part in teaching practice in ordinary-level secondary schools where kinematics graphs are taught. All respondents were sampled randomly from their respective classes and two teachers' training colleges.

3.3   Instruments and Validation

Pre-service teachers were assessed by using the developed and validated kinematics graphs conceptual questions. Some questions were adapted from previous researchers [9] and others were constructed. To ensure validity and reliability, multiple-choice questions were examined by a panel of 11 Physics experts from Iringa, Tanzania. In addition, test items were piloted followed by pre-service Physics teachers' focus group discussion and then re-piloted after 3 months. Also, test re-test reliability revealed a strong correlation with correlation coefficients of .783 and .878 for single and average measures respectively.

3.4   Data Collection Procedures

Two teachers' training colleges (each with 75 participants) were involved, one being the control group and the other as the experimental group. Pre-service Physics teachers in experimental and control groups were taught using YouTube videos and lecture methods respectively. The process used for this study is shown in Fig. 2 below.

In Fig. 2 above, the pre-test was administered to the control and experimental groups. It was given to assess whether groups had a similar level of understanding and misconceptions about kinematics graphs. For 10 weeks, pre-service teachers in the experimental group received interventions through a series of YouTube Videos accessed at www.YouTube.com while in the control group were taught through a series of lectures. A total of 34 carefully selected YouTube Videos were used to teach pre-service teachers kinematics graphs. Four Physics tutors (2 per group) were involved in the teaching process. In the experimental group, pre-service teachers were divided into 5 groups, each with 15 respondents.

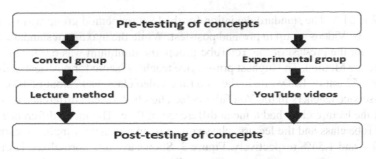

Fig. 2. The process of data collection for control and experimental groups

3.5 Data Analysis Procedures

Simple descriptive and inferential statistics were analysed by using Statistical Package for Social Sciences version 22 and Microsoft Excel. The study also compared the average normalised learning gain scores according to interventions that were given for each group. Average normalised learning gain is the percentage ratio of the mean difference between the post- and pre-test over the mean difference between the highest expected score (100%) and pre-test score [37].

4 Results and Discussions

4.1 Results

The overall pre-service teachers' performance for two interventions indicates that YouTube video interventions improved pre-service teachers' performance more than the lecture method. Table 1 below summarises the mean, standard deviation, minimum, and maximum performance of pre-service teachers for pre-and post-test in all groups.

Table 1. The overall pre-service teachers' achievements from two interventions

S/N	Interventions	N	Test	Pre-service Teachers' Performance			
				Mean	STD	Minimum	Maximum
1	YouTube Videos	75	Pre-test	37.12	10.3	12	44
		64	Post-test	51.50	9.9	32	72
2	Lecture Method	75	Pre-test	36.69	7.9	24	48
		63	Post-test	37.65	6.95	28	52

Table 1 above indicates that all interventions improved pre-service teachers' performance, but performance in the lecture class was not significant i.e., ranging from a mean score of 36.69 to 37.65. In contrast, YouTube Videos interventions provided a considerable improvement in pre-service teachers' achievement which ranged from a mean score of 37.12 to 51.5. The standard deviation for the lecture method group was smaller than the YouTube Videos group in pre-and post-test. While the maximum standard deviation was 10.3 for the pre-test of the YouTube group, the minimum was 6.9 for the post-test in the lecture Method. The highest pre-service teachers' score for post-test in the lecture group was 52 out of 100 while in the YouTube video class was 72 out of 100.

Pre-service teachers in the YouTube video class had a mean difference of 14.38 and those in the lecture class had a mean difference of 0.96. The mean difference between the YouTube class and the lecture, class equate to an average normalized learning gain of 22.87% and 1.54% respectively. Figure 3. Shows average normalized learning gain and mean scores for pre-and post-test according to the interventions given per group.

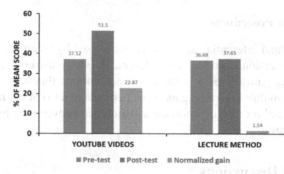

Fig. 3. Teaching interventions and outcome results

The reports from Fig. 3 indicate that there was a significant gain in the YouTube videos class compared to the lecture class. The findings suggest that pre-service teachers who were taught by using YouTube videos significantly outperformed those who were taught by using the lecture method.

Meanwhile, Pre-service teachers' performance across groups (independent samples test) indicates no statistical significance difference in performance between pre-service teachers who were taught by using YouTube videos and those taught through the lecture method during the pre-test. However, findings indicate a statistically significant difference in performance between pre-service teachers in the YouTube videos class and lecture class for the post-test as shown in Table 2 below.

Table 2. Pre-service teachers' performance in pre-and post-test across groups

Test	Groups	N	Mean	STD	T	Df	Sig. (2-tailed)	Effect Size
Pre-test	Control	75	36.69	7.90	−0.285	148	0.776	0.047
	Experimental	75	37.12	10.3				
Post-test	Control	63	37.65	6.95	−9.102	113	<0.001	1.62
	Experimental	64	51.50	9.91				

It is clear from Table 2, that the p-value>0.05 and p-value < 0.001) implies no statistically significant difference between pre-service teachers in the experimental group and those in the control group during the pre-test while there was a statistically significant difference for the post-test. These findings signify that before interventions pre-service teachers in both the control and experimental groups had a similar understanding of kinematics graphs. Similarly, the effect size of 0.047 and 1.62 for the pre-and post-test indicates low and high effect respectively. According to [38], the size effect below 0.2 is regarded as a small effect size, above 0.5 is medium, and above 0.8 is a large effect size. Therefore, YouTube videos produced a higher effect among pre-service teachers than the effect produced by lectures.

4.2 Discussions

YouTube videos as a learner-centred method of teaching and learning were shown to enhance pre-service teachers' conceptual understanding of kinematics graphs more than the lecture method. Therefore, the null hypothesis, which states to be no statistically significant difference between pre-service Physics teachers who were taught through YouTube videos and those through the lecture method was rejected. YouTube videos instructional strategy received a larger effect size, higher mean performance and statistically significant increase (p<0.01) than students taught using lecture methods. This means the lecture method contributed very low in enhancing pre-service teachers to interpret kinematics graphs. Findings align with [23, 39, 40] who reported a positive impact of using YouTube videos for improving performance in Physics. Their findings show that the lecture method resulted in a performance improvement but not so significantly as compared to YouTube videos. However, [21] reported a significant difference in performance in all three groups of students taught through usual teaching methods (lecture), YouTube videos and PhET simulations although the usual teaching method (lecture) had the lowest statistical significance difference.

Similarly, [41, 42] reported that learners' classroom performance is enhanced better when lectures are integrated with technology such as YouTube videos. The study by [43] also found that YouTube videos were the most efficient teaching tool to improve students' performance. Findings concur with [31] who indicated about 94% improved understanding of the subject matter after watching videos. The study by [44] revealed that after making Video a part of the course, many students passed the course. Despite YouTube being one of the learner-centred approaches to teaching and learning, reports from [21] show that many teachers are reluctant to usc lcarner-centred teaching and learning approaches.

5   Summary and Conclusion

This study investigated the impact of YouTube videos as one of the learner-centred strategies that can promote students learning of kinematics graphs in Tanzanian teachers' training colleges. 150 pre-service Physics teachers from two colleges were distributed into two groups and taught through YouTube videos and lecture methods. Developed and validated physics teachers' kinematics graphs conceptual questions were used to measure pre-service teachers' achievement by administering tests before and after interventions. Simple descriptive and inferential statistics were analysed by using Statistical Package for Social Sciences version 22 and Microsoft Excel. The group of pre-service teachers who were taught kinematics graphs by using YouTube videos performed better than those taught through lectures by achieving average normalised learning gains of 22.87% compared to 1.54% for lecture class on the post-test. Also, there was a statistically significant difference (p-value < 0.001) in performance during the post-test between the lecture method and YouTube video classes while there was no significant difference during the pre-test.

The study concludes that YouTube is among of innovative technology tools that teachers can use for classroom instruction in this 21st century of science and technology. The study recommends the following

a) Physics educators should incorporate YouTube videos when teaching kinematics graphs and physics in general and
b) There is a need to conduct similar studies on other topics of physics in which students' performance is weak such as equilibrium and waves to come up with more findings about the use of YouTube videos in teaching and learning.

Acknowledgements. The authors acknowledge the African Centre of Excellence for Innovative Teaching and Learning Mathematics and Science (ACEITLMS) based at the University of Rwanda for the financial support it provides. We are also grateful to Dr. KK Mashood for the useful comments and proofreading. Moreover, the authors wish to extend their appreciation to the owner of the YouTube website and the owners of the YouTube channels watched for their voluntary uploading videos for instructional purposes.

References

1. Ouahbi, I., Darhmaoui, H., Kaddari, F.: Visual block-based programming for ICT training of prospective teachers in Morocco. Int. J. Mod. Educ. Comput. Sci. **14**(1), 56–64 (2022)
2. Adebayo, E.O, Ayorinde, I.T: Efficacy of assistive technology for improved teaching and learning in computer science. Int. J. Educ. Manag. Eng. (IJEME), **12**(5), 9–17 (2022)
3. Nzima, I.: Competence-Based Curriculum (CBC) in Tanzania: Tutors' Understanding and their Instructional Practices. Linnaeus University Press, Sweden (2016)
4. Julius, K.: Nyerere. Education for self-reliance. Dar Es Salaam: Government Printer (1967)
5. Tanzania Institute of Education. Maboresho na mabadiliko ya mitaala toka mwaka 1961 hadi 2010. [Improvements and reforms of curricula from the year 1961 to 2010]. Dar Es Salaam: Tanzania Institute of Education (2013)
6. President's Office - Regional Administration and Local Government. Regional BEST 2019. Dodoma: President's Office - Regional Administration and Local Government (2019)
7. President's Office - Regional Administration and Local Government. Regional BEST 2020. Dodoma: President's Office - Regional Administration and Local Government (2020)
8. Mbwile, B., Ntivuguruzwa, C.: Impact of practical work in promoting learning of kinematics graphs in Tanzanian teachers' training colleges. Int. J. Educ. Pract. **11**(3), 320–338 (2023)
9. Beichner, R.J.: Testing student interpretation of kinematics graphs. Am. J. Phys. **62**(8), 750–762 (1994)

10. Amin, B.D., Sahib, E.P., Harianto, Y.I., et al.: The interpreting ability on science kinematics graphs of senior high school students in South Sulawesi, Indonesia. Jurnal Pendidikan IPA Indonesia **9**(2), 179–186 (2020)
11. Antwi, V.: Impact of the use of MBL, simulation and graph samples in improving Ghanaian SHS science students understanding in describing kinematics graphs. Adv. Life Sci. Technol. **31**(1), 24–33 (2015)
12. Phage, I.B., Lemmer, M., Hitge, M.: Probing factors influencing students' graph comprehension regarding four operations in kinematics graphs. Afr. J. Res. Math. Sci. Technol. Educ. **21**(2), 200–210 (2017)
13. Mbwile, B., Ntivuguruzwa, C., Mashood, K.K.: Development and validation of a concept inventory for interpreting kinematics graphs in the Tanzanian context. Europ. J. Educ. Res. **12**(2), 673–693 (2023)
14. National Examination Council of Tanzania. Students' response item analysis report for form two national assessment: Physics. Dar Es Salaam: National Examination Council of Tanzania (2017)
15. National Examination Council of Tanzania. Students' response item analysis report for form two national assessment: Physics. Dar Es Salaam: National Examination Council of Tanzania (2018)
16. National Examination Council of Tanzania. Students' response item analysis report for form two national assessment: Physics. Dar Es Salaam: National Examination Council of Tanzania (2019)
17. National Examination Council of Tanzania. Students' response item analysis report for form two national assessment: Physics. Dar Es Salaam: National Examination Council of Tanzania (2020)
18. National Examination Council of Tanzania. Students' response item analysis report for form two national assessment: Physics. Dar Es Salaam: National Examination Council of Tanzania (2021)
19. Mbonyiryivuze, A., Yadav, L.L., Amadalo, M.M.: Students' attitudes towards physics in nine years basic education in Rwanda. Int. J. Eval. Res. Educ. (IJERE) **10**(2), 648–659 (2021)
20. Mbwile, B., Ntivuguruzwa, C., Mashood, K.K.: Exploring the understanding of concept inventories for classroom assessment by physics tutors and pre-service teachers in Tanzania. Afr. J. Res. Math. Sci. Technol. Educ. **27**(1), 1–11 (2023)
21. Ndihokubwayo, K., Uwamahoro, J., Ndayambaje, I.: Effectiveness of PhET simulations and Youtube videos to improve the learning of optics in Rwandan secondary schools. Afr. J. Res. Math. Sci. Technol. Educ. **24**(2), 253–265 (2020)
22. Gustafsson, P.: How physics teaching is presented on Youtube videos. Educ. Res. Soc. Change **2**(1), 117–129 (2013)
23. Jones, T., Cuthrell, K.: Youtube: educational potentials and pitfalls. Comput. Sch. **28**(1), 75–85 (2011)
24. Haase, D.G.: The Youtube makeup class. Phys. Teach. **47**(5), 272–273 (2009)
25. Ahmed, N., Nandi, D., Zaman, A.G.M.: Analyzing student evaluations of teaching in a completely online environment. Int. J. Mod. Educ. Comput. Sci. (IJMECS) **14**(6), 13–24 (2022)
26. Tendo, S.N.: multimedia pedagogy among literature lecturers in two universities in Uganda post COVID-19. Int. J. Educ. Manag. Eng. (IJEME), **13**(1), 1–9 (2023)
27. Hassan, M.M., Mirza, T., Hussain, M.W.: A critical review by teachers on the online teaching-learning during the COVID-19. Int. J. Educ. Manage. Eng. (IJEME) **10**(5), 17–27 (2020)
28. Nan, L., Yong, Z.: Improvement and practice of secondary school geography teachers' informatization teaching ability based on the perspective of MOOCs. Int. J. Educ. Manage. Eng. (IJEME) **12**(1), 11–18 (2022)

29. Sweeny, R.W.: There's no I in Youtube: social media, networked identity and art education. Int. J. Educ. Through Art **5**(2/3), 201–212 (2009)
30. Park, J.C.: Video allows young scientists new ways to be seen. Learn. Lead. Technol. **36**(8), 34–35 (2009)
31. Geri, N.: The resonance factor: probing the impact of video on student retention in distance learning. Interdiscipl. J. E-Learn. Learn. Objects **8**(1), 1–13 (2012)
32. Copley, J.: Audio and video podcasts of lectures for campus-based students: production and evaluation of student use. Innov. Educ. Teach. Int. **44**(4), 387–399 (2007)
33. Beheshti, M., Taspolat, A., Kaya, O.S., et al.: Characteristics of instructional videos. World J. Educ. Technol. Curr. Issues **10**(1), 61–69 (2018)
34. McDermott, L.C., Rosenquist, M.L., Van Zee, E.H.: Student difficulties in connecting graphs and physics: examples from kinematics. Am. J. Phys. **55**(6), 503–513 (1987)
35. Planinic, M., Ivanjek, L., Susac, A., et al.: Comparison of university students' understanding of graphs in different contexts Maja. Phys. Educ. Res. **9**, 213–221 (2013)
36. Farghaly, A.: Comparing and contrasting quantitative and qualitative research approaches in education: the peculiar situation of medical education. Educ. Med. J. **10**(1), 3–11 (2018)
37. Richard, R., Hake, R.R.: Interactive-engagement versus traditional methods: a six-thousand-student survey of mechanics test data for introductory physics courses. Am. J. Phys. **66**(1), 64–74 (1998)
38. Cohen, J.: Statistical power analysis for the behavioral sciences (2nd ed.). New Jersey Lawrence Erlbaum (1998)
39. Afriani, T., Agustin, R.R.: Eliyawati. the effect of guided Inquiry laboratory activity with video embedded on students' understanding and motivation in learning light and optics. J. Sci. Learn. **2**(4), 79–84 (2019)
40. Lincoln, J.: Making good physics videos. Phys. Teach. **55**(5), 308–309 (2017)
41. Bwalya, A., Rutegwa, M.: Technological pedagogical content knowledge self-efficacy of pre-service science and mathematics teachers: a comparative study between two Zambian universities. Eurasia J. Math. Sci. Technol. Educ. **19**(2), em2222 (2023)
42. Ljubojevic, M., Vaskovic, V., Stankovic, S., et al.: Using supplementary video in multimedia instruction as a teaching tool to increase efficiency of learning and quality of experience. Int. Rev. Res. Open Distrib. Learn. **15**(3), 275–291 (2014)
43. De la Flor, L.S., Ferrando, F., Fabregat-Sanjuan, A.: Learning/training video clips: an efficient tool for improving learning outcomes in mechanical engineering. Int. J. Educ. Technol. High. Educ. **13**(1), 1–13 (2016)
44. Hsin, W.J., Cigas, J.: Short videos improve student learning in online education. J. Comput. Sci. Coll. **28**(5), 253–259 (2013)

Logical Relationship Analysis and Architecture Design of Logistics Engineering Curriculum Based on ISM Model

Lijun Jiang[1], Wanyi Wei[2(✉)], Lei Zhao[1], and Huasheng Su[1]

[1] Business School of Nanning University, Nanning 530200, China
[2] Saines New Medical College of Guangxi University of Traditional Chinese Medicine, Nanning 530200, China
932370009@qq.com

Abstract. In the context of the rapid development of new technologies such as big data, artificial intelligence, mobile internet, cloud computing, Internet of Things, blockchain, etc., the market's requirements for logistics talents have changed dramatically. In order to cultivate talents suitable for industry enterprises, the logistics engineering specialty has adjusted the curriculum system of talent training, added a curriculum integrating new technology with logistics professional knowledge, and formed a new curriculum system. This paper uses the ISM method to systematically analyze the hierarchical logical relationship between the courses in the new system, and on this basis, combined with the actual situation of the course teaching arrangement, constructs the curriculum topology map of logistics engineering specialty, which makes the talent training ideas clearer, optimizes the connection relationship between the course opening, the semester arrangement and the time planning, and promotes the high quality training of logistics talents.

Keywords: ISM · Logistics engineering · Curriculum system · Course structure

1 Introduction

Modern logistics is the bottom support for the efficient development of modern economy. The development of logistics plays a vital role in the high-quality development of the national economy. According to the statistics of the China Federation of Logistics and Purchasing, the scale of China's social logistics business has maintained an overall growth trend from 2011 to 2021, and the total amount of social logistics will exceed 330 trillion yuan in 2021, setting a new record. The rapid development of new technologies such as big data, artificial intelligence, mobile internet, cloud computing, Internet of Things, blockchain, etc. has promoted the updating and iteration of logistics technology, making manufacturing and service industries face new environment and new challenges in production environment, manufacturing mode, processing method, payment means, and goods circulation methods, which has spawned new ideas, new business forms, and new retail [1, 2]. Under such a market background, the demand of industry enterprises for logistics talents is undergoing subversive changes, and more attention is paid to the application ability of new technology, strong comprehensive ability, logistics professional ability, strong strategic judgment and management ability of logistics talents [3–5]. In order to cope with the changes in the environment and cultivate logistics talents in line with the needs of the new era, colleges and universities have carried out many reforms in the talent training curriculum system and teaching methods of logistics engineering specialty [6]. In the course system, new courses have been set up that combine new technologies such as digitalization and intelligence with logistics and supply chain, such as "big data analysis and application of supply chain", "smart logistics design and technology application" and other courses [7, 8]. How to make the adjusted curriculum system more systematic and scientific is what we need to study and optimize at this stage.

2 Research Trends of Logistics Engineering Curriculum System

The logistics market and logistics professional education research in the United States, Japan, the United Kingdom and other countries started early and have some valuable experience, which has certain reference value for the construction of the logistics engineering professional curriculum system in China [9]. After years of construction and development, the United States, the country with the earliest development of the global logistics industry, has formed a relatively reasonable logistics talent training system [10, 11] with its undergraduate logistics curriculum system highlighting practical teaching and distinctive features. Japan is one of the countries with a high level of logistics modernization, and is also the most effective country to combine industrial production with logistics management. Japan implements "wide caliber" logistics education. The curriculum is not only multifaceted, but also professional, and is committed to cultivating solid basic knowledge of logistics students. The overall setting of the curriculum system makes the knowledge of mathematics, physics, computer, logistics management, information processing, economics, environmental science, geography and other aspects form an organic whole, and the knowledge is integrated with each other to help students better broaden their knowledge, and more importantly, it can adapt to the needs of logistics specialization [12]. However, British logistics education pays more attention to "modularity", and the level of logistics professional teaching and scientific research ranks the world's leading position, which has been recognized by the international academic and industrial circles, and has certain international authority [13]. In addition, the UK has the world's earliest professional organization of logistics and transportation, which is jointly formed by the British Institute of Logistics and the British Institute of Transportation with a history of 100 years. The ILT (Institute of Logistics and Tensport) logistics certification system, standards and corresponding training course modules implemented by it have also been widely adopted by many countries in Europe, America and other regions [14, 15].

China's logistics majors have been established for a short time. In 1994, the logistics management specialty was first established, and a curriculum system based on management, operational research, economics, and information technology was established [16]. In 2001, the Ministry of Education approved some colleges and universities in China to set up logistics engineering undergraduate programs, forming a teaching system that focuses on work and pays equal attention to work management [17]. Since the establishment of logistics engineering undergraduate program, scholars have studied its curriculum system. Wang Qiang (2018), taking the construction of the new engineering discipline as the background, constructed the knowledge, ability and quality structure system for the training of application-oriented logistics engineering professionals. On this basis, he proposed the idea of constructing the curriculum system of application-oriented logistics engineering for the new engineering discipline [18]. Gao Mingjing (2021), taking the logistics engineering major of Dalian Neusoft Information University as an example, explained the course system reform process based on capacity demand research [19]. Luo Darong (2019), on the basis of explaining the problems existing in the logistics curriculum system, analyzed the construction ideas of the logistics curriculum system under the background of engineering education certification, and proposed the construction strategies of the logistics curriculum system under the background of engineering education certification [20]. Guo Zhiying (2022), aiming at the current situation of China's logistics industry development in the context of smart logistics, analyzed the position, education background, experience and other aspects of enterprise logistics talent demand, and put forward relevant suggestions on logistics talent training in colleges and universities [21]. Yan Zhili (2017), taking the logistics major as an example, compared and analyzed the construction of a transformation pilot university in Hebei Province and the University of Applied Science and Technology in Cologne, Germany, from the aspects of talent training objectives, curriculum setting, curriculum arrangement, curriculum implementation and curriculum evaluation, and found that there were significant differences between the two. It is proposed to ensure the accuracy of talent training objectives, the practicality of curriculum arrangement, the modularization of curriculum design, the diversification of teaching methods, and the socialization of curriculum evaluation, so as to accelerate the transformation and development process of local undergraduate universities in China [22]. Only in this way can we continuously promote the continuous progress of professional education in China [23].

Although scholars at home and abroad have done some research on the curriculum system of logistics engineering specialty (or logistics specialty), most of them are based on the background of the times or a certain research perspective on the composition of professional courses, while the research on the logical level and structure of various courses within the professional curriculum system is relatively lacking. Therefore, this paper selects the Interpretative Structure Model (ISM) to analyze the level of logistics engineering professional courses and clarify the logical relationship between the courses in order to improve the systematic nature of the course system and the scientific nature of the course teaching organization.

3 Research on Curriculum Structure of Logistics Engineering

3.1 Introduction to ISM Method

ISM (Interpretative Structural Modeling Method) was developed by American scholar Professor Warfield in 1973 as a method to analyze the structural problems of complex social and economic systems. It is a relatively mature method to reveal the relationship between system elements. ISM can sort out the hierarchical structure of various elements of the system according to the interaction relationship between the system elements, help people decompose the disordered and complex relationship among the system elements into clear and hierarchical structural forms, facilitate in-depth and clear understanding of the whole system, so as to grasp the core problems and find effective solutions to the problems. For the professional curriculum system, there are many subjects, and the relationship between them is complex and disordered. ISM method can effectively sort out the relationship between various courses, form the curriculum hierarchy, and effectively promote the scientific and effective development of professional curriculum teaching organization.

3.2 Research Steps and Process

The general research process of ISM method includes extracting elements and determining the relationship between elements, establishing adjacency matrix and obtaining reachability matrix, and decomposing the reachability matrix to establish a structural model. The application of ISM method to the research of professional curriculum structure has gone through the following three steps.

First, establish an expert group to extract elements and determine the relationship between elements.

Four teachers were selected from logistics engineering and related disciplines, and one industry expert was invited to form an expert group to study the curriculum system. The expert group will discuss the training plan for 2022 talents of logistics engineering specialty, and extract elements (courses) from it. After removing the relatively independent general education courses such as ideological and political courses and university sports from the professional perspective, the expert group included 44 courses into the system research elements, coded them, and confirmed the prerequisite relationship between the courses. See Table 1.

Second, establish the adjacency matrix and find the reachability matrix.

Based on the prerequisite relationship between the above courses, 44 × 44. Let I be the identity matrix of the same order as A, and calculate the reachable matrix M according to the following formula (1).

$$(A+I)^{K-1} \neq (A+I)^K = (A+I)^{K+1} = M \tag{1}$$

Thirdly, decompose the reachable absolute matrix and find out the hierarchical relationship of elements.

Define two sets $R(Si)$ and $A(Si)$.

$R(Si)$: The set of all elements that can be reached from Si is called the reachable set. $A(Si)$: The set of all possible elements is called the antecedent set.

Use $R(Si)$ and $A(Si)$ to find the set of $R(Si)$ and $A(Si)$.

According to the results of $R(Si) = A(Si)$, the hierarchical relationship of curriculum elements is obtained.

Since there are 44 elements in this study, which are large in number and difficult to calculate, the above contents are calculated and analyzed with the help of SPSSAU software. The results show that 44 curriculum elements are divided into 12 levels as shown in Table 2. See Fig. 1 for the relationship between the course elements.

Table 1. Prerequisite relationship between logistics engineering courses

Course No.	Course name	Prerequisites	Course No.	Course name	Prerequisites
(1)	College English		(23)	Logistics system planning a	(2) (8) (10) (16) (17) (18) (19) (20)
(2)	Advanced Math. B		(24)	Logistics system simulation	(23) (28) (28)
(3)	Linear Algebra B	(2) (5)	(25)	International Trade Practice	(1) (32)
(4)	Probability Statistics A	(3)	(26)	Cross-border e-commerce logistics	(1) (10) (25) (32)
(5)	Computer Foundation		(27)	Procurement management	(8) (10) (14)
(6)	Physics B	(2)	(28)	Container multimodal scheme	(10) (15) (20)
(7)	Physical Experiment B	(6)	(29)	Logistics frontier and innovation	(10) (22)
(8)	Principles of Management		(30)	Logistics laws and regulations	(10) (17)
(9)	Engineering drawing	(6)	(31)	Internet of Things project	(5) (13) (19) (21) (33)
(10)	Modern Logistics	(8) (35)	(32)	International Logistics English	(1) (10) (25) (26)
(11)	Engineering mechanics	(3) (6)	(33)	Python programming	(1) (5)
(12)	Fundamental Mechanics	(6)	(34)	New retail marketing	(8)
(13)	Java language programming	(1) (5)	(35)	Technological economics	
(14)	Data statistics and analysis	(1) (2) (3) (4) (33)	(36)	Engineering Training I	(6) (7) (9) (11)
(15)	Operational research	(2) (8)	(37)	Information and experiment	(5) (10) (13) (31)
(16)	Systems engineering	(5) (15)	(38)	Import and export business training	(25) (26)
(17)	Information literacy	(5)	(39)	Supply chain big data analysis	(10) (20) (33)

(*continued*)

Table 1. (*continued*)

Course No.	Course name	Prerequisites	Course No.	Course name	Prerequisites
(18)	Warehousing and inventory	(3) (8) (10) (14) (31) (33) (34)	(40)	Intelligent logistics design/application	(10) (33) (39)
(19)	Transportation scheduling	(8) (10) (14) (15) (31) (33)	(41)	Optimization of logistics equipment	(10) (22)(40)
(20)	Supply chain management	(8) (10) (18) (19) (27) (37)	(42)	Course Design of System Simulation	(23)
(21)	Logistics information	(5) (8) (10) (13) (22) (27)	(43)	Digital supply chain solution	(20) (30) (33) (35)
(22)	Modern logist. Equip	(10) (12)	(44)	Sand table simulation logistics	(8) (10) (18) (19) (20) (21) (27) (30) (35)

Table 2. Hierarchy of curriculum elements of logistics engineering

Hierarchy	Essential factor
The first layer (top later)	(24), (29), (36), (41), (42), (43), (44)
Layer 2	(7), (9), (11), (23), (28), (30), (38), (40)
Layer 3	(16), (17), (25), (26), (32), (39)
Layer 4	(18), (20), (34)
Layer 5	(37)
Layer 6	(19), (31)
Layer 7	(15), (21)
Layer 8	(13), (22), (27)
Layer 9	(10), (12), (14)
Layer 10	(4), (6), (8), (33), (35)
Layer 11	(1), (3)
Layer 12 (ground later)	(2), (5)

According to the above ISM analysis of the hierarchical relationship between the courses of logistics engineering undergraduate course, 44 course elements are divided into 12 levels, among which the basic courses such as "Advanced Mathematics", "College Computer Foundation" and "College Foreign Language" are at the lowest level, and the comprehensive courses such as "Digital Supply Chain Solution Design" and "Sandtable Simulation Experiment of Logistics Enterprise Operation" are at the highest level. The hierarchical structure of the curriculum elements clarifies the logical relationship between the courses, which will be conducive to scientific and reasonable organization and arrangement of curriculum teaching activities.

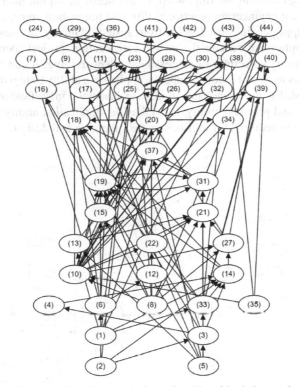

Fig. 1. Schematic diagram of the hierarchical relationship of logistics engineering curriculum elements

4   Improve the Curriculum System Structure of Logistics Engineering

Based on the above analysis, the hierarchical relationship of each course element is obtained. Considering that the undergraduate students of logistics engineering study in school for 6.5 semesters, the courses of advanced mathematics are distributed in 2 semesters, and the courses of college foreign languages are distributed in 4 semesters, and other specific conditions, the curriculum topology map of logistics engineering undergraduate specialty is made through comprehensive exploration and planning, as shown in Fig. 2.

Compared with the original curriculum system, the improved logistics engineering curriculum system is more consistent and prominent in the main line of benchmarking ability training, and the logic of course opening is much clear.

The goal of training logistics engineering professionals is to cultivate high-quality application-oriented talents who have the basic theory, basic knowledge and basic skills necessary for logistics engineering, adapt to the needs of digital international supply chain, have strong engineering practice ability and innovation spirit, and can work in logistics and supply chain-related fields from system planning and design, equipment selection and application, scheme design and optimization, project operation and maintenance. From these objectives of professional talent training, it can be seen that the students' abilities to be cultivated include the application ability of logistics professional knowledge and skills, digital and other modern technologies in the field of logistics, engineering literacy and practical ability, as well as the application ability of mathematics and computer tools required for logistics related planning and design.

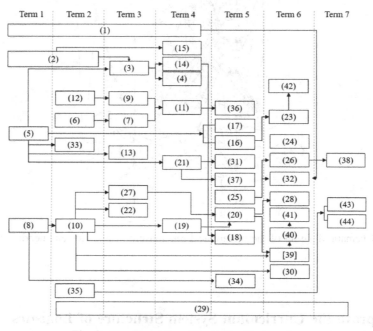

Fig. 2. Topology Map of Logistics Engineering

After the improvement of the system structure of logistics engineering course, it clearly revealed the course content modules corresponding to these abilities, mainly including mathematical methods and tool modules composed of courses such as advanced mathematics, linear algebra, probability theory, operational research, data statistics and analysis, system engineering, etc., modern computer technology knowledge modules composed of courses such as Python, Java language programming, Internet of

Things engineering, etc., college physics, mechanical engineering foundation The engineering basic knowledge and literacy module composed of engineering drawing, engineering mechanics, engineering training and other courses, the logistics professional knowledge and skills module composed of management, modern logistics, warehousing and inventory control and other courses, and the interdisciplinary module of modern technology and logistics integration composed of supply chain big data analysis and application, intelligent logistics design and technology application, digital supply chain solution design and other courses. Among them, the mathematical methods and tools module, modern computer technology knowledge module, engineering basic knowledge and literacy module are mainly concentrated in the first to fifth semesters, while the interdisciplinary module courses of modern technology and logistics integration are concentrated in the sixth and seventh semesters after students master a certain amount of modern technology basic knowledge and logistics professional knowledge. The course module corresponds to the ability training, and has clear logic in the sequence and timing of courses, which can better support the achievement of the training objectives of logistics engineering professionals.

The course team conducted a survey on the students and teachers in the teaching class by means of interviews and answers, and generally responded that according to the new course system, the teaching ideas are clearer, and students' understanding and mastery of knowledge and enthusiasm for learning have also been significantly improved.

5   Conclusion

Based on the new curriculum system established by the logistics engineering specialty according to the change of talent demand in the new environment, the paper better reveals the hierarchical relationship between the courses in the logistics engineering specialty curriculum system by using the ISM method, and effectively helps to sort out the curriculum structure. At the same time, combined with the school's learning time of logistics engineering specialty, as well as the characteristics of some courses and other factors, the curriculum topology map of logistics engineering specialty is explored and constructed. Compared with the original curriculum structure, on the one hand, the improved curriculum topology can clearly see the composition of the curriculum modules, and correspond to the requirements of students' ability training, which helps to clearly grasp the main line of training ideas in the implementation of talent training, and more effectively carry out talent training. On the other hand, more clear and orderly sorting out the logic of the course, promoting the implementation of the course more systematic and scientific, and helping to promote the high-quality development of logistics talent training.

The paper explores the application of ISM model to the research of professional curriculum system structure, scientifically and objectively analyzes the hierarchical relationship among the courses in the system, and makes the various and disordered courses orderly and the mutual relationship clear, which can provide reference for the in-depth study of curriculum system of all majors.

Acknowledgment. This research is supported by the teaching team construction project of Nanning University (2021XJXTD05) and the "Fourteenth Five-Year Plan" project of Guangxi Education and Science (2022ZJY2449).

References

1. Cheng, Q., Yu, L.: Operational mechanism and evaluation system for emergency logistics risks. Int. J. Intell. Syst. Appl. (IJISA) **2**(2), 18–23 (2010)
2. Zhao, X., Feng, L., Lin, R.: Competency-based: strategic thinking on the restructuring of the curriculum system of application-oriented undergraduate logistics engineering in the context of new business. J. High. Educ. **18**, 59–62 (2021). (in Chinese)
3. Zheng, Z., Li, L., Wei, G.: Contradiction between supply and demand of logistics talents training and teaching reform in China. In: International Conference on Artificial Intelligence and Logistics Engineering, vol. 3, pp. 352-359 (2021)
4. Deng, X.: Research on the current situation and strategy of talent training of logistics management specialty from the perspective of intelligent logistics. In: The 4th International Conference on Humanities Education and Social Sciences (ICHESS 2021), vol. 615, pp. 2787–2791 (2021)
5. Yao, W., Fan, Y.: Research on talent cultivation of modern logistics management majors in vocational education based on artificial intelligence. In: The 2nd International Conference on Education, Information Management and Service Science (EIMSS 2022), vol. 7, pp. 331–338 (2022)
6. Benson, G.E., Chau, N.N.: The supply chain management applied learning center: a university-industry collaboration. Ind. High. Educ. **4**, 135–146 (2019)
7. Aliyu, I., Kana, A.F.D., Aliyu, S.: Development of knowledge graph for university courses management. Int. J. Educ. Manage. Eng. (IJEME) **10**(2), 1 (2020). https://doi.org/10.5815/ijeme.2020.02.01
8. Rajak, A., Shrivastava, A.K., Bhardwaj, S., et al.: Assessment and attainment of program educational objectives for post graduate courses. Int. J. Mod. Educ. Comput. Sci. (IJMECS) **11**(2), 26–32 (2019)
9. Muñoz-Martínez, Y., Domínguez-Santos, S., de la Sen-Pumares, S., et al.: Teachers' professional development through the education practicum: a proposal for university-school collaboration. Cypriot J. Educ. Sci. **17**(2), 464–478 (2022)
10. Hua, L., Tian, Y., Wang, J., Yanli, Y.: A comparison between China and the United States in the curriculum system of logistics undergraduate majors - text mining analysis based on 754 majors. Supply Chain Manag. **8**, 75–88 (2022). (in Chinese)
11. Zhu, S., Zhu, R., Zhou, X.: Training mode of logistics talents based on internet era in application colleges and universities. In: Proceedings of the 7th International Conference on Education and Management (ICEM 2017),vol. 1, pp. 22–28 (2018)
12. Yang, Y.: Comparative study of logistics professional education at home and abroad. China Mark. **9**, 150–151 (2017). (in Chinese)
13. Liu, M.: Analysis of the training mode of logistics management talents in British universities. Mark. Weekly **4**, 105–107 (2015). (in Chinese)
14. Mantle, M.J.H.: How different reflective learning activities introduced into a postgraduate teacher training programme in England promote reflection and increase the capacity to learn. Res. Educ. **105**(1), 60–73 (2019)
15. Nawaz, R., Sun, Q., Shardlow, M., et al.: Leveraging AI and machine learning for national student survey: actionable insights from textual feedback to enhance quality of teaching and learning in UK's Higher education. Appl. Sci. **12**(1), 514 (2022)
16. Xu, T.: Undergraduate logistics specialty setting system and training division. High. Eng. Educ. Res. **2**, 23–26 (2002). (in Chinese)
17. Shan, S., Chen, J., Liu, C., et al.: Discussion on the content system of the undergraduate course of logistics engineering. Forest Eng. **20**(6), 26–28 (2004). (in Chinese)

18. Wang, Q., Li, J., Li, W., et al.: Talent training and curriculum system construction of logistics engineering in the context of new engineering. Logist. Technol. **37**(10), 135–138 (2018). (in Chinese)
19. Gao, M.: Research on curriculum reform of logistics engineering specialty based on capacity demand research. Logist. Eng. Manage. **43**(02), 146–148 (2021). (in Chinese)
20. Luo, D.: Construction of logistics curriculum system under the background of engineering education certification. W. Qual. Educ. **5**(22), 3–5 (2019). (in Chinese)
21. Guo, Z.: Demand analysis of logistics talents in the context of smart logistics and suggestions for talent training in colleges and universities. Mod. Commer. Trade Ind. **43**(16), 42–43 (2022). (in Chinese)
22. Yan, Z., Zeng, S., Wang, J.: Comparison of the curriculum system of logistics specialty in Sino-German application-oriented undergraduate universities. Vocat. Tech. Educ. **38**(14), 19–24 (2017). (in Chinese)
23. Robles, A.C.M.O.: The use of educational web tools: an innovative technique in teacher education courses. Int. J. Mod. Educ. Comput. Sci. (IJMECS) **5**(2), 34–40 (2013)

Exploration of the Application of Project Teaching Method in the Course of Urban Public Transport Planning and Operation Management

Lanfang Zhang[✉], Chun Bao, and Jianqiu Chen

School of Transportation, Nanning University, Guangxi, Nanning 530000, China
22452302@qq.com

Abstract. On July 30, 2020, the General Office of the Ministry of Education and the General Office of the Ministry of Industry and Information Technology issued a notice on the "Guide for the Construction of Modern Industrial Colleges (for Trial Implementation)", proposing that cultivating high-quality applied, composite and innovative talents who adapt to and lead the development of modern industries is an inevitable requirement for higher education to support high-quality economic development, and an important initiative to promote the development of classification and characteristics of universities. As a local, application-oriented university, Nanning University should actively explore the development path of modern industrial colleges in the process of training students and conducting course teaching, so that students can quickly adapt to their jobs after graduation, thus promoting the high-quality development of the local economy. This paper explores the application of the project teaching method in terms of the current situation of the construction of the course, the analysis of problems, the specific application of the project teaching method in the course and the evaluation of the course, based on the professional course Urban Bus Planning and Operation Management of Nanning University's transportation major.

Keywords: Project teaching method · Urban public transport planning and operation management · Application · Investigation

1 Introduction

The Urban Public Transport Planning and Operations Management course teaches students about urban public transport, including: introduction to urban public transport, public transport demand forecasting, urban public transport planning, route capacity, travel speed and vehicle utilization, daily dispatching, evaluation and statistics of public transport operation indicators, rapid public transport systems, urban public transport service management, urban public transport service quality management The course provides students with a holistic and systematic understanding of public transport in areas such as In the new educational context, teachers need to improve teaching methods in order to enable students to better meet the employment needs of enterprises after graduation and to meet the requirements of high-quality economic development, and the

Z. Hu et al. (Eds.): ICCSEEA 2023, LNDECT 181, pp. 1121–1131, 2023.
https://doi.org/10.1007/978-3-031-36118-0_95

introduction of the project teaching method plays a crucial role in the teaching reform of the course, transforming the traditional duck-fill teaching to independent exploratory teaching [1].

The project approach sprang up in Europe The earliest prototypes of the idea of work-based education are the 18th century European work-study education and the 19th century United States It was developed and refined in the mid to late 20th century. It is a method in which students are responsible for the collection of information, the design of the project, its implementation and final evaluation. The most significant feature is that "the project is the main line, the teacher is the guide, and the students are the main body", which has changed the passive teaching mode of "teachers speak, students listen" and created a new teaching mode of active participation, independent collaboration and exploration and innovation [2, 3]. At present, the project teaching method has been applied in engineering, software, design, transportation, logistics, English, landscape and other related courses, and the application of this method has improved students' learning motivation, and the teaching effect of the course has been significantly improved, while meeting the requirements of national economic development for higher education [4–6].

Last year, the report of the 20th Party Congress put forward the comprehensive promotion of rural revitalization. As an applied undergraduate school serving local economic development, the School of Transportation of Nanning University needs to keep pace with the times in terms of curriculum reform. The course on public transport planning and operation management mainly corresponds to public transport-related enterprises, and in order to make students quickly applicable to their jobs after graduation, the course teaching and research group has proposed a project through continuous polishing of the course In order to enable students to quickly apply to the workplace after graduation, the course teaching and research group has proposed the project teaching method through continuous polishing of the course [7]. This paper first analyses the current situation and problems of the course construction, then explains the specific application of the project teaching method in the course, and finally evaluates the application effect, showing that the teaching method has a positive effect on the personal growth of students.

2   Analysis of the Current Situation and Problems of Curriculum Development

2.1   Current Status of Course Construction

The course on urban public transport planning and operation management is offered in the second semester of the third year of the transportation major, with a total of 48 h and three credits, two classes per week and two hours per class. In the past, the teaching mode was based on the textbook, and the instructor would introduce case studies for specific content, but the cases were independent of each other and did not form a system [8]. The course has 42 h of theory and 6 h of practice. The course is assessed by 30% of the usual grades, mainly for homework, classroom performance and attendance, and 70% of the final closed-book examination. The course has a resource library, which consists of textbooks, public transport-related specifications and a question bank.

2.2 Analysis of Problems

2.2.1 Students' Knowledge of the Curriculum is not Systematic

Under the old teaching mode, as teachers aim to teach students to master the knowledge of the teaching materials when teaching, this results in students' mastery of the course being in a modular and fragmented state of knowledge, they learn only for the content of the course, but do not understand why they need to learn and what aspects of the course they use in their practical work after learning, although teachers mention the application of knowledge in class, but as students have no practical project experience Therefore they are not clear about it, which also causes students' learning objectives to be weak, their motivation to learn is difficult to mobilize, the effect of learning stays at the knowledge level, it is difficult to form an understanding and application of knowledge, and the learning effect is even more difficult to guarantee [9–11].

2.2.2 Students Lack Practical Application Skills

According to the old teaching method, teachers play a leading role in the whole course learning, they teach and assign homework, and students complete the course tasks according to the requirements [12, 13]. The urban bus planning and operation management course is a professional course with very obvious industry characteristics, if there is no project practice in the teaching process, it is difficult to produce true knowledge, the students can master very limited professional knowledge after the examination, after graduation to the workplace, they need to re-learn, enterprises need to give fresh graduates half a year or even a year of internship, to a certain extent, the waste of national educational resources, not meet the current requirements of the construction of industrial colleges [14, 15]. Therefore, it is the responsibility of schools to train students with practical application skills, so that school education can be seamlessly integrated with the needs of enterprises.

In summary, there is a large gap between the old pedagogy and the new school's objectives for the training of students, and the curriculum needs to be reformed.

3   Application of Project-Based Teaching Methods in the Curriculum

In order to improve the teaching quality and effectiveness of the course and to meet the teaching philosophy of an applied university, especially to meet the requirements of the construction of a modern industrial college, the course teaching and research team has introduced the project teaching method.

3.1 Analysis of the Fit Between the Introduced Project and the Course Content

The projects introduced in this course are actual engineering design projects. In order to improve students' practical ability, the course team invited tutors from design institutes to co-teach. The main consideration of the projects introduced is that they can be linked to the course content in the teaching process, reflecting the principle of linking theory to practice., so that students can apply what they have learnt [16]. On this basis, the course team chose the optimization of public transport lines and stations in the Youjiang District of Baise City as the project for the course (Table 1).

Table 1. Project content and course content matching table

Serial number	Course content	Project content
1	Introduction to Urban Public Transport	Chapter 1 Project Overview
2	Urban public transport demand forecast	Chapter 3 Public Transport Demand Forecast
3	Infrastructure for urban public transport	Chapter 2 Analysis of Public Transport Operations and Service Evaluation, Chapter 4 Analysis of Public Transport Development Patterns, Chapter 5 Optimization of Public Transport Routes, Chapter 6 Optimization of public transport stations (yards)
4	Route capacity, speed and vehicle utilization of public transport operations	Chapter 2 Analysis of the current state of public transport operations and service evaluation
5	Evaluation and statistics of public transport operational indicators	Chapter 2 Analysis of the current state of public transport operations and service evaluation
6	Rapid public transport system	Chapter 4 Analysis of Public Transport Development Patterns
7	City Bus Service Management	Chapter 2 Analysis of the Current Situation of Public Transport Operations and Service Evaluation ,Chapter 7 Study on Integrated Public Transport Development Strategies
8	City Bus Service Quality Management	Chapter 2 Analysis of the Current State of Public Transport Operations and Service Evaluation, Chapter8 Implementation Assessment and Implementation Recommendations

As can be seen from the above matching table, the projects chosen for the teaching process are perfectly matched to the teaching content and, as the project content progresses, students can fully grasp the learning content of the course through teacher-led teaching.

3.2 Methods and Processes for Achieving Course Objectives Through Projects

In order to ensure that the teaching objectives are achieved, teachers need to be familiar with both the course and the project in the process of project teaching method, therefore, the course is taught by dual-teacher teachers, teachers need to have actual project experience working in enterprises, in addition, teachers always keep close contact with the project contact mentor of enterprises, in order to adjust the course teaching according to the actual situation at any time, the following is the project teaching method The following is the implementation process of the project teaching method.

3.2.1 Conducting Lectures on the Introduction to Public Transport

The course begins with an introduction to the background of the project so that students can understand the basics of the project and the teacher goes through the basic concepts of public transport to ensure that students are familiar with the basic terminology. As the students have already completed the pre-requisite courses in transport planning and transport surveys, the teacher groups the students at the end of the introductory lesson and assigns them the task of preparing a survey plan for the project information. After the students have submitted their survey plans in groups, the teacher reviews and summarizes their plans and the class works together to compile a complete and implementable project survey plan and prepare a corresponding questionnaire for use in the survey (Fig. 1).

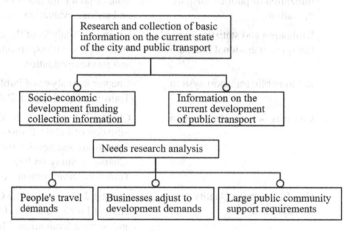

Fig. 1. Research Information Summary Chart

3.2.2 Conducting Project Field Research

Field research is the basis for an in-depth understanding of the project and is the best way for students to link theory with practice. When teaching the project, two days were allocated for field research, mainly on public transport routes and stations, passenger volume surveys, questionnaires on people's travel demands, questionnaires on enterprises' demands for adjustment and development, and questionnaires on the supporting needs of large co-developed communities. The study group leaders, in liaison with their business mentors, held discussions with relevant government departments to gain an in-depth understanding of the current situation of public transport. After the research, the students, under the guidance of their teachers, collate and analyze the research data, mainly in terms of bus routes, bus stations, bus vehicles and the market environment, etc. In the process of analysis, students are guided to carry out calculations of public transport operation indicators in conjunction with textbooks and specifications, and to evaluate the services of public transport (Fig. 2).

Fig. 2. Get onboard video statistics from the bus company on the number of passengers boarding and alighting

3.2.3 Urban Public Transport Demand Forecast

Public transport demand forecasting has always been a difficult part of the course for students to master, so in order for students to learn and apply the project teaching method, this part of the course is specially invited to the enterprise tutor to explain in the computer room, and students follow the enterprise tutor's lecture to carry out the practical operation of transport demand forecasting, and require students to practice the four stages of forecasting before the enterprise tutor teaches, and follow the tutor with questions during the course of the lesson. During the course of the lesson, students followed the tutor with questions to gain a deep understanding and be able to complete the traffic demand forecasting independently.

3.2.4 Public Transport Development Patterns and Route Optimization

In the process of carrying out the analysis of public transport development patterns, the teacher guides the students through the analysis of the project to the content of urban public transport infrastructure. In the analysis of the spatial and traffic characteristics of the city, students are guided to be able to analyze public transport in the city from a global perspective, to understand the morphology of the city in which the project is located and the key nodes of public transport. The teacher takes students through a study of public transport development patterns, important nodes of the line network, the structure of the line network and the types of routes.

The teacher then proceeds to the route optimization part of the project, which is the core content of the course and the basic objective of the course for the students. Through the route optimization process of the specific project, the teacher allows the students to understand the principles and ideas of route optimization, guides them to analyze the current city bus network structure, combines the knowledge of the textbook to analyze the problems of each of the trunk lines, basic lines, branch lines and urban and rural lines, and then comes up with the project to carry out the route network Optimization options, on the basis of which the class will expand on the process of optimizing the structure of the grade line network (Fig. 3).

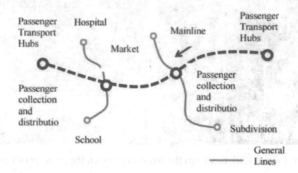

Fig. 3. Schematic diagram of the wire network structure

3.2.5 Optimization of Public Transport Stations (Yards)

Public transport stations are divided into four main categories: repair workshops, parking yards, bus interchange hubs and bus termini. Bus intermediate stations are laid out according to travel road conditions. In the project teaching process, this part of the content combined with the textbook, the specification "urban public transport station, field, plant design specification" (CJJ/T15–2011) for the optimization of the project's bus station analysis, students first according to the teacher's demonstration of the project bus station optimization plan preparation, the teacher then for the plan to explain, so that students clear optimization ideas and principles. At the same time, as the site involves urban land, the teacher also needs to guide the students in the teaching process to connect the site with the higher plan and propose the security of the site (Fig. 4).

3.2.6 Integrated Public Transport Development Strategy

In addition to acquiring the technical knowledge of public transport, students will also need to learn the latest methods of public transport planning and design, understand national policies and regulations, and learn from the experience of other cities' excellent public transport development. In this part of the course, students are given the task of collecting examples of successful public transport development at home and abroad, interpreting them and guiding them to summarize what they can learn from the project

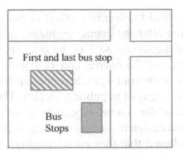

Fig. 4. Platform first and last stop combined with interchange diagram

and propose strategies for the development of public transport in the project, with particular emphasis on the integration of urban and rural public transport and the development of information technology in public transport for students to consider and discuss.

3.2.7 Implementation Assessment and Implementation Recommendations

This is the final stage of the project when the project is being taught. In order for students to understand the reasonableness of the optimization plan, an implementation assessment is required. Students are guided to assess the reasonableness of the optimization plan in terms of capacity allocation, bus mileage, service area, network structure and planned station coverage in relation to the textbook and specifications. Finally, the teacher will recommend a phased implementation of the optimization plan based on the urgency of the public's travel needs, the maturity of the developed sites, the operational pressure on the routes and the ease of adjustment of the bus routes.

3.2.8 Course Assessment

The purpose of the course assessment is to make teachers clear about the learning effect of students. The course mainly assesses students' mastery of knowledge from four aspects. Firstly, basic knowledge is assessed by means of an online classroom test, which accounts for 10% of the grade; secondly, the usual assignments, which account for 30% of the grade, are mainly required to be completed by students in the course of the project teaching method, including the preparation of survey plans, analysis of survey results and calculation of current public transport operation indicators; thirdly, for the understanding and application of knowledge, the teacher assigns a project similar to the fourth is attendance, which accounts for 5% of the grade.

4 Application Effectiveness Evaluation

The original purpose of the project teaching method was to increase students' motivation and improve learning outcomes, to enable students to better adapt to the workplace after graduation in the context of the national promotion of the construction of industrial colleges, to allow companies to reduce the time spent on training, and to enable schools to train students who can be seamlessly integrated with companies [17–19].

In order to fully understand the learning effect of the students, the teacher under-stood the students' situation after the course by means of anonymous questionnaires. By analyzing the results of the survey, 88% of the students were very satisfied with the teaching methods and format of the course, and 12% were satisfied [20]. At the same time, one year after the end of the course, a telephone call back was made specifically to students of transport majors engaged in public transport. The students were all satisfied with the teaching methods of the course, especially those engaged in public transport planning and optimization in design institutes, who were very satisfied with the teaching of the course, and they believed that they could quickly connect to the actual project work content through the course, which improved their work efficiency [21–23] (Fig. 5).

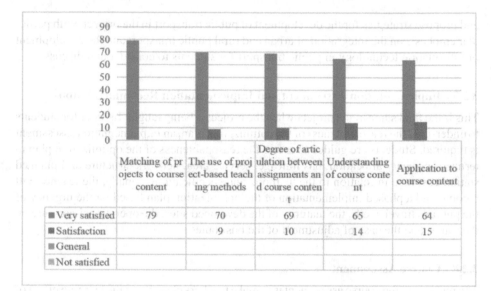

	Matching of pr ojects to course content	The use of proj ect-based teach ing methods	Degree of artic ulation between assignments an d course conten t	Understanding of course conte nt	Application to course content
■ Very satisfied	79	70	69	65	64
■ Satisfaction		9	10	14	15
■ General					
■ Not satisfied					

Fig. 5. Course Questionnaire Results Chart

5 Conclusion

As teachers in the new era, we should keep abreast of the times, pay attention to the latest educational concepts and methods, achieve effective teaching, let the results of teaching promote the high-quality development of the economy, integrate industry and education, and strive to contribute to the comprehensive promotion of rural revitalization.

This paper analyses the problems in the current situation of urban public trans-port planning and operation management courses, combines the new requirements for education in the new era, explores the application of the project teaching method in undergraduate courses in applied universities, and explains the process of realizing the project teaching method and the evaluation of its effects, which is a good reference for similar course teaching reforms in the future.

1) This paper analyses the extent to which the selection of projects for the course fits with the content to be mastered in the course, and shows that the application of the method achieves the content and objectives of the course.
2) This paper elaborates on the process and methods of implementing the project teaching method in the whole process of teaching the course. For the part of the course content that students have not been mastering well enough, they adopt to invite enterprise tutors to co-teach to ensure the effect of the project teaching method.
3) In the process of using the project teaching method, teachers always pay attention to the learning effect of students and the needs of enterprises, adjust the course teaching in time, and follow the development of graduates to evaluate the implementation effect.

Acknowledgment. This project is supported by the second batch of school-level teaching team in 2019 "Applied Effective Teaching Design Teaching Team" (2019XJJXTD10) and Nanning University's second batch of teaching reform project of specialized innovation integration curriculum (2020XJZC04).

References

1. Li, B., Wu, X., Du, J.: Research and application of project-based teaching in the multi-media technology. Adv. Mater. Res. **2385**(694–697), 3692–3695 (2013)
2. Zheng, J.: Guidance on the Application of Teaching Methods. East China Normal University Press, Shanghai (2006). (in Chinese)
3. Hu, Q.: Optimizing Classroom Teaching: Methods and Practices. Beijing: Beijing: Renmin University of China Press (2014) (in Chinese)
4. Li, C.: Practice exploration of term project teaching method for electric elements of power plant course. Curriculum Teach. Method. **5**(12), 137–144 (2022)
5. Li, C.: Research on the application of project teaching method in the course of premiere film and video editing. J. Lanzhou Coll. Arts Sci. (Natl. Sci. Ed.) **36**(06), 107–110 (2022). (in Chinese)
6. Wu, H., Zhang., X.: The application of project teaching method in the teaching of micro-controller principle and interface technology course. Innov. Entrepreneurship Theory Pract. **5**(21), 174–177 (2022). (in Chinese)
7. Zhang, F.: Exploration and practice of project teaching method in the experimental teaching of network service configuration and management. Sci. Technol. **28**, 110–112 (2022). (in Chinese)
8. Dong, G.: Research and application of project teaching method in the curriculum reform of visual communication design. In: Proceedings of 2022 the 6th International Conference on Scientific and Technological Innovation and Educational Development, pp. 949–951 (2022)
9. Huang, L., Chen, Y., Yang, X.: Research on experimental teaching of logistics information management based on project teaching method in the post-epidemic era. Logist. Eng. Manage. **44**(08), 139–141 (2022). (in Chinese)
10. Zhang, L.: Application research of project teaching method in the course of safety management of urban rail transit operation. Front. Educ. Res. **5.0**(9.0), 35–39 (2022)

11. Gu, Y.: The application of project teaching method in computer teaching. Electron Technol. **51**(06), 156–157 (2022). (in Chinese)
12. Ding, J.: The application of project teaching method in teaching electrical and electronics. Integr. Circ. Appl. **39**(04), 168–169 (2022). (in Chinese)
13. Yi, D.: Research on the application of project teaching method in the teaching of landscape sculpture. Art Des. **5**(1), 191–197 (2022)
14. Yao, X.: Exploring the application of project teaching method to the teaching of digital mappin. Mapping Spat. Geogr. Inf. **44**(10), 222–224 (2021). (in Chinese)
15. Chen, J.: Research on the application of project-based teaching reform of professional English course under the cross border e-commerce platform. Adv. Vocat. Tech. Educ. **3**(2), 159–162 (2021)
16. Yin, C., Lin, D., Ren, P.: The application of project driven teaching method in win server 2008. Adv. Mater. Res. **3326**(989–994), 5224–5227 (2014)
17. Sun, X.: Exploration on the project teaching method in fashion design of colleges. In: Proceedings of 2019 5th International Conference on Education Technology, Management and Humanities Science (ETMHS 2019). Francis Academic Press, pp. 373–376 (2019)
18. Zhou, X.: Application of project teaching method in three-dimensional animation teaching. In: Proceedings of 2019 International Conference on Arts, Management, Education and Innovation (ICAMEI 2019). Clausius Scientific Press, pp. 783–787 (2019)
19. Lu, Q., Xia, C.: The Application of project teaching method in hydraulic transmission and control based on CDIO. In: Proceedings of 2018 5th International Conference on Education, Management, Arts, Economics and Social Science (ICEMAESS 2018), pp. 929–932 (2018)
20. Li, Z.: Research on the network teaching mode of PBL project teaching method. J. Comput. Theor. Nanosci. **13**(11), 7762–7767 (2016)
21. Wang, X., Wu, Y.: Exploration and practice for evaluation of teaching methods. Int. J. Educ. Manage. Eng. (IJEME) **3**, 39–45 (2012)
22. Yang, J., Pan, Y.: Analysis on teaching methods of industrial and commercial management based on knowledge transform expansion model of SECI. Int. J. Educ. Manage. Eng. (IJEME) **12**, 76–82 (2011)
23. Adenowo, B.A., Adenle, S.O., Adenowo, A.A.: Towards qualitative computer science education: engendering effective teaching methods. Int. J. Mod. Educ. Comput. Sci. (IJMECS) **9**, 16–26 (2013)

Social Engineering Penetration Testing in Higher Education Institutions

Roman Marusenko[1], Volodymyr Sokolov[2(✉)], and Pavlo Skladannyi[2]

[1] Taras Shevchenko National University of Kyiv, Kyiv 01601, Ukraine
[2] Borys Grinchenko Kyiv University, Kyiv 04053, Ukraine
v.sokolov@kubg.edu.ua

Abstract. Social engineering penetration testing is a complex but necessary tool to test the security of information systems. Such testing requires balancing the organization's benefit and the comfort of an information system user. Penetration testing poses complex ethical concerns affecting people who do not expect it. At the same time, penetration testing is effective only when it mimics the real situation as much as possible, i. e. it is unexpected. The article's authors describe the methodology of social engineering penetration testing of the educational institution information system; substantiate the design of the experiment, which allows for balancing the ethical precautions and the effectiveness of testing. The authors formulate a set of markers that they use to reduce the negative impact of the attack on users of the information system and to help users to identify the true nature of the attack. The experiment conducted by the authors shows the advance of a phishing attack aimed at a large number of system users and its effectiveness. The authors also reveal the challenges such an attack poses to the information system staff, who have to respond to such influence effectively and on time. The experiment shows that half of the responses were received in the first 40 min after mailing. Concluding the research authors analyze the suggested design of social engineering penetration testing experiment, ways to respond to real attacks of this kind, as well as to raise respondents' awareness. The directions for possible future research are outlined. The value of this research is in the object—students of a higher educational institution who constantly work with information. The neglect of personal information indicates the need to introduce information hygiene courses from the very first courses.

Keywords: Penetration testing · Phishing · Social engineering · Sensitive information · Higher education institution · High school

1 Introduction

Protection of information systems (IS) from attacks is based on assessing the resilience of implemented solutions to the actions of attackers. The classic approaches to protection are the use of solutions that have proven their effectiveness in practice compared to others, as well as the audit of a particular IS. The audit aims to make sure that the implemented measures correspond to the plan (statics) and function as planned (dynamics). In this

© The Author(s), under exclusive license to Springer Nature Switzerland AG 2023
Z. Hu et al. (Eds.): ICCSEEA 2023, LNDECT 181, pp. 1132–1147, 2023.
https://doi.org/10.1007/978-3-031-36118-0_96

case, without proper testing, the IS can be vulnerable, which will be known only during the attack [1]. The audit does not involve finding or accounting for vulnerabilities [2].

Protecting IS from attacks of a certain type requires testing the behavior of the system in the context of this type of attack [3]. Penetration testing is a type of security measure designed to verify the readiness of all elements of the IS and security procedures to function properly and effectively and to respond to real threats. It allows you to identify those shortcomings in responding to real attacks, which may not be noticed by the staff of the IS [4]. The tester begins to think "like an attacker" and notices the shortcomings of the security system and the links between them.

Penetration testing can be performed, including social engineering methods [5]. A comparative statistical analysis of Positive Technologies shows that among the surveyed categories of victims of cyberattacks in 2018, the fourth place (7%) was taken by educational institutions [6]. Moreover, for engineering institutions, the main vector of attacks was social engineering (38%). The latter can be a source of information about students, educational systems, and more. Also, the information obtained can be used in a multi-stage attack on other victims' assets. The educational institution has a specific composition of IS users, which are the "weakest link" of any IS. Because of this, the subject of our attention will be social engineering penetration testing in such an institution.

In general, penetration testing dates back to the military in the 1960s [7]. Today, the need for IS penetration testing is directly or indirectly provided by several documents. Thus, the NIST recommendations indicate that "[p]enetration testing is important for determining the vulnerability of an organization's network and the level of damage that can occur if the network is compromised." Intrusion testing involves, in particular, social engineering testing, such as asking an administrator to provide or reset another user's password [8].

Art. 32 of the General Data Protection Regulation (GDPR) provides for the implementation of appropriate technical and organizational measures to ensure a level of security appropriate to the risk, including "...regularly testing, assessing and evaluating the effectiveness of technical and organizational measures for ensuring the security of the processing" [9]. Testing includes the creation of real situations that need to be responded to, i. e. social engineering penetration testing.

Payment Card Industry Data Security Standard (PCI DSS) assessment procedures also require the testing of systems for intrusion, including the verification of the impossibility of modifying authentication credentials by social engineering methods [10].

Phishing remains one of the most common methods of social engineering, so phishing penetration testing will address the greatest threats to IS from social engineering [11–13].

The purpose of the study is to check how vulnerable university students are to phishing attacks in comparison with researchers.

The object of the study is a group of students from one university. None of the participants in the experiment should be warned, as this greatly affects the accuracy of the results of the phishing mailing.

To maintain the accuracy of the results, a script is used to automatically match the results: sent emails and results in the feedback form. This script allows you to exclude the human factor when processing the results of the experiment. It is difficult to repeat this experiment on the same sample.

This work reveals the features of planning and experimenting with penetration testing in higher education institutions by the method of phishing, taking into account ethical constraints. Section 2 analyzes the available research and highlights aspects that need further analysis. Section 3 contains an analysis of the limitations and conditions under which social engineering penetration testing may be conducted. Section 4 describes in detail the experimental procedure, including the collection of information necessary for penetration testing, as well as the results obtained. Section 5 contains an assessment of the results of the experiment and the authors' suggestions for penetration testing, as well as a response to such attacks. The final section outlines the directions for future research in this area.

2 Related Works

The methodology of violating the physical integrity of IS educational institutions by penetration testing is demonstrated in [14], considering the ethical aspects of the attack, although it was a direct impact of the attacker on the staff to physically take over the components of the IS. The paper [15] generally considered the methodology of testing by methods of social engineering and gave examples of its application, but without specifying the different types of such influences.

Social engineering attack scenarios were considered in [16], and training models were also proposed, but the paper contains simulation results that need to be tested in practice.

The methodology of application of social engineering methods was also studied both to understand the behavior of the offender [17] and to develop models to raise awareness of IS personnel [18].

Although this study provides an excellent subject and design of the experiment, it is a development of research [19–21] that substantiates the technology of creating an information security system in an educational institution and conducting an experimental assessment of IS protection against attacks by social engineering methods.

3 Problem Statement

Information systems include software and hardware component and human component (staff, users, administrators, etc.). The latter can be adversely affected as well as software and hardware components. Performing penetration testing of the human component differs from testing the software and hardware components of the IS in several important ways.

First, these are ethical warnings. Social engineering penetration testing is not aimed at demonstrating the absence of vulnerabilities, but at finding existing ones. Therefore, it is always a creative activity. Penetration testing aims to simulate a real situation. The IS user experiences almost the same psychological impact caused by a real attack. However, it cannot be notified in advance of testing, as when checking the hardware and software components of the IS. Awareness of future testing will lead to behavior change, increased attention, etc., and will not allow assessing the normal behavior of the IS user in the real conditions of the IS. On the other hand, deceiving the user, even for testing, will worsen the attitude towards the organizers of testing, because the reactions and emotions will be real.

The difficulty is that testing of software and hardware components can take place with full awareness of IS users about it. Awareness of personnel about the attack does not change the behavior of software or hardware components of the IS. At the same time, testing the behavior of the human component (users, IS administrators) is an impact on them. Influence involves the possibility of exploiting trust, and misleading. This creates an ethical problem, because for such people the attack will look real, and therefore the response will also be real and by security policies. The fact and content of the response actions should also be taken into account when designing the experiment [22].

Second, the user who detected the test "attack" will react realistically, which may lead to legal consequences. Therefore, it is necessary to plan the end of the experiment and the correct notification of the participants that the attack was a test.

Third, there is the question of the possibility of causing non-pecuniary damage to persons who learn that their actions have led to an imaginary breach of IS security.

It is important to find a balance between not testing the damage to the human component of the IS and preserving the essence of penetration testing, which does not involve warning or disclosing the content of the attack before it begins.

Ethical researchers note that in phishing experiments, it is important that communication shows signs of phishing, but does not lead to a real compromise of information. The issue of informed consent given the nature of the study is quite controversial [23]. This is obvious because social engineering by definition includes elements of deception, trickery, exploitation of trust for one's benefit, representation of one entity instead of another, and so on.

Mouton F. et al. note that some studies in the field of social engineering suggest a lack of awareness of the upcoming experiment precisely because a person who knows about it will behave differently, which nullifies the experiment as such [24]. Having prior consent to participate in the experiment distorts the behavior of users, because even if they provide the requested information, it may be because they consider it a manipulation of the experimenter [25]. It will be impossible to interpret such results correctly.

An important issue raised in the literature is the balance between the interests of the penetration testing company and the individual being targeted [26]. The tester must remember that the way to solve such a dilemma will depend on the specific situation.

The literature also indicates one of the possible ways to solve ethical problems—to provide an opportunity to refuse to participate in the experiment [25]. Our study takes into account the possibility of failure at any stage of the experiment. To do this, the authors in the design of the experiment provided several markers that will help users understand that penetration testing is taking place and stop participating in it.

Upon completion of testing, it is important that the results of social engineering penetration testing are summarized and do not identify specific IS users who may suffer from further ridicule or insults from colleagues [27].

Note that penetration testing has certain functional limitations: it covers only the infrastructure that was the subject of testing, allows you to test the human element of the IS completely, illustrates the state of the system only at a certain point in time, may not show all existing system defects [28]. This should be taken into account when interpreting the results and trying to extrapolate them to the behavior of IS users in other situations.

4 Method of Research

For ethical reasons, as already mentioned, penetration testing should be conducted in a way that does not harm the targets of testing. Users of IS with whom social engineering methods interact are naturally prone to react negatively to the fact that their trust has been deceived or exploited. Therefore, even the issue of debriefing after the experiment is debatable [23].

Traditionally, penetration testing of hardware or software components of the IS is directed to these components (Fig. 1a). At the same time, IS users remain observers and may receive negative emotions, especially if the IS that contained the vulnerabilities is built or managed by them.

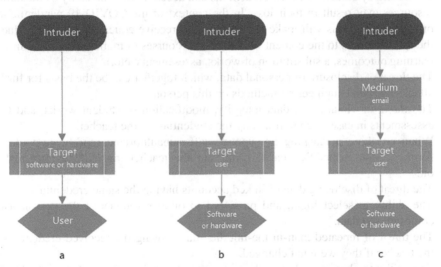

Fig. 1. Options of penetration testing, aiming different components of IS: software/hardware penetration testing model (a); conventional social engineering penetration testing model (b); modified social engineering penetration testing model (c)

Social engineering penetration testing assumes that the object of influence is the human component of the IS (Fig. 1b)—the user of the system. Such influence is exerted within the relationship between the tester acting as the attacker and the victim. In this case, any deception, or tricks can worsen the condition of the victim for a long time, when she learns about the true purpose of interaction with the person who played the role of the attacker.

Because of the above, we decided to choose the model of penetration testing, which provides for the presence of an intermediary between the human component of the IS and the tester, which will be affected (Fig. 1c). In our case, these are phishing emails.

Effective resolution of conflict situations, which are inevitable when a person learns that they have been deceived, should be based on reducing negative emotions towards the opponent. In particular, it is manifested in the transfer of attention and negative emotions from the person to the subject or situation of the conflict. Thus, even in the event of such emotions, they will be directed to the facts, circumstances, and a specific material carrier, in our case—a phishing letter, not a person. This will allow for a more constructive overcoming of the conflict situation during and after the completion of social engineering penetration testing.

Ideally, penetration testing should be realistic, respectful, reliable, repeatable, and reportable to be effective, useful, and not harmful to test participants [14].

The legend of intrusion testing in our study was the presence of an attacker who does not have access rights to the IS and aims to carry out phishing, capturing sensitive data. This required obtaining the necessary information about the IS, its operation, users, and so on. This was also part of the tester's task.

The danger of a successful phishing attack on an educational institution is as follows:

- Disclosure by students of their credentials and access to the university's educational resources may result in their loss. In the context of the COVID-19 pandemic and online learning, this will make it impossible to receive educational services due to the loss of access to the content of educational courses (a malicious compromise of learning outcomes, a substitution of works, assessments, etc.).
- The threat of disclosure of personal data, which together can be the basis for further attacks by social engineering methods on this person.
- The threat of violation of data integrity, modification of student works, and their assessments in case of compromising the credentials of the teacher.
- When attacking a teaching or support staff—unauthorized modification of performance data, schedule, inaccessibility of the teacher, misleading him, and so on.
- The threat of disclosing data of linked accounts having the same credentials.
- The ability to select login, and password to other resources of the person using compromised data.
- The threat of repeated man-in-the-middle attacks using the received credentials of the teacher if they were not changed.
- The ability to change, and reset the password of certain services that are tied to the appropriate mailbox.

The experiment was built on the general scheme of the SE Pyramid [29] and included the following stages:

1) Open-source intelligence (OSINT). Collection of information necessary for the attack (e-mail address, features of the IS mail subsystem, the schedule of IS users, etc.).
2) Development of the content of the phishing letter and its appendix, registration of the sender's address.
3) Sending letters, responding to letters in response to targets. Monitoring attack response measures, correcting behavior in the event of an attack being detected or attempts to involve actors outside the IS in response.

4) Completion of the experiment. Processing of results, and development of recommendations.

In the process of preparing for sending letters, the policy of forming postal addresses
was revealed and posted on the website of the target university. It was found that e-mail
addresses are generated for all teachers and students within the university domain on the
Gmail platform type

< name > . < faculty > < year of admission>@ < high school domain>.edu.ua.

The next step was to find the last names of the students, their faculties, and the
year of entry to generate a pool of addresses. This information could be obtained from
admission orders, which are also contained on the website of the educational institution,
as well as from publicly available national databases of entrants of all higher education
institutions [30, 31].

The correctness of the generated e-mail addresses from the source data and the
existence of the corresponding addresses in the e-mail domain was checked using the
Burpsuite package. It was found that the feature of interaction between the client and
the server part of the Gmail application allows you to enter the address field and check
its existence before entering the password. Therefore, the server responses allow you to
check the existence of a pool of generated addresses.

It was planned to send letters to as many targets as possible at the same time. This
was intended to preclude the possibility of communication between targets if someone
discovered the malicious nature of the letter and wished to inform others about it. Also,
it allowed testing the reaction of system administrators to the attack on a large number
of targets.

According to the content, the phishing letter was supposed to come from the educational department of the university and contained a request to get acquainted with the
rules of academic integrity when taking exams.

The mailing was carried out only to those students who enrolled in the current year
and did not yet have an examination session. This should have justified the sending of a
letter of acquaintance with the rules of academic integrity. The letter contained a link to
the Google Form you had to fill out.

Given the above ethical considerations, several markers were left in the letters sent,
which should help the attentive user to doubt the authenticity of the letter. The user who
noticed one or more markers, according to the plan, should not follow the link from the
letter, and after going—should not enter and send information about themselves.

Researchers note several signs by which recipients of letters can assume that the
letter is phishing [33]. We have selected features from those that affect the success of
phishing [34] but can be noticed by the average user without the use of special knowledge
given the ethical limitations of the experiment (see Table 1).

Our task was to test the vulnerability of targets to a real phishing attack, so the
number of markers that can be contained in one letter corresponded to our idea of what
a real phishing letter might look like.

Table 1. Category and description of the phishing markers implemented into the email

Marker	Generalized description
Familiarity	Sending a letter, not from a corporate email address. In this institution, the teaching and support staff has e-mail addresses provided in the domain of the institution of the type < name > @ < high school domain >.edu.ua. The sender's address was different
Email address policy	To send letters, the address of the type of science. <high school domain> @gmail.com was registered, i. e. with a hint of the scientific department. At the same time, in an educational institution, science and education are separated and have different meanings, as students know. The letter was not about science, but about education. The signature in the letter is formulated as "educational department of the university," not scientific
Visual perception	The letter did not mention the university's name, did not use its logo, and did not provide the name or contact details of the sender, which is not typical for this type of letter
Inconsistency of parts	The title of the letter was formulated to attract the attention of the student, so it contained an instruction to control learning. At the same time, the letter contained a request for information on academic integrity
Atypical structure	The need to send two Google Forms in a row separately is an excessive complication and not justified by the declared purpose of the survey
Attentiveness to details	The text of Google Forms contained a large text of the document, which students usually do not read and only mark the checkbox to read it. There was a disclaimer between the text and the phishing field for the student to enter their data stating that the form was part of a phishing vulnerability study and that the sender of the letter did not collect any personal data. The disclaimer was presented in the same font as the text of the main document, which was required of the student

Indicators of successful phishing of a specific target (recipient of the letter) were selected as follows:

- Go to an external link from a phishing letter (link click). In this case, the letter was sent by an unknown sender and in a real situation, when you follow the link, for example, malware could be downloaded [32].
- Google Forms contained two fields for entering your email addresses and one field for entering a user's password. Correct completion of such forms by the recipient

of the phishing letter and sending of data to an unknown person also testified to the success of phishing (data sent).

An integral indicator of the success of the attack was the time of its detection and the effectiveness of response measures by the school, as well as the cessation of users' sending of their data.

The appearance of the sheet is shown in Fig. 2.

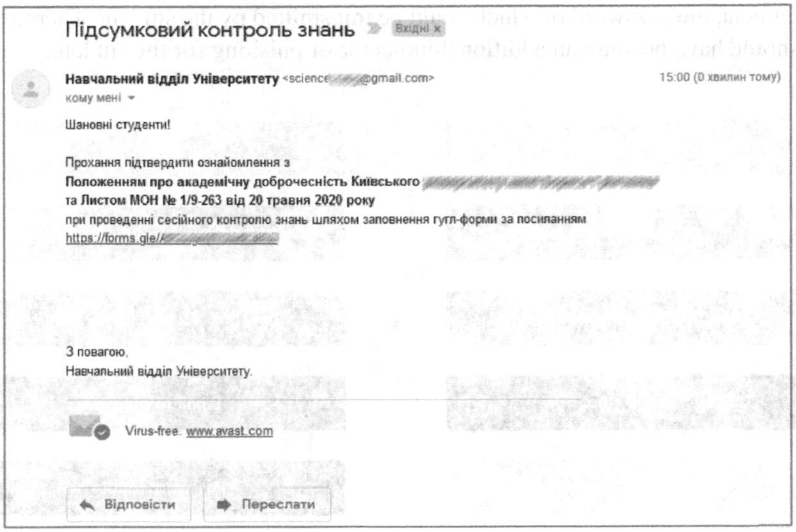

Fig. 2. Sample of the phishing letter with a link to Google Form (in Ukrainian)

Please note that by providing students with additional phishing markers, it was decided to register an address for sending emails to science < high school domain> @gmail.com. At the same time, we tested the possibility of allowing such an address. Thus, it was suggested that despite the vulnerability fix discovered by T. Cotten [35], today similar actions with modifications of the list header by separate HTML tags allow the hiding of the sender's address. Modification of the letter changes the result of its processing by the recipient's browser. While the sender's address still contains the letter in the body, the default browsers do not display it in either the mobile or desktop versions of Gmail. Such a letter is received with a successful check of SPF, DKIM, and DMARC filters and without prior notice to recipients or senders.

A list of mystical links to the Google Form with the first document to learn about the rules of academic volunteering. After filling it out and sending it after switching to other forms. This structure allowed us to test two signs of phishing success. Students who followed the link from the list initially provided only the e-mail address. Based on this fact, we could estimate that (a) the recipient of the list followed the external links from the list from a stranger; (b) the student failed to read the document and sent an unknown personal data file. Another form contained more obvious signs of phishing— the password field, so we predicted a large number of failures to enter it. Indentation, even when sending messages of another form of experimenters, is a small result of filling in the first and can find out that the student goes to a third-party address from the given list. The design of the two Google forms, which contain a link to the letter, was as follows (Fig. 3).

The second form contained an optional field for entering the password. It was not specified what password was meant. First, the need to pass the student's password to an outsider could not be motivated. Secondly, the educational part of the institution has the necessary (increased compared to the student) access rights to the e-learning environment, the password of which could be transmitted by the student. Therefore, this field should have become an additional indicator of phishing for the student.

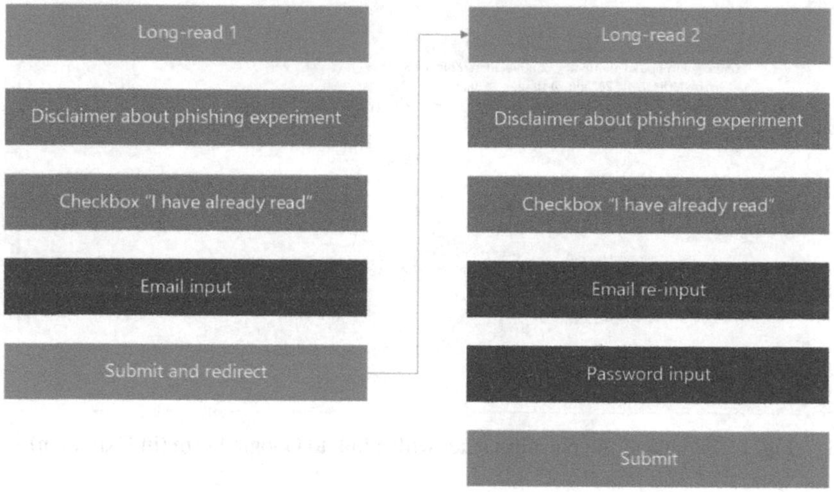

Fig. 3. Structure of the first (a) and the second (b) phishing Google Forms

5   Research Results

For phishing, it has been assumed that the target email domain has basic restrictions on sending emails, which for Gmail is 500 emails per day for a free account [36]. We assume that exceeding the limit could lead to the blocking of the mailing list, the sender's account, and the notification of the domain administrator about unusual activity on the server.

The distribution of letters was planned in a continuous block to a separate faculty. On the one hand, this would allow obtaining a homogeneous sample of responses of targets—by age and type of professional interests. On the other hand, it allowed simulation mailing by the training department, which is usually carried out simultaneously for all users. In general, it was planned to start with the faculties of the humanities and end with the faculty of computer science. Such an order was intended to prevent the rapid detection of the attack, as students of technical specialties have a broader knowledge of the field of IT. University staff did not report the experiment, as it could lead to information leaks about the nature of the experiment and "socially desirable" behavior. The latter could be manifested in the desire to demonstrate better results of student attack recognition or response by IS administrators, and the experiment itself to level.

261 letters were sent. The mailing was carried out by one letter with an attached list of recipients. An unknown factor was the presence or absence of limits in the corporate mail network for the mass distribution of identical letters from one address, which also does not belong to the pool of corporate addresses. It was assumed that the second group of letters to another faculty would be sent the next day.

There was a significant delay in sending the first response to the Google Form, which is not typical of this type of interaction. In a previous study, we found that students respond quickly to a letter they intend to respond to if such a letter does not require additional complex actions. The mode of time of the answers lay within the first hour after the mailing [25]. The experimenters sent a letter to their address confirming the delay, which was probably due to mass mailing from the newly registered address.

We received a total of 99 responses to the first Google Form and 5 responses to the second. The distribution of response time to the first form is interesting (Fig. 4). In the first hour, 59 responses were received. The other 38 responses were received for the next 17 h (including night and morning) until the experiment was stopped. Two more responses arrived two days later. This depends on the number of answers depending on the time is well approximated by the function:

$$N(t) = e^{\pi\left(2 - {}^{t}/100\right)}, \tag{1}$$

where t is the response time, min.

From this, we can conclude that a potential attacker gets most of the desired response from the targets in just one hour. On the other hand, IS protection personnel has significantly less time to protect at least half of the users who could potentially be attacked and compromise the school's IS and themselves. It took only 40 min for the attacking party to get answers from 50% of users (50 people) who were exposed to this phishing.

Mass distribution of phishing emails to users of one IS resembles a reverse DDoS attack. An attacker attacks many targets located outside the physical perimeter of the IS, which are associated with weak and slow information interaction. This poses difficult challenges, as university staff has very little time to detect the full range of targets and warn them of the true nature of the mailing when they detect an attack. Instant messaging tools should be used to prevent this. Because clicking on the link in the letter takes the target outside the conditionally controlled perimeter of the IS, even a complete physical blocking of the network operation of the IS is not able to save the target from the harmful effects of the attacker and the transmission of sensitive data to the attacker.

Analysis of responses to Google Form 1 showed that of the 99 responses received (see Table 2).

Thus, 36.8% of targets responded positively to the phishing letter and fell victim to an experimental attack. One can conclude about the success of the attack design in the form that was proposed.

It is significant that among the addresses there is the address of the university lecturer, which is obvious given the different structure of the teaching addresses of this institution.

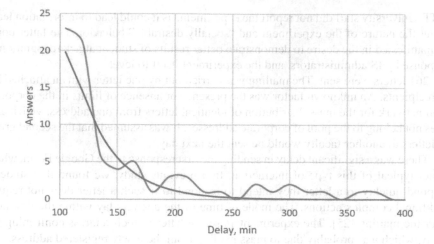

Fig. 4. Response time for the phishing email answer

Table 2. Distribution of active response of addressees for a phishing attack

Answers		Generalized description
Amount	Part, %	
88	33.7	Contained addresses from the mailing pool, i. e. students provided real data
7	2.7	Contained non-corporate e-mail addresses, i. e. students provided their second address. The check showed the authenticity of all seven addresses
2	0.8	Contained a random set of letters. Probably these students recognized phishing
1	0.4	Corresponded to the postal address but is invalid

Note that the e-mail address field did not check the format of the e-mail address, which allowed you to enter any sequence of characters if desired. This was intended to record the responses of those respondents who, when filling out the form, found phishing and wanted to demonstrate it

It was introduced at the stage when the fact of mailing became public and the content of the phishing letter was analyzed by the teaching staff.

Another interesting fact is the receipt of two responses two days after the mailing when its phishing nature was publicly known, and the authors of the experiment have already openly communicated this with stakeholders. This once again confirms our conclusion that a simultaneous attack on distributed targets is a threat given the difficulty of preventing all targets, even if it is detected.

The second form was filled in by five respondents, four of whom entered passwords that looked like real ones, which is 1.5% of respondents. This result demonstrates a high level of document recognition with obvious signs of phishing.

We should also note the passionate position of one respondent who, in response to a phishing letter, sent a message that he did not immediately understand the phishing purpose of the mailing, but now realized it. There was also a question about who is the organizer of the mailing. Since the experiment had not been completed at that time, the experimenters sent a standard response thanking them for participating in the experiment and asking them to wait for it to complete.

Later, the employees of the institution became aware of the mailing. The discovery of the fact of sending phishing letters made it necessary to inform the organizers of its true purpose. Further sending of letters to other faculties was not carried out, as it was no longer possible to ensure non-dissemination of information about the true purpose of the letters.

The response to the experimental attack turned out to be comprehensive, involving employees from various departments and trying to find a mailing list organizer. It should be noted that despite the measures taken, two responses to the letter were received two days after the end of the experiment. Also, the peculiarity of the simultaneous attack on a large number of targets did not allow software and hardware measures to promptly warn all victims of the attack and prevent their compromise.

The results can be summarized as follows. The response and mitigation of an attack require simultaneous and fairly rapid corrective action on the target of the attack. In this case, it could be alerting targets to attack. At the same time, phishing in an institution of this scale could be carried out from different addresses. In turn, this fact, as well as finding the targets of the attack in different geographical locations would not allow us to quickly identify all the victims. As shown in the experiment, the time required to "compromise" the target is short enough to manually notify the target of the attack and prevent the consequences. The software protects the perimeter before the letter hits the target, i. e. before the impact on the latter. If this influence has already occurred, then the software and hardware are powerless.

This study differs from previous studies [20] in the combination of social engineering and man-made attacks. The experiment was preceded by the analysis of information messages that are sent within the corporate network of the university. The masking of the link to a Google form for collecting sensitive information was done taking into account similar emails that are sent to students regularly. This approach gives an order of magnitude better results than mailing to open sources, even taking into account the context [21].

From the experiment, it can be seen that the choice of time and subject of a phishing email stimulates recipients to actively respond. In this case, half of the responses were received in less than an hour, so the speed of response to such incidents should be no more than a quarter of an hour.

6 Conclusions and Future Work

Conducting social engineering penetration testing requires a careful attitude to the ethical side of the design of the experiment, so as not to level its results on the one hand, and the other—not to cause a negative impact and protect the privacy of the targets of the experiment. Finding such a balance will always require consideration of the conditions of a particular IS, the specifics of users of such IS, the incident response system, and the type of information processed in the IS. Phishing is essentially a distributed attack that develops quite rapidly. An effective and timely response is a big challenge. Social engineering, like any attack, exploits the simplest and most effective possible ways to achieve a goal. In this case, it is the sending of letters, which is difficult or impossible to prohibit or restrict the filter of harmful letters.

The primary measure of protection against this type of attack is currently training and raising respondents' awareness. Developing effective and efficient programs in the conditions of the constant evolution of attacks by methods of social engineering requires constant search and improvement of methodology.

In this paper, we have demonstrated a basic version of social engineering penetration testing, which should be noticed by the attentive user. Further development of staff competencies to respond to more complex types of attacks by methods of social engineering variation of testing may be the use of sender addresses with the so-called. Cousin domain, typos in the address, vulnerabilities that allow hiding it, and other techniques. Also, the text of the letter may contain more compelling instructions and may not contain markers that allow you to recognize phishing. However, in this case, researchers will again face the ethics of penetration testing. Therefore, we believe that the study of this area is far from complete.

Acknowledgment. The authors are grateful to Borys Grinchenko Kyiv University administration for assistance in conducting experiments, as well as personally to the deputy head of IT in the Education Laboratory Oksana Buinytska for successfully detecting attack simulation and prompt response.

References

1. Singh, A., Kumar, A., Bharti, A.K., Singh, V.: An e-mail spam detection using stacking and voting classification methodologies. Int. J. Inf. Eng. Electr. Bus. (IJIEEB) **14**(6), 27–36 (2022). https://doi.org/10.5815/ijieeb.2022.06.03
2. Ahraminezhad, A., Mojarad, M., Arfaeinia, H.: An intelligent ensemble classification method for spam diagnosis in social networks. Int. J. Intell. Syst. Appl. (IJISA), **14**(1), 24–31 (2022). https://doi.org/10.5815/ijisa.2022.01.02
3. Fan, W., Kevin, L., Rong, R.: Social engineering: i-e based model of human weakness for attack and defense investigations. Int. J. Comput. Netw. Inf. Secur. (IJCNIS) **9**(1), 1–11 (2017). https://doi.org/10.5815/ijcnis.2017.01.01
4. Smith, J.K., Shorter, J.D.: Penetration testing: a vital component of an information security strategy. Issues Inf. Syst. XI.1, 358–363 (2010). https://doi.org/10.48009/1_iis_2010_3 58-363
5. Jazzar, M., Yousef, R.F., Eleyan, D.: Evaluation of machine learning techniques for email spam classification. Int. J. Educ. Manage. Eng. (IJEME) **11**(4), 35–42 (2021). https://doi.org/ 10.5815/ijeme.2021.04.04
6. Positive Technologies: Cybersecurity threatscape: Q3 2022. https://www.ptsecurity.com/ww-en/analytics/cybersecurity-threatscape-2022-q3/. Accessed 16 Feb 2023
7. The history of penetration testing. https://alpinesecurity.com/blog/history-of-penetration-tes ting. Accessed 16 Feb 2023

8. Scarfone, K., et al.: Technical guide to information security testing and assessment. Recommendations of the National Institute of Standards and Technology. NIST SP800–115. https://nvlpubs.nist.gov/nistpubs/Legacy/SP/nistspecialpublication800-115.pdf. Accessed 16 Feb 2023

9. European Parliament, Regulation 2016/679 on the protection of natural persons with regard to the processing of personal data and on the free movement of such data. https://eur-lex.eur opa.eu/legal-content/EN/TXT/PDF/?uri=CELEX:32016R0679. Accessed 16 Feb 2023

10. Payment Card Industry Data Security Standard. Requirements and security assessment procedures, version 3.0. https://www.pcisecuritystandards.org/minisite/en/docs/PCI_DSS_v3.pdf. Accessed 16 Feb 2023

11. Campbell, N., Lautenbach, B.: Telstra security report. https://www.telstra.com.au/content/dam/shared-component-assets/tecom/campaigns/security-report/Summary-Report-2019-LR.pdf. Accessed 16 Feb 2023

12. Kessel, P.: EY global information security survey. https://assets.ey.com/content/dam/ey-sites/ey-com/en_ca/topics/advisory/ey-global-information-security-survey-2018-19.pdf. Accessed 16 Feb 2023

13. Pescatore, J.: SANS top new attacks and threat report. https://www.sans.org/reading-room/whitepapers/threats/top-attacks-threat-report-39520. Accessed 16 Feb 2023

14. Dimkov, T., et al.: Two methodologies for physical penetration testing using social engineering. In: 26th Annual Computer Security Applications Conference, pp. 399–408 (2010). https://doi.org/10.1145/1920261.1920319

15. Barrett, N.: Penetration testing and social engineering. Inf. Sec. Tech. Rep. **8**(4), 56–64 (2003). https://doi.org/10.1016/s1363-4127(03)00007-4

16. Nguyen, T.H., Bhatia, S.: Higher education social engineering attack scenario, awareness & training model. J. Colloquium Inf. Syst. Secur. Educ. **8**(1), 8 (2020)

17. Indrajit, R.E.: Social engineering framework: understanding the deception approach to human element of security. Int. J. Comput. Sci. Iss. **14**(2), 8–16 (2017). https://doi.org/10.20943/012 01702.816

18. Kelm, D., Volkamer, M.: Towards a social engineering test framework. In: 11th International Workshop on Security in Information Systems, pp. 38–48 (2010). https://doi.org/10.5220/0004980000380048

19. Buriachok, V., et al.: Technology for information and cyber security in higher education institutions of Ukraine. Inf. Technol. Learn. Tools **77**(3), 337–354 (2020). https://doi.org/10.33407/itlt.v77i3.3424

20. Marusenko, R., Sokolov, V., Buriachok, V.: Experimental evaluation of phishing attack on high school students. In: Hu, Z., Petoukhov, S., Dychka, I., He, M. (eds.) ICCSEEA 2020. AISC, vol. 1247, pp. 668–680. Springer, Cham (2021). https://doi.org/10.1007/978-3-030-55506-1_59

21. Marusenko, R., Sokolov, V., Bogachuk, I.: Method of obtaining data from open scientific sources and social engineering attack simulation. Adv. Artif. Syst. Logist. Eng. **135**, 583–594 (2022). https://doi.org/10.1007/978-3-031-04809-8_53

22. Hu, Z., Buriachok, V., Sokolov, V.: Implementation of social engineering attack at institution of higher education. In: International Workshop on Cyber Hygiene, pp. 155–164 (2019)

23. Finn, P., Jakobsson, M.: Designing and conducting phishing experiments, 1–21 (2006)

24. Mouton, F., et al.: Necessity for ethics in social engineering research. Comput. Sec. **55**, 114–127 (2015). https://doi.org/10.1016/j.cose.2015.09.001

25. Resnik, D.B., Finn, P.R.: Ethics and phishing experiments. Sci. Eng. Ethics **24**(4), 1241–1252 (2017). https://doi.org/10.1007/s11948-017-9952-9

26. Faily, S., McAlaney, J., Iacob, C.: Ethical dilemmas and dimensions in penetration testing. In: 9th International Symposium on Human Aspects of Information Security & Assurance, pp. 233-242 (2015). https://doi.org/10.13140/rg.2.1.3897.1360

27. Pierce, J., Jones, A., Warren, M.: Penetration testing professional ethics: a conceptual model and taxonomy. Aust. J. Inf. Syst. **13**(2), 193–200 (2006). https://doi.org/10.3127/ajis.v13i2.52
28. Creasey, J., Glover, I.: A guide for running an effective penetration testing programme. https://www.crest-approved.org/wp-content/uploads/CREST-Penetration-Testing-Guide.pdf. Accessed 16 Feb 2023
29. Hadnagy, C.: Social engineering: The science of human hacking (2018)
30. Introduction.EDUCATION.UA. https://vstup.osvita.ua. Accessed 16 Feb 2023
31. Applicant search service. http://abit-poisk.org.ua. Accessed 16 Feb 2023
32. Kotov, V., Massacci, F.: Anatomy of exploit kits. In: Jürjens, J., Livshits, B., Scandariato, R. (eds.) ESSoS 2013. LNCS, vol. 7781, pp. 181–196. Springer, Heidelberg (2013). https://doi.org/10.1007/978-3-642-36563-8_13
33. Parsons, K., et al.: Do users focus on the correct cues to differentiate between phishing and genuine emails? arxiv:1605.04717
34. Jampen, D., Gür, G., Sutter, T., Tellenbach, B.: Don't click: towards an effective anti-phishing training. a comparative literature review. HCIS **10**(1), 1–41 (2020). https://doi.org/10.1186/s13673-020-00237-7
35. Cotten, T.: Ghost emails: hacking Gmail's UX to hide the sender. https://blog.cotten.io/ghost-emails-hacking-gmails-ux-to-hide-the-sender-46ef66a61eff. Accessed 16 Feb 2023
36. Google: Gmail sending limits in Google Workspace. https://support.google.com/a/answer/166852. Accessed 16 Feb 2023

Author Index